THE MECHANICS OF AEROSOLS

THE MECHANICS OF AEROSOLS

BY

N. A. FUCHS

Karpov Institute of Physical Chemistry
Moscow

TRANSLATED FROM THE RUSSIAN

BY

R. E. DAISLEY

AND MARINA FUCHS

TRANSLATION EDITED BY

C. N. DAVIES

London School of
Hygiene and Tropical Medicine

REVISED AND ENLARGED EDITION

DOVER PUBLICATIONS, INC.
New York

Published in Canada by General Publishing Company, Ltd., 30 Lesmill Road, Don Mills, Toronto, Ontario.

This Dover edition, first published in 1989, is an unabridged and unaltered republication of the edition published in 1964 by Pergamon Press Ltd. It is reprinted by special arrangement with Pergamon Press, Headington Hill Hall, Oxford OX3 OBW, England.

Manufactured in the United States of America
Dover Publications, Inc., 31 East 2nd Street, Mineola, N.Y. 11501

Library of Congress Cataloging-in-Publication Data

Fuchs, N. A.
The mechanics of aerosols.

Translation of: Mekhanika aerozoleĭ.
Reprint. Originally published: Oxford : Pergamon Press, 1964.
Bibliography: p.
Includes index.
1. Aerosols. I. Davies, C. N. (Charles Norman), 1910– . II. Title.
QD549.F913 1989 541.3′4515 89-11840
ISBN 0-486-66055-9

CONTENTS

FOREWORD TO THE ENGLISH EDITION

"MECHANICS OF AEROSOLS" is a new term and it is not yet certain whether it will be accepted by workers in this field. A better title, however, could hardly be suggested for a book which deals with the movements of aerosol particles and clouds, the collisions of particles and hydrodynamic disintegration of aggregates.

The author has tried to cover all the important papers on the subject as completely as possible. They are published in many journals, devoted to widely differing branches of science and technology, and much valuable work is described only in reports and technical papers issued by research organizations which are not available to the author. He was therefore greatly handicapped in his endeavour.

The Russian edition of this book appeared in 1955. When Pergamon Press decided to translate it into English progress had been so rapid that parts of the book were already obsolete. The author was unable to rewrite the whole text, and instead wrote an addendum for the English edition which deals with the work published between 1954 and 1960. This addendum was published in Russian in 1961. Many corrections were also made to the original text for the English edition. In order to make the translation as precise as possible it was sent to the author for correction and approval.

All this made the whole job rather complicated and the author is greatly indebted to the staff of Pergamon Press for the work they have done and especially thankful to Dr. C. N. Davies who undertook the ungrateful task of vetting the translation and has done it perfectly.

Karpov Institute of Physical Chemistry
Moscow

N. FUCHS

FOREWORD

AERODISPERSE systems, or aerosols, consisting of solid or liquid particles suspended in a gas, play a very large part in nature and the life of man. The water cycle in nature involves bulk condensation of water vapour, to form clouds, and subsequent precipitation from them. Clouds make the climate milder by protecting the earth's surface from excessive heating by the sun and by diminishing cooling due to radiation.

The erosion, transport and deposition of solid particles by wind are also of great importance. After many centuries of activity this "pneumatic conveyor" has created enormous deposits of loess, such as those in the basin of the Hwang Ho river. Desert sands carried by the wind are a constant menace to cultivated and irrigated lands, and whole towns have been buried under a thick layer of sand when man has relaxed in his struggle against this terrible enemy. Erosion of light arable soil by wind causes heavy damage, and belts of trees besides providing protection against drying winds, help to prevent fields, gardens and canals from being filled with sand and stop the soil from being carried away by wind. Snowdrifts are discouraged by snow fences in exactly the same way.

Cross-pollination of grasses and many trees is effected by the pollen being scattered as an aerosol by the wind. Considerable numbers of seeds and spores are spread in a similar manner. Many micro-organisms preserve their germinating power over long periods while airborne owing to their association with colloidal matter. The aerial microflora comprise a large assortment of fungi and bacteria which have significance in medicine and in the fermentation industries. Many infectious diseases, including influenza, whooping cough, various lung infections and tuberculosis, are undoubtedly spread through the air.

Dusts containing silica, which are formed during the drilling and mechanical shaping of siliceous rock, are very injurious to the lungs. Silicosis, caused by these dusts, is one of the most dangerous and widespread occupational diseases. Radioactive aerosols formed in the explosion of atomic bombs and in the atomic industry represent an enormous danger. On the other hand, some medicinal preparations are particularly effective when administered by inhalation as aerosols.

Aerosols have become tremendously important in technology. The winnowing of grain, a process used by man from time immemorial, consists of suspending the mixture of grain and chaff as an aerosol and then separating the two constituents with an air stream. Liquid fuel, prior to combustion, is generally converted into a mist by mechanical atomization and the use of pulverized solid fuel has become widespread. Some ores, such as pyrites, are also burnt in the pulverized form. Spray drying of viscous liquids and suspensions has become very important and, fluidized catalysts, consisting of particles suspended in a gas stream, have been receiving much attention. Farm pests and malaria mosquitoes are controlled almost exclusively by aerosols obtained by the atomization of powdered or liquid preparations. Screening smokes and mists are used in military operations.

In contrast to the problem of dispersion is the precipitation of unwanted aerosols, formed in many industrial processes, which is no less important than the conversion of materials into the aerosol state. Examples are smoke and fly ash evolved during the combustion of fuel, particularly pulverized fuel, in the smelting of metals, and in carbon black production; mists formed in sulphuric acid manufacture and electrolysis; dusts generated when materials such as cement are crushed. The need to precipitate aerosols is twofold: like metallic smokes they may contain valuable materials, or they may have a harmful influence on people and the environment, cause corrosion, and make the surrounding locality dirty. Many effective methods, including extremely efficient smoke filters for personal protection, have been devised to deal with industrial aerosols.

Other problems involving undesirable aerosols include the dispersion of fog, which hinders the landing of aircraft, the prevention of dust explosions in such places as coal mines, grinding plants and sugar factories where a detonation wave can put up a dust cloud, and ice formation on aircraft and telegraph wires due to supercooled mists and rain.

Aerosols are also important in science, particularly in experimental physics. Studies of the motion of droplets in a vertical electric field led to the first accurate determination of the electronic charge, of Avogadro's number and hence to the demonstration of the quantum nature of the photoelectric effect. Mist formation by condensation of supersaturated vapour on to ions is exploited in the Wilson cloud chamber, which has been one of the most important instruments in the growth of modern physics. The smoke filament method is widely used in experimental aerodynamics for studying velocity distribution.

The motion of aerosol particles under the action of external forces and molecular collisions controls almost all the phenomena, processes and methods of investigation mentioned above. For rain or snow to fall cloud particles have to come into contact and coalesce and the resulting drops must be large enough to reach the earth before evaporating. The number and distribution of particles of all sizes in the atmosphere are determined, apart from their rate of entry into the atmosphere, partly by their rate of fall and partly by the magnitude of turbulent exchange in the atmosphere.

Aerosol precipitation is determined by the moving particles being able to reach some macroscopic surface and stick to it. Coagulation, another important process taking place in aerosols, results from contact of particles brought about by the thermal or Brownian motion, or by various external influences. The absorption of sound waves by aerosols is caused by relative motion of the particles and the gas in which they are suspended. The combustion of particles, as well as depending on the rate of combustion, is determined by the trajectories of burning particles.

Study of the motion and precipitation of aerosol particles, for which the expression "Mechanics of Aerosols" is here used for the first time, is an important section of aerosol science. It is convenient to include those phenomena, very similar to precipitation and coagulation, which occur when aerosol particles come into contact with one another and with macroscopic objects; it also embraces the interesting but little studied reverse process of detachment of particles from walls and the dissemination of powders as aerosols.

Only a few books and monographs exist which deal with aerodisperse systems and in these comparatively little attention has been paid to the mechanics of aerosols, despite its great practical importance. Papers on this branch of the subject are scattered throughout numerous periodicals devoted to different aspects of science and technology. These articles are often of a strictly applied nature and only in rare cases do they contain data which are also of theoretical interest. Unfortunately most experimental work on aerosols has been carried out with highly polydisperse systems for which the particle size distribution has either not been studied at all or has been determined only very roughly. It is difficult to use such work for checking theoretical results and finding new laws.

The main object of this book is to collect and critically examine all theoretical and experimental material relating to the mechanics of aerosols. Some branches of the subject, for example the theory of steady rectilinear motion of particles of different sizes and the theory of thermal coagulation of aerosols, have been investigated very fully with corresponding effect upon the size of these chapters. On the other hand, important problems like motion, precipitation and coagulation of aerosols in turbulent flow, and detachment and disintegration of particles, are rather blind spots in the mechanics of aerosols, so that the corresponding chapters of this book may leave the reader feeling dissatisfied.

Many theoretical problems in the mechanics of aerosols present great mathematical difficulties; their solution can frequently be obtained only in the form of complex, slowly convergent series, the practical use of which requires an inordinate amount of computation. Sometimes, particularly in problems of diffusion, it has therefore been thought best to deal not with the rigorous solution but to adopt well-known simplifying assumptions in order to obtain an idea, albeit rough, of the phenomenon being examined.

In all sections of the book the author has attempted to quote all the more or less reliable experimental data on the topic under discussion but, in view of the nature of the book, the inclusion of descriptions of measuring techniques and experimental details would not be appropriate. On the other hand, quite a large number of tables and graphs are included which may be useful to people whose work brings them into contact with various aspects of the mechanics of aerosols.

PRINCIPAL NOMENCLATURE

A	coefficient in Cunningham's formula
B	mobility of particles
C_E	electrical capacity
D	diffusion coefficient
D_t	turbulent diffusion coefficient
E	field strength
F	force
F_M	resistance of medium
$F(r)$	cumulative frequency distribution function
G	velocity of thermal motion of a particle
G_g	velocity of thermal motion of a gas molecule
$G(r)$	cumulative weight distribution function
I	current; flow (particles, energy) per $cm^2 \cdot sec$.
K	coagulation constant
L	length
M	molecular weight
N	total number of particles
P	moment of force; dipole moment
R	radius of sphere or cylinder
R_g	gas constant
Re	Reynolds' number for a particle
Re_f	Reynolds' number for a flow
S	area
Sc	Schmidt's number ν/D
Stk	Stokes' number l_i/R
T	absolute temperature
U	flow velocity
U_o	velocity amplitude of oscillations of medium
U^*	friction velocity in turbulent flow
V	velocity of particle
V_s	terminal rate of settling of a particle
V_E	velocity of a particle in an electric field
V_r	velocity of a particle relative to the medium
$W(x,t)$	probability of finding a particle

$W^*(x,t)$	probability that a particle will reach a boundary
a	equatorial semi-axis of ellipsoid of revolution
c	weight concentration; polar semi-axis of ellipsoid of revolution
c_p	specific heat
c_s	velocity of sound
erf	error function
$f(r)$	frequency distribution function
$g(r)$	weight distribution function
g	acceleration due to gravity
k	Boltzmann constant
l	mean free path of gas molecules
l_B	apparent free path of particles
l_i	stop distance of particles
ln	natural logarithm
log	logarithm to base 10
m	mass of particle
m'	mass of medium displaced by particle
m_g	mass of a gas molecule
n	number of particles per cm^3
n_g	number of gas molecules per cm^3
p	pressure
q	charge
r	radius of particle
r_e	equivalent radius
r_s	Stokes' radius
t	time
t_p	period of oscillation
u	electrical mobility; root mean square eddy velocity of medium
v	volume; root mean square eddy velocity of particle
$w(x_o, x, t)$	probability of transition
α	coefficient of accommodation or of sound absorption
γ	density of particle
γ_g	density of medium (gas)
δ	thickness of layer; distance from wall
ε	elementary charge; energy dissipated per sec per gram of medium
ε_k	dielectric constant
η	viscosity of medium
θ	polar angle; phase angle
$\varkappa$	shape factor; Kármán constant; cloud settling factor
Π	potential difference

λ	wave-length; scale of turbulent fluctuations
λ_0	internal scale of turbulence
ν	integer; frequency of oscillations; kinematic viscosity
χ	thermal conductivity
ϱ	distance from centre or axis
σ	specific free surface energy; electric charge density
τ	relaxation time; force of internal friction per cm^2
τ_o	friction force of gas against surface (per cm^2)
φ	coefficient of filling space (volume of disperse phase per cm^3 of the system)
ψ	drag coefficient
$\psi(\varrho)$	potential function
ω	angular velocity; angular frequency
Φ	volume flow rate of gas; number of particles deposited on or passing through unit area per second; strength of point source
Φ'	number of particles deposited per second on unit length of cylinder; strength of a line source per unit length
Γ	gradient of velocity or temperature
Ω	energy
$\ni$	collection or impaction efficiency

TABLE OF CHARACTERISTIC QUANTITIES IN THE MECHANICS OF AEROSOLS

r (cm)	D (cm² sec⁻¹)	$\overline{G}$ (cm sec⁻¹)	τ (sec)	l_B (cm)	$\overline{\Delta x_B}$ (cm)	Δx_s (cm)
10^{-7}	$1{\cdot}28 \times 10^{-2}$	4965	$1{\cdot}33 \times 10^{-9}$	$6{\cdot}59 \times 10^{-6}$	$1{\cdot}28 \times 10^{-1}$	$1{\cdot}31 \times 10^{-6}$
2×10^{-7}	$3{\cdot}23 \times 10^{-3}$	1760	$2{\cdot}67 \times 10^{-9}$	$4{\cdot}68 \times 10^{-6}$	$6{\cdot}40 \times 10^{-2}$	$2{\cdot}62 \times 10^{-6}$
5×10^{-7}	$5{\cdot}24 \times 10^{-4}$	444	$6{\cdot}76 \times 10^{-9}$	$3{\cdot}00 \times 10^{-6}$	$2{\cdot}58 \times 10^{-2}$	$6{\cdot}63 \times 10^{-6}$
10^{-6}	$1{\cdot}35 \times 10^{-4}$	157	$1{\cdot}40 \times 10^{-8}$	$2{\cdot}20 \times 10^{-6}$	$1{\cdot}31 \times 10^{-2}$	$1{\cdot}37 \times 10^{-5}$
2×10^{-6}	$3{\cdot}59 \times 10^{-5}$	55·5	$2{\cdot}97 \times 10^{-8}$	$1{\cdot}64 \times 10^{-6}$	$6{\cdot}75 \times 10^{-3}$	$2{\cdot}91 \times 10^{-5}$
5×10^{-6}	$6{\cdot}82 \times 10^{-6}$	14·0	$8{\cdot}81 \times 10^{-8}$	$1{\cdot}24 \times 10^{-6}$	$2{\cdot}95 \times 10^{-3}$	$8{\cdot}64 \times 10^{-5}$
10^{-5}	$2{\cdot}21 \times 10^{-6}$	4·96	$2{\cdot}28 \times 10^{-7}$	$1{\cdot}13 \times 10^{-6}$	$1{\cdot}68 \times 10^{-3}$	$2{\cdot}24 \times 10^{-4}$
2×10^{-5}	$8{\cdot}32 \times 10^{-7}$	1·76	$6{\cdot}87 \times 10^{-7}$	$1{\cdot}21 \times 10^{-6}$	$1{\cdot}03 \times 10^{-3}$	$6{\cdot}73 \times 10^{-4}$
5×10^{-5}	$2{\cdot}74 \times 10^{-7}$	0·444	$3{\cdot}54 \times 10^{-6}$	$1{\cdot}53 \times 10^{-6}$	$5{\cdot}90 \times 10^{-4}$	$3{\cdot}47 \times 10^{-3}$
10^{-4}	$1{\cdot}27 \times 10^{-7}$	0·157	$1{\cdot}31 \times 10^{-5}$	$2{\cdot}06 \times 10^{-6}$	$4{\cdot}02 \times 10^{-4}$	$1{\cdot}28 \times 10^{-2}$
2×10^{-4}	$6{\cdot}10 \times 10^{-8}$	$5{\cdot}55 \times 10^{-2}$	$5{\cdot}03 \times 10^{-5}$	$2{\cdot}80 \times 10^{-6}$	$2{\cdot}78 \times 10^{-4}$	$4{\cdot}93 \times 10^{-2}$
5×10^{-4}	$2{\cdot}38 \times 10^{-8}$	$1{\cdot}40 \times 10^{-2}$	$3{\cdot}08 \times 10^{-4}$	$4{\cdot}32 \times 10^{-6}$	$1{\cdot}74 \times 10^{-4}$	$3{\cdot}02 \times 10^{-1}$
10^{-3}	$1{\cdot}38 \times 10^{-8}$	$4{\cdot}96 \times 10^{-3}$	$1{\cdot}23 \times 10^{-3}$	$6{\cdot}08 \times 10^{-6}$	$1{\cdot}23 \times 10^{-4}$	1·21
	γ^0	$\gamma^{-1/2}$	γ	$\gamma^{1/2}$	γ^0	γ

THE MECHANICS OF AEROSOLS

CHAPTER I

CLASSIFICATION OF AEROSOLS. SIZE AND SHAPE OF AEROSOL PARTICLES

§ 1. CLASSIFICATION OF AEROSOLS

Disperse systems with a gas-phase medium and a solid or liquid disperse phase are called aerosols or aerodisperse systems. Up to the present time there has existed no generally accepted classification of aerosols and no consistent method of describing different types of aerosol, complete arbitrariness being observed in the literature. It seems rational that classification should differentiate between aerosols formed by dispersion and by condensation while distinguishing systems having solid or liquid disperse phases. In addition, the designation of individual types of aerosol must coincide as far as possible with the names which are given to them in everyday non-technical language (dust, mist, smoke, and so on).

Dispersion aerosols are formed by the grinding or atomization of solids and liquids and by the transfer of powders into a state of suspension through the action of air currents or vibration. Condensation aerosols are formed when supersaturated vapours condense, and as a result of reactions between gases leading to the formation of a non-volatile product such as soot. Apart from their origin, the main difference between these two classes is that dispersion aerosols are in most cases considerably coarser than condensation aerosols; the former contain a wider range of particle size and, when the disperse phase is solid, usually consist of individual or slightly aggregated particles of completely irregular form (fragments). In condensation aerosols solid particles are often loose aggregates of a very large number of primary particles of a regular crystalline or spherical form.

The difference between aerosols with liquid and solid disperse phases is apparent in that, in the former, the particles are spherical and on collision they may fuse together to produce a single spherical particle. Solid particles possess the most varied shapes and coagulate to form more or less loose aggregates, of equally varied shape, the apparent density of which may be much less than the density of the material of which they consist.

The following terms will be used in this book for various kinds of aerosol:

Condensation and dispersion aerosols with liquid particles will be called mists, regardless of particle size; in Russian they are also denoted by one and the same word.

Dispersion aerosols with solid particles will be called dusts, again irrespective of particle size. The view that only coarsely dispersed systems should be described as dust overlooks the fact that dusts with a high degree of dispersion can be formed by separation, either artificially or in the atmosphere.

Finally, condensation aerosols with a solid disperse phase will be called smokes; these include systems of condensation origin containing both solid and liquid particles, the most important of which are smokes formed during the incomplete combustion of fuel. Smokes of hygroscopic materials, such as ammonium chloride, consist of particles which can be solid, semi-liquid or liquid according to the humidity of the medium, and organic substances, which supercool readily, form smokes in which gradual conversion of liquid to crystalline particles takes place. Distinction between smokes and condensation mists and, consequently, strict adherence to the proposed terminology is sometimes difficult; such distinction is nevertheless preferable to the employment of the term smoke for both types of aerosol.

It is not uncommon to be required, in practice, to deal with aerosols containing particles of both dispersion and condensation origin. Thus, in furnace smoke there is always a greater or smaller quantity of mechanically carried ash from the grates; the so-called "atmospheric condensation nuclei" consist partly of dried sea-water spray, and partly of sulphuric acid droplets formed by the oxidation of sulphur dioxide from flue gases. The air of industrial centres contains large numbers of particles of soot and ash, together with products of the destructive distillation of coal and atmospheric moisture, in sizes ranging from tenths of a micron to tenths of a millimetre. Such aerosols cannot be placed in a class according to any of the existing classifications, and the special name "smog" (smoke + fog) has been proposed for them.

Each of the aerosol types indicated above may occur in very different degrees of dispersion; this has a large effect on almost all the properties of disperse systems, and it is therefore desirable to distinguish between highly dispersed and coarsely dispersed aerosols (see p. 4).

The term cloud is given throughout this book to a free aerodisperse system of any type having a definite size and form (rain cloud, dust cloud, cloud of gun smoke, etc.) and is not specific to condensation aerosols with particles greater than 10^{-5} cm in the sense used by some authors [1, 2].

§ 2. PARTICLE SIZE IN AEROSOLS

Let us examine the question of the lower limit to aerosol particle size. Until recently the main method of determining particle size in highly dispersed aerosols ($\sim 10^{-7}$ cm) has been by measurement of their mobility in an electric field (see § 27).

Experiment has shown that in gases there are two types of charged particle known, respectively, as small (gaseous, light) ions and large (heavy, slow) ions. The mobility of the former is about one, and of the latter 10^{-3} to 10^{-4} cm^2 V^{-1} sec^{-1}. It has been established that gaseous ions are molecular aggregates comprising a charged molecule (strictly an ion) with neutral gas molecules attached to it by electrostatic and molecular forces.

Heavy ions differ from light ions in that they are formed only in gases containing suspended particles of solid or liquid; they represent the charged part of highly dispersed aerosols. The existence of "medium" ions, particles with mobilities of 10^{-3} to 10^{-1} cm^2 V^{-1} sec^{-1} has also been established; in the combustion products of illuminating gas and in the sodium flame there are particles with a mobility of 0·2 cm^2 V^{-1} sec^{-1}

[3, 4]. Mobilities of this order are possessed by gaseous ions in the vapour of some organic substances [4], for example amyl alcohol.

It is not possible to distinguish between gaseous ions and charged aerosol particles by mobility but their different behaviour during coagulation may be used. When gaseous ions coagulate, or recombine, the neutral molecular complex that is formed decomposes and the ions are annihilated. Upon the coagulation of aerosol particles, whose existence is not connected with the presence of charge, coarser particles are formed. The mobility of the ions formed in a flame falls several hundred times during a time of the order of 1 sec [3, 4]. The probability of the presence of multiple charges on the particles falls rapidly with reduction in particle size. On the assumption that these ions possess one elementary charge the mobility, $0{\cdot}2\ cm^2\ V^{-1}\ sec^{-1}$, corresponds to a radius of $1{\cdot}5 \times 10^{-7}$ cm (see p. 32).

Electron microscopy is at present the chief method for determining the size of very fine aerosol particles, and with special techniques may be used for particles of liquid and volatile substances. Aerosol particles with r about 10^{-7} cm (for example, silver iodide [5]) have been detected by this method.

Another very important method for measuring the size of particles in highly dispersed aerosols by diffusion (see § 39) has revealed that the radius of the particles in aerosols which have only just been formed in a corona discharge from a metallic point is about $0{\cdot}5 \times 10^{-7}$ cm [22].

Such is the state of affairs from the experimental side of the question. Theoretically it is quite possible that substances with a very stable crystal lattice can produce aerosols with particle sizes of two or three molecular diameters.

Passing on to the question of the upper limit to particle size in aerodisperse systems, it must be remembered that, in a stationary medium, particles with radii of several hundred microns settle so quickly that only with difficulty can they be detected in the suspended state. On the other hand, strongly rising or turbulent air will suspend particles several millimetres in size, as exemplified by sandstorms, snowstorms, the pneumatic lift of friable materials, and the fluidization of granular beds. Such particles should be included in a study of the mechanics of aerosols in view of the essential nature of the problems associated with them.

The subject of aerosols thus involves particles in the very wide range of sizes between 10^{-7} and 10^{-1} cm. It is not surprising that passage from the lower to the upper limit is accompanied by changes not only in nearly all the physical properties of aerosols, but also in the nature of the laws governing them. This can be seen particularly clearly in the resistance which a gas offers to the motion of particles. For very fine particles ($r < 10^{-6}$ cm) the resistance is proportional to the velocity and the square of the particle radius. In the range $10^{-6} - 10^{-4}$ cm there is a gradual change to Stokes' law; the resistance remains proportional to the velocity, but the dependence on radius becomes linear. Further increase in radius brings deviations from Stokes' law in the opposite direction; for velocities that are not very low the proportionality of resistance and velocity ceases while, at sufficiently large velocities and particle sizes, the resistance is more nearly proportional to the square of the radius and the square of the velocity (see p. 31).

Changes in the laws governing some important properties of aerosols are shown in Fig. 1. In all cases the transition region lies, roughly speaking, in the range of particle sizes between 0·5 to 1×10^{-5} and 10^{-4} cm. As regards the properties shown in groups 1 and 2, this is because they are connected either with the ratio of the particle radius to the mean free path of the gas molecules, about 10^{-5} cm in air at atmospheric pressure, or with the average wavelength of visible light ($0{\cdot}55 \times 10^{-4}$ cm). For the rest of the properties the correspondence is fortuitous.

The existence of this transition region enables aerosols to be separated into three groups. The first comprises highly dispersed aerosols with particles of radius less than 0·5 to 1×10^{-5} cm; such particles are characterized by resistance to motion and rates of evaporation and cooling proportional to r^2, the amount of light which they scatter is proportional to r^6, and the coagulation constant of an aerosol made from them

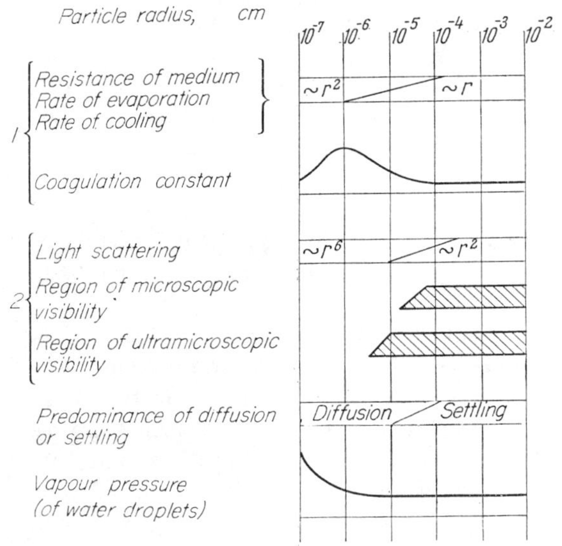

FIG. 1. Some properties of aerosols in relation to particle size.

depends on r. The particles are invisible under the ordinary microscope, and can be detected by ultra-microscopy only under exceptionally favourable conditions. The vapour pressure of the disperse phase in these aerosols is noticeably greater than the normal vapour pressure of the substance, so that distillation of the smaller particles on to the bigger ones can take place. Finally, Brownian movement of the particles predominates considerably over settling under the action of gravity.

In the second group of more coarsely dispersed aerosols, with particles of radius greater than about 10^{-4} cm, both the size and shape of the particles can be determined with the aid of a microscope, the coagulation constant does not depend on r, settling predominates markedly over Brownian movement and the laws stated above are replaced by others, as shown in Fig. 1.

Finally it is convenient to put systems with intermediate particle radii between 0·5 to 1×10^{-5} and 10^{-4} cm into a special group of aerosols with transitional properties. This group plays a very large part in the science of aerosols because the particle size is particularly convenient for ultra-microscopy, one of the principal means of studying aerosols, and predominates in condensation aerosols formed from substances of low vapour tension.

§ 3. PARTICLE SIZE DISTRIBUTION IN AEROSOLS†

The majority of natural and artificially produced aerosols possess quite considerable polydispersity.†† In view of the strong dependence of the physical properties of aerosols on particle size, a mean size is seldom sufficient for the specification of an aerodisperse system; the particle size distribution must be found. Before the invention of the electron microscope this was done with an ordinary microscope, which permits measurement of particles only in coarsely dispersed aerosols ($r > 3$ to 5×10^{-5} cm) or in the coarse fraction of more highly dispersed ones. In the fine fraction of the latter only the total number of particles was usually determined, without measurement of their size; alternatively the fine fraction was completely neglected. Size distribution can now be measured in aerosols having particles down to a radius of 0·5 to 2×10^{-7} cm.

The size distribution can be expressed in several ways. The fraction of the total number of particles having radii which lie between r and $r + dr$ can be written:

$$df = f(r)dr \tag{3.1}$$

with the condition

$$\int_0^\infty f(r)dr = 1. \tag{3.2}$$

The curve representing the function $f(r)$ is called the frequency distribution curve or the differential curve of particle size distribution (Fig. 2); more precisely it should be qualified as the number distribution curve, as distinct from the weight distribution curve which defines the weight fraction dg of particles with radii between r and $r + dr$:

$$dg = g(r)dr \tag{3.3}$$

with

$$\int_0^\infty g(r)dr = 1. \tag{3.4}$$

Note that the area bounded by the differential distribution curve, the axis of abscissae and the two verticals at the points r_1 and r_2 represents the fraction (number or weight) of particles with radii between r_1 and r_2.

The function $g(r)$ can be written

$$g(r) = \beta m_r f(r)$$

where m_r is the mass of a particle with radius r and β is a factor of proportionality which can be determined by means of integration:

$$\int_0^\infty g(r)\, dr = 1 = \beta \int_0^\infty m_r f(r)\, dr = \beta \overline{m}$$

where $\overline{m}$ is the arithmetic mean of the masses of the aerosol particles.

† The problems briefly reviewed in §§ 3 and 4 are set out in detail in Herdan's book [584].

†† Aerosols obtained from a Sinclair–La Mer generator [6] and those containing pollen or spores are comparatively monodisperse: the radius of particles in clover pollen lies between 24·8 and 26·9 μ [7].

Thus the functions $f(r)$ and $g(r)$ are connected by the simple equation

$$g(r) = \frac{m_r}{\bar{m}} f(r). \tag{3.5}$$

Since $\bar{m} = \gamma\bar{v}$, where γ is the density and $\bar{v}$ is the mean volume of the particles then for constant γ, i.e. in the case of an aerosol of uniform composition and free from aggregates, the weight distribution $g(r)$ is identical with the volume distribution $v(r) = (v_r/\bar{v}) f(r)$. For an aerosol such as atmospheric dust, which is not of uniform composition, or for one containing aggregates, the particles have different effective densities (see p. 18); only the volume distribution can then be obtained from microscopical measurement, and may differ considerably from the weight distribution.

The functions expressing the distribution of number and weight over the particle sizes are the most generally used. However, as will be shown below, it is sometimes necessary to use other distribution functions with the general formula $f_\nu(r) = r^\nu f(r)/\overline{r^\nu}$, where ν is a positive or negative integer. The quadratic distribution ($\nu = 2$) is particularly important.

In some aerosol problems it is convenient to express the distribution as a function not of the radius, but of the mass (or volume) of the particles. The mass distribution function of the particles indicates what number fraction $f'(m)\,dm$ or weight fraction $g'(m)\,dm$ of the particles possesses a mass lying between m and $m + dm$. In the case of spherical particles the functions $f(r)$ and $f'(m)$ (also $g(r)$ and g'(m)) are connected by the formula $4\pi\gamma\, r^2 f'(m) = f(r)$.

The differential distribution curves are illustrative, but in working out the results of measurements of aerosol particles and in the solution of some technical problems it is more convenient to use the integral (cumulative) distribution curves, which show what fraction of particles (by number or weight) possesses radii greater or less than a given value r. The appropriate distribution functions are obtained for the number distribution by integrating the function $f(r)$ from r to ∞ in the first case (a) and from 0 to r in the second case (b)

$$F_a(r) = \int_r^\infty f(r)\,dr, \quad F_b(r) = \int_0^r f(r)\,dr \tag{3.6}$$

and similarly for the weight distribution

$$G_a(r) = \int_r^\infty g(r)\,dr, \quad G_b(r) = \int_0^r g(r)\,dr. \tag{3.7}$$

The function $G_a(r)$ is especially frequently used in technology in the investigation of various industrial dusts. The cumulative weight distribution curve defined by it is called the characteristic curve of the given dust and also the residue curve, because it is determined by the weight of dust remaining on a given sieve or at a given air velocity in an air separator. The curve $G_b(r)$ is called the "pass curve". Note that

$$F_a(r) + F_b(r) = G_a(r) + G_b(r) = 1. \tag{3.8}$$

We will illustrate what has been said with the example of the droplet size distribution in a stratus cloud [8], determined by means of a microscope. Let us note that, as opposed to systems with a liquid medium for which sedimentometry yields cumulative distribution curves, in aerosols it is usual to determine the number of particles possessing radii between definite limits; stepped graphs, or histograms, are thus obtained instead of continuous curves.

TABLE 1. DISTRIBUTION OF DROPLET SIZES IN STRATUS CLOUD

Range of droplet radii (μ)	2·5-4	4-5·5	5·5–7	7-8·5	8·5-10	10-11·5
Number of droplets	4	6	15	24	24	12
Range of droplet radii (μ)	11·5-13	13–14·5	14·5–16	16-17·5	17·5–19	
Number of droplet	4	4	4	1	2	

In the case under examination the results shown in Table 1 were obtained from the measurement of 100 droplets.

From these data the histogram illustrated in Fig. 2 was constructed. It can be used immediately to calculate various mean dimensions (see p. 14), but in most cases it is

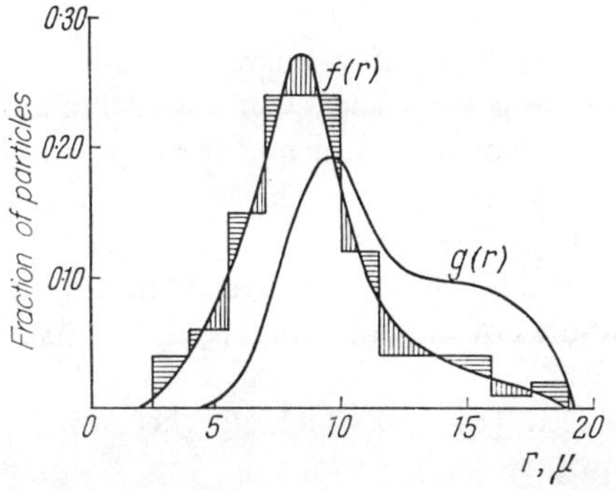

FIG. 2. Frequency distribution curves for number and weight of particles with respect to particle size.

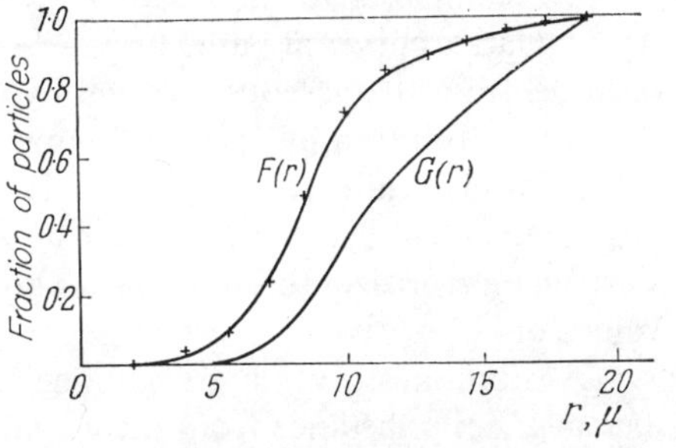

FIG. 3. Cumulative particle size distribution curves.

desirable to convert the histogram into a continuous curve and smooth out irregularities caused by the insufficient number of measurements, for instance the physically improbable rise at the extreme right of the histogram.

Such smoothing out of experimentally determined results is justifiable because the number of particles measured is of necessity limited, with the result that the statistical fluctuations are fairly pronounced. Rather than smoothing the histogram directly, it is better to begin by constructing the smooth cumulative curve $F_b(r)$ from the experimental points (denoted by crosses in Fig. 3). By doing this, it is comparatively easy to smooth out random errors of measurement and fluctuations in the number of particles in each fraction. The differential curve $f(r)$ (see Fig. 2) is then constructed from the integral curve by one of the methods of graphical differentiation.

The weight distribution curves could be obtained by constructing a "weight histogram" directly from the experimental data, although in this case the fluctuations mentioned above would be particularly large. It is better to start from the curve $f(r)$ obtained by smoothing the counts. The axis of r is divided into sufficiently narrow intervals, Δr, for each of which the corresponding average values of the particle mass $m_r = 4\pi\gamma r^3/3$ are calculated. Then, by means of the curve $f(r)$, the mass $m_r f(r) \Delta r$ of the particles in each interval and the mean mass per particle, $\overline{m}$, are determined, whence the required weight distribution function is found from formula (3.5). The curve $g(r)$ in Fig. 2 was constructed in this way. From it, in turn, the cumulative curve $G_b(r)$ (see Fig. 3) is found.

Sieve or sedimentation analysis of dust yields the function $G(r)$ directly and transformation to number distribution is carried out by reversing the procedure above. Construction of quadratic and other size distributions is performed in a similar manner.

The choice of a method for expressing the particle size distribution in an aerosol depends on what properties of the latter are to be characterized. The rate of thermal coagulation of aerosols, for example, depends on the number distribution of particle sizes $f(r)$ (page 295). Evaporation from a coarsely dispersed aerosol at every instant of time is governed by the distribution of particle radius $rf(r)$, because the rate of evaporation of the particles is proportional to their size. The optical density of coarsely dispersed mists is determined by the surface area distribution $r^2 f(r)$ because reflection and scattering of light by large droplets is proportional to the square of their radius. Precipitation of a coarse aerosol by gravity or particle inertia is also controlled by this distribution.

The examples given above involve the differential distribution curves. The integral or cumulative curves are used primarily for calculating the efficiency of removal of the particles from the gaseous medium by various types of separator and for expressing the size distribution by means of empirical equations (see p. 10). The specification of industrial materials, such as insecticidal dusts, in powder form usually includes the percentage of weight which remains on a sieve of a given size and passes through a sieve of a larger size; this amounts to giving the value of the function $G(r)$ at two fixed values of r.

Distribution curves are sometimes encountered, (Fig. 4), which cut the ordinate axis at a finite distance from the origin of coordinates. Curves of this type, giving a false representation of the particle size distribution, are obtained as a result of the limited number of size intervals into which the particles have been divided.

Suppose the curves in Fig. 5 represent the actual size distribution. For sufficiently narrow intervals ($\Delta r = 0{\cdot}1\,\mu$), for example in measuring particles with the aid of an electron microscope, the histogram shown in Fig. 5a would be obtained. On smoothing, this would yield a curve very close to the actual distribution. However, when measuring the particles by means of an ordinary microscope the best that could be done would be to choose $\Delta r = 0{\cdot}2\,\mu$ which would provide the histogram shown in Fig. 5b and a curve of the type A in Fig. 4. Finally, for intervals $\Delta r = 0{\cdot}5\,\mu$, as taken in ordinary industrial hygiene work, the histogram sketched in Fig. 5c and a curve of the type B in Fig. 4 would be obtained.

Such distortion of the true distribution takes place if a significant number of particles are so small that the method used for determining size is not applicable; another cause is very high polydispersity of the aerosol, with particle sizes extending over several orders of magnitude.

As an example, suppose that the radii of the particles of a certain aerosol lie in the range 0·1–200 μ and that the total length of the axis of abscissae is equal to 100 mm. The part of the distribution curve corresponding to the limits 0·1–10 μ receives an interval of only 0·5 mm, and it is impossible to represent the distribution correctly on such a graph.

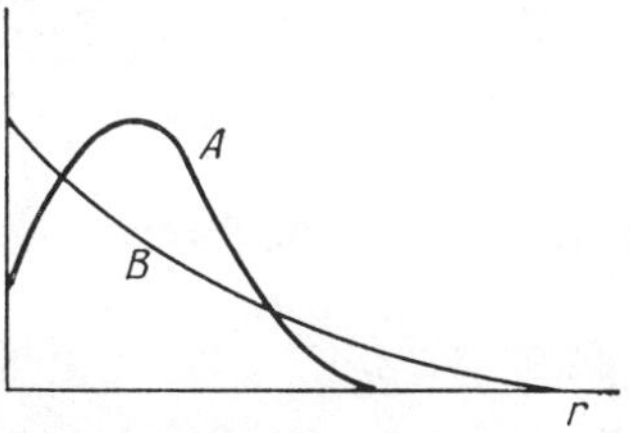

FIG. 4.
Misleading distribution curves.

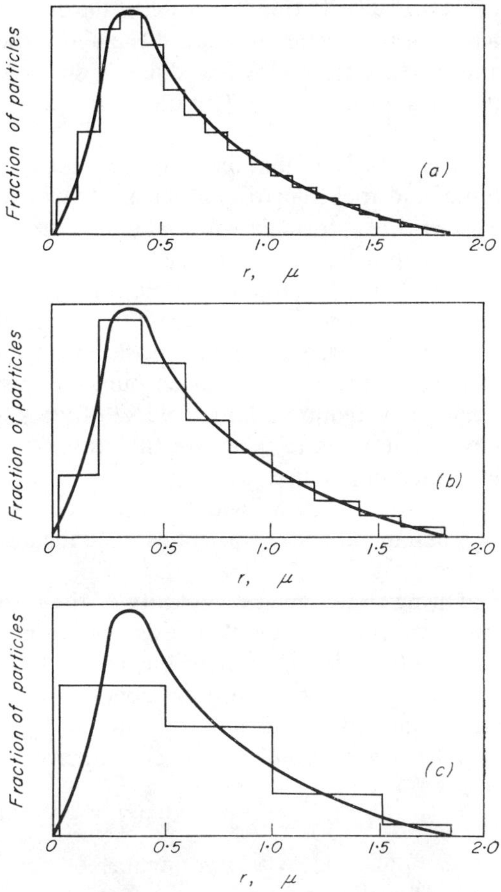

FIG. 5. Relationship of the actual size frequency curve to histograms derived from observations made with different size intervals.

It would seem better in such cases to use a logarithmic scale along the axis of abscissae, thus avoiding the crowding together of very fine particles and devoting to each order of magnitude an equal length of scale. However, a new difficulty now arises. If a logarithmic scale is adopted for the radii and the distribution curve is constructed by simply plotting values of the function $f(r)$ along the ordinate axis, then the area bounded by this curve, the axis of abscissae, and the verticals at the points r_1 and r_2 will. be equal to

$$\int_{r_1}^{r_2} f(r)\, d\log r = \int_{r_1}^{r_2} \frac{1}{r} f(r)\, dr,$$

i.e., this area is no longer proportional to the fraction of particles with radii from r_1 to r_2 and the curve loses its illustrative character. In order to preserve the significance of the area, $rf(r)$ must be plotted along the ordinate axis instead of $f(r)$, but for very polydisperse aerosols the curve $rf(r)$ practically coincides with the axis of abscissae in the region of small r. Thus we again arrive at the inadequacy of visual representation of the size distribution by means of the curve. Recently in the study of aerosols there has been a tendency to discard the above methods of representing size distribution and to replace them by coordinate systems which cause the distribution to be displayed as a straight line. This question is closely connected with others which will now be examined.

The use of distribution curves for specifying industrial aerosols and solving theoretical and applied aerosol problems is inconvenient. It is simpler to express the curves by means of a formula with the minimum number of coefficients, the magnitude of which is peculiar to a given distribution. It is desirable that such a formula should be applicable to the largest possible number of aerodisperse systems so that in going from one system to another only the values of the coefficients change. By employing a sufficiently large number of coefficients it would be possible to represent by a single formula all the distributions encountered in practice. However the choice of coefficients would require a large volume of work and it would be difficult to assign any physical meaning to them; for this reason formulae with many coefficients have not been used in practice.

As a rule, formulae with only two coefficients are chosen, this number being the very minimum, one coefficient defining the mean particle size and the other the spread in size.

Owing to the great complexity of the formation processes of both condensation and dispersion aerosols, there exists, as yet, no theoretical derivation of a distribution formula (with one exception: see below), but there is a series of empirical formulae applicable mainly to aerosols obtained by the mechanical disintegration of solid and liquid bodies. The most common of these are:

(1) Roller's formula [9] which, with symbols as used above, has the form†

$$G_b(r) = ar^{1/2} \exp(-s/r) \tag{3.9}$$

and is applicable to a large number of powdered industrial materials over a wide range of particle sizes.

(2) The Rosin–Rammler formula [10]

$$G_a(r) = \exp(-ar^s), \tag{3.10}$$

applicable to comparatively coarsely dispersed dusts and mists obtained by mechanical atomization.

A better formula for the latter has been proposed by Nukiyama and Tanasawa [11]

$$f(r) = ar^2 \exp(-br^s), \tag{3.11}$$

in which a and b are not independent but are functions of s and the mean size of the drops. These functions have been worked out and tabulated by the authors.

† For $r \to \infty$, $G_b(r)$ in this formula also $\to \infty$. Therefore the integral distribution curve must in this case be terminated at such a value, r_1, that $G_b(r_1) = 1$.

Formula (3.9) can be rewritten in the form

$$\log [G_b(r)/r^{1/2}] = \log a - 0{\cdot}434\, s/r. \tag{3.12}$$

If values of $1/r$ are plotted as abscissa and $\log (G_b/r^{1/2})$ as ordinate then, if this formula applies, the experimental points should fall on a straight line from which a and s can easily be determined.

Similarly from formula (3.10) we obtain

$$\log G_a(r) = -0{\cdot}434\, ar^s. \tag{3.13}$$

In this case the value of s must be chosen so that the experimental points lie on a straight line when r^s and $\log G_a$ are chosen as coordinates. Similar treatment of experimental data is carried out for formula (3.11) which takes the form

$$\log [f(r)/r^2] = \log a - 0{\cdot}434\, br^s. \tag{3.14}$$

Putting the formula into such a form that the size distribution is expressed as a straight line greatly simplifies the task of choosing the coefficients in these formulae and smoothing the experimental data.

In a very few aerosols, such as those formed by plant spores [12], the distribution curves are symmetrical and close to the shape of the Gaussian curve or normal distribution:

$$f(r) = \frac{1}{\beta\sqrt{2\pi}} \exp [-(r - \bar{r})^2/2\beta^2] \tag{3.15}$$

where $\bar{r}$ is the mean particle radius; $\beta^2 = \overline{(r - \bar{r})^2}$ is the mean square of the deviation of the radius from $\bar{r}$. This is termed the variance and its square root is the standard deviation of the radii.

Introducing the auxiliary variable

$$\xi = (r - \bar{r})/\beta\sqrt{2}. \tag{3.16}$$

gives the fraction of particles with radii $\leqslant r_1$ as

$$\int_0^{r_1} f(r)\, dr = \frac{1}{\beta\sqrt{2\pi}} \int_0^{r_1} e^{-(r-\bar{r})^2/2\beta^2}\, dr = \frac{1}{\sqrt{\pi}} \int_{-\frac{\bar{r}}{\beta\sqrt{2}}}^{\frac{r_1-\bar{r}}{\beta\sqrt{2}}} e^{-\xi^2}\, d\xi. \tag{3.17}$$

The meaning of the function $f(r)$ makes it different from 0 only for $r \geqslant 0$, but it is convenient to take the lower limit of the integral as $-\infty$ which introduces a negligible error provided that the mean radius is large compared with the standard deviation.

$$\int_0^{r_1} f(r)\, dr = \frac{1}{\sqrt{\pi}} \int_{-\infty}^{\frac{r_1-\bar{r}}{\beta\sqrt{2}}} e^{-\xi^2}\, d\xi = \frac{1}{2}\left[1 + \operatorname{erf}\left(\frac{r_1 - \bar{r}}{\beta\sqrt{2}}\right)\right] = \frac{1}{2}(1 + \operatorname{erf} \xi_1), \tag{3.18}$$

where

$$\operatorname{erf} \xi_1 = \frac{2}{\sqrt{\pi}} \int_0^{\xi_1} e^{-\xi^2} d\xi \quad \text{(Error function)} \tag{3.19}$$

and ξ_1 is the value of ξ corresponding to r_1.

Let us plot ξ on an arbitrary scale along the ordinate axis (Fig. 6) and set against it the corresponding values of $0{\cdot}5\,[1 + \operatorname{erf}(\xi)]$, i.e., the fraction of particles in the normal distribution under investigation for which $r < \bar{r} + \beta\sqrt{2}\,\xi$. As before, we will plot r along the axis of abscissa. If the integral curve $F_b(r)$, expressing the fraction of particles with radii less than r, is constructed in this "probability" coordinate system,

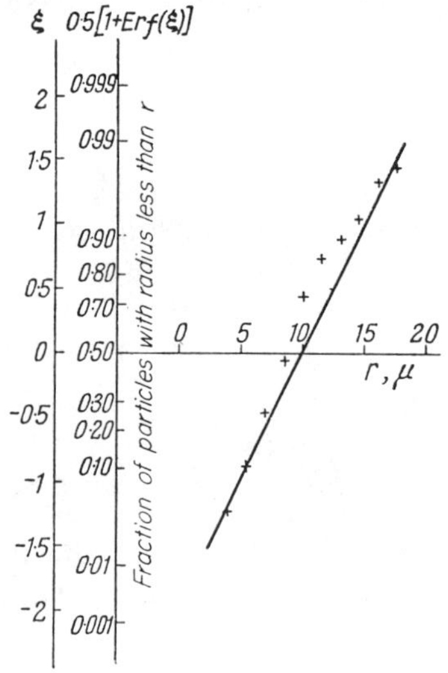

FIG. 6. Cumulative distribution curves on a normal probability plot.

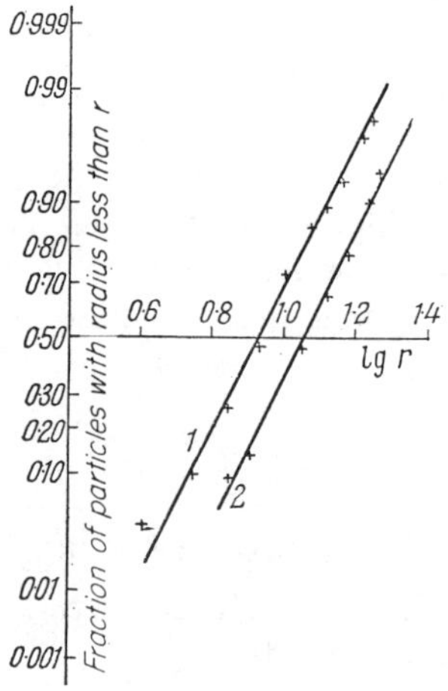

FIG. 7. Cumulative distribution curves on a logarithmic probability plot.

then in the case of the normal distribution of sizes, in agreement with equation (3.16), a straight line cutting the axis of abscissa at $r = \bar{r}$ must be obtained. The tangent of the angle of inclination of this straight line to the axis of abscissa is proportional to $1/\beta$.

The data of Table 1, for a water mist, do not give a straight-line graph on the probability plot, Fig. 6, on which size distribution of droplets is shown by crosses. This is because the frequency distribution is skew (see Fig. 2). The overwhelming majority of condensation and dispersion aerosols possess such asymmetrical distributions, the steeper slope being on the side of small r. This is the reason for the crowding together of points representing fine particles, which was mentioned above, when the size is used on a linear scale of abscissa in the distribution curves. If the logarithm of the radius is taken along the abscissa then the curves aquire a more symmetrical form and often approximate to a Gaussian curve. In this case the distribution is lognormal and is expressed by the formula

$$f(r)\,dr = \frac{1}{\log \beta_g \sqrt{2\pi}} \exp\left[-\frac{(\log r - \log r_g)^2}{2(\log \beta_g)^2}\right] d\log r. \tag{3.20}$$

Here $\log r_g = \overline{\log r}$ and consequently r_g is the geometrical mean of the particle radii. $(\log \beta_g)^2 = \overline{(\log r - \log r_g)^2}$ is the variance of the logarithms of the radii. The standard deviation of the logarithm of the sizes, β_g, is called the standard geometric deviation. As can be seen in Fig. 7, the mist taken as an example (stratus cloud) gives a straight-line graph† on the logarithmic probability plot. The distribution of droplet sizes in natural clouds has been shown by Levin to be lognormal for a large number of observations on mount Elbrus [13]. The distribution has also been found in other aerosols including rock [14] and uranium [15] dust formed by mechanical grinding, mists produced by a disk atomizer [16], and aerosols of NH_4Cl and H_2SO_4 formed by mixing gaseous reagents [17]. The lognormal distribution may have theoretical significance [18]. In particular, Kolmogorov has shown [19] that simple hypotheses about the process of grinding solid particles, result in a particle size distribution which tends asymptotically to the lognormal distribution with progressive size reduction [20]. It would be extremely interesting to find out under what conditions this distribution results from condensation processes.

When the number of particles in a given size range is distributed lognormally [formula (3.20)] the weight and other distributions of a power of the radius are also lognormal with the same value of β_g; all such derived distributions are therefore represented by parallel straight lines [21].

The distribution corresponding to the ν-th power of the radius is expressed by

$$a \exp(2{\cdot}302\, \nu \log r) \frac{1}{\log \beta_g \sqrt{2\pi}} \exp\left[-(\log r - \log r_g)^2/2\log^2 \beta_g\right]$$

$$= \frac{a}{\log \beta_g \sqrt{2\pi}} \exp\left[-\frac{(\log r - \log r_g)^2 - 2{\cdot}302\, \nu \log r \,.\, 2\log^2 \beta_g}{2\log^2 \beta_g}\right]$$

$$= \frac{1}{\log \beta_g \sqrt{2\pi}} \exp\left\{-\frac{[\log r - (\log r_g + 2{\cdot}302\, \nu \log^2 \beta_g)]^2}{2\log^2 \beta_g}\right\}.$$

where a is normalising constant.

The weight distribution of droplet sizes in a mist having the number distribution represented by the straight line 1 in Fig. 7 is expressed by the parallel straight line 2.

It is characteristic of each of the distribution formulae examined in this section that points towards the ends of the graph fail to conform with the straight line. This has no great practical significance because points at the extremes of the integral curves represent only a small fraction of the particles. Some authors [9] claim that the formulae proposed by them are correct over the whole range and assign the deviations to experimental errors, premature settling of particles, etc. It is difficult to agree with this since the theoretical significance which can be attached to these formulae is doubtful. They are more or less successful empirical approximations to the actual distribution. The fact that each formula has been applied with success to several groups of aerosols suggests that general significance is lacking. Nevertheless, the practical value of these formulae is undoubted, because they make it possible to express the properties governed by particle size in terms of two parameters and so to compare one aerosol with another.

† In drawing straight lines on log–probability plots it should be borne in mind that experimental points lying far from the median axis have lower statistical weight because they refer to a relatively small number of particles.

The labour expended in finding a suitable formula and determining the values of the two coefficients is usually considerably less than that required to measure a sufficiently large number of particles; the further development of aerosol studies, particularly the theory of aerosol formation, would be assisted by analysing in this manner measurements carried out on as large a number of aerodisperse systems as possible.

One of the most important tasks in experimental study of aerosols is the establishment of the relationship between their properties, e.g. retention in filters, respiratory tract etc., and the particle size. If such a study is made with polidisperse aerosols, it is necessary to determine the fraction of particles of each size retained on the filter etc., which involves measurements of a great number of particles. These measurements are so cumbersome that usually insufficient number of particles are measured and therefore the results are much less reliable than those obtained with fairly monodisperse aerosols.

Since the time of Aitken, particles in the atmosphere have been counted, after condensing water upon them, in special counters for condensation nuclei, as these particles are termed. Presumably, only a very small portion of the particles contained in the atmosphere are poorly wetted by water and cannot be detected by this means. Junge [585], who studied the size distribution of condensation nuclei by means of an impactor for large sizes, and from their electric mobility in the case of small ones, found that for particles greater than 0·1 μ the distribution was

$$f(r) = C/r^4 \tag{3.21}$$

where C is a constant. This formula was also obtained by Junge by a rather rough deduction from the theory of coagulation.

Condensation nuclei having $r > 1\mu$ are called gigantic nuclei, those with $1\mu > r > 0{\cdot}1\,\mu$ are large and those with $r < 0{\cdot}1\,\mu$ are termed Aitken nuclei.

§ 4. MEAN SIZES OF AEROSOL PARTICLES

Knowledge of the particle size distribution is necessary for the complete specification of an aerosol. However, an indication of a mean size may be all that is available when, instead of the size-distribution, some property depending on it has been measured, such as the diffusion coefficient, line broadening on an X-ray photograph of deposited particles, or the diameter of a diffraction ring. An indication of the dispersity of aerosols is often obtained from the ratio of the weight concentration (mass of the disperse phase in unit volume) to the number concentration (number of particles in unit volume); this is equal to the mean mass of a particle and, if the density is known, the mean particle size can be obtained.

The mean particle sizes determined by different methods can differ considerably from one another. Corresponding to distributions of number, weight and other functions of the radius in respect of particle size there exist a series of mean values of the radius.

(1) Arithmetic mean radius

$$r_1 = \bar{r} = \int_0^\infty r f(r)\, dr \approx \sum_\nu r_\nu N_\nu / N, \tag{4.1}$$

where N_ν is the number of particles in the ν-th interval of radii;
r_ν is the radius at the middle of this interval;
N is the total number of particles;

(2) Mean square (mean surface area) radius

$$r_2 = \sqrt{\overline{r^2}} = \left[\int_0^\infty r^2 f(r)\, dr\right]^{1/2} \approx \left[\sum_\nu r_\nu^2 N_\nu/N\right]^{1/2}; \tag{4.2}$$

(3) Mean cube (mean volume or weight) radius

$$r_3 = \sqrt[3]{\overline{r^3}} = \left[\int_0^\infty r^3 f(r)\, dr\right]^{1/3} \approx \left[\sum_\nu r_\nu^3 N_\nu/N\right]^{1/3} \tag{4.3}$$

Other mean radii can be similarly defined.

These means are derived from distributions with respect to size so that the coarse and fine particles are equally represented. Means derived from $g(r)$, the distribution with respect to weight, are very important in practice, for example,

$$r_1' = \int_0^\infty r g(r)\, dr \approx \sum_\nu r_\nu g_\nu/G, \tag{4.4}$$

where g_ν is the weight of the particles in the ν-th interval of particle weight and
G is the total weight of the particles.

In addition, use is frequently made of:

(4) the geometric mean radius r_g, defined by the formula

$$\log r_g = \overline{\log r} = \int_0^\infty \log r \,.\, f(r)\, dr \approx \sum_\nu N_\nu \log r_\nu/N; \tag{4.5}$$

(5) the number median radius r_m, determined from the condition $F_a(r_m) = F_b(r_m) = 0{\cdot}5$, i.e., half the particles have radii exceeding r_m and the other half are smaller;

(6) the weight median radius r_m', determined from the similar condition $G_a(r_m') = G_b(r_m') = 0{\cdot}5$, i.e., the weight of the particles with radii greater than r_m' is half the total weight of the aerosol.

In normal distributions $r_m = \bar{r}$; in lognormal $r_m = r_g$ and $\log r_m' = \log r_m + 6{\cdot}908\,(\log \beta_g)^2$ [14].

As an illustration the various mean particle sizes in the mist examined above will be calculated. It is best to start directly from the experimental data (see Table 1), without smoothing, and to carry out the calculation by formulae

$$r_1 = \sum_\nu r_\nu N_\nu/N \tag{4.6}$$

and so on.

In this way it is found that $r_1 = 8{\cdot}9\,\mu$, $r_2 = 9{\cdot}4\,\mu$ and $r_3 = 9{\cdot}9\,\mu$.

The curves for F_b and G_b of Fig. 3 give $r_m = 8{\cdot}5\,\mu$ and $r_m' = 11{\cdot}1\,\mu$. Curves 1 and 2 of Fig. 7 give the values $r_m = 8{\cdot}6\,\mu$ and $r_m' = 11{\cdot}5\,\mu$.

Means obtained from the experimental determination of particle size depend on the method of measurement. The ratio of weight to number of particles gives r_3; the corona method (diameter of diffraction rings) r_1, and so on.

Choice of a mean to characterize the operative particle size of an aerosol, like the choice of a distribution curve, is governed by the properties which are important. The optical density of coarse aerosols and their rate of precipitation in gravitational or inertial force fields depend on the mean square radius, r_2; the rate of evaporation is related to r_1, and so on. In some cases more complex means have to be constructed. Thus, the specific surface of an aerosol, or the surface area possessed by unit mass or volume of the disperse phase, can be visualized from a particle having the same specific surface as the whole aerosol. The radius of such a particle, r_s, is found from the equation

$$\int_0^\infty 4\pi r^2 f(r)\,dr \Big/ \int_0^\infty \tfrac{4}{3}\pi r^3 f(r)\,dr = 4\pi r_2^2/\tfrac{4}{3}\pi r_3^3 = 4\pi r_s^2/\tfrac{4}{3}\pi r_s^3 \tag{4.7}$$

or

$$r_s = r_3^3/r_2^2.$$

In the mist examined $r_s = 11{\cdot}0\,\mu$. The same value is relevant to the absorption of light by unit volume of material dispersed as a coarse aerosol.

In the distribution which has been worked out, the means differ appreciably from one another. The wider the range of particle size in an aerosol the greater is the difference between the various means, and when several orders of magnitude are present the idea of a mean radius tends to lose physical meaning.

The mean radii and distribution of sizes in aerosols with particles of irregular form will be considered later.

§ 5. SHAPE AND STRUCTURE OF AEROSOL PARTICLES

The simplest aerodisperse systems are mists, the particles of which are, as a rule, of spherical shape and form spherical particles on coagulation. In mists of very viscous liquids, such as of Apiezon grease, the union of a pair of droplets which come into contact may be slow, so that particles of irregular shape [23] are sometimes observed. More complex phenomena have been observed in mercury mists which are described in the numerous papers on the determination of the electronic charge by measuring the rate of movement of aerosol particles in a vertical electric field (see p. 52). Whilst in oil mists the particles have normal density, equal to that of the oil from which they have been produced, particles in mercury mists often have a density substantially less than that of mercury [24–26]. It was noticed that particles of normal density decreased in size with the passage of time because of the evaporation of mercury, whereas those of abnormal density did not evaporate [25]. Normally evaporating particles are obtainable by the mechanical atomization of very pure mercury [24] or by evaporation at not too high a temperature, under conditions which do not promote oxidation. On the other hand, mists formed by means of electrical discharges [25, 27] or from mercury contaminated with lead [28], yield, as a rule, particles which do not evaporate.

Non-evaporating particles are covered with a more or less thick oxide film, the density of which is considerably less than that of metallic mercury; this circumstance alone may noticeably reduce the density of the particles. In addition, a further phenomenon takes place: mercury droplets covered with an oxide film do not coalesce on contact with one another but form aggregates like solid particles. The apparent density of these aggregates, determined from measurements in a vertical electric field, is sometimes a tenth of the density of mercury. Such flocculent aggregates are typical of aerosols with solid particles, and oxidised mercury mists are fully analogous to smokes.

In aerosols with solid particles distinction should be made between the shapes of the primary particles and of the aggregates formed from them. The primary particles in smokes formed by direct vapour-crystal transition usually have a regular crystalline form, but if the vapour condenses into liquid droplets, which subsequently solidify,

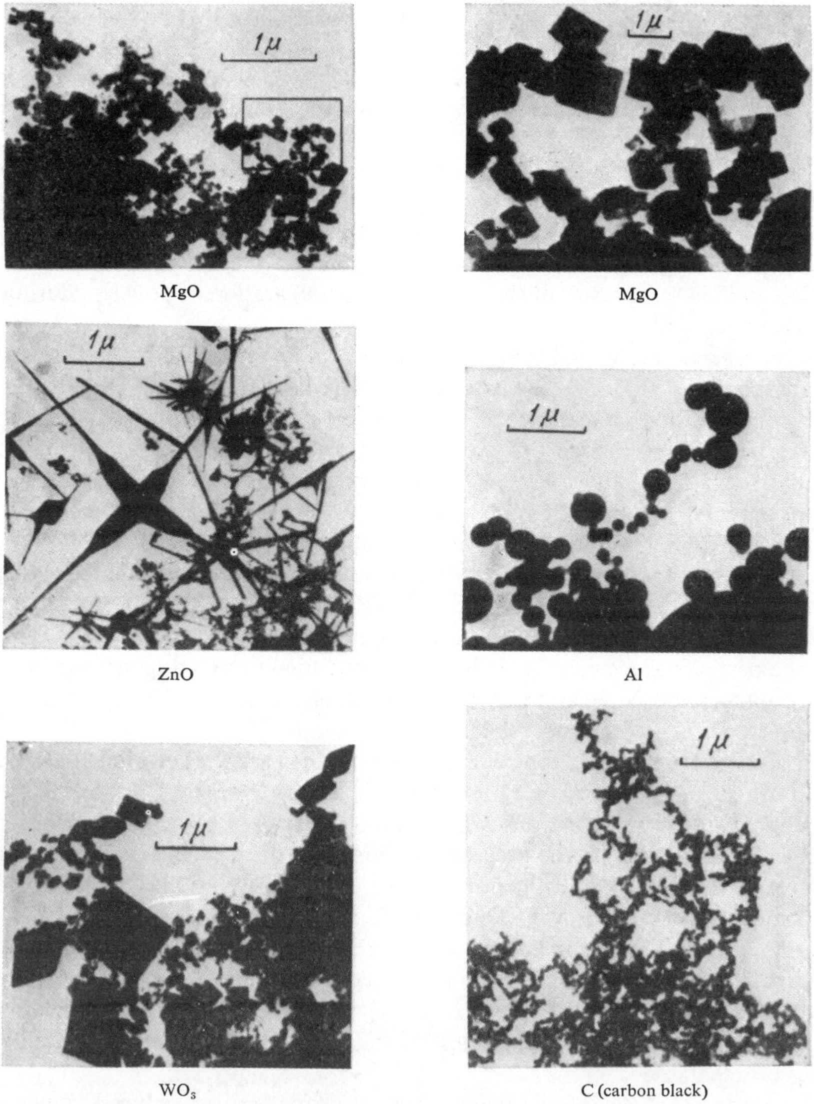

FIG. 8. Electron photomicrographs of aerosol particles.

the smoke particles may be either crystalline or spherical. Primary smoke particles are often so small that their shape and size can hardly be determined by electron microscopy, although they are sometimes easily distinguishable under the ordinary microscope. Some idea of the variety of shapes and sizes of primary particles in smokes is given by the electron photomicrographs shown in Fig. 8 [29] and 73 (p. 314).

Particles of dust are usually irregularly shaped, being fragments of solid bodies, but their crystalline nature is often evident. Even the finest particles may present a lamellar form, as in mica, slate and graphite dust, while fibrous asbestos and textile particles similarly reveal the crystal habit of the parent material.

It is convenient to divide aerosol particles, according to their shape, into three classes determined by their relative dimensions along three axes perpendicular to one another:

(1) isometric particles, in which all three dimensions are roughly the same. Spherical or regular polyhedral shapes and particles approximating to these forms belong here;

(2) particles having much greater lengths in two dimensions than in the third—platelets, leaves, scales etc.;

(3) particles with great length in one dimension—prisms, needles, threads etc.

Aggregates in aerosols arise from the coagulation of individual smoke or dust particles, and also as a result of the incomplete disintegration of powders during their dispersion (see § 58).

The number of individual particles contained in an aggregate is variable, ranging from two to many millions. The rate of thermal coagulation of aerosols for a given weight concentration increases rapidly as particle size decreases; the stability of the aggregates and their capacity to resist the dispersing action of turbulence is also enhanced (see § 58). This is why aggregation of aerosols is more pronounced if the primary particles are fine. Systems derived from highly dispersed primary particles consist, as a rule, of aggregates and rarely contain individual particles.

In the ultramicroscopy of aerosols individual particles and aggregates can be differentiated by the fact that, for the same settling rate, the former execute a much more violent Brownian motion and are less bright than the others [27]. However, when the particles are very small it is extremely difficult to distinguish the two types with the ultra-microscope, and this has been one of the reasons for the origination of the "sub-electron" hypothesis (see page 53).

Aggregates may be either approximately isometric in shape or else linear and thread-like (fibrous). Linear aggregates are chains of primary particles (see page 314). Very often aggregates consist of several such chains. The type of aggregate depends on the disperse phase and also on the gaseous medium [30]; the presence of an electric field, and other factors, may also influence aggregation. The formation of linear aggregates is examined in more detail in Chapter VII.

It has already been pointed out that the apparent density of aggregates can be much less than the true density of the substance of which they are made. An investigation, specially set up by Whytlaw-Gray [31], using the vertical electric field method (see page 54) gave the following values for the apparent density of particles in some smokes (Table 2).

Considering that deviations from spherical shape lead to reduced values of the effective density (page 54), it is likely that the upper limits to the densities of the particles in the Table relate to primary particles. The very small values of effective density (0·07 for Hg; 0·2 for Au) undoubtedly refer to aggregates of thread-like shape and are therefore much reduced. Actually the ratio of apparent to true densities of smoke particles obviously oscillates within the limits of approximately 0·1–0.7, depending on the nature of the packing of primary particles, in a similar way to the ratio of the bulk and true densities of various powders.

Mention must be made of the suggestion advanced by a few authors [32, 33] that the low apparent density of aerosol particles is caused by the presence of a stationary gaseous envelope on their surface. In order to explain observed deviations from the true density of the material concerned, it has been necessary to ascribe a thickness of a few tenths of a micron to this envelope. This concept has neither theoretical nor experimental foundation. In the past this hypothetical envelope has often been resorted to, to explain incomprehensible facts. We will return to this question below.

TABLE 2. DENSITY OF PARTICLES IN SMOKES

Material	Density		Method of producing smoke
	True	Apparent	
Au	19·3	0·2 – 8·0	Vaporization in electric arc
Ag	10·5	0·64 – 4·22	Vaporization in electric arc
Hg	13·6	0·07–10·8	Heating in boat
MgO	3·6	0·24 – 3·48	Burning metallic magnesium
$HgCl_2$	5·4	0·62 – 4·3	Heating in boat
CdO	6·5	0·17 – 2·7	Vaporization in electric arc

Sometimes it is desirable to have a numerical expression for the degree of irregularity of particles. Since all laws describing the properties of aerosols are expressed particularly simply for particles of spherical shape, the degree of irregularity is usually assessed by the deviation of the particle shape from spherical. It is simplest of all to take as the "coefficient of sphericity" [34] the ratio of the surface area of a sphere with the same volume as the given particle to the surface area of the particle. The coefficient of sphericity $\varkappa_s$ is equal to one for spheres and is smaller for any other particle shape. Isometric particles have $\varkappa_s$ close to 1; for an octahedron it is 0·846, for a cube 0·806, and for a tetrahedron 0·670. Particles markedly extended in one or two dimensions, for example crystalline skeletons such as snowflakes, have values considerably less than 1.

The problem of specifying the size distribution of aerosols with non-spherical particles presents great difficulties. Even for particles which are regular polyhedra, the question of what to take as their size can be interpreted in more than one way. Still greater arbitrariness is possible in defining the "size" of extended particles. The specification of particle sizes by two or three numbers is too complex. Some averaged dimension must be used, for which purpose the equivalent radius and the Stokes or sedimentation radius are frequently employed. The equivalent particle radius r_e is the radius of a sphere with the same volume as the given particle; the Stokes or sedimentation particle radius r_s is the radius of a sphere with the same density and rate of settling. For spherical particles $r_e = r_s$, but in general these radii are different, and the difference increases as the coefficient of sphericity of the particles decreases. The question of the value of r_s for particles of various shapes and the relationship between r_s and r_e is examined in § 12. Methods of finding the size of aerosol particles by their rate of settling obviously give the quantity r_s; pneumatic separators and elutriators fractionate powders on the same basis.

Determination of particle size by the number-weight method yields the average value of the equivalent radius r_e, which also results if individual particles are suspended in a vertical electric field (page 51) or when they are measured with an optical or electron microscope.

Determination of particle size in three dimensions under the microscope, in the case of coarsely dispersed aerosols, is possible by focussing alternately upon the upper and lower surfaces or by a stereoscopic method proposed by Ran'ko [35]; these measurements can only be made if the particles exceed 2–3 μ. A more general method consists in "shadowing" the particles in vacuum at a known angle to the plane of support and in measuring the length of the shadow.

In general, determination of equivalent radii under the microscope is a laborious and sometimes impractical task; in practice it is rarely carried out. Measurement is usually confined to the average size of the particle projection in the field of view of the microscope taking two mutually perpendicular directions [36], preferably the directions of the maximum and minimum lengths (Fig. 9); the arithmetic mean of these is equal to twice the mean particle radius. Such a mean radius can differ markedly from r_e.

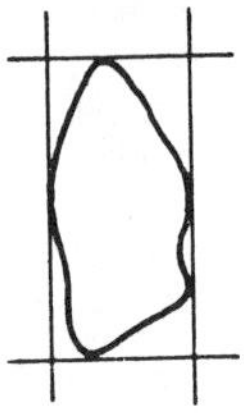

FIG. 9. Dimensions of a particle of irregular shape.

The chief industrial method for determining the size of solid particles and for the fractionation of powders is screen or sieve analysis. The semi-aperture, or half-width of the openings in the sieves, coincides with r_s and r_e for spherical particles and is not much different for isometric ones. The passage of disk-like and plate-like etc. particles through a sieve is governed by their width which exceeds r_s and r_e; for acicular, prismatic, and fibrous particles the sieve semi-aperture is equal to the thickness of the needle which is less than r_s and r_e. Sieve and sedimentometer methods of analysing powders containing particles of irregular shape (such as coal dust) often give completely different results. In connection with the recent appearance of photoelectric devices for automatic measurement of particles deposited on a transparent support [586], the term "projected radius" introduced by Heywood [587], i.e. the radius of the circle whose area is equal to that of the particle projection upon the surface of the support, has become of particular importance, as it is this radius which is determined in most of the devices mentioned. Moreover, in a well-known visual method of microscopic determination of particle sizes by means of eye-piece circular graticules, a graticule is chosen whose area is closest to that of the image of the particle.

Observations made by Dallavalle and Orr [588] show that the sedimentation rate of MgO aerosols, and particularly of those of NH_4Cl, is greatly increased in a damp atmosphere. Microscopic examination shows that the aggregates become more compact, probably by capillary tightening by a condensed water film.

CHAPTER II

STEADY RECTILINEAR MOTION OF AEROSOL PARTICLES

§ 6. RESISTANCE OF A GAS TO THE MOTION OF VERY SMALL PARTICLES

A characteristic feature of the mechanics of aerosols is that of the three types of force acting on the particles—external forces (gravitational, electrical etc.), resistance of the medium†, and interaction of the particles—the third is in most cases considerably less than the others and can be neglected; the particles can be considered independent of one another. Study of the mechanics of aerosols can thus be reduced to the theoretical and experimental investigation of the motion of individual particles in a resisting medium under the influence of various external forces; when necessary a correction is made for interaction of particles. The situation is similar for other aerosol properties, such as scattering of light and rate of evaporation, where it is also considered that each particle evaporates or scatters light as if it were alone and a corresponding correction for secondary scattering etc. is introduced. This method of investigation is used very widely in the theory of aerosols.

It is advantageous to start the study of the dynamics of aerosols with the simple case of steady motion under the action of a constant force. The motion of spherical particles has been studied most, both theoretically and experimentally, and this will be examined first.

As already remarked (page 4), equations expressing some aerosol properties, including the resistance of a gas to the motion of particles, depend on the ratio of the particle radius to the mean free path, l, of the gas molecules. For $r \ll l$, in finely dispersed aerosols, or at low gas pressures, the motion of the particles is of a molecular character; it does not upset the Maxwellian distribution of the velocities of the gas molecules, either in magnitude or in direction, and creates no currents in the gas. The resistance offered by the gas is due to a greater number of molecules impinging on the surface of the moving particle from in front than from behind. The resistance is proportional to the surface area or to the square of the particle radius. If the mass, m, of the particle is much greater than the mass m_g of one of the gas molecules, that is for radii exceeding $0{\cdot}5 \times 10^{-7}$ cm, the resistance for the molecular regime of particle motion is expressed by the formula

$$F_M = -\frac{4}{3}\pi\delta n_g m_g G_g r^2 V, \qquad (6.1)$$

† The forces acting on a particle due to the medium surrounding it, whether it is at rest or in motion, are not considered to be external forces.

where n_g is the number of gas molecules in 1 cm^3,
G_g is their mean velocity,
V is the velocity of a particle,
δ is a factor, whose magnitude depends on the way the gas molecules rebound from the surface of the particle (Epstein [37]).

For specular reflection $\delta = \delta_1 = 1$; for diffuse reflection with conservation of the absolute velocity of the incident molecules $\delta = \delta_2 = 13/9 = 1{\cdot}442$; for diffuse reflection in which the gas molecules assume a velocity distribution in equilibrium with the surface temperature of the particle, that is for perfect accommodation, $\delta = \delta_3 = 1 + \pi/8 = 1{\cdot}393$. Other molecular reflection mechanisms, which are sometimes postulated, are incompatible with the principles of statistical mechanics.

By means of the well known expression for the viscosity of a gas

$$\eta = \varphi\, n_g\, m_g\, G_g\, l \tag{6.2}$$

where φ is a numerical coefficient, the resistance can be expressed in the following way [38]:

$$F_M = -\frac{4/3\pi\delta r^2 \eta V}{\varphi l} = -\frac{6\pi\delta r^2 \eta V}{\varphi 4{\cdot}5 l} = -\frac{6\pi\eta r^2 V}{(A+Q)\,l}\,. \tag{6.3}$$

The quantity $4{\cdot}5\,\delta$ is denoted by $A + Q$. For the three types of reflection discussed $A + Q$ has the following values: $(A+Q)_1 = 1{\cdot}175$; $(A+Q)_2 = 1{\cdot}091$; $(A+Q)_3 = 1{\cdot}131$.

The motion of oil droplets has received the most careful experimental study by Millikan; in this case a value of 1·154 was obtained for $A + Q$ [39].

Other experiments (page 26) lead to the conclusion that reflection of gas molecules from the surface of droplets is diffuse with a small persistence of velocity, that is, a tendency for the tangential component of velocity of the molecules to be conserved. Mathematically this is equivalent to specular reflection of a definite proportion of the molecules (about 10 per cent in the case of oil droplets) and diffuse reflection of the rest. On this assumption the theoretical value of the coefficient $A + Q$ becomes equal to 1·125 with conservation of the gas molecule velocities and to 1·164 with complete accommodation. From comparison with the experimental value $A + Q = 1{\cdot}154$ it follows that diffuse reflection is evidently accompanied by fairly complete accommodation of the molecules.

The range of validity of equations (6.1) and (6.3) is discussed on page 29.

The resistance offered by air at very high particle velocities is of interest to such new branches of technical science as rocket flight etc. Tsien [589] and Krzywoblowski [590] deduced the following formula for particles which are much smaller than the mean free path at velocities comparable to the mean molecular velocity G_g

$$F_M = -\frac{1}{2}\pi r^2 n_g m_g V^2 f(V/G_g) \tag{6.4}$$

where f is a rather complicated function. At

$$V \gg G_g \tag{6.5}$$

$$F_M \approx -\pi r^2 n_g m_g (V^2 + G_g^2). \tag{6.6}$$

§ 7. STOKES' FORMULA

When $r \gg l$ and certain additional conditions discussed below are fulfilled, the resistance of a gas to the motion of spherical particles is expressed by the well-known Stokes' formula:

$$F_M = -6\pi\eta r V. \tag{7.1}$$

The resistance is proportional to the radius and the motion of the particles is of a hydrodynamic nature. The isotropic character of the velocity distribution of the molecules is destroyed and a bulk flow is created. The resistance of the medium is caused by hydrodynamic forces.

The most important application of Stokes' formula is the settling of aerosol particles under the action of gravity. The force acting on a particle is then equal to

$$F = \frac{4}{3}\pi r^3 g(\gamma - \gamma_g) \approx \frac{4}{3}\pi r^3 g\gamma, \tag{7.2}$$

where g is the acceleration due to gravity; γ and γ_g are the densities of the particle and the gas. In view of its smallness, the latter can be neglected. From (7.1) and (7.2) we obtain for the steady rate of settling of the particle

$$V_s = \frac{2}{9}\frac{r^2 g\gamma}{\eta} = g\tau, \tag{7.3}$$

where $\tau = \dfrac{2r^2\gamma}{9\eta}$, a quantity with the dimension of time, which will be seen to play an important role in the mechanics of aerosols.

The derivation of Stokes' formula from the basic dynamical equations of viscous flow depends upon the following conditions [40]:

(1) incompressibility of medium;
(2) infinite extent of medium;
(3) very small rate of movement;
(4) constant rate of movement;
(5) rigidity of particle;
(6) absence of slipping at its surface.

A brief examination of each will now be made.

(1) As is well known, the compressibility of a medium begins to have effects only at velocities comparable to the rate of propagation of mechanical impulses, which is equal to the speed of sound in the medium. Provided that phenomena such as the motion of aerosols formed in an explosion are excluded, the gaseous medium can be considered incompressible in the mechanics of aerosols.

(2) The condition of infinite extent of the medium in all directions is never observed in practice; at a greater or lesser distance from aerosol particles there are always some macroscopic bodies, such as the walls of a vessel, besides other particles. The influence of walls on the motion of small particles according to Stokes' law has been investigated very thoroughly, both theoretically and experimentally; a liquid medium was used and good agreement between theory and experiment was obtained [41]. The effect of walls is to increase the resistance of the medium by a factor $1 + b\dfrac{r}{x}$, where x is the

distance of the particle centre from the wall, and b is a coefficient depending on the shape and disposition of the walls. It is supposed that r/x is not greater than 0·1. For a particle moving parallel to a plane wall b is about 9/16, for motion at right angles to the wall it is 9/8, and for motion along the axis of an infinitely long cylinder $b = 2{\cdot}1$†. Correction for the effect of the walls becomes appreciable only when they are at a distance, x, less than 10 r. However, in view of their comparatively high rate of movement on account of Brownian motion, convection and sedimentation, aerosol particles are unable to remain for any length of time in close proximity to a wall without coming into contact with it, and its effect on resistance can usually be neglected. The more complicated question of the mutual effect of aerosol particles on their motion is examined below (see § 13).

(3) In the derivation of Stokes' law the inertia terms are omitted from the equations of motion; this is permissible only for very small rates of motion, so that Stokes' formula is a first approximation. The second approximation, which partly allows for inertia forces, was obtained by Oseen [42] and has the form

$$F_M = -6\pi\eta rV\left(1 + \frac{3}{8}\frac{r\gamma_g V}{\eta}\right) = -6\pi\eta rV\left(1 + \frac{3}{16}\,\mathrm{Re}\right), \tag{7.4}$$

where γ_g is the density of the medium, and

$$\mathrm{Re} = 2\,r\gamma_g V/\eta = 2\,rV/\nu \tag{7.5}$$

is the Reynolds number of the particle. Stokes' law can be used only at low Reynolds' number. It will be shown below (see page 31) that the error in Stokes' law is approximately proportional to Re and is about 1·7 per cent at Re = 0·1.

(4) The effect of non-uniform movement of particles upon the resistance is examined in Chapter III.

(5) In the derivation of Stokes' formula it is assumed that the particle is rigid. For liquid droplets two new factors appear:

(a) A droplet can deform under the action of the medium. This is significant only for large drops and is discussed below (page 44).

(b) Circulation develops in a moving droplet being directed, near the surface, towards the lee side; this reduces friction and, consequently, the resistance offered by the medium. The resistance of the medium to the motion of a spherical particle is in this case given by the formula [43, 44]

$$F_M = -6\pi\eta rV\,\frac{1 + (2\eta/3\eta_p)}{1 + (\eta/\eta_p)}, \tag{7.6}$$

where η is the viscosity of the medium;

η_p is the viscosity of the liquid of which the droplets consist.

† The correction factor derived by Faxen [454] for a sphere moving along the axis of a cylindrical tube

$$\left[1 - 2{\cdot}104\,\frac{r}{R} + 2{\cdot}09\left(\frac{r}{R}\right)^3 - 0{\cdot}95\left(\frac{r}{R}\right)^5\right]^{-1},$$

where $\frac{r}{R}$ is the ratio of the radii of the sphere and the tube, holds at Re $\leqslant$ 0·02 down to values of $\frac{R}{r} \approx 3$, according to Bacon's experiments [111].

Since the viscosity of gases is much less than the viscosity of liquids, this correction to Stokes' law is negligible. Mists in air require corrections of 0·7 per cent for water and hundredths or thousandths of one per cent in the case of oil.

(6) In the derivation of Stokes' law it is assumed that there is no velocity discontinuity at the surface of the sphere; the infinitely thin layer of medium immediately adjacent to the surface is stationary with respect to the particle. Should velocity discontinuity, or slip along the surface of the particle, exist, the resistance of the medium must, of course, be reduced. If the tangential forces acting on the particle are taken as proportional to this velocity jump, and the factor of proportionality, called the coefficient of external friction, is denoted by η_e, then the resistance is given by the formula [45]

$$F_M = -6\pi\eta r V \frac{2\eta + r\eta_e}{3\eta + r\eta_e}. \tag{7.7}$$

Stokes' formula is obtained from this when η_e is infinitely large.

Denoting the ratio η/η_e, called the slip coefficient, by β we obtain for small values of β/r

$$F_M = -6\pi\eta r V \frac{1 + (2\beta/r)}{1 + (3\beta/r)} \approx -6\pi\eta r V \bigg/ \left(1 + \frac{\beta}{r}\right). \tag{7.8}$$

Experiment shows that in a liquid no slip occurs at the surface of moving particles, and the correction discussed is not necessary. But it is different in a gas. Here the phenomenon of slip, associated with the small amount of space filled by matter in a gas, plays a big part.

§ 8. RESISTANCE OF A GAS TO PARTICLES OF A SIZE COMPARABLE WITH THE MEAN FREE PATH

In all transfer processes considered in the kinetic theory of gases — heat conduction (transfer of thermal energy), viscosity (momentum transfer) and diffusion (material transfer) — a discontinuity in the corresponding parameter (temperature, velocity, concentration) may occur at the surface of a solid body across which transfer is taking place. The size of the discontinuity is, roughly speaking, equal to the product of the gradient of the parameter and the mean free path of the gas molecule l. The effect produced by the discontinuity reaches a noticeable magnitude either due to the value of l being comparable with the dimensions of the space occupied by the gas, or when the gradient of the parameter close to the surface is large, particularly at the surface of small particles. In this case the gradient is proportional to y/r for all transfer processes, where y is the difference of temperature, velocity or vapour pressure between the particle and the surrounding medium at an infinitely great distance from the particle†. Thus the discontinuity in the parameter is Ayl/r, where A is a numerical factor depending on the nature of the reflection of gas molecules from the particle surface. As a result of the discontinuity the effective difference in the parameter (temperature, velocity or vapour pressure) decreases from y to $y(1 - Al/r)$. Hence

† In particular the gradient of tangential velocity at the surface of moving spheres at low Re is equal to $\frac{3}{2} V \sin\theta/r$, where θ is the angle between the radius vector at a given point on the surface and the direction of motion, and V is the velocity of the particle.

it follows that change of rate of a transfer process as a result of the discontinuity is expressed by the correction factor $1 - Al/r \approx 1/(1 + Al/r)$.

Rigorous analysis of the effect of discontinuity in the tangential velocity at the surface of a particle (Epstein [37]) shows that the slip coefficient, β, in equation (7.8) equals

$$\beta = 0{\cdot}7004\left(\frac{2}{\alpha} - 1\right)l \tag{8.1}$$

where α denotes the fraction of gas molecules undergoing diffuse reflection at the surface and $1-\alpha$ the fraction undergoing specular reflection. Hence for the resistance of the medium we obtain

$$F_M = -6\pi\eta rV\Big/\left(1 + A\frac{l}{r}\right) \tag{8.2}$$

where

$$A = \beta/l = 0{\cdot}7004\left(\frac{2}{\alpha} - 1\right) \tag{8.3}$$

and for the rate of settling of the particles

$$V_s = \frac{2}{9}\frac{r^2\gamma g}{\eta}\left(1 + A\frac{l}{r}\right) = \frac{mg}{6\pi\eta r}\left(1 + A\frac{l}{r}\right). \tag{8.4}$$

Because (7.8) is derived for small values of $\beta/r = A\frac{l}{r}$ and A is of the order of unity, formula (8.2), first proposed by Cunningham [46], is true only for small l/r.

The value of A found experimentally by Millikan [38]† for oil droplets in air is 0·864 which, according to (8.3), corresponds to $\alpha = 0{\cdot}895$. Thus about ten per cent of the molecules are reflected specularly and the rest diffusely.

More precisely, the ten per cent specifies the degree of persistence of the tangential velocity of molecules on reflection (see page 22). The slip coefficient for air on the surface of oil has also been determined with a Couette viscometer†† covered with a film of oil [47]. The very close value $A = 0{\cdot}870$ was obtained. For droplets of oil, mercury and aqueous solutions of $BaHgI_4$ in various gases, values of A lying in the range 0·820–0·900 (see [48]) were found by several authors, which corresponds to a comparatively small change in the nature of molecular reflection: from 8 to 12 per cent of the molecules are reflected specularly. Similar results were obtained for solid spherical particles of selenium.

Considerably larger variations in the slip coefficient have been observed with other solids [47]: for glass spheres it was found that $\beta = 0{\cdot}82 \times 10^{-5}$; for machined brass cylinders $0{\cdot}66 \times 10^{-5}$; for cylinders covered with shellac $0{\cdot}97 \times 10^{-5}$; for the same cylinders when the shellac has aged, $0{\cdot}68 \times 10^{-5}$. After substitution of $l = 0{\cdot}94 \times 10^{-5}$ cm in equation (8.3) it follows that the quoted values of β correspond to the following values of α: 0·92, 1·00, 0·81 and 0·99. This indicates that on the freshly formed and very smooth surface of an amorphous body, reflection is of the same nature as at a liquid surface. A surface obtained by mechanical working, or one which is old and cracked, is so rough that reflection of molecules from it is completely diffuse. In

† Millikan and other authors mentioned in § 8 took a value of $0{\cdot}942 \times 10^{-5}$ cm for l in air at 760 mm and room temperature, which corresponds to a value $\varphi = 0{\cdot}350$ in equation (6.2).

†† The Couette viscometer consists of two coaxial cylinders, one of which rotates. The torque transmitted by fluid viscosity to the other cylinder is measured.

practice $A = 0{\cdot}86$ may be taken for liquid droplets and very smooth solid spheres, and $A = 0{\cdot}70$ for rough spheres. According to (8.3) the minimum possible value for A is 0·70. Systematic errors may be suspected in the occasional investigations where smaller values of A have been found.

We have analysed the question of the resistance of a gas in two cases—under molecular ($r \ll l$) and hydrodynamic ($r \gg l$) particle motion. It remains to examine the resistance when r and l are of the same order of magnitude, and to define the applicability of the limiting laws described above. This problem has proved too difficult for theoretical solution and has been dealt with by the introduction of an empirical formula which agrees satisfactorily with experiment [39, 49]

$$F_M = -6\pi\eta r V \Big/ \left(1 + A\frac{l}{r} + Q\frac{l}{r}e^{-br/l}\right). \tag{8.5}$$

This expression approaches (8.2) for $r \gg l$ and (6.3) for $r \ll l$. For oil droplets in air Millikan [39] found $A = 0{\cdot}864$, $Q = 0{\cdot}29$, $b = 1{\cdot}25$, and Mattauch [26] $A = 0{\cdot}898$, $Q = 0{\cdot}312$ and $b = 2{\cdot}37$. For droplets of an aqueous $BaHgI_4$ solution in air [50] $A = 0{\cdot}879$, $Q = 0{\cdot}23$, $b = 2{\cdot}61$. For glass spheres in air† $A = 0{\cdot}77$, $Q = 0{\cdot}40$, $b = 1{\cdot}62$. It must be borne in mind that the accuracy in the experimental determination of Q and b is much less than in the determination of A.

Millikan's data should probably be considered the most reliable. In particular, according to Mattauch $A + Q = 1{\cdot}210$, which is contrary to the theoretical considerations given at the end of § 6. In this book, therefore, Millikan's values for the coefficients A, Q and b are used. They depend also on the value assigned to the mean free path, l, of the gas molecules which is usually calculated by means of equation (6.2) from the viscosity of the gas; the values of A, Q and b thus depend on φ in this equation. A value $\varphi = 0{\cdot}499$ is now accepted which makes $l = 0{\cdot}653 \times 10^{-5}$ cm. The coefficients A, Q and b then become 1·246, 0·42 and 0·87.

The ratio of the velocity of a particle, V, to the force, F, causing steady motion is called the mobility B. According to (8.5)

$$B = \left(1 + A\frac{l}{r} + Q\frac{l}{r}e^{-br/l}\right) \Big/ 6\pi\eta r. \tag{8.6}$$

Values of log B as a function of log r, calculated from this equation for oil droplets in air at 23° and 760 mm mercury are presented in Table 3 taking the viscosity of air to be $18{\cdot}3 \times 10^{-5}$ poise [51]. The same relationship is illustrated in Fig. 10. The curves for particles of different materials practically coincide when plotted on the scales used for this figure.

Formula (8.6) shows that the error in using the molecular kinetic formula (6.3) is 1 per cent for $r = 2 \times 10^{-7}$ cm and 10 per cent for $r = 2 \times 10^{-6}$ cm. The error in using Cunningham's formula (8.2) is 1 per cent for $r = 1{\cdot}8 \times 10^{-5}$ cm and 10 per cent for $r = 5 \times 10^{-6}$ cm. Finally Stokes' formula (7.1) gives an error of 1 per cent for $r = 8 \times 10^{-4}$ cm and 10 per cent at $r = 0{\cdot}8 \times 10^{-4}$ cm.

Table 4 shows the regions of applicability of these formulae for errors of 1 per cent and 10 per cent. To complete the picture, deviations from Stokes' law due to inertia

† In an article by Knudsen and Weber [49] rather higher values of A and Q are given, because in expression (6.2) for the viscosity of the gas the authors took $\varphi = 0{\cdot}310$.

TABLE 3. MOBILITY OF OIL DROPLETS IN AIR AT 23°C AND 760 mm Hg

$\log r$	$\log B$	Δ	$\log u$[1]	$\log r$	$\log B$	Δ	$\log u$[1]
$\bar{8}.3$	12.898		1.102	$\bar{6}.9$	7.893		$\bar{4}.097$
		200				156	
$\bar{8}.4$	12.698		0.902	$\bar{5}.0$	7.737		$\bar{5}.941$
		200				148	
$\bar{8}.5$	12.498		0.702	$\bar{5}.1$	7.589		$\bar{5}.793$
		200				141	
$\bar{8}.6$	12.298		0.502	$\bar{5}.2$	7.448		$\bar{5}.652$
		200				135	
$\bar{8}.7$	12.098		0.302	$\bar{5}.3$	7.313		$\bar{5}.517$
		199				129	
$\bar{8}.8$	11.899		0.113	$\bar{5}.4$	7.184		$\bar{5}.388$
		199				122	
$\bar{8}.9$	11.700		$\bar{1}.904$	$\bar{5}.5$	7.062		$\bar{5}.266$
		199				118	
$\bar{7}.0$	11.501		$\bar{1}.705$	$\bar{5}.6$	6.944		$\bar{5}.148$
		200				115	
$\bar{7}.1$	11.301		$\bar{1}.505$	$\bar{5}.7$	6.829		$\bar{5}.033$
		199				113	
$\bar{7}.2$	11.102		$\bar{1}.306$	$\bar{5}.8$	6.716		$\bar{6}.920$
		199				111	
$\bar{7}.3$	10.903		$\bar{1}.107$	$\bar{5}.9$	6.605		$\bar{6}.809$
		200				109	
$\bar{7}.4$	10.703		$\bar{2}.907$	$\bar{4}.0$	6.496		$\bar{6}.700$
		199				107	
$\bar{7}.5$	10.504		$\bar{2}.708$	$\bar{4}.1$	6.389		$\bar{6}.593$
		198				106	
$\bar{7}.6$	10.306		$\bar{2}.510$	$\bar{4}.2$	6.283		$\bar{6}.487$
		197				105	
$\bar{7}.7$	10.109		$\bar{2}.313$	$\bar{4}.3$	6.178		$\bar{6}.382$
		196				103	
$\bar{7}.8$	9.913		$\bar{2}.117$	$\bar{4}.4$	6.075		$\bar{6}.279$
		195				103	
$\bar{7}.9$	9.718		$\bar{3}.922$	$\bar{4}.5$	5.972		$\bar{6}.176$
		194				102	
$\bar{6}.0$	9.524		$\bar{3}.728$	$\bar{4}.6$	5.870		$\bar{6}.074$
		193				101	
$\bar{6}.1$	9.331		$\bar{3}.535$	$\bar{4}.7$	5.768		$\bar{7}.972$
		192				100	
$\bar{6}.2$	9.139		$\bar{3}.343$	$\bar{4}.8$	5.667		$\bar{7}.871$
		190				101	
$\bar{6}.3$	8.949		$\bar{3}.153$	$\bar{4}.9$	5.567		$\bar{7}.771$
		187				101	
$\bar{6}.4$	8.762		$\bar{4}.966$	$\bar{3}.0$	5.466		$\bar{7}.670$
		183				101	
$\bar{6}.5$	8.579		$\bar{4}.783$	$\bar{3}.1$	5.365		$\bar{7}.569$
		180				101	
$\bar{6}.6$	8.399		$\bar{4}.603$	$\bar{3}.2$	5.264		$\bar{7}.468$
		175				100	
$\bar{6}.7$	8.224		$\bar{4}.428$	$\bar{3}.3$	5.164		$\bar{7}.368$
		169				100	
$\bar{6}.8$	8.055		$\bar{4}.259$	$\bar{3}.4$	5.064		$\bar{7}.268$
		162					

[1]) u = electric mobility see page 113

forces [formula (7.4)] have been taken into account and the upper limit of applicability of formula (7.3) has been indicated for particles of unit density falling freely in air under the action of gravity. Departure (magnitude of order m_g/m) from formula (6.3) is indicated for particles of unit density in air when they are comparable in size with the molecules of air.

TABLE 4. REGIONS OF APPLICABILITY OF VARIOUS FORMULAE FOR THE RESISTANCE OF A MEDIUM

Formula	Permissible Error	
	1%	10%
Stokes (7.1)	$8 \times 10^{-4} < r < 15 \times 10^{-4}$ cm	$0{\cdot}8 \times 10^{-4} < r < 35 \times 10^{-4}$ cm
Cunningham (8.2)	$1{\cdot}8 \times 10^{-5} < r < 8 \times 10^{-4}$ cm	$5 \times 10^{-6} < r < 8 \times 10^{-5}$ cm
Molecular kinetic (6.3)	$1 \times 10^{-7} < r < 2 \times 10^{-7}$ cm	$5 \times 10^{-8} < r < 2 \times 10^{-6}$ cm

Examining water drops of $r = 0{\cdot}4 - 1{\cdot}0\,\mu$ in air saturated with water vapour in a Millikan condenser, Gokhale and Gatha [591] found $A = 1{\cdot}032$ at $l = 0{\cdot}653 \times 10^{-5}$ cm.

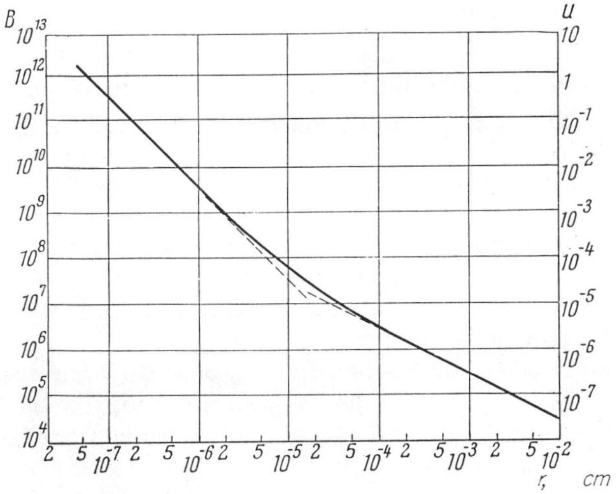

FIG. 10. Size dependence of mobility of aerosol particles.

Schmitt [592] studied droplets of silicone oil of $r = 0{\cdot}7 - 1{\cdot}2\,\mu$ in various gases at pressures of 10–760 mm Hg and obtained in nitrogen $A = 1{\cdot}45 \pm 3$ per cent, $Q = 0{\cdot}40 \pm 10$ per cent and $b = 0{\cdot}9 \pm 20$ per cent, using in his computation of l the formula (6.2) with $\varphi = 0{\cdot}499$. The values of Q and b are very close to the ones obtained by Millikan, but those of A and of $A + Q$ are much greater than his; according to § 6 this means that the proportion of gas molecules which is reflected specularly from the surface of the particles is very large ($\sim$ 50 per cent).

§ 9. EXPERIMENTAL VERIFICATION AND ACCURACY OF STOKES' FORMULA

Conditions for the experimental verification of Stokes' formula by measuring the rate of settling of solid spheres are particularly favourable in a liquid. By using a liquid of high viscosity a large size of sphere can be combined with a low rate of settling, and high accuracy achieved at fairly low Reynolds' numbers. The only difficulty in working with large spheres is the need to introduce a fairly large correction for the effect of the walls. Experiments carried out in liquids have shown that deviation of the measured rate of settling of spheres from that calculated by Stokes' law does not exceed the experimental error [52]. Hence, by making use of the principal of dynamical similarity it could be concluded that Stokes' law was also applicable to motion in gases, although this conclusion required direct experimental confirmation.

The experiments are much more difficult in a gas than in a liquid because a sufficiently small Re can be obtained only by employing very small spheres ($r < 10\,\mu$) for which the effect of slip is manifesting itself. In addition, it is rather difficult to obtain solid particles of this size which are exact spheres† and, in the case of liquid droplets, it is difficult to determine their size accurately under the microscope. Direct experiments have, therefore, been able to provide only approximate verification of Stokes' formula for gaseous media [12].

The accurary of the formula has, however, been established by an indirect method. The value of $4{\cdot}77 \times 10^{-10}$ e.s.u., determined by Millikan [53] (page 52) on the basis of Stokes' law (with correction for slipping), for the charge of the electron was very close to the value of $4{\cdot}80 \times 10^{-10}$ e.s.u. calculated from the dimensions of unit cells in crystals, determined by X-ray diffraction. It was found that the reason for the small (0·6 per cent) difference between these figures was inaccuracy of the value used for the viscosity of air [54]; when new, more thorough measurements of viscosity had been made the difference decreased to 0·1 per cent. Physicists working in this field are of the unanimous opinion that the residual discrepancy is still due to a slight error in the viscosity measurements [51]. Thus Stokes' formula expresses one of the most accurate physical laws for both liquids and gases.

The importance of Stokes' law in the theory of aerosols is very great. Although, as shown above, the range of particle sizes over which the formula holds accurately is small, its validity can be extended by means of appropriate corrections, or by doing without great accuracy, to the range $10^{-5}\text{ cm} < r < 5 \times 10^{-3}\text{ cm}$ which embraces many aerosols of practical significance. It will be seen from what follows that Stokes' law is the basis of almost the entire mechanics of aerodisperse systems, including the theory of the motion and deposition of suspended particles in all sorts of industrial apparatus.

§ 10. RESISTANCE OF A MEDIUM BEYOND THE STOKES REGION

As indicated above, the increasing effect of inertia forces results in the resistance of a medium calculated by Stokes' law deviating more and more below the true value as the Reynolds' number increases; at Re = 0·5 the error exceeds 5 per cent. Oseen's

† This difficulty can now be overcome by the use of industrial polystyrene latex.

formula (7.4) and the somewhat more exact formula of Goldstein [55]

$$F_M = -6\pi\eta r V\left(1 + \frac{3}{16}\,\text{Re} - \frac{19}{1280}\,\text{Re}^2 + \cdots\right), \tag{10.1}$$

deduced with partial consideration of inertia forces, give a rather better approximation with the calculated resistance bigger than the actual. The widespread opinion that Oseen's formula provides a much better approximation than Stokes' rather exaggerates the improvement. It derived largely from Schmiedel's experiments [56], which were accepted as the most accurate at small Re. These values for the resistance, however, are undoubtedly too large and noticeably exceed even the figures calculated according to Oseen in contrast to the data of all other investigators. This is evidently explained by errors, reaching 1–2 per cent, in measuring the viscosity of the liquids in which Schmiedel performed his experiments. As a result the experimental points approached Oseen's curve and departed from Stokes' curve. In Möller's later measurements [57], which should be more accurate, the following results were obtained; at Re = 0·1 Stokes' formula gives a deviation of − 1·5 per cent, Oseen's formula + 0·4 per cent; at Re = 0·2 the deviations are − 3 per cent and + 0·8 per cent, at Re = 0·5, − 6·5 per cent and + 1·5 per cent. Thus Oseen's formula gives a four times better approximation than Stokes' formula. By using formula (10.1) it is possible to obtain a slightly better approximation which is insufficient to make up for its complexity. In practice there is no particular point in using these formulae; it is simpler to use the empirical ones given below.

In contrast to molecular and viscous motion, where the resistance is proportional to the velocity of the particle, the region of large Re values, where inertia forces cannot be neglected, is characterised by the resistance varying as V^s, where s steadily increases with Re. The concept of the mobility of a particle, independent of its velocity, loses its meaning and the non-linear dependence of resistance on not one, but two variables – size and velocity of the particle – has to be expressed. Use of the dynamical similarity principle enables this complex problem to be simplified considerably. The dimensionless quantity

$$\psi = F_M \Big/ \frac{\gamma_g V^2}{2} \cdot \pi r^2, \tag{10.2}$$

called the drag coefficient of a sphere, is a unique function of the Reynolds' number, which is also dimensionless.

Where Stokes' law applies the drag coefficient is equal to

$$\psi = 24/\text{Re} \tag{10.3}$$

while Oseen's formula is expressed by

$$\psi = 24/\text{Re} + 4{\cdot}5. \tag{10.4}$$

The relationship between ψ and Re has been determined in many experimental studies [58], in which measurements have been made either of the rate of settling of solid spheres in liquids or of the force acting on a stationary sphere situated in a flowing liquid or gas. Apart from the wall effect, the force of interaction between the sphere and the medium depends only on their relative velocity and has the same value whether the sphere moves through a stationary medium or vice versa.

TABLE 5. VALUES OF THE DRAG COEFFICIENT ψ AND THE FUNCTIONS $Re^2\,\psi$ AND ψ/Re FOR SPHERICAL BODIES IN THE RANGE OF Re 0·01–1000

log Re	log ψ	Δ	log $Re^2\psi$	log (ψ/Re)	log Re	log ψ	Δ	log $Re^2\psi$	log (ψ/Re)
$\bar{2}$·0	3·380	99	$\bar{1}$·380	5·380	0·6	0·928	76	2·128	0·328
$\bar{2}$·1	3·281	100	$\bar{1}$·481	5·181	0·7	0·852	74	2·252	0·152
$\bar{2}$·2	3·181	99	$\bar{1}$·581	4·981	0·8	0·778	72	2·378	$\bar{1}$·978
$\bar{2}$·3	3·082	100	$\bar{1}$·682	4·782	0·9	0·706	70	2·506	$\bar{1}$·806
$\bar{2}$·4	2·982	99	$\bar{1}$·782	4·582	1·0	0·636	68	2·636	$\bar{1}$·636
$\bar{2}$·5	2·883	100	$\bar{1}$·883	4·383	1·1	0·568	66	2·768	$\bar{1}$·468
$\bar{2}$·6	2·783	99	$\bar{1}$·983	4·183	1·2	0·502	64	2·902	$\bar{1}$·302
$\bar{2}$·7	2·684	99	0·084	3·984	1·3	0·438	63	3·038	$\bar{1}$·138
$\bar{2}$·8	2·585	99	0·185	3·785	1·4	0·375	61	3·175	$\bar{2}$·975
$\bar{2}$·9	2·486	99	0·286	3·586	1·5	0·314	59	3·314	$\bar{2}$·814
$\bar{1}$·0	2·387	98	0·387	3·387	1·6	0·255	57	3·455	$\bar{2}$·655
$\bar{1}$·1	2·289	98	0·487	3·189	1·7	0·198	55	3·598	$\bar{2}$·498
$\bar{1}$·2	2·191	98	0·591	2·991	1·8	0·143	53	3·743	$\bar{2}$·343
$\bar{1}$·3	2·093	98	0·693	2·793	1·9	0·090	51	3·890	$\bar{2}$·190
$\bar{1}$·4	1·995	97	0·795	2·595	2·0	0·039	48	4·039	$\bar{2}$·039
$\bar{1}$·5	1·898	96	0·898	2·398	2·1	$\bar{1}$·991	46	4·191	$\bar{3}$·891
$\bar{1}$·6	1·802	95	1·002	2·202	2·2	$\bar{1}$·945	43	4·345	$\bar{3}$·745
$\bar{1}$·7	1·707	94	1·107	2·007	2·3	$\bar{1}$·902	41	4·502	$\bar{3}$·602
$\bar{1}$·8	1·613	92	1·213	1·813	2·4	$\bar{1}$·861	38	4·661	$\bar{3}$·461
$\bar{1}$·9	1·521	91	1·321	1·621	2·5	$\bar{1}$·823	36	4·823	$\bar{3}$·323
0·0	1·430	89	1·430	1·430	2·6	$\bar{1}$·787	33	4·987	$\bar{3}$·187
0·1	1·341	87	1·541	1·241	2·7	$\bar{1}$·754	31	5·154	$\bar{3}$·054
0·2	1·254	85	1·654	1·054	2·8	$\bar{1}$·723	28	5·323	$\bar{4}$·923
0·3	1·169	83	1·769	0·869	2·9	$\bar{1}$·695	24	5·495	$\bar{4}$·795
0·4	1·086	80	1·886	0·686	3·0	$\bar{1}$·671	21	5·671	$\bar{4}$·671
0·5	1·006	78	2·006	0·506	3·1	$\bar{1}$·650		5·850	$\bar{4}$·550

The results of these investigations are given in Table 5. In compiling this table Möller's results [57] were used for Re < 0·4 and Davies' data [59], obtained by averaging the more reliable experimental results, for higher values of Re; the probable error in the values of ψ given in Table 5 is about 1 per cent for Re < 0·5, gradually increasing with increasing Re to reach 3–4 per cent at Re = 500.

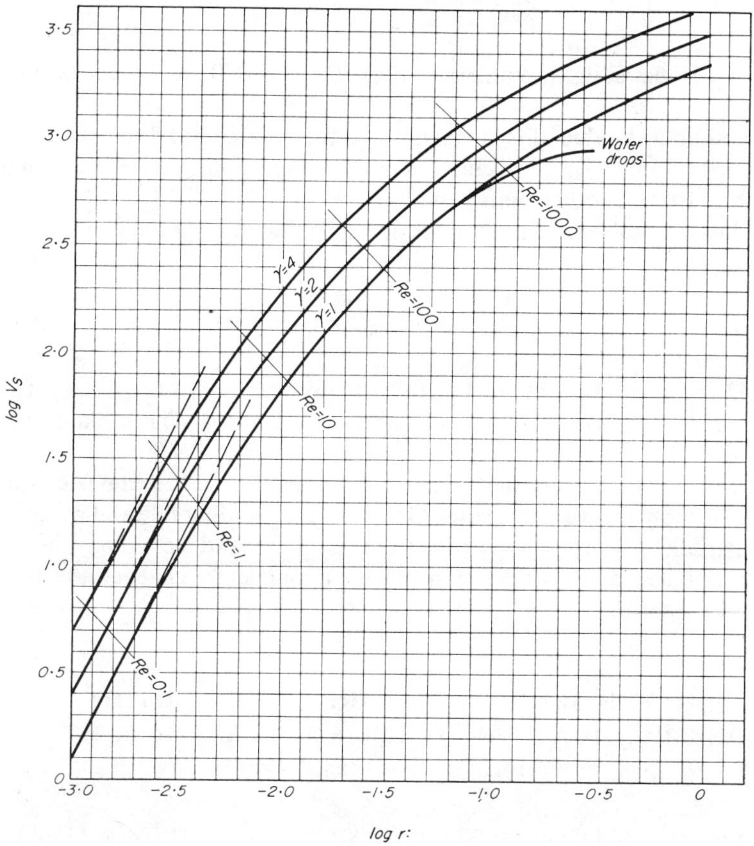

FIG. 11. Rate of settling of aerosol particles in air.

Quite a large number of empirical formulae relating ψ and Re have been proposed by different authors. The most successful from the standpoint of simplicity and accuracy seems to be the one proposed by Klyachko [60]

$$\psi = \frac{\mathrm{Re}}{24} + \frac{4}{\sqrt[3]{\mathrm{Re}}}. \tag{10.5}$$

In the range 3 < Re < 400 this agrees within 2 per cent with the values of ψ given in Table 5.

Table 5 can be used to obtain the rate of settling of particles having radii > 10^{-3} cm [for smaller particles it is simpler and more accurate to use formula (8.4)] in the following manner. From (7.5) and (10.2)

$$\mathrm{Re}^2\psi = \frac{8F_M\gamma_g}{\pi\eta^2}. \tag{10.6}$$

Replacing F_M by the gravitational force on the particle,

$$\mathrm{Re}^2\psi = \frac{32}{3}\frac{r^3\gamma\gamma_g g}{\eta^2}. \tag{10.7}$$

Since the particle radius and all other quantities in (10.7) are known, $\mathrm{Re}^2\psi$ can be calculated.

The corresponding value of Re is then found from Table 5 and the rate of settling, V_s, determined from (7.5). The calculation is carried out similarly if the external force acting on the particle is other than gravitational.

For the inverse problem of finding the size of particles from their rate of settling, the equation

$$\mathrm{Re}/\psi = \frac{3V_s^3\gamma_g^2}{4\gamma g\eta} \tag{10.8}$$

is used as a starting point.

Figure 11 shows graphs obtained in this way for the rate of fall in air, at 20° and 760 mm, of spherical particles having densities of 1, 2 and 4. The broken lines were calculated by Stokes' law.

It must be emphasized that the formulae and tables presented above can be used directly only in comparatively rare cases to calculate the settling rate and other ordered motions of particles in the aerosols encountered in nature and technology, because the presence of turbulence and convection currents completely changes the nature of settling (see Chapter VI).

§ 11. THE GENERAL NATURE OF THE MOTION OF NON-SPHERICAL PARTICLES. ROTATION OF PARTICLES IN A SHEAR FLOW

So far we have examined the motion only of spherical particles, because these have been most widely studied on both the experimental and theoretical sides, although solid aerosol particles are usually non-spherical. The problem of resistance to the motion of non-spherical particles has been solved theoretically only for ellipsoids, which include as limiting shapes thin elliptical plates and cylinders possessing an infinitely large ratio of length to diameter. As far as experimental study of this problem is concerned, like the experimental verification of Stokes' formula it is much easier to work in a liquid than in a gas; nearly all the experimental material available was obtained from experiments in liquids. These results can be used for the mechanics of aerosols with the aid of the principle of dynamical similarity.

When non-spherical particles fall through a resisting medium phenomena are observed which do not take place with spheres. At low velocities, in the viscous flow regime, ellipsoidal particles may be oriented in any way relative to their direction of fall [61, 62] and, depending on their size, can either keep the orientation originally given to them or adopt all sorts of orientations as a result of Brownian rotation. This firmly established experimental fact is in complete agreement with the findings of hydrodynamics, according to which the torque acting on an ellipsoid of revolution due to viscous motion through a fluid produced by a force acting through its centre is equal to zero [63]. This is not true for shapes which lack the symmetry of the ellipsoid of revolution.

A characteristic feature of the motion of non-spherical particles is that the directions of particle motion and of resistance due to the fluid, instead of lying in the same straight line, may be separated by an angle θ. Motion parallel to an axis of symmetry, however, is always opposed along the axis. For particles shaped as ellipsoids of revolution the force of resistance acts through the centre and the angle θ at low Re has been calculated by Gans [63] with the aid of formulae (12.2–12.5) (see page 37). The value of θ depends on the angle of inclination θ of the axis of rotation of the ellipsoid to the direction of motion and reaches a maximum at $\theta = 40–45°$ for oblate ellipsoids and $\theta = 45–55°$ for prolate ellipsoids. The angle θ increases as the ratio β of the major and minor axes of the ellipsoid increases. For oblate ellipsoids the maximum value of θ is 4·6° at $\beta = 2$ and 9·4° at $\beta = 10$, while for prolate ellipsoids it is 3·9° at $\beta = 2$ and 10·4° at $\beta = 10$.

When a certain value of Re, of the order 0·05–0·1, is exceeded the nature of the motion begins to change. Elongated particles strive to orient themselves, when falling freely, in a way which depends on the balance between the resistance of the fluid and net gravitational force; for disks, needles etc. this will usually be the position of maximum resistance in which their most developed boundaries and longest edges are disposed perpendicularly to the direction of motion [62, 64]. Homogeneous particles in the shape of regular polyhedra, cubes and tetrahedra tend to position themselves with one face perpendicular to this direction [61]. The orienting force increases as Re is increased, orientation becoming complete at some value of Re between 10 and 100. This phenomenon is also in agreement with theory: under potential flow conditions, corresponding to large values of Re, a couple also acts on moving non-spherical particles trying to turn them across the stream [65].

As Re gets bigger the angle θ increases and the trajectory of a freely falling particle deviates further from the vertical, the bigger the size of the particle. This has been established by experiments with coal ash particles in the absence of air currents [66]. The average deviation from the vertical during fall from a height of 100 cm was 0·45 cm for a particle radius $r = 0·04$ mm and 1·4 cm at $r = 0·15$ mm. For sufficiently large Re the path of the particles becomes spiral or zigzag [61, 66]. This phenomenon is particularly strongly pronounced for elongated, needle-like and plate-like particles. As particles orient themselves across the direction of flow their mobility in the direction of the external force acting on them becomes much less than in the perpendicular direction, so that while settling they slip sideways or glide [67]; this is easy to detect by observing the motion of dust particles in the sun's rays. The extent of gliding, perceived as the ratio between the horizontal and vertical rate of movement, depends on the shape and size of the particles. For Re $< 0·1$, i.e. for $r < 10\,\mu$, this phenomenon is not observed in the settling of particles under gravity but it may take place at high speeds, for example in cyclones, explosions etc.

In conclusion, the orientation of elongated particles in a shear flow will be examined (Fig. 12). If in a rectilinear laminar flow directed along the x axis with a gradient Γ along the z axis there is placed a prolate ellipsoid of revolution with its major axis lying in the xz plane, then it can be shown that the ellipsoid, in addition to moving forward, will rotate about its minor axis which lies in the y direction. The rate of rotation [68, 69] is

$$\frac{d\theta}{dt} = \Gamma \cdot \frac{a^2 \cos^2\theta + c^2 \sin^2\theta}{a^2 + c^2} \tag{11.1}$$

where a is the minor and c the major semi axis of the ellipsoid and θ the angle between the major axis and the x axis. Because $c > a$ the rate of rotation is minimal at $\theta = 0$, when the major axis coincides with the direction of flow, and maximal at $\theta = \pi/2$. The major axis of the ellipsoid is thus directed close to the direction of flow for most of the time. If this axis lies in the xy plane there is no orientation. Finally, in intermediate cases when the major axis is more or less inclined to the xy plane, the greater the angle of inclination the more fully is the ellipsoid oriented. Such is the picture for orientation of particles elongated in one direction. The theory of the phenomenon has been verified in model experiments carried out in viscous liquids [70].

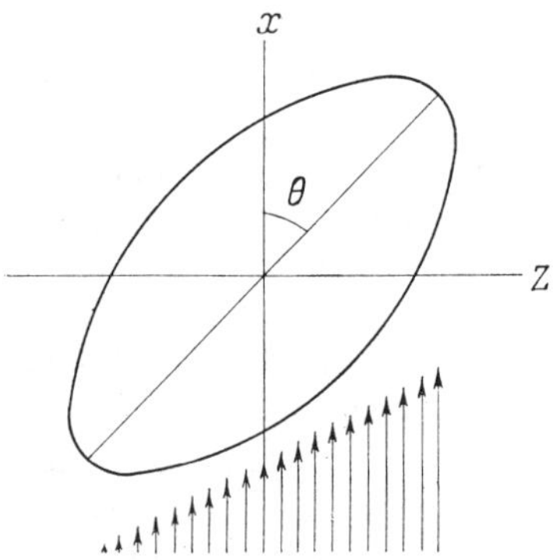

FIG. 12. Orientation of an ellipsoid in a shear flow.

Particles elongated in two directions (platelets, scales) are similarly oriented with their broad sides parallel to the direction of flow. Since some velocity gradient is practically always present in a fluid undergoing laminar flow, elongated particles which are suspended in it are almost always oriented. The phenomena have been studied in detail in colloidal solutions and suspensions but never in aerosols, although they can easily be observed and measured by changes in the intensity and polarization of light scattered by the aerosol.

Brownian rotation of the particles counteracts the orienting action of the flow gradient. As a result, a definite statistical distribution of the directions of the particle axes is set up and can be calculated [68], but the formulae derived are so cumbersome that they are not reproduced here.

Equation (11.1) shows that for spherical particles

$$\frac{d\theta}{dt} = \frac{\Gamma}{2}. \tag{11.2}$$

This equation was confirmed by model experiments with glass spheres [593] having $r = 75 - 140\,\mu$ in corn syrup at $\Gamma = 0 - 1{\cdot}8\ \text{sec}^{-1}$.

According to Sharkey [594], when a thin isokinetic jet of a carbon black suspension was introduced into water in laminar flow along a tube, the jet moved gradually towards the axis, the shift being greater the farther the jet was from the axis and the greater the flow rate. Sharkey thought this phenomenon was due to the Magnus force arising from the rotation of the particles in the velocity gradient. It must be remembered, however, that the Magnus effect requires that the rotating particles should have translational velocity relative to the medium, which was not the case in Sharkey's experiments. In fact, no transverse forces were found to be acting on spherical particles in a velocity gradient in the experiments of Manley and Mason [595]. Moreover, the Magnus formula is valid only when Re is large. According to theoretical work by Garstang [596] the drag upon a spherical particle rotating about an axis normal to the direction of the translational motion at small Re is the same as in the absence of rotation, while the transverse force is either absent or in a direction opposite to that of the Magnus force.

Tollert [597] also refers to the Magnus force in an attempt to explain his observation that the density of a sediment, formed at the bottom of a vertical tube containing liquid or air by settling spherical particles (at Re = 500–7000), increases from the periphery to the centre. In this case, only those particles which fall near the walls rotate, due to the wall effect mentioned on p. 34; the direction of this rotation, however, is such that the Magnus force is directed towards the walls of the tube rather than towards the axis.

The orientation of moving, non-spherical particles is of great importance for microscopic examination of aerosol sediments. According to the observations of Cruize [598], which are also confirmed by other authors [599], deposits in a thermal precipitator (see § 16) had the particles oriented at random. However, when the particles settled under gravity most of them were deposited on the surface in the most stable orientation, that is, flat to the surface. Unfortunately, the author does not indicate the size of the particles examined. On the other hand, Timbrell [600] observed a random orientation of settled particles. The lack of agreement is presumably due to larger particles being used in one case, and smaller ones in the other. The fact that the particles are lying flat [601] in the deposit obtained in a conifuge (see § 28) may be accounted for by the great velocity of sedimentation.

§ 12. RESISTANCE OF A MEDIUM TO THE MOTION OF NON-SPHERICAL PARTICLES

Purely viscous flow at low Re will be examined first. Experiment shows that the resistance offered by a fluid is given by Stokes' law, even for particles of non-spherical shape, if the numerical coefficient is adjusted according to the shape of the particle. A theoretical expression for the coefficients has been obtained only for ellipsoidal particles, and for ellipsoids of revolution the integrals contained in this expression reduce to simple functions. If the equatorial semi-axis of an ellipsoid of revolution is denoted by a and the ratio of the major to the minor axis by β, the resistance of the medium [71] is given by the formula

$$F_M = -6\pi\eta V a \varkappa', \tag{12.1}$$

in which the numerical coefficient $\varkappa'$ is expressed by one of the following formulae:

For the motion of a prolate ellipsoid along the polar axis:

$$\varkappa_c' = \frac{4}{3}(\beta^2 - 1) \Bigg/ \left\{ \frac{2\beta^2 - 1}{\sqrt{\beta^2 - 1}} \ln\left(\beta + \sqrt{\beta^2 - 1}\right) - \beta \right\}. \tag{12.2}$$

For the same transverse to the polar axis:

$$\varkappa_a' = \frac{8}{3}(\beta^2 - 1) \Bigg/ \left\{\frac{2\beta^2 - 3}{\sqrt{\beta^2 - 1}} \ln\left(\beta + \sqrt{\beta^2 - 1}\right) + \beta\right\}. \tag{12.3}$$

For the motion of an oblate ellipsoid along the polar axis:

$$\varkappa_c' = \frac{4}{3}(\beta^2 - 1) \Bigg/ \left\{\frac{\beta(\beta^2 - 2)}{\sqrt{\beta^2 - 1}} \operatorname{arc\,tan} \sqrt{\beta^2 - 1} + \beta\right\}. \tag{12.4}$$

For the same transverse to the polar axis:

$$\varkappa_a' = \frac{8}{3}(\beta^2 - 1) \Bigg/ \left\{\frac{\beta(3\beta^2 - 2)}{\sqrt{\beta^2 - 1}} \operatorname{arc\,tan} \sqrt{\beta^2 - 1} - \beta\right\}. \tag{12.5}$$

TABLE 6. VALUES OF THE SHAPE FACTOR $\varkappa'$ FOR THE ELLIPSOIDAL PARTICLES

Ratio of the axes β	Prolate ellipsoids during motion			Oblate ellipsoids during motion		
	along the polar axis	across the polar axis	statistical average	along the polar axis	across the polar axis	statistical average
2	1·20	1·38	1·32	0·90	0·79	0·82
3	1·40	1·73	1·62	0·88	0·72	0·77
4	1·60	2·06	1·91	0·87	0·68	0·74
6	1·97	2·68	2·44	0·86	0·64	0·72
8	2·31	3·26	2·94	0·85	0·62	0·70
10	2·65	3·81	3·42	0·85	0·61	0·69
20	4·16	6·38	5·64	0·85	0·59	0·68
∞	—	—	—	Infinitely thin circular plate		
				$\frac{8}{3}\pi \approx 0{\cdot}850$	$\frac{16}{9}\pi \approx 0{\cdot}566$	0·66

Values of $\varkappa'$ calculated from these formulae are given in Table 6. If the polar axis of a particle makes an angle θ with the direction of motion, the motion of the particle can be resolved into components parallel and perpendicular to the polar axis on account of the linearity of equation (12.1). The respective components of the resistance of the medium are equal to $-6\pi\eta Va\varkappa_c' \cos\theta$ and $-6\pi\eta Va\varkappa_a' \sin\theta$ and in the general case their resultant, as we have seen above, is at an angle to the direction of motion. The projection of the resultant on the direction of motion of the particle is equal to:

$$F_M = -6\pi\eta Va(\varkappa_c' \cos^2\theta + \varkappa_a' \sin^2\theta). \tag{12.6}$$

Because of Brownian rotation the orientation of the particles is constantly changing. Averaging the resistance of the fluid over all directions of the polar axis, we obtain for F_M the expression

$$F_M = -6\pi\eta Va\left(\frac{1}{3}\varkappa_c' + \frac{2}{3}\varkappa_a'\right). \tag{12.7}$$

The average resistance is thus the same as it would be if the polar axis were one third of the time oriented with the particle motion and two thirds of the time perpendicular to the motion. The statistical mean values of the resistance given in the table were calculated from this.

By putting $\beta \to \infty$ in formulae (12.4) and (12.5) the following expression for the resistance of a medium to the motion of an infinitely thin circular disk of radius a is obtained:

$$F_M = -16\eta a V, \tag{12.8}$$

when the disk is perpendicular to the direction of motion, and

$$F_M = -\frac{32}{3}\eta a V \tag{12.9}$$

for the parallel position. Formula (12.8) with Oseen's correction introduced

$$F_M = -16\eta a V\left(1 + \frac{\text{Re}}{2\pi}\right) \tag{12.10}$$

gives very good agreement with experimental data† [72].

In a similar way the following expressions are obtained from formulae (12.2) and (12.3) for the resistance of a medium to the motion of "ellipsoidal needles" of length $2L$:

$$F_M = -\frac{4\pi\eta VL}{\ln 2\beta} \tag{12.11}$$

when the needle moves along its axis, and

$$F_M = -\frac{8\pi\eta VL}{\ln 2\beta} \tag{12.12}$$

when the needle moves trasverse to its axis (β is the ratio of the length of the needle to its thickness).

A cylindrical needle with a very large ratio of length to radius (R), moving transverse to its axis, is opposed by a force per unit length which is given by Lamb's formula [73]:

$$F_M = -\frac{4\pi\eta V}{2{\cdot}002 - \ln \text{Re}}$$

$$\left(\text{Re} = \frac{2RV}{\nu}\right). \tag{12.13}$$

This agrees well with experiments [74] for Re $\leq 0{\cdot}5$. In this case the resistance is not proportional to the speed.

In the formulae given above the resistance is related to the least (or greatest) diameter of an ellipsoidal particle. Another method of expressing the dependence of the resistance of a medium on the shape of a particle has greater theoretical significance. The ratio $\varkappa$ of the resistance of a given particle to that of a spherical particle having the same volume will be called the dynamic shape factor of the particle. The

† In the recently published work of Aoi [561] the resistance of a medium to the motion of ellipsoids with various values of β at Re ≤ 4 has been accurately calculated.

radius of an equal sphere called the equivalent radius, r_e, is $a\beta^{\frac{1}{3}}$ for a prolate and $a\beta^{-\frac{1}{3}}$ for an oblate ellipsoid (a is the equatorial semi-axis). Thus

$$\varkappa = \frac{6\pi\eta Va\varkappa'}{6\pi\eta Vr_e} = \varkappa'\beta^{\mp\frac{1}{3}}, \tag{12.14}$$

in which the upper sign refers to prolate and the lower to oblate ellipsoids. If the densities of sphere and ellipsoid are the same the coefficient $\varkappa$ is equal to the ratio of the rates of settling of the two particles under the action of gravity. The rate of settling of a non-spherical particle will thus be given by the formula

$$V_s = \frac{2r_e^2\gamma g}{9\eta\varkappa}. \tag{12.15}$$

From the definition of the sedimentation radius† (see page 19) it follows that

$$V_s = \frac{2r_s^2\gamma g}{9\eta} \tag{12.16}$$

and, consequently,

$$\varkappa = r_e^2/r_s^2 \tag{12.17}$$

hence the dynamic shape factor equals the square of the ratio of the equivalent and sedimentation radii.

Values of $\varkappa$ for ellipsoidal particles are given in Table 7.

TABLE 7. VALUES OF THE DYNAMIC SHAPE FACTOR $\varkappa$ FOR THE ELLIPSOIDAL PARTICLES

Ratio of the axes β	Prolate ellipsoids during motion			Oblate ellipsoids during motion		
	along the polar axis	across the polar axis	statistical average	along the polar axis	across the polar axis	statistical average
1·1	0·994	1·005	1·001	—	—	—
1·3	0·970	1·027	1·008	—	—	—
1·5	0·940	1·044	1·010	1·072	0·958	0·996
2	0·95	1·09	1·05	1·14	0·99	1·04
3	0·97	1·20	1·12	1·26	1·04	1·11
4	1·01	1·30	1·20	1·38	1·08	1·18
6	1·08	1·47	1·34	1·56	1·17	1·30
8	1·15	1·62	1·47	1·71	1·25	1·40
10	1·22	1·76	1·58	1·83	1·32	1·49
20	1·54	2·34	2·08	2·31	1·59	1·83

Particles of regular ellipsoidal shape are very rarely encountered; the recently published results of measurements of the rate at which particles of various shapes fall in viscous liquids at low Re are, therefore, of great value [62, 75]. These experiments

† In this book the term "sedimentation radius" has been used in place of "Stokes' radius" for, generally speaking, this radius is not connected with Stokes' formula and can be used for particles of each size (very large or very small) for which it is not valid. In some cases, for example, aerosols containing particles of different or unknown densities, it is convenient to use another term, the "reduced sedimentation radius", which is the radius of a spherical particle of unit density and having the same falling velocity.

were conducted with ellipsoids of revolution, circular cylinders, rectangular parallelepipeds of square cross-section and bodies consisting of two cones joined at their bases (Table 8).

TABLE 8. DYNAMIC SHAPE FACTOR

Ratio of the height to the diameter (or to the base, or the ratio of the axes)	Cylinders[1]		Parallelepiped with square base[1]		Ellipsoids of rotation[2]		Bodies consisting of two circular cones joined by the bases[2]	
					Prolate	Oblate		
	Position of the axis		Position of the normal to the base		Position of the polar axis		Position of the axis	
	Horizontal	Vertical	Horizontal	Vertical	Horizontal	Vertical	Horizontal	Vertical
0·25	1·09	1·31 (1·34)	1·15	1·39 (1·40)	—	—	—	1·48
0·50	1·04	1·16	1·07	1·18	—	—	—	—
1·00	1·06	1·04 (1·00)	1·08	1·08 (1·04)	—	—	—	1·07
2·00	1·14	1·02	1·16	1·04	—	—	—	—
3·00	1·24	1·04	1·22	1·03	—	—	—	—
4·00	1·32 (1·30)	1·07	1·31 (1·29)	1·09	1·28	1·36	1·27	—

[1] Heiss's experiments [75]; within brackets, experiments of McNown and Malaika [62].
[2] Experiments of McNown and Malaika [62].

There is quite satisfactory agreement between the experimental values of $\varkappa$ for ellipsoids (1.28 and 1.36) and the theoretical values (1.30 and 1.38). It is also demonstrated by these results that the values of $\varkappa$ for elongated bodies of different shape but with the same axial ratio are very close to one another. Hence it follows that in calculating the mobility of rod-like particles they may be taken as prolate ellipsoids, and disk-shaped particles may be taken as oblate ellipsoids; the resulting errors are of the order of a few per cent.

The value of $\varkappa$ has also been determined for regular polyhedra, with the aid of models [61], when the following values were obtained; 1·06 for an octahedron, 1·07 for a cube and 1·18 for a tetrahedron. At $\mathrm{Re} < 0{\cdot}05$ the value of $\varkappa$ is almost independent of the orientation of the polyhedra.

Finally, measurements of $\varkappa$ at low Re have also been made on model aggregates consisting of glass spheres stuck together [64]. The aggregates, shaped like chains or disks, were disposed horizontally during fall. For chains of two particles it was found that $\varkappa = 1{\cdot}16$, of three particles 1·31, four particles 1·70, eight particles 2·14. For flat aggregates of three particles $\varkappa = 1{\cdot}26$, seven particles 1·70. For octahedra of six particles it was 1·31. These figures show clearly why the apparent density of particles determined, without reference to shape, from their falling speed is particularly low in the case of linear aggregates (see page 18).

An important conclusion follows from the results above. As Table 7 shows, $\varkappa$ can have values rather less than 1 for the motion of ellipsoidal particles in the direction of their major axes. It is possible that profiles exist which give still smaller values of $\varkappa$

when favourably oriented. However, such orientation is unstable, either on account of Brownian motion or of hydrodynamic forces. Large particles oriented with their long axes perpendicular to the direction of motion always have $\varkappa$ greater than 1. Nor in practice, as the table shows, does the statistical mean value of $\varkappa$ for small particles fall below 1. Hence it is possible to make the general statement that particles of spherical shape have greater mobility than particles of any other shape which have the same volume. In other words the sedimentation radius of non-spherical particles is always less than their equivalent radius [see formula (12.17)]. This explains why determinations of the apparent density of smoke particles (see Table 2), may yield maximum values undoubtedly relating to individual particles, which are appreciably less than the true density.

The values taken by $\varkappa$ at large Re have been given little attention. It remains more or less constant up to a certain Re and then begins to increase rapidly [61, 62]. This increase, associated with the creation of vortices by protruding edges, begins sooner the sharper these edges. Thus it commences when Re is about 100 for cubes and octahedra and 10 for tetrahedra and thin disks falling in the horizontal position.

The following practical conclusions can be drawn. In dynamic calculations with non-spherical particles the data presented above must first be used to determine the probable mean value of $\varkappa$ for particles of the given shape. For large particles held in a definite orientation during their motion it is only necessary to know the corresponding value of $\varkappa$. For small particles changing their orientation because of Brownian rotation, the statistical mean value of $\varkappa$ should be taken and Stokes' formula used in the following form

$$F_M = -6\pi\eta V r_e \varkappa. \tag{12.18}$$

For values of Re at which Stokes' law is no longer applicable, use is made of the method of calculation described in § 10, replacing ψ by $\psi\varkappa$ and remembering that r in all formulae means the equivalent radius r_e. By this procedure round, cubical and octahedral particles up to Re = 100, and particles with sharper angles and edges up to Re = 10 can be dealt with.

An important application of the theory of steady motion of aerosol particles is found in air-elutriation of powdered materials. Fractionation by sieves is applicable only to particles of diameter exceeding 40 μ; for smaller sizes recourse must be had either to sedimentation or elutriation (air separation). An elutriator for the size analysis of powders is illustrated in Fig. 13 [76, 77]. A weighed quantity of powder is placed in the constricted lower part A of the apparatus. Air admitted through the tube B disperses the powder and carries the particles upwards. The conical part called the diffuser serves to smooth the air flow and the hood E acts as a trap. The air velocity in the separator must be such that the flow is laminar and $\mathrm{Re}_f = 2R\overline{U}\gamma_g/\eta$ ($\overline{U}$ is the mean air velocity, R is the separator radius) must not exceed the critical value (about 2000).

The air velocity in the cylindrical part of the apparatus corresponds to a definite particle size. If the velocity were constant over the whole cross-section of the separator, all particles having $V_s < U$ would be carried away and all those with $V_s > U$ would be left behind. However the air velocity, U_1, near the axis of the separator is greater than at the walls. Therefore particles with V_s close to U_1 are carried out only if they get to the middle of the separator, and their removal proceeds fairly slowly. When the blow-through has been completed at one speed and the percentage of powder carried out has been determined, the operation is repeated at a greater air flow, corresponding to bigger particles, and so on.

Another factor affecting the working of the separator is the permanence of aggregates of small particles. For complete disruption of aggregates prolonged blowing through is sometimes required while with powders like kaolin complete disaggregation cannot be achieved at all; small aggregates, instead of individual particles, are therefore blown out of the elutriator.

It will be seen from what has been stated previously that the pneumatic separator fractionates particles according to the sedimentation radius and in the case, for example, of plate-like particles the results of sieve and pneumatic fractionation must differ markedly. Microscopic examination of nominally the same fractions of coal

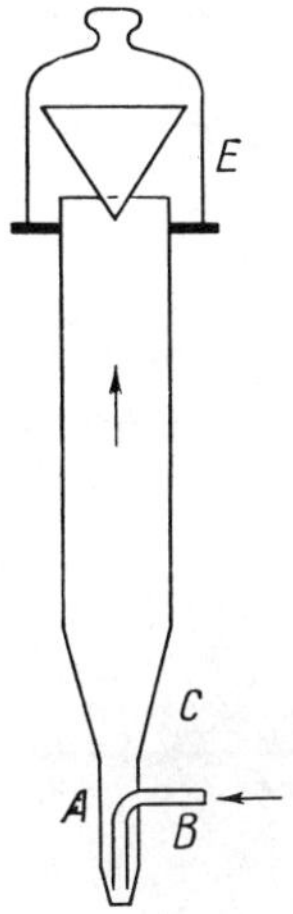

FIG. 13. Elutriator.

dust obtained by two methods have shown that particles are encountered in pneumatic separation which are two or three times bigger than in the corresponding sieve fraction [78]. These particles, as would be expected, are comparatively thin flakes of coal.

In conclusion we will say a few words about the resistance of a gas to the motion of large drops. This is of great importance in the theory of rainfall from clouds. Experiment shows that large drops are appreciably flattened when they are falling (Fig. 14) which causes increased resistance [579].

Recent experimental data on the terminal velocity of water drops in air [79, 80] are given in Table 9 and Fig. 11. It will be seen that appreciable deviation from the rate of fall of equal-sized solid spheres begins at about $r_e = 0{\cdot}4$ mm. At $r_e > 2$ mm there is scarcely any further increase in rate of fall with drop size because increase in drop weight is compensated by a corresponding increase in deformation.

When the equivalent radius of a drop reaches a value of 2·5–3 mm the deformation of the falling drop is so great that it ruptures. Raindrops larger than this are not observed but, in tests carried out in a wind tunnel, it has been found possible to suspend drops of radius up to 4·5–5 mm in an ascending air stream [81]. The degree of turbulence of the flow was of great importance: it was sufficient to disturb the air

TABLE 9. FINAL DROP VELOCITY OF WATER DROPS IN AIR AT 20° AND 760 mm Hg

r_e, cm	0·005	0·01	0·015	0·02	0·025	0·03	0·035	0·04	0·045	0·05
V_s, cm/sec	27	72	117	162	206	247	287	327	367	403

r_e, cm	0·06	0·07	0·08	0·09	0·10	0·11	0·12	0·13	0·14	0·15	0·16	0·17
V_s, cm/sec	464	517	565	609	649	690	727	757	782	806	826	844

r_e, cm	0·18	0·19	0·20	0·21	0·22	0·23	0·24	0·25	0·26	0·27	0·28	0·29
V_s, cm/sec	860	872	883	892	898	903	907	909	912	914	916	917

flow by moving the hand beneath a drop to cause its immediate rupture. The same effect was produced by a sudden small increase in the flow velocity. The rupture process can be seen from Fig. 15 obtained by photographing a suspended drop.

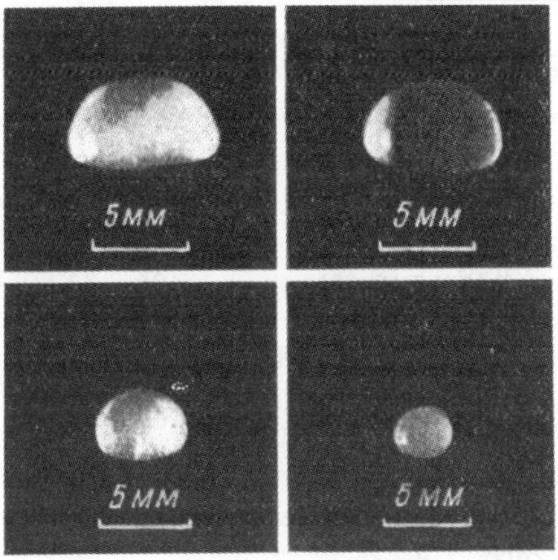

FIG. 14. Shape of falling water drops.

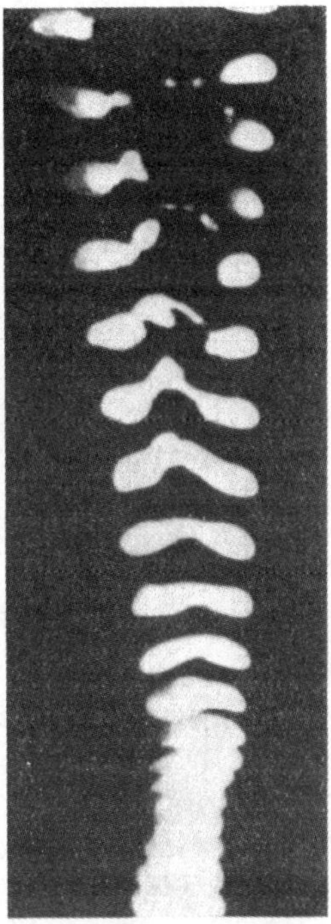

FIG. 15. Rupture of falling drops.

Timbrell [602] described a very convenient method for measuring the sedimentation radius of aerosol particles of irregular shape. Spherical particles of glass, polystyrene etc. were mixed with the aerosol and a thin jet was introduced isokinetically into the centre of a horizontal rectangular channel through which laminar flow of air was maintained. A transparent strip was placed on the bottom of the channel on which the particles were collected and examined. The sedimentation radii of particles of irregular shape, aggregates etc. were determined from the size of the spheres deposited nearby. Instead of the very long channel necessary for small particles, a channel which gradually widened laterally was used. The method may be unsuitable for very small particles as it is extremely difficult to eliminate thermal convection in the channel. In this apparatus Timbrell investigated the relationship between the projected area radius, r_{pr}, which is the radius of a circle equal in area to the particle, and the sedimentation radius, r_s, of glass and coal particles. The majority of particles in the sediment were oriented in a random manner. The ratio r_s/r_{pr} varied greatly from one particle to another (in the case of glass particles from 0·5 to 1·1). The mean value of r_s/r_{pr} was equal to 0·67 for glass and to 0·74 for coal particles.

Timbrell also determined r_s for doublets of glass spheres. When the particles forming the doublet were of the same radius, r, the value of r_s for the doublet, on an average, was 1·13 r and the orientation of the doublets in the experiments seemed to be horizontal. The dynamic shape factor $\varkappa$ of the doublets was found to be equal to 1·24, a value which is clearly high.

In a paper by Ludwig [603], who measured the sedimentation velocities of cylinders in a viscous fluid at small Re, the orientation is not indicated and the results differ considerably from those given in Table 8.

In model experiments of Eveson *et al.* [604] the sedimentation velocities of doublets composed of identical spheres in a viscous fluid at Re $<$ 0·02 were measured. The dynamic shape factor of these doublets $\varkappa$ was found to be equal to 1·15 for horizontal and to 1·04 for vertical orientation of the doublets. Theoretical computation by Faxén [605] in this latter case gives $\varkappa = 1{\cdot}023$.

Gurel *et al.* [606] proposed an empirical formula for the sedimentation velocities of particles of more or less isometric shape in terms of their volume and surface area. However, the formula is obviously unsuitable, even for the simplest case of spherical particles.

Chowdhury and Fritz [607], from their model experiments with spheres, cubes, octahedrons and tetrahedrons at Re $<$ 0·2, obtained the formula:

$$\varkappa = 1/(1 + 0{\cdot}862 \log \varkappa_s) \tag{12.19}$$

where $\varkappa_s$ is the coefficient of sphericity (see p. 19). The values of $\varkappa$ computed from this formula are close to those found by other authors.

Van der Leeden *et al.* [608] measured the velocities attained by water drops, aqueous solutions of surface active substances and some organic liquids after falling from different heights. Considering the air resistance to be dependent only on the velocity of the drops and not on their acceleration, the authors determined the value of the drag coefficient ψ in the formula $F_M = -\psi\pi r^2\gamma_g V^2/2$ (where r is the radius of the undeformed drop). Since $r = 1{\cdot}8 - 3$ mm an appreciable deformation of these drops, falling with Re, between 1000–3000 (see Fig. 11), must have occurred. In this range of Re ψ for a sphere has a constant value $\sim 0{\cdot}4$, and ψ for falling drops is a function of the shape of the drops only. Using a well-known graph of pressure distribution on the surface of a sphere in a potential flow, Hinze [609] calculated the change in the shape of falling drops. To a first approximation, assuming the deformed drops to be oblate ellipsoids, he found the relative decrease in the radius of the drop in the direction of the motion to be equal to 0·069 We, where We $= \gamma_g z V^2/\sigma$ is the Weber number and σ the surface tension of the liquid. Using the values of ψ for ellipsoids and Hinze's formula, Van der Leeden *et al.* obtained a theoretical curve

(ψ, We), which at We $= 1 - 5$ is in fair agreement with the experimental data of the authors and other investigators (Fig. 15a).

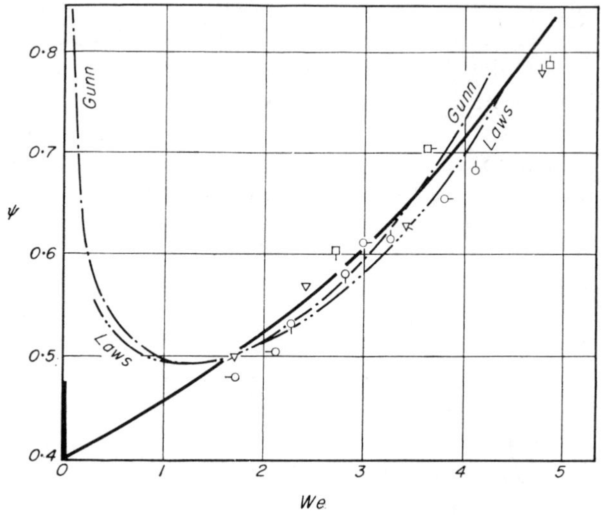

FIG. 15a. Resistance of a gas to the motion of liquid drops at large Re (continuous curve-theoretical).

§ 13. THE SETTLEMENT OF CLOUDS OF PARTICLES

So far only the motion of isolated aerosol particles has been examined. In the mechanics of clouds of particles the hydrodynamic interaction between individual particles must also be considered (see page 95). Two essentially different cases can be distinguished; an unbounded, free cloud and an aerosol filling a space bounded by walls.

In the first place the unbounded cloud will be discussed. Imagine a constant force such as gravity, mg, acting on the particles. Their motion under the action of this force to some extent entrains the surrounding medium with the result that the resistance of the medium to the motion of the cloud is reduced. Theoretical calculation of this reduction has been performed only for a system of two spherical particles [82, 83], in which case the speed of both particles under the action of a force F equals

$$V = (F/6\pi\eta r)\left[1 + \frac{3}{4}(1 + \cos^2\theta)\frac{r}{\varrho} + \cdots\right], \tag{13.1}$$

where ϱ is the distance between the centres of the particles and θ is the angle between the line of centres of the particles and the direction of motion.

Higher terms of the series which are omitted contain powers of r/ϱ from three upwards. Comparison of (13.1) with the expression (34.4) for the speed of flow of a liquid round a sphere at low Re_f shows that the additional velocity received by each particle is simply equal to the flow velocity caused by the other particle at the point occupied by the centre of the first particle. At a rather larger value of Re the effect on the rear particle exceeds that on the leading one (page 100).

For a cloud of many particles Burgers [83] assumed that the additional velocities imparted to a given particle by all the others could be computed from formula (13.1) and added together. Calculation of the resultant velocity of the particles in a spherical cloud on this assumption indicated that the particles and the medium moved with the same speed; the medium was thus completely entrained by the disperse phase and the cloud moved as one entity according to the laws of motion of a liquid sphere in a medium having the same viscosity. The centre of the cloud moves with a velocity

$$V = \frac{4}{3}\pi R^3 nF/5\pi\eta R. \tag{13.2}$$

The numerator is the total force acting on all the cloud particles (R is the radius of the cloud and n the number of particles per cm^3), and the denominator is the resistance of the medium at unit velocity according to formula (7.6) in which η_p has been made equal to η. Simultaneously with the forward motion of the cloud circulation arises in it but the shape and size of the cloud (see Fig. 63, page 241) are preserved.

Burgers concludes that a cloud can move in this way at any aerosol concentration regardless of its size, but it is impossible to agree with this. To understand the problem more easily suppose that the cloud particles are fixed in space while the fluid moves past them with velocity U. The force exerted on each particle is assumed to be given by Stokes' law without any corrections. The total force acting on the cloud is then

$$F_1 = 6\pi r\eta U \,.\, nv = 6\pi r\eta U\frac{4}{3}\pi R^3 n, \tag{13.3}$$

where v is the volume of the cloud. This force is the drag on the cloud when it is being blown through. If the medium flows round the cloud the drag is

$$F_2 = \psi\pi R^2 \,.\, \gamma_g U^2/2. \tag{13.4}$$

If $F_1 \gg F_2$ then the cloud cannot be blown through and the medium completely flows round it. In the case of a freely moving cloud this means that the cloud particles completely entrain the medium and are stationary with respect to it, as in the case examined by Burgers. When $F_1 \ll F_2$ the medium blows through the cloud; this means that in a free cloud, entrainment of the medium by the particles is negligible and the speed of the particles relative to the medium is given by Stokes' formula without corrections. If F_1 and F_2 are of the same order the medium blows partly through and partly round the cloud, and the disperse phase moves relative to the medium with a speed less than that given by Stokes' formula.

Because the velocity U in formulae (13.3) and (13.4) is not fixed but is determined by the external force acting on the cloud, it is more convenient to compare not the forces F_1 and F_2 but the velocities of the cloud under different conditions of motion when a given external force acts on the particles. Conditions corresponding to a higher velocity occur when the difference between these velocities is large, and intermediate conditions when the difference is small. Thus when a cloud is settling under the action of gravity the velocity corresponding to blowing through is

$$V_{S_1} = \frac{2\,r^2\gamma g}{9\eta}. \tag{13.5}$$

The velocity corresponding to flowing-round is obtained by equating the weight of the disperse phase of the cloud, $\frac{4}{3}\pi R^3 n \frac{4}{3} r^3 \gamma g$, to the resistance of the medium [see formula (13.4)]

$$V_{S_2} = \sqrt{\frac{32 R n \pi r^3 \gamma g}{9 \psi \gamma_g}}. \tag{13.6}$$

The character of the motion is determined by the value of the ratio

$$\varkappa = V_{S_2}/V_{S_1} = \sqrt{\frac{72 \pi R n \eta^2}{\psi \gamma_g \gamma r g}}. \tag{13.7}$$

For $\varkappa \gg 1$ flowing-round occurs, and for $\varkappa \ll 1$ blowing-through. If Stokes' formula for a liquid sphere (13.2) is applicable to the motion of the whole cloud, then

$$V_{S_2} = \frac{4 R^2 n}{15 \eta} \cdot \frac{4}{3} \pi r^3 \gamma g, \tag{13.8}$$

$$\varkappa = V_{S_2}/V_{S_1} = 1{\cdot}6 \pi R^2 r n. \tag{13.9}$$

This case has been examined by Smoluchowski [82].

In most clouds encountered in nature, industry and everyday life, at least in the initial period of their existence, $\varkappa \gg 1$ and they move as one entity. Later, when the cloud is dispersed by air currents and diffusion, the product $R^3 n$, proportional to the total number of particles in the cloud, remains constant, and the products Rn and $R^2 n$ which occur in formulae (13.7) and (13.9) decrease as the cloud volume increases; with them $\varkappa$ also decreases until, eventually, the cloud can be blown through.

It must be borne in mind that in practically all clouds of condensation origin, and also in those obtained by the dispersion of volatile substances, the gases inside and outside the cloud have somewhat different compositions and temperatures, and hence, also, different densities. This difference in density is usually considerably greater than the difference due to the weight of the disperse phase.

In natural clouds, for example, the liquid water content is of the order 1 gm/m^3 and an equivalent difference in air density is reached at a temperature difference between the inside and outside air of the order of 0·2°C, or an absolute humidity difference of the order 1 mm Hg. The weight of the disperse phase therefore plays a secondary role in controlling the vertical movement of natural clouds, which is mainly decided by differences in temperature and humidity between the air inside and outside the cloud.

The situation is similar for clouds formed in explosions, by the combustion of fuel in stoves, furnaces and internal combustion engines, for those created with the aid of special generators of screening and insecticidal mists, by smoke cannisters, and so on. Being at a high temperature, such clouds rise initially. If they manage to cool before being scattered in the atmosphere, at a high CO_2 content or high concentration of the disperse phase they begin to settle towards the earth. In all these cases the cloud moves as a whole, independently of its size and the aerosol concentration, because the cloud of gas and particles cannot be blown through and is only washed away at the surface by diffusion and air currents.

Insufficient attention to these considerations sometimes leads to erroneous conclusions. Thus to explain the shape of the path of a stream of tobacco smoke injected horizontally into a smoke chamber Prosad [84], starting from the assumption that smoke particles settle individually, had to take the absurd value of $24\,\mu$ for the mean particle radius. In fact the shape of the path was undoubtedly determined by the rate of settling of the stream as a whole because its density exceeded that of air on account of the carbon dioxide contained in it. A striking example of rapid settling of a cloud is furnished by the "fire cloud" which descended with tremendous speed from the volcano Mont Pelée in 1902 and turned the town of Saint Pierre into cinders. Evidently the concentration of the disperse phase (volcanic ash etc.) was so great that the density of the cloud, despite its high temperature, was much higher than that of air.

A very complicated system of movement exists in cumulo-nimbus clouds containing droplets of all sizes from $r = 10\,\mu$ to $r =$ 2–3 mm. In this case, under the influence of the higher temperature of the cloud in comparison with the surrounding air, a rapid rise of the whole cloud takes place at a rate up to 10 m sec^{-1} while the drops of water in it are falling individually at speeds between 0·01 and 8–9 m sec^{-1}.

The resultant velocity of some droplets is therefore directed upwards while others move downwards. These phenomena play an essential role in the process of precipitation from clouds (see page 319).

§ 14. THE MOTION OF AN AEROSOL IN A CONFINED SPACE

For aerosols in an enclosure the motion of the particles includes that of the medium caused by convection currents, artificial agitation, etc., as well as their own motion relative to the medium. Just now we are interested only in the latter, and shall examine it for particles settling under gravity. If the particles of an aerosol occupying a space confined by walls settle with a velocity V the medium moves in the reverse direction with a mean velocity φV, where φ is usually a very small fraction of the total volume of the disperse phase. Since the medium is entrained in the vicinity of the particles, then in the spaces between them the velocity of the counterflow is greater then φV. Thus the rate of settling of particles in the present case, unlike the motion of a free cloud, is less than that of isolated particles in an infinite volume by the factor $1 + \varkappa\varphi$ where $\varkappa > 1$.

According to Cunningham [46] still another factor should be taken into account; in the derivation of Stokes' formula one of the boundary conditions is that the velocity of the medium is zero at an infinitely great distance from a particle. When a cloud of particles settles in a confined space, however, the velocity of the medium is zero at a distance ϱ from the centre of a particle, where $2\varrho \approx n^{-1/3}$ is the mean distance between adjacent particles. Thus each particle experiences the same resistance which it would experience at the centre of a closed spherical vessel of radius ϱ. According to Cunningham's calculations this resistance, on a Stokes approximation, is equal to $6\pi r V\eta \times (1 + 1{\cdot}25 r/\varrho)$. Following Oseen, the correction becomes less the greater the Reynolds number $V\varrho\gamma/\eta$. All other authors occupied with this problem have arrived by way of fairly complicated, but not rigorous, considerations at a correction factor of $1 + \varkappa\varphi$ with values of $\varkappa$ equal to 5·5 [85], 7·0 [83] and 4·5 [86]. Rigorous solution of the problem is obviously extremely difficult.

The difference between correction factors of the type $1 + \varkappa\varphi$ (I) and $1 + \varkappa\, r/\varrho \approx 1 + \varkappa\varphi^{\frac{1}{3}}$ (II) is of importance because at the usual values of φ in aerosols the factor (I) is practically equal to 1 while the factor (II) may be a few per cent greater than 1. For the small values of φ which are of interest this problem has been investigated experimentally only by Kermak [85] who measured the rate of settlement in monodisperse suspensions of various animal erythrocytes with radii 2·4, 3·0, 3·7 and 4·4 μ in water. It turned out that for $\varphi \leqslant$ 0·04–0·08 the experimental results agree well with a correction factor $1 + \varkappa\varphi$ and $\varkappa$ has values lying in the range 4·8–6·9 for various erythrocytes. Unfortunately the settling rate of isolated particles was not measured in this work but was determined by means of extrapolation.

Thus from the rather scanty data available it may only be said that in the settling of aerosols in a confined space the resistance of the medium at low φ is probably equal to $6\pi r V\eta\,(1 + \varkappa\varphi)$ and $\varkappa$ is close to 5 or 6.

The rate of settling of concentrated suspensions has become important recently in connection with the fluidization of powders (see page 367). In the fluidized state a concentration of particles for which the settling rate is equal to the flow velocity is automatically established. Experiments on fluidization have led to the formula

$$V_s' = V_s(1 - \varphi)^\alpha, \tag{14.1}$$

where V_s' is the settling rate of the entire system of particles and V_s is that of an isolated particle.

For spherical particles Lewis and Bowerman [87] and Richardson and Zaki [88] obtained the same value 4·65 for the coefficient α. An approximate theoretical calculation of the settling rate was made by Richardson and Zaki who started from two models for the distribution of spheres in space; they obtained two curves (V_s', φ) one of which lies about 40 per cent higher and the other 20 per cent lower than the experimental curve.

In conclusion a phenomenon will be mentioned which is familiar to everyone working with aerosols. When concentrated aerosols settle the upper boundary is usually flat and horizontal, a phenomenon which is exhibited both in the laboratory and in natural mists. The explanation is that, for an aerosol density exceeding that of the gas adjoining it, hydrostatic forces counteract any disturbance of the horizontal position of the upper boundary of the aerosol by convection, just as in liquids. Such stabilization of the upper surface will be observed only when the particles move as a whole with the medium, which necessitates a sufficiently high concentration (see preceding section).

The surface of aerosols dispersed in dense gases like chlorine or carbon dioxide etc., is particularly stable [89].

Many theoretical and experimental papers have been devoted lately to the sedimentation of particles in a limited space, or hindered settling. Only equations which refer to very small values of the volume fraction of the disperse phase, φ, (the fraction of the total volume which is filled by the disperse phase) will be given here.

Following Cunningham's idea (see p. 49), but allowing for backwards flow, Happel [610] and Kuwobara [611] obtained at $\varphi \to 0$ the formula $V_s'/V_s = 1 - \varkappa\varphi^{\frac{1}{3}}$ with $\varkappa = 1{\cdot}5$ [610] and 1·62 [611]. Brinkman [612] deduced the formula $V_s'/V_s = 1 - 2{\cdot}1\,\varphi^{\frac{1}{2}}$ and Hawksley [613] $V_s'/V_s = 1 - 4{\cdot}5\,\varphi$. Experiments [614] confirmed the expression $V_s'/V_s = 1 - 2{\cdot}1\,\varphi^{\frac{1}{2}}$ for almost isodisperse liquid suspensions. The results of all other investigations can be expressed by the formula $V_s'/V_s = 1 - \varkappa\varphi \approx (1 + \varkappa\varphi)^{-1}$ with $\varkappa = 4{\cdot}0$ [615, 616], 4·5 [617] and 5·4 [618]. The conclusion (see p. 50) that the velocity of hindered settling at small concentrations depends on φ raised to the first power seems to be confirmed, but no theoretical basis for this is apparent.

The principal difficulty encountered in precision measurements of V_s'/V_s, at small φ, is convection. Wilson [619] using very dilute aqueous suspensions of glass spheres of $r = 1$–$5\,\mu$, found that it was impossible to obtain strictly vertical trajectories of particles at room temperature, although they were realized at 4 °C, when the thermal coefficient of water was equal to zero. Only in more concentrated suspensions and aerosols is the downward gradient of concentration high enough to suppress convection. Unfortunately, this is often ignored. It is difficult to combine the two conditions, a high enough weight concentration and a low enough particle concentration, which is necessary for the neglect of coagulation in not very coarse aerosols. It seems, therefore, that much sedimentation analysis of aerosols is erroneous.

§ 15. MOTION OF PARTICLES IN VERTICAL AND HORIZONTAL ELECTRIC FIELDS. PRACTICAL APPLICATIONS

The motion of aerosol particles in an electric field is no different in principle from motion in the earth's gravitational field. The force acting on a particle in an electric field is qE where q is the charge on the particle and E the field strength. The velocity of the particle given by formula (8.2) is

$$V_E = qEB = qE\left(1 + A\frac{l}{r}\right)\Big/ 6\pi r\eta. \tag{15.1}$$

The movement of particles in a vertical field is very interesting on account of the practical advantage obtained by the electric field being superimposed upon the earth's gravitational field. The vertical electric field method developed by Millikan [90] and Ehrenhaft [91] is one of the most fruitful methods of studying aerosols and has played a very large role in advancing knowledge in this field.

Aerosol particles are introduced into a chamber formed by two horizontal condenser plates and having side walls of insulating material provided with windows for the observation, illumination and charging of the particles. Observations are made with a horizontal microscope having an eyepiece graticule. The field strength $E = \Pi/h$, where Π is the potential difference and h the distance between the condenser plates. The strength and sense of the electric field can be varied as desired. The rate of fall of a particle V_s is determined first with the field switched off, and then under the simultaneous influence of the electric and gravitational fields, $V_s + V_E$ or $V_s - V_E$ depending on the sense of the electric field. Hence V_E is found. In addition, the field intensity E_B which exactly balances the gravitational force on the particle is sometimes determined

$$E_B = mg/q = \frac{4}{3}\pi r^3 \gamma g/q. \tag{15.2}$$

In some chambers provision is made for varying the pressure between wide limits both above and below atmospheric pressure. The technique of working with the vertical field method has been well set out in the literature [53, 92]; it permits the following problems to be solved.

A. Determination of the value of the elementary electric charge and the law of resistance of a gas to the motion of small particles

The first application of the vertical field method was to the solution of these important problems [53]. Writing the expression for V_s:

$$V_s = mgB = mg\left(1 + A\frac{l}{r}\right)\Big/ 6\pi r\eta \tag{15.3}$$

or

$$V_s = \frac{2r^2 g\gamma}{9\eta}\left(1 + A\frac{l}{r}\right), \tag{15.4}$$

taking the square root of (15.4), and multiplying by (15.1) gives

$$q = \left[\frac{18\pi V_E}{E}\left(\frac{\eta^3 V_s}{2\gamma g}\right)^{1/2}\right]\left(1 + A\frac{l}{r}\right)^{-3/2} = q_{St}\left(1 + A\frac{l}{r}\right)^{-3/2}. \tag{15.5}$$

The expression in square brackets is denoted by q_{St} which is the magnitude of the charge on the particle determined from experiment on the assumption that Stokes' formula applies. In the case of liquid (oil) droplets all the quantities entering into this expression – the viscosity of air, density of the particles, field strength, and velocities V_E and V_s – can be determined by experiment. In charging a particle by X- or γ-rays the change in its charge Δq is equal to a small positive or negative whole number of elementary charges ε:

$$\Delta q = (\nu_1 - \nu_2)\varepsilon, \tag{15.6}$$

where $\nu_1 - \nu_2$ is usually ± 1, less often ± 2, and so on. Similarly $\Delta q_{St} = (\nu_1 - \nu_2)\varepsilon_{St}$. Thus by taking the highest common factor of several Δq_{St} it is easy to determine ε_{St}.

It follows from equation (15.5) that

$$\varepsilon = \varepsilon_{St}\Big/\left(1 + A\frac{l}{r}\right)^{3/2}, \tag{15.7}$$

hence

$$\varepsilon_{St}^{2/3} = \left(1 + A\frac{l}{r}\right)\varepsilon^{2/3} = \varepsilon^{2/3} + A\varepsilon^{2/3}\cdot\frac{l}{r}. \tag{15.8}$$

When values of $\varepsilon_{St}^{2/3}$ are plotted for various droplets as a function of l/r the points must lie on one straight line according to (15.8), provided that Cunningham's formula is applicable. The point of intersection of the straight line with the ordinate axis gives the value of $\varepsilon^{2/3}$ and the tangent of the angle of inclination to the axis of the abscissa gives the value of $A\varepsilon^{2/3}$. In fact the quantity determined directly from experiment is not the true radius r but

$$r_{St} = \left(\frac{9V_s\eta}{2\gamma g}\right)^{1/2}, \tag{15.9}$$

that is the radius calculated on the assumption that Stokes' formula holds. However, at small values of the correction term Al/r, r_{St} differs little from r so that replacing r in (15.8) by r_{St} has little effect on the results; in particular the experimental points lie on a straight line. The first accurate determination of the electronic charge was made in this way and the applicability of Cunningham's correction at small values of l/r was demonstrated. To obtain accurate results in the determination of ε the effect of Brownian motion should be excluded as far as possible by working with comparatively large droplets ($r = 2$–$5\,\mu$).

The use of small drops is likewise avoided when studying the law of resistance of a medium at large values of l/r by making the measurements at low pressures, that is at large l. In these experiments [39] reduction of the pressure to 0·5 mm Hg made it possible to measure the resistance of the gas up to a value of $l/r = 134$. The empirical formula (8.5) was established in this manner. The most thorough method of studying the law of resistance for small particles consists in measuring the speed of the same particle at various pressures [94, 95, 28].

The first experiments using a vertical electric field were carried out not with individual droplets but with an aqueous mist obtained by the condensation of steam on gaseous ions. The mist droplets carried one elementary charge each. V_s and V_E were determined by the movement of the upper boundary of the cloud and the value obtained for ε was about 30 per cent below the true value [96]. This is not surprising since the velocity of the upper boundary is always determined by the slowest particles (the smallest in the measurement of V_s and the largest in the measurement of V_E) so that, according to (15.1), a low value is obtained for the charges on the particles.

So far only spherical particles have been considered. In passing on to non-spherical particles and aggregates the following points must be taken into account. For aggregated particles γ denotes the apparent density, and if the true density is inserted in (15.5) in place of γ, values of q, and hence of ε also, several times less than the true ones may be obtained. The "sub-electron" hypothesis [97] arose in this way. If a particle is not spherical $r_e^2/\varkappa$ instead of r^2 must be put in (15.4) (see 12.15) and $r_e\varkappa$ instead of r in (15.1) (see 12.18). As a result the factor $\varkappa^{3/2}$ appears in the denominator of the expression for q [see formula (15.5)]. Because $\varkappa > 1$ for non-spherical particles (see page 40), too low a value will again be obtained for the elementary charge if this factor is not taken into account.

Knowing the value of ε and the law of resistance of a medium the following problems can be solved by the vertical field method.

B. Measurement of charge and mobility of particles

By measuring the speed of a particle in a vertical electric field, first moving upwards and then downwards, and taking the arithmetic mean of the two speeds, $V_E - V_s$ and $V_E + V_s$, V_E is obtained and is proportional to the number, ν, of elementary charges on the particle. By charging it repeatedly, as indicated above, ν can be evaluated and the total charge $q = \nu\varepsilon$ found. Knowing V_E and q the mobility B of the particle can be determined from formula (15.1).

The mobility of a particle can also be calculated from its rate of fall, V_s, in the absence of the field [see formula (15.3)], and the balancing field E_B which depends on its mass (15.2).

The shape and density of the particle are not involved explicitly in either of these methods. It is very likely, however, that the mobility of a particle in an electric field, determined by the first method, and the mobility in the gravitational field, determined by the second, may differ appreciably for irregular shapes on account of the orienting action of the electric field, which tends to turn the long axis in line with the field (see § 43). This does not seem to have been borne in mind by any of those who have worked in this field and may explain many discrepancies between theory and experiment in the motion of aerosol particles in a vertical electric field.

C. Determination of particle size

Two difficulties are encountered in determining the size of aerosol particles from their rate of settling under gravity by applying formula (15.4)

(1) The apparent density of aggregated particles in (15.4) is unknown.

(2) Small particles ($r < 10^{-4}$ cm) have a large Brownian motion which results in fluctuations of the measurements.

These difficulties are largely abolished by the following variant of the method [98]. V_E and q are determined as indicated above and hence r, or $r_e \varkappa$ for non-spherical particles is, found from (15.1). The density of the particle need not be known. The effect of Brownian motion can be reduced by making the velocity V_E sufficiently large by increasing the field strength E. This, of course, cannot be done for V_s.

The size of particles of unknown density can be found from their rates of fall at various pressures [99] in the following way.

The mean free path of gas molecules is inversly proportional to the gas pressure, hence (15.4) can be written in the form

$$V_s = \frac{2\gamma r^2 g}{9\eta}\left(1 + \frac{A'}{pr}\right) \tag{15.10}$$

where $A' = Apl = Ap_0 l_0$, l_0 being the mean free path at atmospheric pressure p_0. Thus, if V_{S_1} and V_{S_2} are the rates of fall at pressures p_1 and p_2,

$$\frac{V_{S_1}}{1 + (A'/p_1 r)} = \frac{V_{S_2}}{1 + (A'/p_2 r)}, \tag{15.11}$$

whence

$$r = \frac{V_{S_2}p_2 - V_{S_1}p_1}{V_{S_1} - V_{S_2}} \cdot \frac{A'}{p_1 p_2}. \tag{15.12}$$

This method cannot give very accurate results since r is determined by the ratio of the small differences between nearly equal pairs of quantities. Another uncertainty is that for sizes of the same order as the mean free path the theory of the motion of small particles in a gas (§ 8) has only been worked out for spheres. It is not known what shape factor ought to be applied to the correction term $1 + Al/r$.

D. Determination of apparent density and dynamic shape factor of particles

Having determined the radius, as indicated above, and the mass, from the value of the balancing field, these quantities can be used to find the apparent density, γ, of an aggregated particle. Whytlaw-Gray [31] found the values given on page 19 for the density of some smoke particles in this way. Correct results can be obtained only for spherical particles. For particles of other shapes the coefficient $\varkappa$ must be introduced into (15.1), otherwise a high value for the equivalent radius r_e and an excessively low value for the apparent density will be obtained.

The equivalent radius, r_e, of individual non-spherical particles of known density can be determined by using (15.2) after measuring their charge and the magnitude of the balancing field. In addition, the value of $\varkappa$ can be found either from the velocity V_E in an electric field or from V_s for motion under gravity. It has already been pointed out that different values for $\varkappa$ may be found in the two cases if the particles orient in the electric field.

Particle density can be determined by a modification of the vertical field method proposed by Placzek [24]; an inhomogeneous electric field with diverging lines of force is employed. The force, F, acting on an uncharged particle in such a field is

$$F_E = \varkappa_E v \operatorname{grad} E^2, \tag{15.13}$$

where v is the volume of the particle and $\varkappa_E$ is a coefficient depending on the shape and dielectric constant ε_k of the particle. For an insulating sphere $\varkappa_E$ is equal to $(3/8\pi)(\varepsilon_k - 1)/(\varepsilon_k + 2)$ while for a conducting one it is $3/8\pi$. In the latter case $\varkappa_E$ is the same for solid particles and for aggregates provided that electrical contact exists between the primary particles of which the aggregates are composed. If E and grad E are directed vertically at the point occupied by the particle, the force acting on it is

$$F_E = 2\varkappa_E v E\, dE/dz. \tag{15.14}$$

If the gradient is upwards the particle will be held stationary when $F_E = mg$ or

$$E dE/dz = \frac{mg}{2\varkappa_E v} = \frac{\gamma g}{2\varkappa_E}. \tag{15.15}$$

Equation (15.15) shows that uncharged spherical particles having the same density and dielectric constant are in equilibrium at the same value of $E\, dE/dz$, whatever their size may be. This has been verified by experiments with droplets of oil and mercury, an inhomogeneous field being produced by suitably shaping the upper plate of the condenser. In this case $E\, dE/dz$ is proportional to the square of the voltage on the condenser plates.

During these experiments it was found that the values of $E\, dE/dz$ corresponding to the equilibrium of both oil and mercury droplets were such that the coefficient $\varkappa_E$ was practically the same in each case. Oil droplets must therefore be polarized in a steady electric field like conductors of electricity, evidently because of ionizing contaminants contained in them. Since mineral oils as a class are very poor conductors of electricity, a similar phenomenon probably occurs in particles of most other materials. In future, therefore, it will be supposed that the aerosol particles behave like conductors in an electric field.

The value of $E\, dE/dz$ at various points of an electric field produced by a condenser can be measured accurately by observing individual droplets and applying equation (15.15). The apparent density of aggregated particles of metallic aerosols can then be found from similar observations on the particles. Such particles come to equilibrium at various values of $E\, dE/dz$ because they have different apparent densities. This made it possible to establish that the droplets in mercury mists obtained by mechanical atomization at low air pressure have normal density, but those in mists atomized at high pressures or formed by volatilization possess apparent densities which are one fifth to one-tenth of the density of mercury (see page 19).

The vertical field method has also been used to investigate the kinetics of the evaporation of droplets, the photoelectric effect on particles, Brownian motion, motion in a temperature gradient and a number of other physical problems.

The motion of aerosol particles in the earth's gravitational field with a horizontal electric field superimposed can also be used to determine ε and for some other problems discussed above. Motion then takes place along an inclined straight line and is described by the same equations (15.1) and (15.3), the only difference being that V_s is here the vertical and V_E the horizontal component of the particle velocity. Equation (15.5) and all the other results still hold here. Measurement of both velocity components is carried out by a photographic method using intermittent illumination [100]. This method can be used with coarser droplets (up to $r = 10\,\mu$) than the vertical field method, which is an advantage in the precise determination of ε. For another variant of this method see page 109.

It is impossible to suspend large particles in a Millikan condenser since the powerful electric field required results in ionization. According to Straubel [620] droplets up to $100\,\mu$ in diameter could be suspended and observed for long periods at the centre of a horizontal metal ring connected with an a.c. source of several kilovolts; small droplets executed vertical oscillations, and large ones remained practically stationary.

A similar, but a more elaborate device for suspending large particles was designed by Wuerker *et al.* [621]. They used a condenser with convex plates separated by an insulated convex ring. Controlled voltages on plates and ring with d.c. and a.c. components produced an inhomogeneous field in the condenser. A pressure of 10^{-4} to 10^{-2} mm Hg was maintained in the condenser. Charged aluminium particles ($r = 10\,\mu$) of which had certain ratios of charge to mass (q/m) were retained in the condenser at definite voltages, while all others were rejected. By controlling the voltage and frequency it was possible to control the location and mode of oscillation of the particles and to determine the ratio q/m. With rise of pressure the amplitude of oscillation decreased due to increase in air resistance. This device appears to be very convenient for the study of the resistance of gases at small values of r/l during oscillation at large Re, as well as for a number of other problems in the mechanics of aerosols.

§ 16. RADIOMETRIC FORCES IN AEROSOLS. THERMOPHORESIS, PHOTOPHORESIS AND DIFFUSIOPHORESIS

The repulsion of aerosol particles from hot surfaces was known in the eighteen eighties but has only recently been interpreted correctly. It can easily be observed by side illumination of an aerosol in which a heated object is situated, when a "black" layer containing no particles is revealed, the thickness of which increases with the temperature of the object.

The phenomenon arises from the so-called radiometric forces exerted by a gas on non-uniformly heated objects, in this case aerosol particles, which it surrounds.

The radiometric force may be caused either by a temperature gradient in the gas (thermophoresis) or by illumination of particles from one side (photophoresis).

Radiometric forces, like other interactions between particles and a gas, depend essentially on the ratio of the particle radius to the mean free path of the gas molecules.

For $r \ll l$ thermophoresis is the result of gas molecules impinging on the particle from opposite sides with different mean velocities. The net force caused by difference in momentum received by the particle from each side is equal to [93, 252]

$$F_T = -\frac{\pi p l r^2 \Gamma_a}{2T} \tag{16.1}$$

where Γ_a is the temperature gradient in the medium. The coefficient of accommodation of gas molecules on the surface of a particle has been taken as unity. In this case the radiometric force is proportional to the square of the particle radius and is independent of gas pressure because pl is constant.

Photophoresis is due to the gas molecules rebounding from the hotter, illuminated side of the particle with greater velocities than from the unilluminated side. The resultant force is equal to [101]

$$F_T = -\frac{\pi \alpha p r^3 \Gamma_i}{3T}, \tag{16.1'}$$

where Γ_i is the temperature gradient inside the particle, and α is the coefficient of accommodation.

In this case, therefore, the radiometric force is proportional to the pressure of the gas and the cube of the particle radius.

For $r \gg l$ the mechanism of the radiometric force is rather more complicated [102]. Imagine a non-uniformly heated wall in contact with a gas. In the layer of gas adjacent to the wall approximately the same tangential temperature gradient will be established. If the temperature is higher to the left, the gas molecules impinging on the element of surface ds from the left have on the average a higher velocity than those coming from the right. As a result the wall receives an impulse directed to the right, against the temperature gradient, and an equal impulse directed to the left is imparted to the gas, making the gas slip along the surface towards the higher temperature with a velocity

$$U = \frac{3\eta \Gamma_i}{4\gamma_g T}. \tag{16.2}$$

The nature of radiometric flow of a gas round a non-uniformly heated object is essentially different from normal viscous flow, is which examined in § 34. In the latter case the tangential velocity of the gas is zero at the solid surface and increases with the distance from it so that the frictional force acting on the surface is in the direction of flow. In radiometric flow, however, the velocity of the gas is maximal (16.2) at a distance l from the surface and then decreases with distance; the force acting on the body is thus directed against the flow towards the lower temperature. Hydrodynamic calculation shows [103] that the force on a spherical particle in the case of photophoresis is

$$F_T = -\frac{3\pi\eta^2 r R_g \Gamma_i}{pM}, \tag{16.3}$$

where R_g is the gas constant and M is the molecular weight of the gas.

The radiometric force in this case is inversely proportional to the pressure of the gas (as long as it is not too low and the viscosity of the gas can be considered constant) and proportional to the radius of the sphere.

When the radius is comparable with the mean free path Hettner [103] proposed the empirical interpolation formula

$$F_T = -\frac{\pi r^2 \eta \sqrt{\dfrac{\alpha R_g}{MT}}\,\Gamma_i}{\dfrac{p}{p_0} + \dfrac{p_0}{p}}, \tag{16.4}$$

where
$$p_0 = \frac{3\eta}{r}\sqrt{\frac{R_g T}{M\alpha}}. \tag{16.5}$$

It can easily be verified that formula (16.4) reduces to (16.1′) for $l \gg r$ and to (16.3) for $l \ll r$. The absolute value of the force F_T reaches a maximum at $p = p_0$. In air at normal temperature, for $\alpha = 1$

$$p_0 \approx 1{\cdot}5 \times 10^{-5}/r \quad \text{atm}. \tag{16.6}$$

The dependence of radiometric force on pressure, expressed by (16.4), has been confirmed by experiments on the photophoresis of selenium particles [104]. The fact that some authors have not found any effect of pressure on the rate of photophoresis is explained by their having worked at pressures close to p_0, the maximum value of the radiometric force, in the vicinity of which the force varies only slightly with pressure.

The main difficulty in calculating the radiometric force on a particle is the determination of the temperature gradient in the particle itself. This problem is solved comparatively simply if the gradient is caused by non-uniformity of the temperature of the gas itself. In the absence of convection currents (when the temperature gradient of the medium is directed upwards) and if the emission of heat from the particle by radiation is neglected, the gradient inside a spherical particle is given for $r \gg l$ by the formula [105]

$$\Gamma_i = \frac{3\chi_a \Gamma_a}{2\chi_a + \chi_i}, \tag{16.7}$$

where Γ_a is the temperature gradient of the medium, χ_i is the thermal conductivity of the material of the particle and χ_a is the thermal conductivity of the gas.

The radiometric force is then equal to

$$F_T = -\frac{9\pi\chi_a}{2\chi_a + \chi_i}\,\frac{\eta^2 r R_g \Gamma_a}{pM} = -\frac{9\pi\chi_a}{2\chi_a + \chi_i}\,\frac{\eta^2 r \Gamma_a}{\gamma_g T}. \tag{16.8}$$

This formula, derived by Epstein [105], has been checked experimentally by Rosenblatt and La Mer [106] on tricresylphosphate droplets of radius 0·4–1·6 μ by the vertical electric field method. The temperature of the upper plate of a condenser was held higher than that of the lower plate so as to eliminate convection in the condenser. The rate of vertical movement of a particle, both with and without the temperature gradient, was measured and the velocity of the particle due to the radiometric force was determined from the difference; the radiometric force was then evaluated by means of (8.2). The force was strictly proportional to the temperature gradient. Further, experiments conducted with a single particle at various pressures showed that at a large value of r/l the radiometric force increases linearly with $1/p$ according to formula

(16.8) but as l approached r the graph began to turn off below the straight line, as would be expected from formula (16.4). At a high pressure (760 mm Hg) the radiometric force, in agreement with formula (16.8), is proportional to the particle radius so that F_T/r does not depend on r; at lower pressures F_T/r increases somewhat as r decreases in contradiction to formula (16.4). The absolute value of the force was found to be 50 per cent higher than the theoretical value, possibly because the value of the thermal conductivity of tricresyl phosphate used in the calculation was unreliable.

In the later work of Saxton and Ranz [107], carried out by the same technique with droplets of paraffin and castor oil having radii 0·5–2 μ, the pressure was atmospheric and the temperature gradient 18–140° cm^{-1}. The results obtained are shown in Fig. 16, on which the straight line has been plotted according to formula (16.8). In view of the appreciable experimental errors inherent in these measurements, which lead to marked scatter of the experimental points on the graph, the formula can, therefore, be considered verified by experiment.

The effect of a temperature gradient upon the motion of aerosol particles can be seen in everyday life since it causes thick dust deposits† to be formed on walls facing sources of heat. The smoky flame of an oil lamp quickly covers the lamp glass with a layer of soot; similar deposition of soot and ash occurs on chimney walls, the temperature of which is below that of the flue gases. Whenever an aerosol is passed through a series of surfaces at various temperatures some particles will be deposited on the

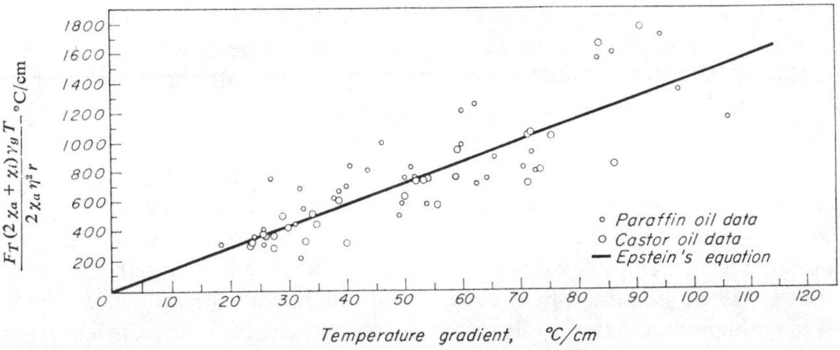

FIG. 16. Radiometric forces acting on droplets.

colder areas. Especially effective is the deposition in a narrow gap, formed by two massive metallic blocks, in the middle of which a heated metallic filament is taughtly stretched perpendicular to the flow direction. Instruments of this type (thermal precipitators) have found wide application in the study of aerosols [110]. The great advantage of the method is that, in agreement with (16.26) and (16.27), the velocity of particles in a temperature field depends comparatively little on their size, unlike their velocity under the action of gravitational and inertia forces which drops rapidly with diminishing size. Other advantages of thermal precipitation over inertial de-

† When rooms are heated by steam or water pipes the air is warmer than the walls, which in time become covered with an even layer of dust [108].

position are that breaking up of aggregates and blowing away of deposited particles are avoided.

According to Watson [112] the thickness of the "black" layer surrounding a heated solid (seen by side illumination), or the "dust-free space", is roughly constant over the whole surface of the solid and is given by the formula

$$h = K(T - T_0)^{\alpha} p^{-\beta} \tag{16.9}$$

where K is a constant, $T - T_0$ is the temperature difference between the solid and the medium and p is the pressure of the gas. The values of the indices are $\alpha \approx 0{\cdot}5$ and $\beta \approx 0{\cdot}6$.

The motion of aerosol particles illuminated from one side (photophoresis), being a special example of thermophoresis, has attracted particular attention on the part of research workers. The phenomena observed are of a more complex nature than those caused by non-uniformity of the temperature of the medium, because the temperature distribution in the illuminated particle can be very different, depending on its shape, size, transparency and refractive index. In transparent particles the rear side may be heated by rays refracted inside the particle more than the side which faces the source of light, with the result that the particle moves towards the light (negative photophoresis). Particles of some materials, such as selenium, exhibit negative photophoresis while their size is less than a certain critical value, but at larger sizes the sign of the photophoresis is reversed [101]. The explanation of this phenomenon is that, as the particle becomes larger, the attenuation of the light passing through it increases and consequently the heating of the rear side of the particle decreases.

This has been demonstrated particularly well in experiments with monochromatic illumination of coloured naphthalene smokes [113]: positive photophoresis is observed in illumination with rays strongly absorbed by the particles, and negative photophoresis for weakly absorbed rays. In the latter case the negative photophoresis changes to positive when the concentration of the dye is sufficiently large.

Calculation of the temperature gradient in an illuminated particle is a very complicated task. In order to find even a rough approximation to the value of the photophoretic force in the simplest case of an absolutely black spherical particle for $r \gg l$, the problem will be simplified by assuming that a temperature gradient Γ_i, constant in both magnitude and direction, is set up in the particle, and the heat lost from the particles by radiation will be neglected. ΔT denotes the temperature difference between an element dS of the particle surface and the gas at a great distance from the particle. The normal component of the temperature gradient of the medium at the surface of the particle, when the temperature distribution is established, is equal to

$$dT/d\varrho_{\varrho=r} = -\Delta T/r \tag{16.10}$$

[see analogous problem on page 202, formula (38.35)]. Hence it follows that the amount of heat given out in 1 sec by the element dS of surface is

$$d\Omega = \frac{\chi_a \Delta T dS}{r}. \tag{16.11}$$

The amount of light energy absorbed by the particle in one second is $\Omega = \pi r^2 I$, where I is the intensity of luminous flux (erg cm^{-2} sec^{-1}).

In accordance with the assumed temperature distribution in the particle, the temperature of the surface element dS will be taken as

$$T = T_0 + \Gamma_i r \cos \theta, \tag{16.12}$$

where T_0 is the temperature at the equatorial plane of the particle and θ is the angle between the radius vector to the surface element and the direction of the light rays.

The heat balance for the front half of the particle is expressed by the equation

$$\pi r^2 I - \frac{\chi_a}{r} \int_F \Delta T \, dS - \pi r^2 \chi_i \Gamma_i = 0, \tag{16.13}$$

where the integral is taken over the front half of the surface. The first term of this equation expresses the flow of light energy to the particle, the second is heat given out to the gas and the third, the heat which flows to the rear half of the particle. For the rear half of the particle we have similarly

$$-\frac{\chi_a}{r} \int_R \Delta T \, dS - \pi r^2 \chi_i \Gamma_i = 0. \tag{16.14}$$

Expressing T by (16.12), carrying out the integration and subtracting (16.14) from (16.13) we get

$$\Gamma_i = \frac{I}{2(\chi_i + \chi_a)}. \tag{16.15}$$

Thus the radiometric force is in this case equal to

$$F_T = -\frac{3\pi\eta^2 r R_g I}{2pM(\chi_a + \chi_i)}. \tag{16.16}$$

If the particle is neither transparent nor absolutely black, I must be replaced by $I\alpha$, where α is the light absorption coefficient.

For particles having the shape of an oblate ellipsoid of revolution, with a small ratio of the polar semi-axis c to the equatorial semi-axis a, Epstein derived the following similar formula [105] when the polar axis is directed parallel to the light rays

$$F_T = -\frac{3\pi\eta^2 a R_g I}{2pM\left(\dfrac{\pi}{4}\chi_a + \dfrac{a}{4c}\chi_i\right)}. \tag{16.17}$$

The explanation of negative photophoresis given above has been contested on the grounds that the temperature gradient inside the particle in the direction of the light rays cannot be established because of Brownian rotation of the particle [114]. This objection concerns not only negative photophoresis but also all other cases of the radiometric effect in aerosols.

To find the effect of Brownian rotation on photophoresis an estimate will be made of the value of the temperature relaxation time τ for an illuminated particle, that is the time required to establish the stationary temperature gradient (16.15). For this we will use the same simplified temperature distribution as before.

Suppose that at $t = 0$ the particle has a constant temperature T_0 throughout its volume, this temperature being determined from the heat balance equation

$$\pi r^2 I = 4\pi \Delta T_0 \chi_a r \tag{16.18}$$

whence

$$\Delta T_0 = \frac{rI}{4\chi_a}. \tag{16.19}$$

The temperature difference between the front and rear halves of the particle in a steady gradient is considered small in comparison with ΔT_0, and the difference in heat emission of the two halves is neglected. This is equivalent to discarding χ_a in formula (16.15) and is permissible because the thermal conductivity of solid and liquid materials is usually much greater than that of gases. On this assumption one half of the light energy absorbed by the particle is given out to the gas by the front half of the particle, and the other half of the energy is transmitted to the rear half of the particle. In the presence of a gradient Γ_i the heat content of the front half of the particle is

$$\Omega = \gamma c_p \int T dv = \gamma c_p \int_0^r (T_0 + \Gamma_i x)\pi y^2 \, dx = \gamma c_p \left(\frac{2}{3}\pi T_0 r^3 + \frac{\pi}{4}\Gamma_i r^4\right), \tag{16.20}$$

where γ is the density and c_p the specific heat of the material of the particle. The heat balance in the rear half of the particle is expressed by the equation

$$\frac{d\Omega}{dt} = \frac{\pi}{4}\gamma c_p r^4 \frac{d\Gamma_i}{dt} = \frac{\pi}{2} r^2 I - \pi r^2 \chi_i \Gamma_i, \tag{16.21}$$

a solution of which, reducing to zero at $t = 0$, will be

$$\Gamma_i = \frac{I}{2\chi_i}\left[1 - \exp\left(-\frac{4\chi_i t}{\gamma c_p r^2}\right)\right]. \tag{16.22}$$

Thus the temperature relaxation time of an illuminated particle is equal to

$$\tau = \frac{\gamma c_p r^2}{4\chi_i}. \tag{16.23}$$

Since the thermal diffusivity $\chi_i/\gamma c_p$ of most solids and liquids has a value of the order 10^{-3} or greater†, it follows that $\tau < 2{\cdot}5 \times 10^{-6}$ sec for $r = 1\,\mu$ and $\tau < 2{\cdot}5 \times 10^{-8}$ sec for $r = 0{\cdot}1\,\mu$.

The mean square of the angle of rotation of a spherical particle in a time τ (see § 43) is equal to

$$\overline{\theta^2} = \frac{kT\tau}{4\pi\eta r^3}. \tag{16.24}$$

Hence for air at room temperature and atmospheric pressure it follows that $\sqrt{\overline{\theta^2}} < 0{\cdot}4°$ at $r = 1\,\mu$ and $\sqrt{\overline{\theta^2}} < 1{\cdot}1°$ at $r = 0{\cdot}1\,\mu$. Thus during the temperature relaxation time τ the particle is able to rotate through a very small angle, and, consequently the effect of Brownian rotation on the photophoresis of aerosol particles

† The thermal diffusivity of water is $1{\cdot}5 \times 10^{-3}$, mineral oils $0{\cdot}7 \times 10^{-3}$, sulphur 3×10^{-3}.

with sizes of the order $10^{-4} - 10^{-5}$ cm is negligible. It is not difficult to verify that this conclusion remains valid for transparent particles.

The pressure of light, which in the case of opaque black particles is equal to $\pi r^2 I/c$, c being the velocity of light, does not play an important role in the phenomenon of photophoresis because it is considerably less than the radiometric force.

The simplest case of photophoresis – that of uniform spherical particles – has been examined above. When the particles are non-spherical or non-uniform the phenomenon becomes more complex. The surface temperature of the particle at protruding edges and corners, or on areas of the surface which do not absorb the light, is less than on flat faces or in strongly absorbing areas. The direction of the resultant radiometric force will in general no longer coincide with the direction of the light rays, and the resultant moment of these forces will not be equal to zero. The motion of illuminated particles takes on, therefore, a rather complicated character [562].

Complex motions are observed also when radiometric forces are combined with other kinds of force (electric, magnetic, gravitational etc.). Magnetophotophoresis has been studied most. Solid particles in a uniform magnetic field move, when illuminated, along a spiral, the axis of which is parallel to the field strength vector. The velocity component of a particle in the direction of this vector, V_H, first increases as the field strength H increases, reaches a maximum value (at a given illumination) and then remains constant. The theory of magnetophoresis has been developed by Rohatschek [563]. The magnetic field orients the particle in a definite way. Radiometric forces compel the particle to move at a definite angle to the direction of the field and rotate it about an axis which passes through the particle parallel to the field. Owing to the rotation the direction of the resultant radiometric force changes continuously and the particle describes a spiral. Since Brownian rotation counteracts orientation of the particle the degree of orientation, and hence also V_H, increases with increasing H until complete orientation is achieved. The relationship between V_H and H derived by Rohatschek agrees satisfactorily with the results of measurements.

Using modern theory of transfer phenomena in gases, a more accurate equation for the thermophoretic force at $r \ll l$ was deduced by Waldmann [622] and by Derjaguin and Bakanov [623]. According to Waldmann

$$F_T = -\frac{32 r^2 \chi_a \Gamma_a}{15 G_g} \approx -\frac{4 r^2 \Gamma_a l p}{T} \tag{16.25}$$

a value $8/\pi$ times greater than formula (16.1) deduced in a rather elementary way by Einstein.

The coefficient of accommodation of gas molecules, at the surface of the particles, is assumed to be equal to 1. The radiometer force is therefore proportional to the square of the particle radius and does not depend on gas pressure since lp = const. In (16.25) χ_a is the translational part of the thermal conductivity of the gas, equal to $(15/4)\, k\eta/m_g$ and (6.2) with $\varphi = 0{\cdot}499$ is used for η. The thermophoresis velocity at $r \ll l$ is equal to:

$$V_T = -\frac{\chi_a \Gamma_a}{5p(1 + \pi\alpha/8)} = -\frac{3 l G_g \Gamma_a}{8(1 + \pi\alpha/8) T}, \tag{16.26}$$

where α is the fraction of gas molecules reflected diffusely by the particle.

The velocity, V_T, does not depend on the size of the particles at all, and only very slightly, through α, on their nature; F_T and V_T are related by (6.1).

When $r \gg l$, allowing for the fact that generally $\chi_a \ll \chi_i$ the velocity of thermophoresis is given by (16.8) as

$$V_T = -\frac{3\chi_a \eta \Gamma_a}{2(2\chi_a + \chi_i)\gamma_g T} \approx -\frac{0{\cdot}75\chi_a l G_g \Gamma_a}{\chi_i T}. \tag{16.27}$$

Experiments by Schmitt [592], using the method of [106] and [107], were performed on the thermophoresis of droplets of silicone oil and paraffin wax with $r = 0{\cdot}7 - 1{\cdot}2\,\mu$ in various gases at pressures of 10–700 mm Hg; the range of the values of r/l was from 0·3 to 20. The author found that F_T was practically independent of pressure for $r/l \leqslant 0{\cdot}4$. According to (16.25) this marks the commencement of thermophoresis "at small r/l". The measurements showed that at $r/l \approx 0{\cdot}33$ F_T was strictly proportional to r^2, in agreement with (16.25).

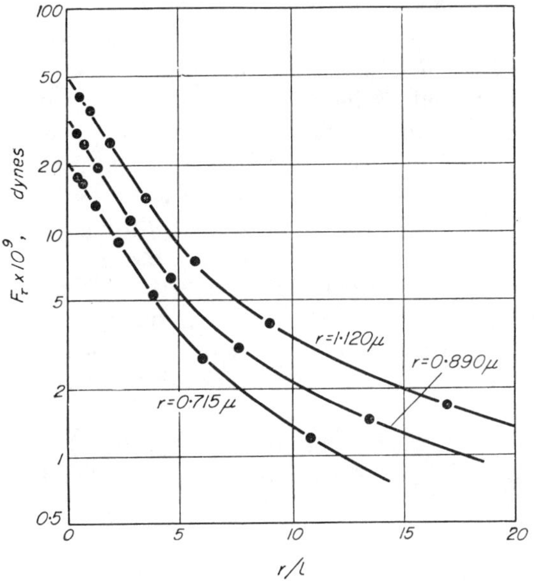

FIG. 17. Thermophoretic force in argon (droplets of silicone oil). $\Gamma_a = 49{\cdot}4°$ C/cm.

On Fig. 17 log F_T is plotted against r/l for silicone oil droplets of different sizes in argon. Since at $r/l < 5$ the graphs are linear, $F_T = F_{T0}e^{-br/l}$, where F_{T0} corresponds to $r/l \to 0$; $b \approx 0{\cdot}38$ in all gases except H_2. The ratio of theoretical (16.25) to experimental values of F_{T_0} is equal to 1·05 in Ar, 1·21 in N_2, 1·22 in CO_2 and 1·36 in H_2 (measurements in H_2 are not very reliable since the thermophoresis velocity is large). Still better agreement with theory is obtained for the velocity of thermophoresis according to (16.26), in which Schmitt assumes $\alpha = 1$, which is, however, at variance with his data for the mobility of particles (see page 29).

A graph of $V_T^* = V_T p/760$ (p is the pressure in mm Hg) against $r^* = rp/760 = 0{\cdot}065 . 10^{-4}\, r/l$ for silicone oil droplets is given on Fig. 17a, which shows that from $r/l \approx 15$ onwards, the thermophoresis velocity no longer depends on the particle size;

the range of large r/l begins here. The transition range therefore extends, approximately, from $r/l = 0{\cdot}4$ to 15. The effect of the thermal conductivity of the particles upon F_T begins to show at $r/l = 3$. The fact that V_T for large particles in H_2 is higher than that of the smaller ones is accounted for by the extremely large thermal conductivity of H_2.

At large r/l the ratio of theoretical to experimental values of V_T is equal to 0·83 in Ar, 1·09 in N_2, 0·71 in CO_2 and 1·07 in H_2.

It follows from (16.26) and (16.27) that the ratio of the thermophoresis velocities of particles having $r \ll l$ and $r \gg l$ is about $0{\cdot}4\,\chi_i/\chi_a$. In air ($\chi_a = 0{\cdot}10^{-5}$ cal. deg^{-1}. cm^{-1}. sec^{-1}) this ratio is 2–4 for particles of organic substances, for NaCl ≈ 100, for Fe ≈ 1000. Taking into consideration the work of Schadt and Cadle (see below) the two last figures ought to be reduced to about 4 and 20, respectively. The data of Schmitt, at atmospheric pressure in air indicates that the transition from (16.26) to (16.27) proceeds when r is about 0·025–1 μ. In this range the thermophoresis velocity decreases continually with rise of r.

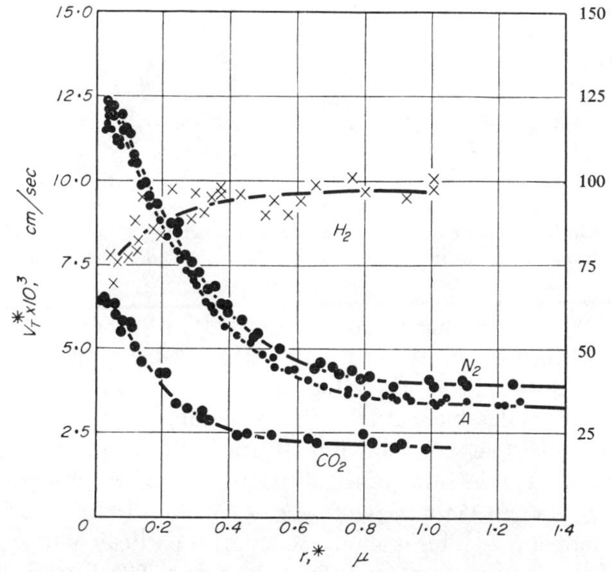

FIG. 17a. Dependence of the thermophoretic velocity on the radius of the drops. $\Gamma_a = 49{\cdot}9$ deg/cm (on the right—the scale for H_2).

In samples obtained in a thermal precipitator, the average particle size must therefore continually increase from the front edge, nearest the intake, towards the back edge, a fact which has already been noted in the literature [624]. To illustrate this, electron micrographs taken at different points of the sediment of a highly polydisperse aerosol of NaCl in a thermal precipitator are shown in Fig. 17b [625]. Similar micrographs were recently obtained by Thürmer [626] with MgO smoke. It should be noted, however, that size segregation of particles in the thermal precipitator is continued also in the range of $r > 1\,\mu$, where theoretically it should cease. The efficiency of the thermal precipitator begins to fall with an increase in r above $2\,\mu$, probably for this reason [627]. It is impossible to explain these facts by inertia of the particles, by their rotation in the flow or their hydrodynamic repulsion from the walls of the precipitator [625]. It is probable that large particles are carried away in the air stream, but this has not yet been shown.

Various aerosols (stearic acid having fairly uniform particles with $\bar{r} = 0{\cdot}15$–$2{\cdot}5\,\mu$, and polydisperse aerosols of NaCl with $\bar{r} = 0{\cdot}1$ and $0{\cdot}2\,\mu$, of glycerol with $\bar{r} = 0{\cdot}25$ and $0{\cdot}5\,\mu$, and of Fe with $\bar{r} = 0{\cdot}7\,\mu$) were drawn in the experiments of Schadt and Cadle [628] through a thermal precipitator with a heated ribbon 1·4 mm wide. A flat slit 0·15 mm wide was formed between the ribbon and the cold plate of the precipitator.

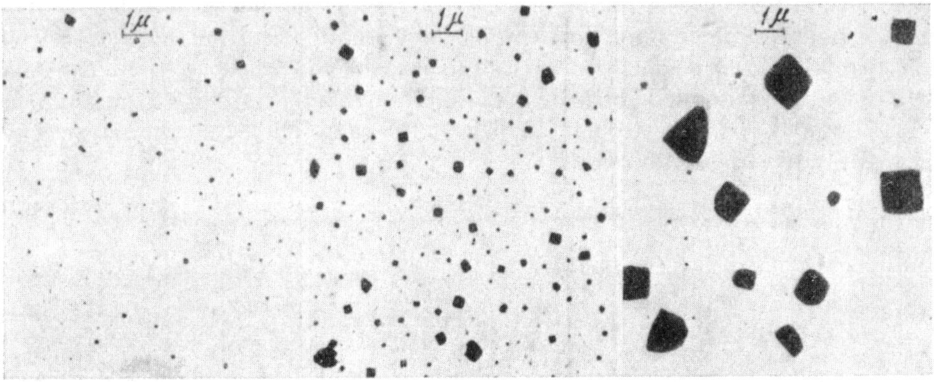

(a) (b) (c)

FIG. 17b. Segregation of particles in a thermal precipitator. (a) At the front edge of the deposit. (b) In the middle of the deposit. (c) At the rear edge of the deposit.

Having measured the ribbon temperature and the width of the deposit, while neglecting the thermal convection, the authors estimated the value of the ratio $\beta_e = F_T/r\,\Gamma_a$ and compared it with the theoretical value of this ratio, β_t, calculated from (16.8). Although there is a considerable scatter in the values of β_e, they show clearly that for particles with low thermal conductivity (stearic acid $\chi_i = 3{\cdot}0 \times 10^{-4}$, glycerol, $6{\cdot}4 \times 10^{-4}$) β_e and β_t have the same order of magnitude; on the other hand for NaCl ($\chi_i = 0{\cdot}5 \times 10^{-2}$) β_e is some 25 times and for Fe ($\chi_i = 1{\cdot}6 \times 10^{-1}$) some 50 times larger than β_e. The thermophoretic force is about the same for particles of organic substances as for NaCl, and for Fe it is about one-fifth of this value. The divergence from theory cannot be explained by electrical effects since the ribbon was connected with the body of the precipitator. In all the experiments with a Millikan condenser, droplets of organic liquids with low χ_i were studied, which explains the good agreement with theory. If Epstein's theory were valid in all cases, it would be practically impossible to use thermal precipitators for collecting metal aerosols.

In using a thermal precipitator to sample aerosols for examination by electron microscopy, a difficulty is encountered. The temperature of the thin organic film is much lower where it is in contact with the wire gauze than at the unsupported parts so that particles are precipitated predominantly near the wires of the gauze and cannot be photographed. Some authors [629, 630] believe that the dependence of the thermophoresis velocity on particle size may give rise to errors not only in the number of the particles but also in their size distribution. This difficulty can be evaded if the deposit on an unsupported area of the film is obtained, cut out and transferred upon a gauze [630].

A theoretical estimate of the thickness h of the dust-free space surrounding heated bodies was made by Zernik [631]. He considers h to be determined by thermophoresis and by a relatively small component, U_n, of the free convection of the gas normal to the surface and directed towards it. U_n was calculated theoretically by Schmidt and Beckmann [632] and determined experimentally by Kraus [633] for a vertical plate and a horizontal cylinder. Assuming in (16.8) $\chi_i = 0$ (which is inadmissible as $\chi_i > \chi_a$), Zernik found that at the middle of the plate and in the horizontal plane passing through the axis of the cylinder the values of h were close to those found by Watson [634]. According to his theory, however, $h = 0$ at the bottom of the heated body and gradually widens upwards, while the drawings of Watson show h as more or less constant. Zernik's calculations, as he points out himself, are valid only at Grashoff numbers of 10^4–10^9, whereas in Watson's experiments $\mathrm{Gr} = 5 \times 10^2 - 5 \times 10^4$.

The theory does not explain the very sharp boundary of the dust-free space observed in all experiments, even with a highly polydisperse aerosol of MgO having $r = 0{\cdot}05$–$1\,\mu$, nor does it account for the thickness of the layer being identical for different aerosols. It rather seems that the agreement between theory and experiment obtained by Zernik may be accidental.

The type of thermophoresis most often encountered in industry is the reverse of the above, solid aerosol particles from hot gases being deposited on the relatively cold walls of boilers and heat exchangers to form a layer of low thermal conductivity which greatly reduces heat transfer and the efficiency of the equipment.

A number of experimental studies have been published lately [635, 636] of the various optical factors determining the direction of photophoresis (positive or negative). The theory of various types of complex photophoresis, involving magnetism, gravitation etc., is thoroughly discussed by Rohatschek [637].

Aitken [638] noticed that at the surface of bodies wetted with a volatile liquid a dust-free space was formed, even when a temperature difference was lacking. This was later corroborated by Watson [634]. Facy [639] described the formation at the surface of an evaporating water droplet of a dust-free space which gradually increased in thickness. Conversely, the condensation of super-saturated vapour on to a droplet caused aerosol particles to move towards it and be absorbed. Facy's explanation of these phenomena was incorrect. He supposed that a diffusion movement of the particles, similar to thermophoresis and due to concentration gradients of the components of the gas mixture was taking place. As will be shown later, this process really exists but in the experiments of Aitken, Watson and Facy the principal role was played by another phenomenon, also due to gas diffusion, the so-called Stephan flow.

This is a hydrodynamical flow of a vapour-gas mixture, normal to the surface of the evaporating liquid, which compensates for the gas diffusion towards the surface [640]. Derjaguin and Dukhin [643] were the first to appreciate this. The rate of flow is

$$U = -\frac{D_g \operatorname{grad} c'}{c'} \tag{16.28}$$

where c' is the gas concentration, D_g the coefficient of diffusion of the gas (or vapour) in the gas-vapour mixture. For a spherical drop of radius R

$$U = \frac{D_g(c_s - c_\infty)\, R\, M'}{c' \varrho^2 M} \tag{16.29}$$

where M' is the molecular weight of the gas, M that of the vapour, ϱ is the distance from the centre of the drop, c_s the vapour concentration at the surface of the drop and c_∞ that at $\varrho = \infty$.

The motion of the particles due to Stephan flow is sometimes of essential importance. In the precipitation of aerosols by water sprays, in a Venturi scrubber for instance, the Stephan flow hinders the capture of particles by water drops if the gas is unsaturated and facilitates it when the gas is supersaturated. Heating the water sprayed into a scrubber increases the evaporation rate of the drops and lowers the efficiency of the scrubber appreciably [642]. Facy [639] pointed out that an intensive annihilation of the condensation nuclei takes place by the condensation of vapour on the drops of natural clouds and fogs. On the other hand, the view expressed by Aitken [638] and others [643], that the Stephan flow may diminish deposition of aerosols in the lungs, appears to be wrong since inhaled air is saturated with water vapour and heated in the windpipe [644].

A theory of the diffusion motion or diffusiophoresis was advanced by Derjaguin and Bakanov [645] in 1957 and by Waldmann [622] in 1959 for $r \ll l$. In a binary gas mixture the velocity of an aerosol particle according to Waldmann is equal to

$$V_D = \frac{\left(m_1^{\frac{1}{2}}\,\delta_1 - m_2^{\frac{1}{2}}\,\delta_2\right) D_g \operatorname{grad} n_1}{n_1 m_1^{\frac{1}{2}}\,\delta_1 + n_2 m_2^{\frac{1}{2}}\delta_2} \tag{16.30}$$

where m_1, m_2 are the masses and n_1, n_2 the concentrations of molecules of the two components; $\delta = 1 + \pi\alpha/8$.

The velocity of the particle is referred to coordinates such that the total number of all gas molecules crossing an area normal to the concentration gradient is zero. If the gas is contained in a closed vessel and the concentration gradient is the same everywhere, the system is stationary†. The direction of V_D coincides with that of diffusion of the heavier gas (provided the differences in values of δ are small). The cause of diffusiophoresis is that a greater proportion of the heavier molecules, which have greater momentum, bombard the particles on one side than on the other.

In the formula deduced by Derjaguin and Bakanov V_D is expressed in terms not of D_g but of the mass and the radii of the gas molecules. This makes it appear more complex.

The difference between diffusiophoresis and the motion of particles in Stephan flow can be seen from (16.30); when $m^{1/2}\delta_1 = m_2^{1/2}\delta_2$ the diffusiophoresis ceases and only Stephan flow remains.

The rate of deposition Φ of an aerosol with concentration n_0 on a spherical drop on which vapour is condensing, so that the diffusion of the particles themselves and the Stephan flow act together, was calculated by Dukhin and Derjaguin [643]

$$\Phi = \frac{4\pi D_g(c_\infty - c_s) n_0 R}{c'\{\exp\left[D_g(c_\infty - c_s)/Dc'\right] - 1\}} \tag{16.31}$$

where D is the coefficient of diffusion of aerosol particles. It is seen that the effect of Stephan flow upon the aerosol deposition on the drop is negligible only at small values of the ratio $D_g(c_\infty - c_s)/Dc'$. From (16.31) it is possible to estimate the effect of Stephan flow upon the absorption of condensation nuclei by clouds.

† Bakanov (Thesis, 1957) derived a similar equation for the case of a system moving with the centre of gravity of the gas mixture.

A direct measurement of repulsive forces acting on aerosol particles in the vicinity of an evaporating drop was made by Prokhorov [646]. He measured the force acting upon a silver coated glass sphere of $r \approx 1$ mm located at various distances from a water drop of $r \approx 2$ mm. The chief difficulty in these experiments was convection resulting from the decrease in temperature of the evaporating drop. In order to eliminate convection the drop was heated to air temperature by means of a thin metal filament. In spite of these precautions, the forces measured proved to be 2–3 times as large as those calculated from Stokes' formula and (16.29).

Up to this point we have considered diffusiophoresis of particles which do not interact with the gas. Dukhin and Derjaguin [647, 648] investigated the behaviour of a drop of a volatile liquid while a concentration gradient of its vapour was maintained around it. The drop travels in the direction of the gradient with a velocity (at $c \ll c'$) $V_D = D_g \operatorname{grad} c/c'$, approximately. The motion was called by the authors "diffusion polarization" of the drop and is due to the evaporation rate, and hence the velocity of the Stephan flow and its resulting recoil force, being greater on the side of the drop towards the decrease in the vapour concentration than on the opposite side. The particles travel in the direction of the increase in the vapour concentration. If the concentration gradient is established by evaporation from the same liquid, the diffusion polarization is compensated by the Stephan flow so that the drop will remain almost stationary. This is not the case if the temperature of the drop differs from that of the evaporating surface. If the drop assumes the psychrometric temperature T_p, as usually occurs in practice, it will be attracted by the surface when the temperature of the latter is less than T_p, and repelled in the alternative case. Water droplets must therefore be repelled by freezing water drops in natural clouds and attracted by thawing icicles.

Even when the temperature of evaporating drops is the same, a second order interaction force exists between them of a sign dependent on the natures of gas and liquid. For water drops in the air the force is repulsive.

Fedoseev and Poliansky [649] measured the repulsion between two evaporating water drops of $r \approx 1$ mm at 20° at 60 per cent relative humidity, finding that when the drops were 0·1 mm apart the force was $1{\cdot}6 \times 10^{-4}$ dynes. No precautions having been taken to eliminate convection, it is hardly possible to draw conclusions from these data.

The study in reference [650] is concerned with the thermophoretic effect caused by the cooling of evaporating drops in respect of evaporating and non-evaporating aerosol particles. The calculations show, however, that the natures of the drops, particles and gas, as well as particle size, determine whether diffusion (repulsive) or thermal (attractive) forces predominate; in all the experiments described as yet in the literature a repulsion of the particles was always observed, so that diffusion effects must have been in the ascendancy.

CHAPTER III

NON-UNIFORM RECTILINEAR MOTION OF AEROSOL PARTICLES

§ 17. NON-UNIFORM MOTION OF PARTICLES AT LOW Re IN A STATIONARY MEDIUM

The motion of an aerosol particle with a constant velocity, which was examined in the preceding chapter, must be regarded as an ideal; in reality the velocity is always changing, both in magnitude and direction. It suffices to think of convection and turbulent currents in a gas, which can only be eliminated under artificial laboratory conditions in very small volumes.

The accelerated motion of aerosol particles is, of course, more complex than motion at constant velocity and the relevant differential equations can be solved in comparatively few cases; for others one must resort to numerical approximation. Only a few of the more important examples of unsteady motion will be discussed, assuming for simplicity that the particles are spherical, and attention will be restricted, for the most part, to low Reynolds' numbers so that the resistance of the medium is proportional to the velocity of the particle.

Non-uniform rectilinear motion of particles will be taken first. The general differential equation for a spherical particle travelling in a straight line through a resisting medium at low Re [115, 116] is

$$m\frac{dV}{dt} = F(t) - \frac{2}{3}\pi\gamma_g r^3\frac{dV}{dt} - 6\pi\eta r V - 6\,r^2\sqrt{\pi\eta\gamma_g}\int_0^t \frac{dV}{dx}\frac{dx}{\sqrt{t-x}}. \tag{17.1}$$

The first term on the right hand side of the equation is the external force† acting on the particle which, in general, varies with time. The third term is the resistance of the medium for a constant velocity equal to the instantaneous value of the velocity of the particle. The second and last terms express that part of the resistance which is due to the energy expended in setting the medium itself in motion. The second term is the resistance of an ideal fluid to the accelerated motion of a sphere. It is equivalent to an increase in the mass of the sphere by half the mass of the medium displaced††; in view of the low density of a gas compared with that of a particle, this term may be neglected.

The integral term in the equation has also been neglected in most published work on aerosols, no arguments to justify the simplification having been put forward.

† As before, forces on a particle due to a stationary or moving medium will not be regarded as "external".

†† The resistance of an ideal fluid to a sphere moving with a constant velocity is zero.

Before an analysis of this question is embarked upon, the equation of motion of a particle settling under the action of gravity from rest at $t = 0$ will be solved. The integral term in (17.1) and also the second term, in accordance with what has been said above, will be neglected, giving the equation

$$\frac{dV}{dt} + \frac{V}{\tau} - g = 0 \tag{17.2}$$

where

$$\tau = m/6\pi\eta r = \frac{2\,r^2\gamma}{9\eta}. \tag{17.3}$$

The solution of equation (17.2) reducing to zero at $t = 0$ is

$$V = V_s(1 - e^{-t/\tau}) \tag{17.4}$$

where $V_s = \tau g$ is the terminal (steady) velocity of settling of the particle. The distance travelled by the particle is

$$x = V_s t - V_s t\ (1 - e^{-t/\tau}) \tag{17.5}$$

and the acceleration is

$$\frac{dV}{dt} = ge^{-t/\tau} = \frac{V_s}{\tau}\,e^{-t/\tau}. \tag{17.6}$$

If the particle moves under the action of any other constant force F, g in these formulae must be replaced by F/m.

The mathematically equivalent case of motion of a particle possessing an initial velocity V_0 in the absence of external forces can easily be seen to lead to the formulae

$$V = V_0 e^{-t/\tau} \tag{17.7}$$

$$x = V_0\tau(1 - e^{-t/\tau}). \tag{17.8}$$

In this case the expression (17.6) is again obtained for the acceleration, but with the opposite sign.

The equations of non-uniform motion of particles are simplified if τ is taken as the unit of time. According to equations (17.4) and (17.7), at the instant $t = \tau$ the velocity of a particle is $1/e$ of the initial velocity in the case of (17.7) or differs by $1/e$ from the terminal velocity in the case of (17.4). Thus τ is a kind of relaxation time for the moving particle. As shown below, the character of unsteady motion is always governed by the value of τ, which is therefore a most important quantity in the mechanics of aerosols.

Time will be denoted by t' and particle velocity by V' in the new units, so that $t' = t/\tau$, $V' = V\tau$ and $dV'/dt' = \tau^2 dV/dt$. t' is thus dimensionless while V' has the dimension length. Equations (17.4) and (17.6) now become

$$V' = V_s'(1 - e^{-t'}), \tag{17.9}$$

$$dV'/dt' = V'_s e^{-t'}, \tag{17.10}$$

and (17.1), with the second term of the right hand side discarded, transforms to

$$\frac{1}{\tau^2}\frac{dV'}{dt'} + \frac{1}{\tau^2}V' - \frac{F'}{m\tau^2} + \frac{6r^2\sqrt{\pi\eta\gamma_g}}{m}\int_0^{t'} \tau^{-3/2}\frac{dV'}{dx'}\frac{dx'}{\sqrt{t'-x'}} = 0, \quad (17.11)$$

where $x' = x/\tau$ and $F' = \tau^2 F$ – the value of the external force in the new units.

Replacing m by $\frac{4}{3}\pi r^3\gamma$ and τ by expression (17.3) leads to

$$\frac{dV'}{dt'} + V' - \frac{F'}{m} + \frac{3\sqrt{\gamma_g}}{\sqrt{2\pi\gamma}}\int_0^{t'}\frac{dV'}{dx'}\frac{dx'}{\sqrt{t'-x'}} = 0. \quad (17.12)$$

The solution of equation (17.12) for the case of a force which operates from $t = 0$ upon a particle initially at rest has been given by Boggio [117, 118] as

$$V'(t') = \frac{1}{\lambda_1 - \lambda_2}\int_0^{t'} \Phi(x)\,[e^{\lambda_1(t'-x)} - e^{\lambda_2(t'-x)}]\,dx + C_1' e^{\lambda_1 t'} + C_2' e^{\lambda_2 t'}, \quad (17.13)$$

where $C_1' = -C_2' = g'(0)\,(\lambda_1 - \lambda_2)$; $g' = F'/m$; λ_1 and λ_2 are the roots of the characteristic equation $\lambda^2 + (2 - \alpha^2)\,\lambda + 1 = 0$;

$$\Phi(x) = g'(x) + \frac{dg'(x)}{dx} - \frac{\alpha}{\sqrt{\pi}}\frac{d}{dx}\int_0^{x}\frac{g'(z)\,dz}{\sqrt{x-z}}; \quad \alpha = \sqrt{\frac{9\gamma_g}{2\gamma}}.$$

Equation (17.13) is the general solution for the acceleration of a particle from rest at $t = 0$. When the force acting on the particle is constant, $g' =$ const.,

$\Phi(x) = g'\left(1 - \frac{\alpha}{\sqrt{\pi x}}\right)$ and (17.13) becomes

$$V'(t') = g' + \frac{g'}{\lambda_1 - \lambda_2} \times$$

$$\times\left\{e^{\lambda_1 t'}\left(\frac{1}{\lambda_1} + 1\right) - e^{\lambda_2 t'}\left(\frac{1}{\lambda_2} + 1\right) - \frac{\alpha}{\sqrt{\pi}}\int_0^{t'}\frac{e^{\lambda_1(t'-x)} - e^{\lambda_2(t'-x)}}{\sqrt{x}}\,dx\right\}. \quad (17.14)$$

Since α is small the approximate expressions $\lambda_1 = -1 + i\alpha$ and $\lambda_2 = -1 - i\alpha$ can be employed with

$$\frac{1}{\lambda_1} + 1 = \frac{\alpha^2}{2} - i\alpha \quad \text{and} \quad \frac{1}{\lambda_2} + 1 = \frac{\alpha^2}{2} + i\alpha.$$

Introducing these expressions into (17.14) and replacing g' by V_0', the terminal velocity of the particle in the new units based on τ, yields finally

$$V'(t') = V_0'\left\{1 - e^{-t'}\left[\cos(\alpha t') - \frac{\alpha}{2}\sin(\alpha t') + \frac{1}{\sqrt{\pi}}\sin(\alpha t')\int_0^{t'}\frac{e^x\cos(\alpha x)\,dx}{\sqrt{x}}\right.\right.$$

$$\left.\left. - \frac{1}{\sqrt{\pi}}\cos(\alpha t')\int_0^{t'}\frac{e^x\sin(\alpha x)}{\sqrt{x}}\,dx\right]\right\}. \quad (17.15)$$

Particles with a density of unity in air at 20° and 760 mm Hg have $\alpha = 0{\cdot}0736$. For this value the function $V'(t')$ has been calculated by means of graphical integration and is shown in Fig. 18 by the continuous line. Without the integral term equation (17.12) reduces to

$$V'(t') = V_0'(1 - e^{-t'}) \tag{17.16}$$

which is shown on the diagram by the broken line.

The integral term thus leads to a small reduction in the velocity, V', which depends neither on the size of the particles nor on the viscosity of the medium but exclusively on the value of α, which is a function of the ratio of the densities of particle and fluid. The effect is therefore large in liquids but negligible in gases. For the distance travelled by a particle in a definite time, the maximum value of the correction due to the integral term does not exceed 4 per cent at a particle density equal to unity, 2 per cent at a particle density of 4, and so on.

Thus, in the case examined, neglecting the integral term does not lead to appreciable error, regardless of particle size, and the resistance can in practice be considered non-inertial and the same as the resistance at a constant velocity equal to the value of the velocity at the given instant. A similar conclusion is reached for the oscillatory motion of a particle in a gas.

For decelerated motion of a particle having an initial velocity V_0 in the absence of external forces the equation of motion is

$$\frac{dV'}{dt'} + V' + \frac{3\sqrt{\gamma_g}}{\sqrt{2\pi\gamma}} \int_0^{t'} \frac{dV'}{dx'} \frac{dx'}{\sqrt{t' - x'}} = 0 \tag{17.17}$$

in place of (17.12).

The initial condition is $V'(0) = V_0'$. If the function $V^*(t) = V_0' - V'(t)$ is introduced an equation identical to (17.12) with F'/m replaced by V_0' and an initial condition $V^*(0) = 0$ is obtained. Thus the solution (17.15) is valid for $V^*(t)$ and, by turning Fig. 18 through 180°, we obtain the accurate and the approximate curves $V'(t)$ for the second problem. The correction arising from the integral term, however, now has a larger value than for accelerated motion, as will be seen below.

Putting $t = \infty$ in (17.8) we find that the distance l_i traversed by the particle in an unlimited time is equal to

$$l_i = V_0\tau = \frac{2V_0 r^2\gamma}{9\eta}. \tag{17.18}$$

The length l_i is called the stop-distance of a particle with initial velocity V_0, and will be seen to play a large role in the curvilinear motion of particles. From Fig. 18 it is clear that the stop-distance l_i, represented by the area under the continuous curve V_t which includes the integral term, may be appreciably greater than the value of $V_0\tau$ corresponding to the area under the approximate broken curve. The velocity of a decelerating particle, owing to the form of the integral term, continues for quite a long time to differ from zero by a small but finite amount (as seen from Fig. 18). This is a result of the flow produced by the moving particle in the surrounding fluid (assumed infinite in all equations of this section) which dies out much more slowly than the motion relative to the fluid of the particle itself. However, since in the first place the equations of flow of a viscous fluid round a spherical particle, derived by

Stokes, are valid only in the vicinity of the particle and, in the second place, in all practical cases the motion of a particle takes place in a space bounded by walls or other particles, the last stage of damping of the particle motion probably proceeds more rapidly than is suggested by the curve in Fig. 18. For this reason the simple expression (17.18) for l_i will be used henceforth.

After a time much longer than τ has elapsed the second term on the right hand side of equation (17.5) may be neglected which is equivalent to supposing that the particle moved from the very beginning with a velocity V_s. This idea can be applied in a general way, to motions having a duration which is large compared with τ, by considering either that the particle is stationary with respect to the medium, if no external forces act upon it, or that it moves with a velocity $V(t) = BF(t)$ where $F(t)$ is the instantaneous value of the external force. This kind of particle motion is called *quasi-stationary*.

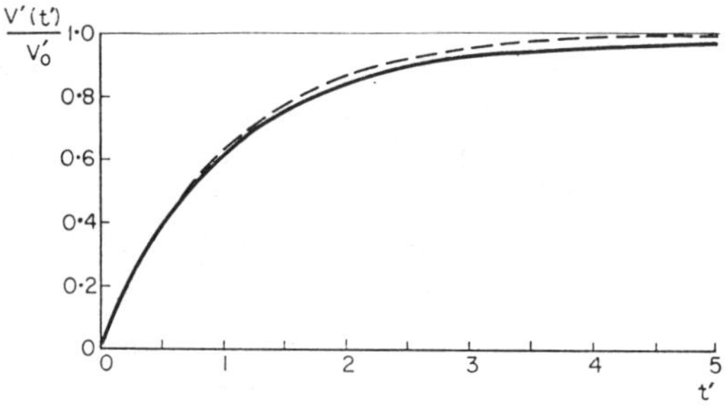

FIG. 18. Motion of particles under a constant force.

An experimental investigation of non-uniform rectilinear motion has been conducted by Berkowitch [119]. A charged particle of selenium was suspended in a condenser M_1 shunted by a high resistance R_E, after which another condenser M_2 of capacity C_E and initial potential difference Π_0 was discharged through this resistance. Variation of the voltage across condenser M_1 with time is of quite a complex nature owing to the electrial oscillations which occur. However, the period of these oscillations is considerably less than the relaxation time of the particles investigated ($r \approx 10^{-4}$ cm, $\gamma = 4{\cdot}4$, $\tau \approx 2 \times 10^{-4}$ sec) and if they are neglected the discharge of condenser M_2 is expressed by the differential equation

$$-C_E \frac{d\Pi}{dt} = \frac{\Pi}{R_E} \tag{17.19}$$

and hence if follows that

$$\Pi = \Pi_0 e^{-t/C_E R_E}. \tag{17.20}$$

In these experiments $C_E = 5{\cdot}74 \times 10^{-9}$ farad, $R_E = 1{\cdot}5 \times 10^6\,\Omega$, $\Pi_0 = 2500$ V. The relaxation time of the condenser $C_E R_E = 8{\cdot}6 \times 10^{-3}$ sec is thus considerably greater than τ, and the motion of the particle can be considered quasi-stationary. The velocity of the particle is given by the equation

$$\frac{dx}{dt} = \frac{\Pi_0 q B}{h} e^{-t/C_E R_E} \tag{17.21}$$

where h is the distance between the plates of the condenser M_1, q is the charge on a particle, and the expression obtained for the path traversed by the particle is

$$x_\infty = \frac{\Pi_0 q B}{h} \int_0^\infty e^{-t/C_E R_E} dt = \frac{\Pi_0 q B C_E R_E}{h}. \tag{17.22}$$

A more accurate derivation with account taken of electrical oscillations in the condenser circuit and inertia of the particles (but without the integral term) leads to the same expression. The maximum velocity of the particle in these experiments did not exceed 20 cm/sec, which corresponds to Re $\leqslant$ 0·03 within the region of applicability of equation (17.1).

The values of x_∞ measured by Berkowitch agreed excellently with formula (17.22); they were strictly proportional to each of the quantities Π_0/h, C_E, R_E and B deviations not exceeding $\pm$1–1·5 per cent. The mobility B of a particle was determined by the rate of fall and by the strength of the balancing field (see § 15B). The absolute value of x_∞ differed by not more than ± 3 per cent from the theoretical value and the deviations were completely random in character, probably being caused by the Brownian motion of the particle. It is important to note that the selenium particles used in these experiments were spherical so that orientation in the field could not take place. Correction for the integral term, not considered by Berkowitch, must have been negligible owing to the high density of selenium; in addition, in the first stage of discharging the condenser the particle accelerated and in the second it slowed down so that the effect of the integral term in the two stages was cancelled to some extent.

For non-uniform motion of small spherical particles in a non-uniformly moving medium, eq. (17.1) becomes

$$\frac{4}{3}\pi r^3(\gamma\dot{V} - \gamma_g\dot{U}) + \frac{2}{3}\pi r^3(\dot{V} - \dot{U})$$

$$+ 6\pi\eta r\left[V - U + \frac{r}{\sqrt{\pi\nu}}\int_0^t \frac{\dot{V}(\theta) - \dot{U}(\theta)}{\sqrt{t-\theta}}\,d\theta\right] - \frac{4}{3}\pi r^3 g(\gamma - \gamma_g) = 0. \tag{17.23}$$

Gravity is the extraneous force in this case. $\dot{U}$ and $\dot{V}$ are derivatives with respect to the time. Assuming

$$\alpha = 9\nu\gamma_g/r^2(2\gamma + \gamma_g) \approx 1/\tau$$

$$\beta = 3\gamma_g/(2\gamma + \gamma_g) \approx \frac{3}{2}\frac{\gamma_g}{\gamma}$$

eq. (17.23) reduces to

$$\dot{V} - \beta\dot{U} + \alpha(V - U - V_s) + \sqrt{\frac{3\alpha\beta}{\pi}} \int_0^t \frac{\dot{V}(\theta) - \dot{U}(\theta)}{\sqrt{t-\theta}}\,d\theta = 0. \tag{17.24}$$

Tchen [651] gave the following solution to this integro-differential equation:

$$V(t) = \frac{1}{\vartheta}\int_0^\infty e^{-ky}\sin(\vartheta y)\,F(t-y)\,dy \tag{17.25}$$

where $$k = \alpha\left(1 - \frac{3}{2}\beta\right) \approx 1/\tau;$$

$$\mathscr{E}^2 = \alpha^2 - k^2 = 3\alpha^2\beta\left(1 - \frac{3}{4}\beta\right) \approx 9\gamma_g/2\tau^2\gamma;$$

$$F(t) = \alpha^2[V_s + U(t)] + \alpha(1 - 2\beta)\,\dot{U} + \beta\ddot{U} - \sqrt{\frac{3\alpha\beta}{\pi}}(\beta - 1)\int_0^\infty \frac{\ddot{U}(t-x)}{\sqrt{x}}\,dx$$

$$\approx \frac{V_s + U}{\tau^2} + \frac{\dot{U}}{\tau} + \frac{3\gamma_g}{\gamma}\ddot{U} + \sqrt{\frac{9\gamma_g}{2\gamma\tau\pi}}\int_0^\infty \frac{\ddot{U}(t-x)}{\sqrt{x}}\,dx. \qquad (17.26)$$

Pearcey and Hill [652] computed the effect of the integral term in (17.23) upon the motion of a particle in a medium at rest, after removal of the extraneous force. To a first approximation the additional path traversed by the particle due to this effect is proportional to the square root of the time and therefore tends to infinity at $t \to \infty$. However, to a second approximation, this path and therefore the stop distance l_i prove to be finite. An accurate evaluation of l_i presents difficulties. Unfortunately, the rough estimate made by the authors holds only for the values of γ_g/γ corresponding to the motion in air of particles of density $< 0{\cdot}15$.

§ 18. NON-UNIFORM MOTION OF PARTICLES AT LARGE Re

Accelerated rectilinear motion at high Reynolds' numbers such as are encountered in the mechanics of aerosols (up to about Re = 1000) has not been studied theoretically, and experimental data on this problem are insufficient and contradictory. In Schmidt's work [120] the rate of rise in air of hydrogen-filled balloons of radius 8 and 26 cm (Re = 5000–35,000) was measured by a photographic method (as in all the other investigations mentioned in this section). Analysis of the measurements shows that the air resistance at any instant appreciably exceeds the resistance at a corresponding constant velocity. In addition, the velocity of the balloon does not increase steadily but falls after a time and then again increases. Similar behaviour was observed for the rise of wax spheres in water; their radii were 1·0–1·5 cm (Re = 500 to 1500), and study of the flow near the spheres, by the coloured streamer method, led the author to conclude that the fluctuations of velocity were due to the creation, growth and breakaway of vortices from the spheres.

Lunnon [121] measured the rate of fall of solid spheres of various materials with r = 1–10 mm in a deep shaft and also found that the resistance of a medium for accelerated motion is considerably greater than at constant velocity. At Re = 10,000 and an acceleration of 5 m sec^{-2} ψ is 15 per cent greater and, at an acceleration of 8 m sec^{-2}, 35 per cent greater than at constant velocity; at Re = 35,000 the figures are 25 and 100 per cent respectively.

In the work of Khudyakov and Chukhanov [122, 123] the motion of sand particles with radii 35 μ, 100 μ and 420 μ was studied in a vertical tube through which air was passed from top to bottom at a velocity of 10–25 m sec^{-1}. The sand particles were introduced into the tube with no appreciable initial velocity so that their motion in the first stages was slower than that of the air. The resistance of the medium at Re = 20–600 was found to average about half that at constant velocity. However, in these experiments, because of strong turbulence and "twist" of the flow, the particles did not fall vertically but were squeezed out to the walls of the tube and the values found for ψ are not completely reliable.

On the other hand, the experimental data of Laws [124], who measured the rate of fall in air of water droplets with r = 0·6–3 mm at various distances from the point of release, lie well on the theoretical curves [125] based on values of ψ for constant rate of fall (Re $\leq$ 1500). The calculations are inexact because the deformation of the droplet, which increases ψ, depends on the velocity, whereas the values of ψ employed are those for an equal sphere at constant velocities. Laws' data for droplets with r = 0·6 mm (Re $\leqq$ 200), for which the flattening effect is negligible, have been analysed by the present author who found values for ψ 5–10 per cent below those given in Table 5. Thus for Reynolds' numbers not exceeding a few hundred it is possible, without making a large error, to assume that the resistance of air does not depend on the acceleration; this is often done in engineering calculations.

On this assumption the motion of an aerosol particle in the absence of external forces can be calculated in the following way. Differentiation of formula (7.5) $V = \mathrm{Re}\, \nu/2r$ gives

$$\frac{dV}{dt} = \frac{\eta}{2r\gamma_g} \cdot \frac{d\,\mathrm{Re}}{dt}. \tag{18.1}$$

Substitution of the expressions for F_M (10.2), V and dV/dt in the equation of the motion of a particle in the absence of external forces

$$m\frac{dV}{dt} = -F_M \tag{18.2}$$

leads [126], after tidying, to

$$\frac{d\,\mathrm{Re}}{\psi\,\mathrm{Re}^2} = \frac{\pi\eta\,dt}{4m}, \tag{18.3}$$

whence

$$\int_{\mathrm{Re}_1}^{\mathrm{Re}_2} \frac{d\,\mathrm{Re}}{\psi\,\mathrm{Re}^2} = -\frac{\pi\eta r(t_2 - t_1)}{4m} = -\frac{3\eta(t_2 - t_1)}{16r^2\gamma}. \tag{18.4}$$

The integral in (18.4) can be calculated graphically from the values of the function $\psi\,\mathrm{Re}^2$ given above (page 32). The change in Re, and consequently the change in velocity of the particle, can thus be found as a function of time. A similar derivation leads to the equation

$$\int_{\mathrm{Re}_1}^{\mathrm{Re}_2} \frac{d\,\mathrm{Re}}{\psi\,\mathrm{Re}} = -\frac{\pi r^2\gamma_g(x_2 - x_1)}{2m} = -\frac{3\gamma_g(x_2 - x_1)}{8r\gamma} \tag{18.5}$$

giving the relationship between the distance covered by the particle and the velocity.

As an example of the use of these equations the stop-distance, l_i, for particle sizes and velocities at which (17.18) is not applicable will be found. It is necessary first to substitute $\mathrm{Re}_1 = 2rV_0/\nu$, where V_0 is the initial velocity of the particle, and $\mathrm{Re}_2 = 0$

in equation (18.5); $x_2 - x_1$ is equal to l_i. The value of $\int \frac{d\,\mathrm{Re}}{\psi\,\mathrm{Re}}$ was determined graphically and the graph shown in Fig. 19 was obtained, giving the quantity $\frac{3\gamma_g l_i}{8 r \gamma}$ as a function of $\frac{2 r V_0}{\nu} = \mathrm{Re}_0$. The dashed straight line on the graph represents formula (17.18), based on Stokes' law of resistance.

The relaxation time τ for particles of radius 1 mm and density 1 having an initial velocity of 30 cm sec^{-1} will next be calculated with the aid of formula (18.4). This quantity will be needed in § 45. With $\mathrm{Re}_1 = 2 \times 0{\cdot}1 \times 30/0{\cdot}17$ and $\mathrm{Re}_2 = \mathrm{Re}_1/e$ graphical integration gives $\tau = 6{\cdot}3$ sec (instead of 12 sec according to Stokes). For $r = 0{\cdot}1$ mm, $\tau = 0{\cdot}10$ sec instead of 0·12 sec according to Stokes.

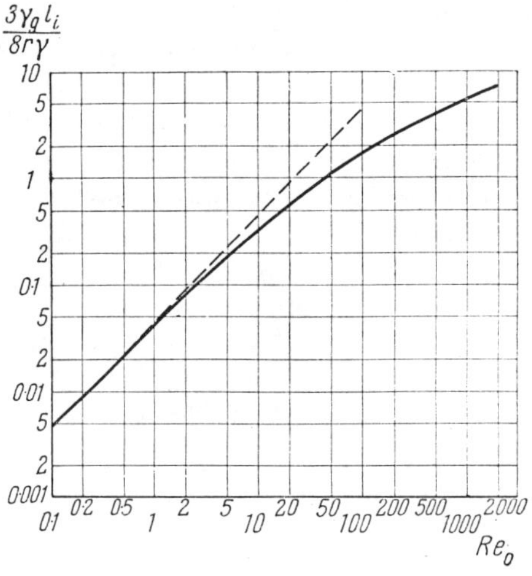

FIG. 19. Stop-distance of particles beyond the Stokes region.

Problems of non-uniform motion of particles under the action of a constant force, such as gravity, are solved in a similar fashion. In this case the equation of motion is

$$m \frac{dV}{dt} = -F_M + mg \tag{18.6}$$

and, instead of (18.4),

$$\int_{\mathrm{Re}_1}^{\mathrm{Re}_2} \frac{d\,\mathrm{Re}}{a - \psi\,\mathrm{Re}^2} = -\frac{3\eta(t_2 - t_1)}{16\,r^2\gamma}, \tag{18.7}$$

where $a = 32\,g\gamma\gamma_g r^3/3\,\eta^2$.

Equation (18.5) is replaced by

$$\int_{\mathrm{Re}_1}^{\mathrm{Re}_2} \frac{\mathrm{Re}\,d\,\mathrm{Re}}{a - \psi\,\mathrm{Re}^2} = -\frac{3\gamma_g(x_2 - x_1)}{8 r \gamma} \tag{18.8}$$

and graphical integration can again be used to find the relationship between the velocity of the particle and time or distance traversed.

Serafini [653] found an analytical expression for the motion of a particle in a gas at large Re, in the absence of external forces, a problem which had previously been tackled by graphical methods. His expression is derived from a more general form of (10.5), $\psi = \frac{24}{\mathrm{Re}}(1 + \varepsilon\,\mathrm{Re}^{2/3})$, where ε is a constant. This leads to the following formula for the velocity of a particle at time t:

$$V = \frac{V_0}{\mathrm{Re}_0\,\varepsilon^{3/2}}\left[(\mathrm{Re}_0^{-2/3}\,\varepsilon^{-1} + 1)\,e^{\theta} - 1\right]^{-3/2} \tag{18.9}$$

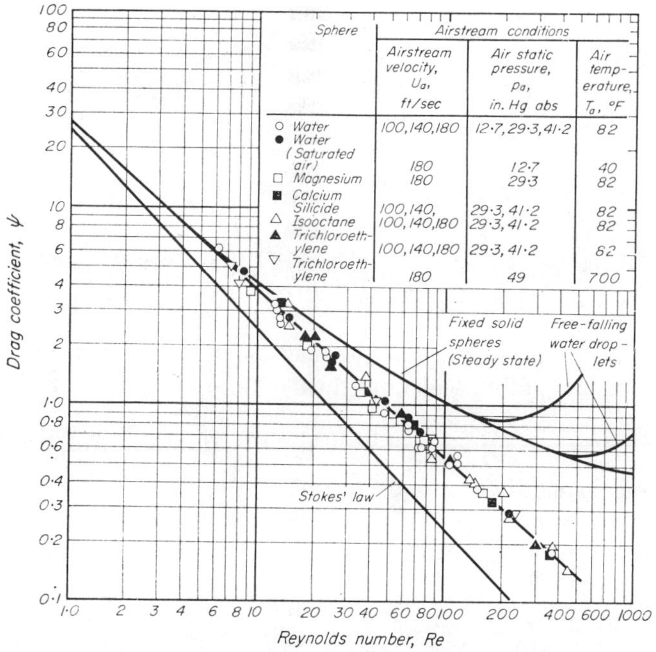

FIG. 19a. Resistance of a gas to particles in non-uniform motion.

where $\theta = 3\,\eta t/\gamma r^2$; V_0 and Re_0 are the initial values of quantities concerned. The path traversed is equal to

$$x = \frac{r\,\varepsilon^{-3/2}\gamma}{3\gamma_g}\{\mathrm{Re}_0^{1/3}\varepsilon^{1/2} + \arctan(\mathrm{Re}_0^{-1/3}\varepsilon^{-1/2})$$

$$- [(\mathrm{Re}_0^{-2/3}\varepsilon^{-1} + 1)e^{\theta} - 1]^{-1/2} - \arctan[(\mathrm{Re}_0^{-2/3}\varepsilon^{-1} + 1)\,e^{\theta} + 1]^{1/2}\} \tag{18.10}$$

whence it follows that the maximal value of $x(\theta \to \infty)$, or the stop-distance, is

$$l_i = \frac{r\varepsilon^{-3/2}\gamma}{3\,\gamma_g}\left[\mathrm{Re}_0^{1/3}\varepsilon^{1/2} + \arctan(\mathrm{Re}_0^{-1/3}\varepsilon^{-1/2}) - \frac{\pi}{2}\right]. \tag{18.11}$$

Ingebo [654] experimented with aerosols of spherical particles having $r = 10$–$60\,\mu$ in a wind tunnel at flow rates from 30 to 54 m/sec. The size and the velocity of particles at different distances from the entrance were found photographically and the acceleration and drag calculated as functions of the relative velocity of particles and air. As shown in Fig. 19a, the drag coefficient, ψ, is the same as for uniform motion only at small Re; it differs considerably at larger Re. The experiments of Schmidt and Lunnon (see page 76) show the resistance to be greater for accelerated motion of particles than for uniform. A lower resistance would be expected for decelerated motion, as in Ingebo's experiments. The experimental results of Ingebo are expressed by an empirical formula

$$\psi = 27\,\mathrm{Re}^{-0.84} \quad (6 < \mathrm{Re} < 400). \tag{18.12}$$

§ 19. OSCILLATION OF AEROSOL PARTICLES DUE TO A PERIODIC EXTERNAL FORCE

The following equation, derived by Stokes [127], describes the oscillation of a spherical particle in a resisting medium caused by a periodically varying external force F; it is only valid at low Re:

$$m\frac{dV}{dt} = F - \frac{9}{4}m'\omega\beta(1+\beta)V - \left(\frac{1}{2} + \frac{9}{4}\beta\right)m'\frac{dV}{dt}, \tag{19.1}$$

m is the mass of the particle; m' is the mass of the medium displaced by it; ω is the angular frequency of the oscillation (equal to the number of oscillations per second multiplied by 2π); $\beta = \frac{1}{r}\sqrt{\frac{2\nu}{\omega}}$. The magnitude of β is of the order $1/r\sqrt{\omega}$ and in air, under normal conditions, $\beta = 0{\cdot}55/r\sqrt{\omega}$.

Since $m \gg m'$ the term $\frac{1}{2}m'(dV/dt)$ will be neglected as before, reducing equation (19.1) to

$$m\frac{dV}{dt} + \frac{9}{4}m'\beta\frac{dV}{dt} + \frac{9}{4}m'\omega\beta V + \frac{9}{4}m'\omega\beta^2 V - F = 0. \tag{19.2}$$

Although β, in the second term of the equation, tends to infinity as $\omega \to 0$, the oscillatory motion, to which (19.2) applies, has $V = V_0 \sin \omega t$ so that $dV/dt = \omega V_0 \cos \omega t$; hence all terms of (19.2) except the last two tend to zero when $\omega \to 0$. The penultimate term then becomes equal to $6\pi\eta r V$, which is the resistance of the medium at constant velocity V. Thus when $\omega \to 0$ equation (19.2) becomes the normal Stokes formula.

If the "reduced" mass of the particle, $m + \frac{9}{4}m'\beta$, is denoted by m_r and $\frac{9}{4}m'\omega\beta(1+\beta) = \frac{9}{4}m'\omega\beta + 6\pi\eta r$ is denoted by $1/B_r$ (B_r is the reduced mobility), then (19.2) becomes

$$m_r\frac{dV}{dt} + \frac{V}{B_r} - F = 0. \tag{19.3}$$

If F varies sinusoidally with time, $F = F_0 \sin \omega t$, the solution of (19.3), reducing to 0 at $t = 0$, is

$$V = \frac{F_0 B_r \sin(\omega t - \varphi)}{\sqrt{1 + B_r^2\omega^2 m_r^2}} + \frac{F_0 B_r e^{-t/B_r m_r}\sin\varphi}{\sqrt{1 + B_r^2\omega^2 m_r^2}}, \tag{19.4}$$

and

$$\tan\varphi = B_r \omega m_r. \tag{19.5}$$

The second term in (19.4), corresponding to the transient stage of oscillation, quickly tends to zero and under steady conditions oscillation of the particle proceeds according to the formula

$$V = V_0 \sin(\omega t - \varphi) \tag{19.6}$$

with a velocity amplitude

$$V_0 = \frac{F_0 B_r}{\sqrt{1 + B_r^2 \omega^2 m_r^2}}. \tag{19.7}$$

Replacing m_r and B_r by the full expressions given above,

$$V_0 = \frac{F_0 f}{\omega m \sqrt{f^2 + 3\beta f + \frac{9}{2}\beta^2 + \frac{9}{2}\beta^3 + \frac{9}{4}\beta^4}}, \tag{19.8}$$

$$\tan\varphi = (\tfrac{2}{3}f + \beta)/\beta(1 + \beta), \tag{19.9}$$

where

$$f = \frac{2m}{3m'} = \frac{2\gamma}{3\gamma_g}. \tag{19.10}$$

These equations show that the amplitude and phase of the oscillation is determined by the ratio of the densities of particle and fluid and by the parameter β, i.e. for a given fluid, by the product $r\sqrt{\omega}$.

If the second and third terms in equation (19.2), representing the inertial part of the resistance of the medium and equivalent to the integral term in equation (17.1), are omitted, m_r becomes m, B_r becomes B, and instead of (19.8) and (19.9) two simpler formulae are obtained:

$$V_0 = \frac{F_0 B}{\sqrt{1 + \omega^2\tau^2}} = \frac{F_0 B}{\sqrt{1 + (2\pi\tau/t_p)^2}} = F_0 B \cos\varphi, \tag{19.11}$$

$$\tan\varphi = \tau\omega = 2\pi\tau/t_p, \tag{19.12}$$

where $t_p = 2\pi/\omega$ is the period of the oscillations, and the amplitude is

$$a = F_0 B/\omega\sqrt{1 + \omega^2\tau^2}. \tag{19.13}$$

In order to estimate the error introduced by this simplification, the ratio of the velocity amplitudes of particles oscillating in a gas, V_0, and in a vacuum V_0^0 will be deduced. The latter comes from the differential equation for oscillation in a vacuum

$$m\frac{dV}{dt} = F_0 \sin\omega t, \tag{19.14}$$

a solution of which is

$$V^0(t) = V_0^0 \sin\left(\omega t - \frac{\pi}{2}\right) = \frac{F_0}{\omega m}\sin\left(\omega t - \frac{\pi}{2}\right). \tag{19.15}$$

According to the exact formula (19.8)

$$V_0/V_0^0 = f\Big/\sqrt{f^2 + 3\beta f + \frac{9}{2}\beta^2 + \frac{9}{2}\beta^3 + \frac{9}{4}\beta^4}. \tag{19.16}$$

According to the approximation (19.11)

$$V_0/V_0^0 = \frac{B\omega m}{\sqrt{1 + \omega^2\tau^2}} = \frac{\omega\tau}{\sqrt{1 + \omega^2\tau^2}}. \tag{19.17}$$

The curve expressed by formula (19.17) is illustrated in Fig. 20 (broken line 2). The continuous line shows the equation (19.16), when $\gamma = 1$, in air under normal conditions. The error introduced into the amplitude of oscillation by omitting the terms depending on the inertia of the fluid from the basic equation (19.2) is small and can be neglected, almost always. It is easy to see that when the density of the particles increases the error is still reduced, as in the case analysed in § 17.

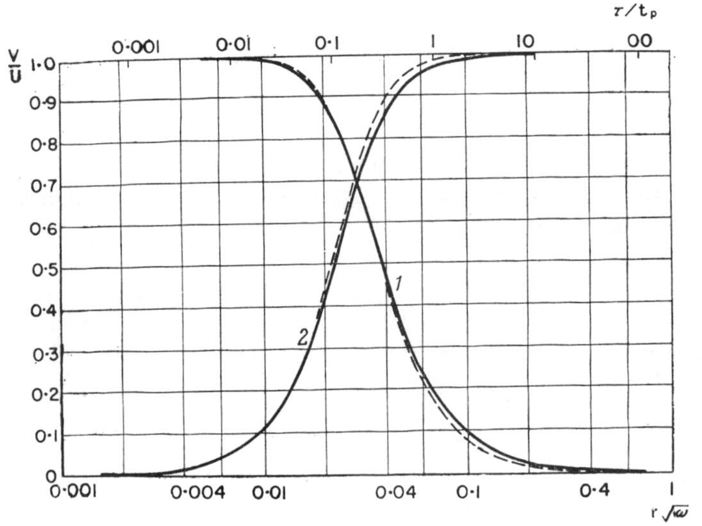

FIG. 20. Oscillations of aerosol particles.

Equations (19.11) and (19.12) show that the nature of the oscillation is determined by the ratio of the relaxation time τ of the particle to the period t_p of the oscillations. For a very large value of this ratio (coarse particles, high frequencies) $\tan\varphi \to \infty$, $\varphi \to \frac{\pi}{2}$, $V_0 \to F_0/\omega m$, and the equation reduces to the one for oscillation in a vacuum. Thus at large τ/t_p the resistance of the medium can be neglected (inertial regime of the oscillations).

At very small values of τ/t_p, $\tan\varphi \to 0$, $\varphi \to 0$, $V_0 \to F_0 B$ and equation (19.6) becomes

$$V = F_0 B \sin\omega t = FB. \tag{19.18}$$

Thus, in this case, the motion is of a quasi-stationary nature; the inertia terms disappear from the oscillation equation and the particle moves, at every instant, with the same velocity that it would have possessed under the action of a constant force equal to the instantaneous value of the variable force (viscous regime of oscillations).

As seen from Fig. 20, transition from the viscous to the inertial regime of oscillation occurs in the range of values of τ/t_p between 0·002 and 1. In the viscous region the particle and the force are in phase; as τ/t_p increases, the particle begins to lag behind until, in the inertial region, the lag reaches a quarter period.

The most important example of the oscillatory motion of aerosol particles under the action of an external force is provided when they are subjected to a periodic electric field. In this case $F = Eq$, where E is the field intensity and q is the charge on the particle. The most careful experimental investigation of such oscillations has been carried out by Yü-Cheng-Yang [128]. A charged particle was suspended in a constant vertical electric field, after which a variable vertical field of 50 c/s was superimposed. The radius of the particle was of the order 10^{-4} cm so that τ was of the order 10^{-5} sec and the oscillations were purely viscous. Re in these experiments did not exceed 0·03.

The oscillating particles under these conditions had the appearance of vertical streaks having a length equal to twice the amplitude of oscillation. The amplitude was strictly proportional to the field strength and to the charge of the particle, in agreement with equation (19.18). The absolute magnitude of the amplitude, however, was higher than calculated by (19.18), in some experiments by as much as 7 per cent.

The mobility of the particle in these experiments was determined by the rate of settling in the absence of an electric field and by the magnitude of the balancing field (§ 15B). In calculating the mobility it was therefore unnecessary to introduce corrections for the shape and density of the particles. The author ascribed the discrepancy in the value of the amplitude to imperfect sinusoidal variation of the electric field. Had this been the case, the error would have been constant instead of varying from one experiment to another. The discrepancy is more likely to be due to the employment of particles, formed by sparking between metallic electrodes, which were undoubtedly aggregates. Aggregates of extended form orient themselves with the long axis parallel to an electric field (see § 43) and therefore possess greater mobility in a field than when settling under gravity.

Measurement of the amplitude of oscillation in a periodic electric field offers a convenient method of determining the charge on a particle (see page 109). Oscillation of particles in an electric field could be used for the artificial coagulation of aerosols, but an acoustic field (see following section) is more effective. A curious phenomenon – the "flashing" of smoke leaving the funnel of a steam train passing below the overhead conductors of an electrified railway – can probably be explained by oscillation of the smoke particles in the electric field between the conductors and the funnel [129].

§ 20. OSCILLATION OF AEROSOL PARTICLES INDUCED BY ACOUSTIC WAVES

A particle situated in a vibrating medium is itself set in oscillation. Let V be the velocity of a particle, U the velocity of the medium. The force acting on a spherical particle situated in an ideal fluid moving with acceleration dU/dt is

$$F_M = m'\frac{dU}{dt} - \frac{m'}{2}\left(\frac{dV}{dt} - \frac{dU}{dt}\right), \qquad (20.1)$$

where m' is the mass of fluid displaced by the particle.

In order to understand the physical meaning of this equation suppose, for a moment, that the densities of the particle and the medium are equal. The particle will then move with the medium, provided it is small compared with the wavelength, and $V = U$. Since the only force acting on the particle is the reaction of the medium F_M, then this is equal to the product of the mass of the particle and its acceleration:

$$F_M = m' \frac{dV}{dt} = m' \frac{dU}{dt}. \tag{20.2}$$

The first term on the right hand side of (20.1) therefore represents the reaction of the medium on the particle moving with it due to the pressure gradient in the medium $m' \frac{dU}{dt} = v \gamma_g \frac{dU}{dt} = v \operatorname{grad} p$. The term $-\frac{m'}{2}\left(\frac{dV}{dt} - \frac{dU}{dt}\right)$ expresses that part of the reaction of the medium which depends on the acceleration of the particle with respect to the medium. In the case of an ideal fluid this reaction amounts to an apparent increase in mass of the particle by half the mass of the medium displaced by it. Thus, in an ideal fluid the equation of motion of a particle has the form

$$\begin{aligned} m \frac{dV}{dt} &= m' \frac{dU}{dt} - \frac{m'}{2}\left(\frac{dV}{dt} - \frac{dU}{dt}\right) \\ &= \frac{3}{2} m' \frac{dU}{dt} - \frac{m'}{2} \frac{dV}{dt}. \end{aligned} \tag{20.3}$$

In a real medium terms depending on viscosity should be added on the right hand side. Then V in (19.1) denotes the relative velocity of the particle and the medium, equal to $V - U$ in the case now under consideration. Finally, for the motion of a particle in an oscillating viscous medium the equation becomes

$$\begin{aligned} m \frac{dV}{dt} &= \frac{3}{2} m' \frac{dU}{dt} - \frac{1}{2} m' \frac{dV}{dt} \\ &\quad - \frac{9}{4} m' \omega \beta (1 + \beta)(V - U) - \frac{9}{4} m' \beta \left(\frac{dV}{dt} - \frac{dU}{dt}\right) \end{aligned} \tag{20.4}$$

or, discarding the term $-\frac{m'}{2}\frac{dV}{dt}$ as before and using the notation of the previous section,

$$m_r \frac{dV}{dt} + \frac{V}{B_r} = \frac{9}{4} m' \omega \beta (1 + \beta) U + \left(\frac{9}{4} m' \beta + \frac{3}{2} m'\right) \frac{dU}{dt}. \tag{20.5}$$

If the vibration of the medium is expressed by the equation

$$U = U_0 \sin \omega t \tag{20.6}$$

then (20.5) takes the form

$$\begin{aligned} m_r \frac{dV}{dt} + \frac{V}{B_r} &= \frac{3}{2} m' \omega U_0 \times \\ &\quad \times \sqrt{1 + 3\beta + \frac{9}{2}\beta^2 + \frac{9}{2}\beta^3 + \frac{9}{4}\beta^4} \cdot \sin(\omega t + \theta), \end{aligned} \tag{20.7}$$

where
$$\tan\theta = \left(\frac{2}{3} + \beta\right) \Big/ \beta(1 + \beta), \tag{20.8}$$

similar to the equation for a particle oscillating under the influence of a periodic force (19.3). The steady-state solution to (20.7) is

$$V = U_0 \mu \sin(\omega t + \theta - \varphi), \tag{20.9}$$

where
$$\mu = \frac{V_0}{U_0} = \sqrt{\frac{1 + 3\beta + \frac{9}{2}\beta^2 + \frac{9}{2}\beta^3 + \frac{9}{4}\beta^4}{f^2 + 3f\beta + \frac{9}{2}\beta^2 + \frac{9}{2}\beta^3 + \frac{9}{4}\beta^4}}, \tag{20.10}$$

$$\tan(\varphi - \theta) = \frac{\frac{3}{2}(f - 1)\beta(1 + \beta)}{f\left(1 + \frac{3}{2}\beta\right) + \frac{3}{2}\beta + \frac{9}{2}\beta^2 + \frac{9}{2}\beta^3 + \frac{9}{4}\beta^4}. \tag{20.11}$$

Formulae (20.10) and (20.11) for the ratio of the amplitudes of the oscillations† of the particle and the medium, $\mu = V_0/U_0$, and for the corresponding difference in phase, $\varphi - \theta$, were first derived by König [130]; his rather complicated derivation has here been replaced by a simpler one. The amplitude of the relative oscillatory motion of particle and medium, $V_r = V - U$, will next be found by discarding the first two very small terms on the right hand side of (20.4) and subtracting $m\dfrac{dU}{dt}$ from both sides.

This converts the equation to

$$m_r \frac{dV_r}{dt} + \frac{V_r}{B_r} = -m\frac{dU}{dt}. \tag{20.12}$$

Putting $U = U_0 \sin \omega t$ and solving the resulting equation gives, as before

$$V_r = -\frac{U_0 f \sin\left(\omega t + \frac{\pi}{2} - \varphi\right)}{\sqrt{f^2 + 3f\beta + \frac{9}{2}\beta^2 + \frac{9}{2}\beta^3 + \frac{9}{4}\beta^4}}. \tag{20.13}$$

Comparison with formula (19.16) shows that the ratio of the amplitude of the relative oscillation of particle and medium to that of the oscillation of the medium itself is equal to the ratio of the amplitude of the oscillation of a particle under the action of a periodic external force in a gas to that of a particle in a vacuum (Fig. 20, curve 2). In calculating the amplitude of the relative oscillation of particle and medium the simplified formula (19.17) can therefore be used instead of the exact one (19.16).

† The ratio of the amplitudes of vibration and that of the amplitudes of the velocities of particle and medium are equal in simple harmonic motion.

Thus

$$\frac{V_{r_0}}{U_0} = \frac{\omega\tau}{\sqrt{1+\omega^2\tau^2}}, \tag{20.14}$$

$$V_r = -V_{r_0}\sin\left(\omega t + \frac{\pi}{2} - \varphi\right), \tag{20.15}$$

$$\tan\varphi = \omega\tau. \tag{20.16}$$

If all terms on the right hand side of (20.4) are discarded, except

$$\frac{9}{4}m'\omega\beta^2(V-U) = 6\pi\eta r(V-U),$$

that is if the resistance of the medium is considered non-inertial, then in place of (20.7) we obtain the equation

$$m\frac{dV}{dt} + \frac{V}{B} = \frac{U}{B} = \frac{U_0\sin\omega t}{B}. \tag{20.17}$$

For the ratio of the amplitudes of the oscillations of particle and medium the following expression is obtained instead of (20.10)

$$\mu = \frac{V_0}{U_0} = \frac{1}{\sqrt{1+\omega^2\tau^2}} = \frac{1}{\sqrt{1+(2\pi\tau/t_p)^2}}, \tag{20.18}$$

and for the difference in phase, instead of (20.11)

$$\tan(\varphi-\theta) = \omega t = 2\pi\tau/t_p. \tag{20.19}$$

The broken curve 1 in Fig. 20 illustrates the relationship (20.18) and the continuous curve (20.10); the latter is for $\gamma = 1$ in air under normal conditions. The difference between the two curves is so insignificant that the simplified formulae (20.18) can be used unconditionally in practice.

From (20.14) and (20.18) follows the important formula

$$V_{r_0}^2 = U_0^2 - V_0^2. \tag{20.20}$$

The vibration of aerosol particles in an acoustic field is thus dependent on the value of τ/t_p. When this ratio is very small $(\varphi-\theta)\to 0$, $\mu \to 1$ and the particle oscillates with the same amplitude and phase as the medium, thus moving with it. At very large values of τ/t_p, μ is very small,† but in this case the complete formula (20.10) should be used giving $\mu = \dfrac{1}{f} = \dfrac{3\gamma_a}{2\gamma}$. The same expression holds for the oscillations of a particle in an ideal fluid [865]. The transition from complete entrainment of particle to near immobility occurs when τ/t_p varies from 0·02 to 20. The way in which vibration takes place in an acoustic field will be shown later to have a decisive effect on the interaction of aerosols and acoustic waves.

† This has been confirmed by photomicrographs of coarse and fine particles exposed to an ultrasonic field [131].

The equations given above are not applicable to very coarse particles at high sound frequencies because of the large value of Re, but in this case, as indicated, the particles remain practically stationary.

Experimental verification of the theory of the vibration of particles in an acoustic field is hampered by the difficulty of measuring the amplitude of vibration of the gas at the point occupied by the particle. This has been overcome in a method devised by Zernov [132] for producing a uniformly oscillating mass of air by attaching to the stem of a tuning fork a closed cylindrical vessel provided with windows for illumination and observation; the dimensions of the vessel must be considerably less than the wavelength. The air inside the vessel oscillates as a whole, and the amplitudes of oscillation both of the vessel itself and of the particles inside can easily be measured. Measurements made on *Lycopodium* spores ($r = 16\,\mu$, $t_p = 0{\cdot}012$ sec) by Wagenschein [133], using this method, gave satisfactory agreement with theory. However, the accuracy of such experiments is low and *Lycopodium* spores are rather unsuitable because of their ridged surface.

The following method of determining the sizes of aerosol particles [134] is based on the variation of the oscillation with τ/t_p. The amplitude of vibration of the particles is determined photographically, first for particles with $t_p \gg \tau$, which vibrate with the medium, and then for particles with t_p comparable to τ; τ and r can then be found from formula (20.18).

Experiments have shown that large particles or aggregates of flat shape in sound waves of high frequency and large amplitude, hence at large Re, are oriented perpendicularly to the direction of vibration [135] in agreement with § 11.

An important result in the mechanics of aerosols follows from § 17, 19 and 20. In the region where Stokes' formula applies the resistance, without introducing appreciable error, can be considered non-inertial and set equal to $6\pi\eta r V_r$, where V_r is the instantaneous value of the velocity of the particle relative to the medium; this makes theoretical investigation of the motion of particles very much easier.

The theory of particle oscillation in a sonic field has been verified by Gucker and Doyle [655]. The mean radii of droplets of dioctylphthalate in isodisperse mists were found by measuring the sedimentation velocity. The value of $\bar{r}$ ranged from 0·8 to 4μ. Using (20.18) the theoretical value of the ratio V_0/U_0 was estimated. The amplitude of sonic oscillations U_0, at a frequency of 4850 c/s, was measured by means of very small droplets. Then V_0 was measured for a large number of mist droplets and the mean value of V_0/U_0 in each mist was calculated. The scatter of experimental points, due mainly to variations in the sound intensity, was rather large but a fair agreement between experimental and theoretical values of V_0/U_0 was obtained.

§ 21. PRESSURE ON AEROSOL PARTICLES DUE TO SOUND

The action of sound waves on suspended particles is not limited to the forces discussed above, which are due to the viscosity of the medium and are variable in direction. The particles also experience hydrodynamic sound pressure which acts always in the same direction and depends not on the viscosity but on the density of the gas. If the radius of a particle is small compared with the wavelength, λ, of sound the pressure upon a stationary spherical particle due to travelling sound waves produces a force (King [136])

$$F_M = 2{\cdot}4\pi U_0^2\gamma_g r^6\left(\frac{2\pi}{\lambda}\right)^4, \tag{21.1}$$

in the direction of propagation of the waves. In standing waves the sound pressure has the considerably greater value

$$F_M = \frac{5}{3\lambda}\pi^2\gamma_g r^3 U_0^2 \sin(4\pi x/\lambda), \tag{21.2}$$

where x is the distance of the particle from the nearest node. This force acts towards the nearest antinode. It reaches its maximum value half way between the nodes and antinodes at $x = (1/8)\,\lambda$, $(3/8)\,\lambda$, $(5/8)\,\lambda$ and $(7/8)\,\lambda$.

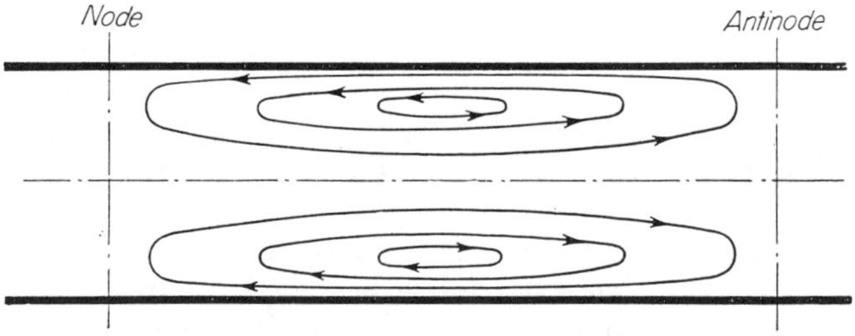

FIG. 21. Circulation of air in an acoustic field.

At the nodes and antinodes $F_M = 0$, the particle being in unstable equilibrium at the former and in stable equilibrium at the latter. King's formula has been confirmed experimentally with cork spheres ($r = 1$ mm) suspended on threads in standing waves of frequency 400–2800 c/s [137].

Westervelt has shown [138] that the viscous energy losses in the boundary layer of gas near the surface of the particle and the lack of symmetry of the flow field on both sides of the particle, which follows from Oseen's theory, are not taken into account in King's derivation. The first of these effects gives a force proportional to the first power of the particle radius and the second, one proportional to the square of the radius; these forces can exceed that calculated according to King by a full order of magnitude. Their importance increases with decrease in particle size, which probably explains why they were not evident in the experiments mentioned. It is highly desirable that measurements of the pressure of sound waves on fine particles should be made.

In sufficiently intense ultrasonic waves sound pressure can exceed the force of gravity and suspend particles in a vertical tube in which standing waves are generated.

Sound pressure is the principal reason for the formation of Kundt dust figures which are periodic deposits of dust in horizontal tubes in which standing sound waves are produced. These deposits have the appearance of narrow ridges disposed perpendicularly to the axis of the tube at a short distance from one another. The ridges are biggest near the antinodes and disappear at the nodes. The mechanism of their formation is probably the following. Standing sound waves in tubes are accompanied by circulation of the gas caused by friction against the wall of the tube and having the form shown in Fig. 21 [139]. Besides a circulatory motion, dust stirred up by the gas flow acquires, under the action of sound pressure, a velocity component directed towards the antinodes. It then falls out of the circulatory flow and is deposited on the walls of the tube in the planes of the antinodes.

The formation in the distance of half a wavelength of not one but several ridges is explained by the presence inside the tube of harmonics, or waves with frequencies an integral number of times the fundamental frequency. It follows from equation (21.2) that sound pressure grows quickly with increase in particle size. In addition, it is necessary for the production of dust figures that the particles should readily be stirred up by the gas flow. This phenomenon is therefore particularly well produced with coarse, light particles such as those of cork dust.

In formulae (21.1) and (21.2) it is assumed that particles remain stationary in an acoustic field. In the alternative case, U_0 must be understood to represent the amplitude of vibration of the particle relative to the medium. As entrainment of particles by the oscillating medium increases, the sound pressure decreases in proportion and becomes zero for complete entrainment.

Sound pressure is obviously of great importance in the coagulation of aerosols by ultrasonic waves (see § 53). Dukhin [656] drew attention to the fact that in a standing sonic wave the oscillations of aerosol particles which are incompletely entrained by the gas are not harmonic and are somewhat asymmetric. As a result, the particles gradually drift to the nodes, this effect being much greater in some cases than that due to acoustic pressure.

§ 22. SCATTERING AND ABSORPTION OF SOUND WAVES BY AEROSOLS

When sound or electromagnetic waves are transmitted through an aerosol they undergo reflection, scattering, absorption and diffraction. In the case of sound waves these phenomena are comparatively weak, and only a brief exposition of the problem will be presented.

The theory of the scattering of waves by an aerosol is simplified considerably if the particles are stationary, spherical and small compared with the wavelength, λ. Let U_0 denote the velocity amplitude and ω the angular frequency of sound with a plane wave front falling on a spherical particle. The energy flux carried per second by the waves through 1 cm^2 of the wave front is

$$I = \frac{U_0^2 \gamma_g c_s}{2}, \tag{22.1}$$

where c_s is the velocity of sound. The energy carried away per second in waves scattered by the particle is shown by theory [140] to be

$$\Omega = \frac{7 \pi \gamma_g U_0^2 \omega^4 r^6}{18 c_s^3}. \tag{22.2}$$

Thus, when sound waves pass through an aerosol layer of thickness dx containing n such particles in 1 cm^3, the ratio of the scattered energy to the incident energy is

$$\alpha_d dx = \frac{7 \pi \omega^4 r^6 n}{9 c_s^4} dx = \frac{7 \pi (2\pi)^4 r^6 n}{9 \lambda^4} dx. \tag{22.3}$$

Hence it follows that the energy flux per cm², in the original direction after the sound has passed through an aerosol layer of thickness x is

$$I = I_0 e^{-\alpha_d x} \tag{22.4}$$

where I_0 is the initial energy flux of the waves per cm² and

$$\alpha_d = \frac{7\pi(2\pi)^4 r^6 n}{9\lambda^4} \tag{22.5}$$

is the coefficient of scattering of the sound waves by the aerosol. If the particles do not remain stationary in an acoustic field but are entrained to some extent by the oscillations of the medium, the scattering coefficient is reduced accordingly.

When ordinary sound waves ($\lambda \approx 1$ m) pass through natural clouds or fog ($r \approx 10^{-3}$ cm, $n \approx 10^3$), α_d is a negligibly small quantity of the order 10^{-20}.

Even for ultrasonic waves and the densest aerosols encountered in nature or industry α_d still remains extremely small, so that scattering of sound by aerosols can practically always be neglected.

The absorption of sound by aerosols, caused by the viscosity of the gas, will be explained by an elementary derivation of the basic equation, assuming that the resistance obeys Stokes' law. For a particle moving in a viscous medium all the work performed against the resistance at low Re,

$$\Omega = \int_0^x F_M dx = \int_0^x \frac{V}{B} dx \tag{22.6}$$

is converted into heat. If the medium also moves with velocity U then V in equation (22.6) should be replaced by the relative velocity $V_r = V - U$ and the path differential dx by $V_r\, dt$, so that

$$\Omega = \frac{1}{B}\int_0^t V_r^2 dt. \tag{22.7}$$

Expressing the motion of the particle by equation (20.15),

$$\Omega = \frac{V_{r_0}^2}{B}\int_0^t \sin^2\left(\omega t + \frac{\pi}{2} - \varphi\right) dt. \tag{22.8}$$

When t is sufficiently large $\Omega \to V_{r_0}^2 t/2B$ so that in unit time the particle absorbs an amount of energy equal to $V_{r_0}^2/2B$. Hence, as before, it follows that when sound waves pass through an aerosol layer of thickness dx the ratio of absorbed to transmitted energy is equal to

$$\alpha_a dx = \frac{V_{r_0}^2 n dx}{U_0^2 B \gamma_g c_s} = \frac{V_{r_0}^2 6\pi\nu r n\, dx}{U_0^2 c_s}, \tag{22.9}$$

and the absorption equation, like the scattering equation, is

$$I = I_0 e^{-\alpha_a x}, \tag{22.10}$$

with the coefficient of absorption of sound waves by the aerosol equal to [see formula (20.14)]

$$\alpha_a = \frac{\omega^2\tau^2}{1+\omega^2\tau^2}\frac{6\pi\nu r n}{c_s}. \tag{22.11}$$

A more rigorous derivation (Sewell [141]) leads to the expression

$$\alpha_a = \frac{\omega^2\tau^2}{1+\omega^2\tau^2}\frac{6\pi\nu r n}{c_s}\left(1+\sqrt{\frac{r^2\omega\gamma_g}{2\nu}}\right). \tag{22.12}$$

Putting in (22.12) values of ν and c_s for air at normal pressure and temperature, we obtain

$$\alpha_a \approx 8{\cdot}3\times 10^{-5}\times rn\left(1+r\sqrt{\frac{\omega}{0{\cdot}3}}\right)\frac{\omega^2\tau^2}{1+\omega^2\tau^2}. \tag{22.13}$$

As can be seen from the foregoing, absorption of sound by aerosols depends largely on the extent to which the particles follow the vibration of the sound waves.

In the results presented above the particles have been taken to be rigid spheres. In the work of Epstein and Carhart [143] deformation, particularly of liquid spheres, was taken into consideration, and shown to be negligible. A more important factor considered by them was the dissipation of energy by heat conduction, which increases the absorption coefficient for aqueous mists by about 40 per cent (see Table 10).

The first measurements of absorption of ultrasound by aerosols (tobacco smoke) were made by Altberg and Goltzmann [144]. Unfortunately, in these experiments the particle size was not determined and, in addition, tobacco smoke contains much carbon dioxide, which is strongly absorbing at the wavelengths used for the measurements. The results obtained are therefore difficult to compare with theory.

In the experiments of Laidler and Richardson [145] an aerosol of *lycoperdon* spores ($r = 2{\cdot}5\,\mu$), containing $1{\cdot}5\times 10^6$ particles/cm³ was taken. At the high frequencies used $\omega\tau$ had values between 10 and 100 so that the particles remained practically stationary. The following values were obtained for the absorption coefficient: at a frequency of 42,000 c/s, $\alpha_a = 0{\cdot}029$; at 98,000 c/s, $\alpha_a = 0{\cdot}031$. Theoretical values of α_a are equal to 0·038 and 0·042 respectively.†

In the work of Knudsen, Wilson and Anderson [146] the rate of decay of acoustic vibrations in a resonant cavity containing an aerosol was measured. The component, β, of the attenuation coefficient of the sound caused by the presence of aerosol particles and expressed in db sec⁻¹, is related to α_a by the formula

$$\beta = 10\log e\cdot\alpha_a c_s. \tag{22.14}$$

By substituting for α_a from (22.13) the following expression for β is obtained:

$$\beta = \frac{2{\cdot}94\,\varphi}{r^2}\left(1+r\sqrt{\frac{\omega}{0{\cdot}3}}\right)\frac{\omega^2\tau^2}{1+\omega^2\tau^2}, \tag{22.15}$$

where φ is the volume of disperse phase contained in 1 cm³ of the aerosol.

Unfortunately these experiments were carried out with very polydisperse water and oil mists obtained by mechanical atomization. For a water mist with $\varphi = 2\times 10^{-6}$ and $\bar{r} = 6{\cdot}25\,\mu$ the following results were obtained (Table 10).

† Theoretical values of α_a given by the authors in their paper are erroneous.

Before experimental and theoretical values of β are compared the effect on the theoretical calculations of the presence of a range of particle sizes must be considered. If for simplicity only the main factor $6\pi\nu\, rn/c_s$ in formula (22·12) is taken into consideration, the absorption coefficient is proportional to $\sum_i n_i r_i = n\bar{r}$ and $\varphi = \frac{4}{3}\pi \sum_i n_i r_i^3 = \frac{4}{3}\pi n r_3^3$ where $r_3 = \sqrt[3]{\overline{r^3}}$ (see page 15). This makes n proportional to φ/r_3^3 so that $1/r^2$ in (22.15) should be replaced by $\bar{r}/r_3^3$.

In polydisperse aerosols $r_3 > \bar{r}$ so formula (22.15) yields a value of β which is too great and increases with the degree of polydispersity. Knudsen *et al.* did not take this into account and, putting $r = \bar{r}$ in formula (22.15) obtained the too high values of β given in Table 10. Epstein and Carhart [143] made use of known size distributions of the mist to which the data of Table 10 refer and calculated β for size fractions. Although they allowed for heat conduction in their calculations the values of β, except at a frequency of 500 c/s, are considerably below the experimental values. It is possible that evaporation and condensation of vapour on the droplets (see below) played an appreciable role in the absorption of sound.

TABLE 10. ABSORPTION OF SOUND IN A WATER MIST

Frequency of vibrations (c/s)	Attenuation coefficient β (db sec^{-1})		
	Experimental	Theoretical	
		Knudsen	Epstein
500	5	10·1	5·0
1000	7	13·8	5·7
2000	9·4	16·0	6·3
4000	10·1	17·1	6·9
6000	12·0	18·2	7·5
8000	13·2	18·8	7·7

Passing on to the absorption of sound by natural atmospheric fog, $r = 5\,\mu$ and $n = 2000$ may be taken as typical values, so that

$$\alpha_a \approx 8 \times 10^{-5} \frac{10^{-7}\omega^2}{1 + 10^{-7}\omega^2} \approx \frac{0{\cdot}34}{\lambda^2 + 4200}, \tag{22.16}$$

where λ is the wavelength of the sound. The absorption coefficient is thus not very sensitive to wavelength and for metre waves it is one third of the value for centimetre waves. At the same time the coefficient of absorption of sound in gases is, to a first approximation, proportional to $1/\lambda^2$. At $\lambda = 5$ cm the absorption coefficient of sound in air is equal to $0{\cdot}8 - 1{\cdot}6 \times 10^{-4}$ so that the gas and the disperse phase attenuate the sound to roughly the same extent. For longer waves the absorption of sound by water droplets predominates over absorption by air. Theory therefore indicates that dense atmospheric fog, contrary to the opinion [148] held since the time of Tyndall's experiments [147], must attenuate sound appreciably. In the fog discussed the energy of sound waves with a frequency of 500 c/s, according to formula (22·16), is decreased by a factor of about 2/3 in a distance of 100 m. According to Sieg's observations [149], sound of this frequency in fog is reduced to 1/1·3 of its intensity in the distance quoted. The concentration of the fog and sizes of the droplets were not measured.

Besides hydrodynamic absorption of sound by aerosols, discussed above, mists of volatile substances exhibit additional absorption as a result of the periodically recurring processes of evaporation and condensation of vapour on the droplets, caused by temperature oscillations in the acoustic field [151]. Absorption of sound by this mechanism is negligible at very low frequencies, since equilibrium can then be established between the droplets and the vapour, and at very high frequencies when only very small quantitites of liquid can evaporate. The theory of this phenomenon shows that the greatest absorption of sound due to the processes of evaporation and condensation in water mists would be expected at frequencies close to the lower limit of audibility of sound (~ 16 c/s) and observations appear to confirm this result [146].

Experimental investigation of the absorption of sound by fog is attended by great difficulties and all attempts in this direction have so far been of a qualitative nature. The difficulties are caused by attenuation of sound with increasing distance from the source as a result of expansion of the wave front and by atmospheric inhomogeneities. The first can be excluded by placing the source of sound and the receiver, or observer, at fixed points and making measurements in the presence and absence of fog, but it is impossible to eliminate atmospheric irregularities of density and humidity which cause sound waves to be reflected, refracted and scattered.

These phenomena reduce the intensity of sound much more than the absorption by air and fog which we are discussing [152]. Many contradictory observations can be accounted for in this way. The temperature and humidity of the gas phase in fog frequently differ from that of the surrounding air; thus visible clouds are simultaneously "acoustic clouds", and the presence of several such clouds between the source of sound and the observer may completely damp out the sound. On the other hand, thick continuous masses of fog are often very homogeneous, so that if the whole space between the source of sound and the observer is filled with fog, attenuation may be less than in the absence of fog but in the presence of atmospheric inhomogeneities. The idea that atmospheric fog does not absorb sound is probably due to this. The matter requires further and more exact study.

The important paper by Epstein and Carhart [143] on the absorption of sound by aerosols (page 91) merits detailed consideration; an excellent account of this work has been given by Zink and Delsasso [657].

The coefficient of sound absorption due to viscous loss is

$$\alpha_a^v = \frac{6\pi\nu r n}{c_s}\left(1 + \frac{1}{\beta}\right)\mu_\nu \tag{22.17}$$

where

$$\mu_\nu = \frac{V_{r_0}^2}{U_0^2} = \frac{f^2}{f^2 + 3f\beta + \frac{9}{2}\beta^2 + \frac{9}{2}\beta^3 + \frac{9}{4}\beta^4} \tag{22.18}$$

is an accurate expression for the ratio $V_{r_0}^2/U_0^2$ (see 20.13), for which an approximate value was used in (22.12).

The coefficient of sound absorption due to the periodic irreversible heat transfer to and fro between the particle and the gas, which leads to an increase in entropy and a decrease in free energy, is given by

$$\alpha_a^\lambda = \frac{4\pi\lambda r n}{c_s}\left(\frac{c_p}{c_v} - 1\right)\left(1 + \frac{1}{\beta}\right)\mu_\lambda \tag{22.19}$$

$$\mu_\lambda = \frac{f^2}{f^2 + 3f\theta + \frac{9}{2}\theta^2 + \frac{9}{2}\theta^3 + \frac{9}{4}\theta^4} \tag{22.20}$$

where c_p/c_v is the ratio of the specific heats of the gas at constant p and v, λ the heat diffusivity of the medium and $\theta = (2\lambda/\omega)^{1/2}/r$. μ_λ expresses the degree of equalisation of the temperatures of particle and medium. At large θ these temperatures are almost identical, the decrease in free energy is negligible and $\mu_\lambda = 0$. At very small θ the temperature of the particles is almost constant, the temperature difference and the decrease in free energy attain their maximum possible values and $\mu_\lambda \approx 1$.

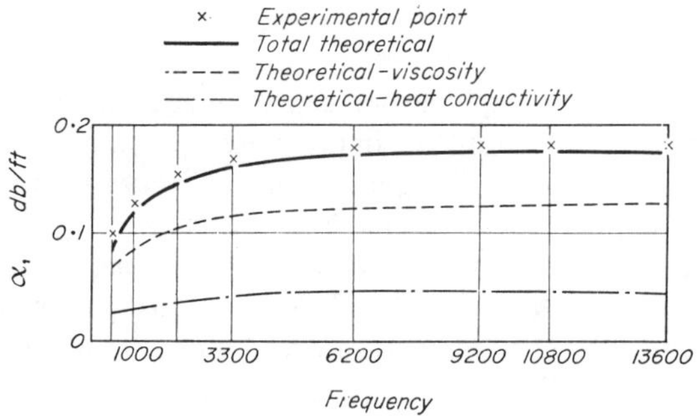

FIG. 21 a. Absorption of sound by an aerosol.

The coefficients α_a^v and α_a^λ are additive only at moderate particle concentrations ($n < 10^6$) and provided that $\left(\frac{c_p}{c_v} - 1\right)\omega\lambda/c_s \ll 1$, which, in the case of the air, means $\omega \ll 4 \times 10^5$.

In Zink and Delsasso's experiments [657] a sound impulse consisting of a score of sinusoidal waves was passed through an aerosol of spherical particles of Al_2O_3 with $r = 2{\cdot}5 - 7{\cdot}5\,\mu$ and $n \approx 3 \times 10^4$. Oscillograms of the waves entering the aerosol and emerging from it were obtained on the same screen. The absorption coefficient was determined from the ratio of the amplitudes, and the time taken by the sound to pass through the aerosol, and hence the velocity of the sound, from the distance between the sinusoids on the screen. The aerosol particles were counted in five fractions, α_a^v and α_a^λ being calculated for each fraction from the formulae given above and then added together. The very good agreement between theory and experiment obtained in air and helium is shown on Fig. 21 a. In argon and oxygen the experimental points were 5–10% above the theoretical curves.

A formula was deduced by the authors for the decrease in the velocity of sound, Δc_s, caused by suspended particles; it is valid only at small Δc_s:

$$\Delta c_s = \tfrac{1}{2} c_s \left[\frac{M_r(1 - \mu_v)}{M_g} + \frac{R_g c_r M_r (1 - \mu_\lambda)}{c_v M_g C_p}\right] \tag{22.21}$$

M_r/M_g is the ratio of the masses of the disperse phase and of the medium, c_r/c_v – the ratio of their specific heats, C_p the molar thermal capacity of the gas at constant pressure and R_g is the gas constant. The right-hand side of (22.21) must be summed up over all fractions. The first term in (22.21) determines the decrease in the sound velocity due to an increase in its density, the second term that due to an increase in its thermal capacity, owing to the presence of suspended particles.

For these factors to have an effect on the sound velocity it is necessary that the aerosol particles should participate in the oscillation of the medium, as well as absorb and give up the heat. This is taken care of by the factors μ_ν and μ_λ, respectively. The effects of sound absorption and decrease in sound velocity are thus complementary. Formula (22.21) is also in good agreement with experiment.

§ 23. HYDRODYNAMIC INTERACTION BETWEEN AEROSOL PARTICLES

A particle moving relative to a fluid produces currents which in turn act upon other particles and give rise to hydrodynamic interaction between them. In aerosols such forces develop mainly when particles are moving on parallel courses and discussion will be restricted to such cases.

The magnitude and direction of the forces of hydrodynamic interaction depend largely on the nature of the flow created. In the viscous flow regime, at low Re, interaction between two similar particles moving at constant velocity in the same direction leads, as shown in § 13, to a reduction in the resistance of the medium according to formulae (13.1). In the presence of a large number of particles the forces of interaction are summed and, as a result, an aerosol cloud can move considerably faster than isolated particles.

The component of hydrodynamic interaction which tends to change the distance between the particles is equal to zero in the viscous flow regime at very low Re; there is neither attraction nor repulsion between particles. Oseen's theory, with partial consideration of inertial forces [153] shows that when particles move in a single front they repel, and when they move one behind the other the front one is repelled and the rear one attracted (see below). A particle moving parallel to a wall at low Re is repelled by the wall with a force [154]

$$F_M = \frac{9}{16}\pi\gamma_g r^2 V^2, \tag{23.1}$$

which does not depend on the distance x of the centre of the particle from the wall. This formula is true only for small values of the ratios r/x and Vx/ν.

A completely different picture is observed at large Re. The theory of hydrodynamic interaction at large Re was worked out by Kirchhoff for an ideal fluid, without consideration of any vortices created during the motion [155]. According to this theory, a spherical particle of radius r_1 moving with velocity V acts on another particle of radius r_2 (Fig. 22) with a force which does not depend on the motion of the second particle. If the distance ϱ between the centres of the particles is much greater than the radii, the component of this force along the line of centres is

$$F_\varrho = \frac{3\pi\gamma_g r_1^3 r_2^3 V^2}{\varrho^4}\left(\frac{3}{2}\cos 2\theta + \frac{1}{2}\right), \tag{23.2}$$

and perpendicular to the line of centres

$$F_\tau = \frac{3\pi\gamma_g r_1^3 r_2^3 V^2}{\varrho^4}\sin 2\theta, \tag{23.3}$$

where θ is the angle between the line of centres and the direction of motion. The force field of the hydrodynamic interaction, created by a moving particle, is illustrated in Fig. 23. It is the same as the force field of the interaction between two similarly directed electric dipoles (see § 52), but the direction of the forces is reversed.

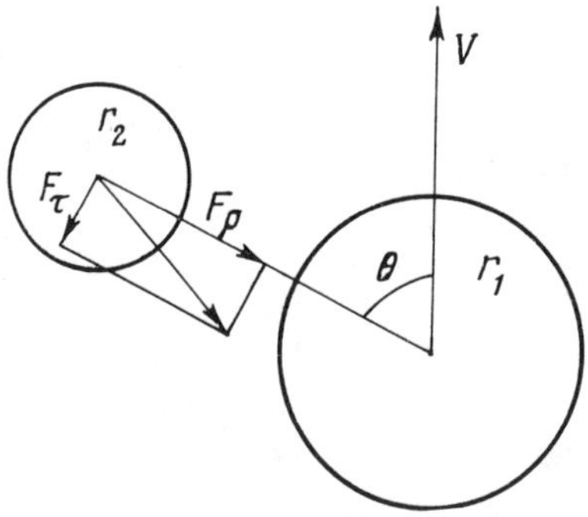

FIG. 22. Hydrodynamic forces between particles.

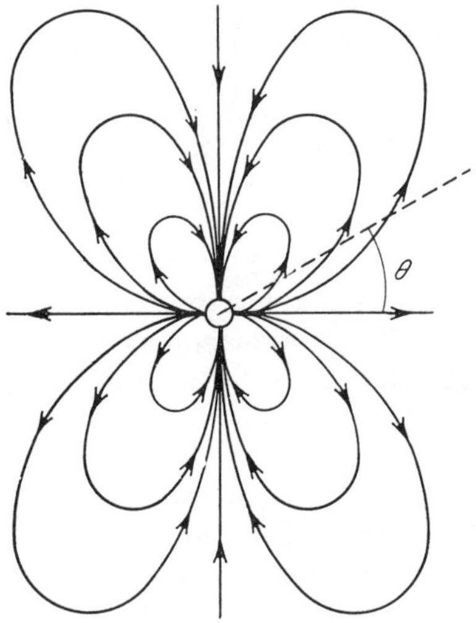

FIG. 23. Hydrodynamic field created by a moving particle.

It follows from formula (23.2) that for $\cos 2\theta < -1/3$, $F_\varrho < 0$, so that there is an attraction, and for $\cos 2\theta > -1/3$ a repulsion. The attraction reaches a maximum value at $\theta = \pi/2$ and $3\pi/2$, this being

$$F_\varrho = -\frac{3\pi\gamma_g r_1^3 r_2^3 V^2}{\varrho^4}. \tag{23.4}$$

The maximum value of the repulsion (when θ is equal to 0 or π) is twice as large.

Only in these two cases is the force of interaction directed along the line of centres of the particles ($F_\tau = 0$). As already indicated, the force with which the first particle acts on the second depends only on the motion of the first particle relative to the medium and not on the motion of the second particle. The action and reaction, therefore, are not equal to one another except when the velocities of both particles are equal in magnitude and direction. Similar forces act on stationary spheres situated in a medium flowing uniformly with velocity V.

At large Re a particle moving parallel to a wall in a fluid at rest with respect to the wall is attracted by it with a force

$$F = -\frac{3\pi\gamma_g r^6 V^2}{16 x^4} \tag{23.5}$$

where x is the distance from the centre of the particle to the wall [156]. This formula is only true for $r/x \ll 1$.

The modification of the equations presented above to take account of vortices formed in a real fluid has not been investigated. Since vortices are only formed behind moving spheres, and the flow in the remaining volume differs little from the potential flow of an ideal fluid, it is probable that in the most important case for the physics of aerosols – attraction between particles – the equations above remain valid even in real fluids.

Another important question, also unanswered, is at what values of Re does attraction or repulsion between particles occur according to equations (23.2) and (23.15)† (see below). The experimental verification of the theory of hydrodynamic interaction between forward-moving bodies has been limited to the observation of attraction or repulsion but the magnitude of the interaction force has not been measured. Experiments have been carried out in liquids with fairly coarse spheres at large Re. Gorbatchev and Severnyi [157] detected hydrodynamic attraction (at $\theta = \pi/2$) and repulsion (at $\theta = 0$) between liquid drops of radius 25 μ suspended on fine threads in a gas stream moving with a velocity of 15 cm sec^{-1}, at Re = 0·5. The problem needs further experimental investigation.

Hydrodynamic interaction is most important when aerosol particles are set in oscillation, especially in an acoustic field. The forces acting between two particles oscillating with the same amplitude and in the same phase are given according to Bjerknes [158], by equations (23.2) and (23.3), but V^2 is in this case equal to the mean square velocity of the particles $\overline{V_r^2}$ relative to the medium. Thus

$$F_\varrho = \frac{3\pi\gamma_g r_1^3 r_2^3 V_{r_0}^2}{2\varrho^4}\left(\frac{3}{2}\cos 2\theta + \frac{1}{2}\right), \tag{23.6}$$

$$F_\tau = \frac{3\pi\gamma_g r_1^3 r_2^3 V_{r_0}^2}{2\varrho^4}\sin 2\theta, \tag{23.7}$$

where V_{r_0} is the amplitude of oscillation of the particle relative to the medium that is $\sqrt{2\overline{V_r^2}}$.

If the particles oscillate in antiphase (for example oppositely charged particles in an alternating electric field), a minus sign must be inserted on the right hand sides of formulae (23.6) and (23.7). In the interaction of an oscillating and a stationary particle the former always attracts the latter, but in this case the force is appreciably less than when both particles are oscillating. If the oscillations take place along the line of centres the attractive force is

$$F_\varrho = \frac{3\pi\gamma_g r_1^6 r_2^3 V_{r_0}^2}{\varrho^7}. \tag{23·8}$$

The influence of the viscosity of the fluid upon hydrodynamic interaction has remained unexplained for vibrating particles. The theory of oscillation in a viscous fluid [159] indicates that the effect of viscosity on the waves formed by the oscillations

† Remember that for $\theta = \pi/2$ attraction between particles occurs at large Re and repulsion at low Re.

of a sphere becomes negligible at a distance ϱ from the centre such that $\varrho\sqrt{\frac{\omega}{2\nu}} \gg 1$ $\left(\varrho \gg \sqrt{\frac{0{\cdot}3}{\omega}} \text{ in air}\right)$. Thus for particles of size $r \leqslant 10^{-4}$ cm, where hydrodynamic attraction in an acoustic field is of particular interest in connection with ultrasonic coagulation (see § 51), and at a frequency of 10 kc/s, ϱ/r must be $\gg 22$. According to (23.6), however, when particles are so far apart the force of interaction is vanishingly small. The theory presented above clearly needs substantial correction for the effect of viscosity.

Hydrodynamic interaction between particles suspended in an acoustic field has usually been studied experimentally with solid or liquid spheres supported on fine threads in a tube in which standing sound waves are generated [160, 161, 162]. The attractive or repulsive force was measured by the deviation of the threads from the vertical. Qualitative agreement with theory has been obtained, spheres arranged along the tube repelling one another and those arranged across the tube attracting†. Circulation of air in the tube must have strongly influenced the results (see page 88).

These experiments are obviously of little use for checking the theory quantitatively. The more accurate measurements made by Thomas [163] using Zernov's method (see page 87) with spheres of radius 3 – 7 mm in a closed cylindrical vessel oscillating along its axis at a frequency 60 c/s gave good agreement with theory. The force of interaction was strictly proportional to the square of the amplitude of the oscillations at any distance between the centres of the spheres. The dependence of force on distance was in full agreement with theory even at small distances in the case of attraction, and at distances greater than $4\,r$ in the case of repulsion. At smaller distances the repulsive force was less than the theoretical value, probably because of vortex formation. The dependence of force on the frequency of the oscillations was also verified, the absolute value of the force for spheres 3 – 4 mm in radius being the same as the theoretical value. The condition mentioned above for the effect of viscosity leads in the case considered to $\varrho/r \gg 0{\cdot}07$, and may be considered fulfilled. Since the amplitude of oscillation in these experiments was about 0·1 cm (Re = 70 – 150) the question as to the magnitude of the hydrodynamic forces between oscillating aerosol particles at low Re remains open.

In order to calculate the hydrodynamic interaction between two spherical particles moving in a viscous medium on Stokes' approximation, it is first necessary to find the field of flow satisfying the equations: at an infinite distance from the particles the flow velocity is zero and on the surface of a particle it is equal to the velocity of the particle. The solution is obtained by the method of successive approximation.

The sum of the Stokes flow fields of each particle (34.4) provides the first approximation which for parallel motion of two identical particles is equation (13.1). This fails to satisfy rigorously the condition at the surface of the particles, but by the successive addition of correction terms a closer approach is obtained. The solutions involve a function of the ratio r/ϱ, usually a power series the convergence of which, and the accuracy of the solutions, rapidly deteriorates as r/ϱ approaches 0·5, when the particles touch one another.

† It should be noted that, when the spheres arranged across the tube are brought close enough together (of the order $\frac{1}{10}$ diameter), attraction is under some conditions replaced by repulsion, which increases as the spheres are moved still closer [564].

Since the fields of flow are linear with respect to the velocities of each particle, the hydrodynamic interaction need only be calculated for the motion of particles along the line of their centres, when the force $F_{||}$ is also directed along this line, and for motion normal to it, $F_{\perp}$. The most important case is when the particles move in the same plane, when $F_{\perp}$ lies in this plane, too; owing to the symmetry of Stokes' flow past a sphere $F_{\perp}$ is also normal to the line of centres. The particles also rotate so that, in the case of falling particles, the left-hand one rotates clockwise and the right anti-clockwise.

In general it is necessary to resolve the velocities of the particles into their components $V_{||}$ and $V_{\perp}$, in order to calculate the forces $F_{||}$ and $F_{\perp}$, and then to add them. Owing to the symmetry of Stokes' flow past a sphere, if the radii, r_1 and r_2, as well as the velocities of the particles, V_1, V_2, are equal, the force of interaction is the same upon each particle, both in magnitude and direction.

Several investigations have been devoted to the hydrodynamic interaction between two spherical particles. Hocking [658], Kynch [142] and Faxén [659] deduced formulae (the former two authors for $F_{||}$ and $F_{\perp}$, the latter only for $F_{||}$) for the general case $r_1 \neq r_2$, $V_1 \neq V_2$; Smoluchowski [660] for $F_{||}$ and $F_{\perp}$ at $r_1 \neq r_2$ and $V_1 = V_2$, Burgers [83] the same with $r_1 = r_2$ and $V_1 = V_2$, Stimson and Jeffery [577] for $F_{||}$ with $r_1 = r_2$ and $V_1 = V_2$.

According to Burgers:

$$F_{||} = 6\pi\eta r V \left[\frac{3}{2}\frac{r}{\varrho} - \left(\frac{r}{\varrho}\right)^3 - \frac{15}{4}\left(\frac{r}{\varrho}\right)^4 + \cdots\right] \tag{23.9}$$

$$F_{\perp} = 6\pi\eta r V \left[\frac{3}{4}\frac{r}{\varrho} + \frac{1}{2}\left(\frac{r}{\varrho}\right)^3 + \cdots\right]. \tag{23.10}$$

These formulae show that the ratio of the sedimentation velocity of two identical interacting particles to that of an isolated particle is

$$V_s'/V_s = 1 + \frac{3}{4}(1 + \cos^2\theta)\frac{r}{\varrho} + \frac{1}{2}(1 - 3\cos^2\theta)\left(\frac{r}{\varrho}\right)^3 - \frac{15}{4}\cos^2\theta\left(\frac{r}{\varrho}\right)^4 + \cdots \tag{23.11}$$

where θ is the angle between the line of centres and the vertical. The particles fall vertically only when $\theta = 0$ or $\pi/2$; for other values their trajectories are inclined towards the line of centres at an angle φ where

$$\sin\varphi = \frac{3}{4}\frac{r}{\varrho}\left[1 - \frac{2}{3}\left(\frac{r}{\varrho}\right)^2 - \frac{1}{5}\left(\frac{r}{\varrho}\right)^4 + \cdots\right]\sin\theta\cos\theta. \tag{23.12}$$

In Smoluchowski's paper the expression in square brackets is replaced by $[1 - 3r/2\varrho]$.

The values of V_s/V_s' calculated from different formulae for the sedimentation of identical spherical particles along the line of their centres, as well as the experimental data of Eveson *et al.* (see below), are presented in Table 10a. Faxén's formula is not included in the table because his power series converge very slowly when r/ϱ approaches 0·5.

TABLE 10A. THEORETICAL AND EXPERIMENTAL VALUES OF V_s/V_s' IN SEDIMENTATION OF TWO IDENTICAL SPHERICAL PARTICLES ALONG THE LINE OF THEIR CENTRES, TO STOKES' APPROXIMATION

r/ϱ	0·0	0·1	0·2	0·3	0·4	0·5
Stimson and Jeffery	1·0	0·870	0·776	0·712	0·677	0·645†
Smoluchowski (complete formula)	1·0	0·872	0·790	0·752	0·760	0·813
Smoluchowski (abridged formula)	1·0	0·870	0·770	0·690	0·625	0·571
Kynch	1·0	0·870	0·777	0·714	0·673	0·602
Burgers	1·0	0·870	0.778	0·718	0·695	0·718
Hocking	1·0	0·871	0·777	0·710	0·637	0·260
Experimental value	1·0	0·900	0·800	0·730	0·686	0·655

† This value was obtained by Faxén [605] from the formula of Stimson and Jeffery by a simple mathematical procedure.

As seen from the table, the formula of Stimson and Jeffery, which is very cumbersome, agrees best with experiment up to the contact between the particles and must be considered to be the most accurate. The next in accuracy is the formula of Kynch, followed by the relatively simple one of Burgers which is given here.

An experimental study of the interaction between two particles at small Re ($\sim$ 0·01) was made by Eveson *et al.* [604]; the sedimentation velocity of plastic spheres of $r = 1{\cdot}4 - 2{\cdot}4$ mm was measured in castor oil with the angle θ varying from 0 to 90°; the vessel was 25 cm in diameter so that the wall effect was insignificant. In all cases the results of the experiments were similar to those presented in Table 10a for $\theta = 0$. At small r/ϱ the ratio V_s/V_s' was somewhat larger than the theoretical value, but when the spheres approached one another the discrepancy diminished. The fact that both spheres moved with identical velocities confirms that the pattern of flow was that of Stokes. The angle of the slope of the trajectories to the vertical was in good agreement with Smoluchowski's calculations (see page 99). It is surprising that no rotation of particles was observed by the authors†. The reason for small discrepancies between theory and experiment at small r/ϱ, where the theoretical deductions are undoubtedly accurate, is hard to explain.

In similar experiments Happel and Pfeffer [641] investigated only $\theta = 0$ at Re $= 0{\cdot}008 - 0{\cdot}03$. The results were in excellent agreement with the formula of Stimson and Jeffery in the range of r/ϱ from 0·09 to 0·5.

Pshenai-Severin [661] examined hydrodynamic interaction between particles of different size settling along the line of centres. In this case, on Stokes' approximation,

$$V_{s_1}' \approx V_{s_1} + \frac{3}{2}\frac{r_2}{\varrho} V_{s_2}; \quad V_{s_2}' \approx V_{s_2} + \frac{3}{2}\frac{r_1}{\varrho} V_{s_1}. \tag{23.13}$$

Let the index 1 refer to the particle moving in front and suppose $r_2 > r_1$. The second particle therefore overtakes the first. Interaction, as seen from (23.13), reduces the rate of approach of the particles and the effect is still larger at small distances between the particles for which formula (23.13) is no longer valid.

† In a later communication [881] the authors report a slow rotation of the particles.

The same author [662] also studied the problem by the method of Stimson and Jeffery. A table given in his paper shows that the velocities of water droplets, for instance, of $r_1 = 4\mu$ and $r_2 = 6\mu$ in air, for which $V_{s_1} = 0{\cdot}20$ and $V_{s_2} = 0{\cdot}46$ cm/sec, are equal to 0·49 and 0·53 cm/sec respectively at the distance $\varrho = 16\mu$.

It should be emphasized that the calculation of the hydrodynamic interaction to Stokes' approximation is admissible only if $\beta = V/v \ll 1$ and appreciable departures may occur when the condition $Vr/v \ll 1$ is still observed, i.e. when Stokes' formula for the motion of one particle is still valid. According to Pshenai-Severin the error is as large as 9 per cent at $\beta = 0{\cdot}2$ and 21 per cent at $\beta = 0{\cdot}5$.

Turning to the theory of hydrodynamic interaction, using Oseen's approximation, the expressions deduced by Oseen himself [153] will first be considered. Imagine two spherical particles of radii r_1 and r_2 moving (e. g. sedimenting) along the x-axis with the same velocity V_s, the first particle being in advance of the other. In addition to the usual Stokes' force, $-6\pi\eta r_1 V_s$, the leading particle is also being acted upon in the direction of the motion by the force:

$$F_x^{(1)} = \frac{9\pi r_1 r_2 \eta V_s}{2\varrho} \exp\left[-\frac{V_s}{2v}(\varrho + x_1 - x_2)\right] \tag{23.14}$$

where x_1 and x_2 are the coordinates of the particles; along the line of centres another force acts,

$$F_\varrho^{(1)} = \frac{9\pi r_1 r_2 \eta v}{\varrho^2}\left\{1 - \left(1 + \frac{V_s\varrho}{2v}\right)\exp\left[-\frac{V_s}{2v}(\varrho + x_1 - x_2)\right]\right\}. \tag{23.15}$$

The forces acting upon the second particle are obtained from these formulae by transposing the indices. The forces $F_x^{(1)}$ and $F_x^{(2)}$ have the same direction as the vector V_s. The force $F_\varrho^{(1)}$ is directed along the line of centres towards the side opposite to the second particle and is a repulsive force. The force $F_\varrho^{(2)}$ at small angles θ, as in the case of motion along the line of centres, is directed towards the first particle, while at large θ, for instance, $\theta = \pi/2$, it acts in the opposite direction (repulsion).

At $r_1 = r_2$, for motion along the line of centres, the total interaction force is

$$F^{(1)} = F_x^{(1)} + F_\varrho^{(1)} = \frac{9\pi\eta v r^2}{\varrho^2}\left[1 - \exp\left(-\frac{V_s\varrho}{v}\right)\right] = \frac{9\pi\eta r^2 V_s}{\varrho}\,\frac{1 - e^{-\beta}}{\beta} \tag{23.16}$$

$$F^{(2)} = F_x^{(2)} + F_\varrho^{(2)} = \frac{9\pi\eta r^2 V_s}{\varrho} \tag{23.17}$$

where $\beta = \varrho V_s/v$. Since $(1 - e^{-\beta})/\beta < 1$, the first particle moves more slowly than the second, as is confirmed by experiments (see p. 322). The ratio of the velocity of their mutual approach to V_s is equal to $\vartheta = \frac{3}{2}\frac{r}{\varrho}\left[1 - \frac{1 - \exp(-\beta)}{\beta}\right]$, the motion of the particles being supposed to be quasistationary (see page 74). In the case of water drops of $r = 30\mu$, ϑ is equal to 0·056 at $r/\varrho = 0{\cdot}05$ and to 0·112 at $r/\varrho = 0{\cdot}2$.

Thus, on Oseen's approximation, settling particles of equal size approach one another rather more quickly than on the Stokes basis. This is due to the asymmetry in Oseen's field of flow past a sphere: the flow velocity behind a moving particle decreases with distance much more slowly than in front (see Fig. 74a). When particles of different size settle along the line of centres and $r_2 > r_1$, so that the second particle overtakes the first, the Oseen interaction, unlike that of Stokes, markedly increases (according to Pshenai-Severin) the rate of mutual approach of the particles in this case as well.

Additional discussion of this question will be found in § 54 in connection with the theory of gravitational coagulation.

The effect of viscosity on the hydrodynamic interaction between particles of equal size in a sonic field was also dealt with by Pshenai-Severin [663] who calculated the velocity of mutual approach of particles, on Oseen's approximation, by the same method used for uniform motion. Mutual approach of particles was assumed to occur only when the relative velocity of the particles and the medium corresponded to Re > 1. It was shown that at the velocity amplitude of the oscillations of the gas $U_0 = 400$–1000 cm/sec, the sonic energy density being 50–200 erg/cm³, the time of the approach of particles with $r = 1$–$15\,\mu$ from $\varrho = 100\,r$ to $10\,r$, for favourable relationships between the period of oscillations t_p and the relaxation time of the particles, τ, amounted to some fractions of a second.

This effect can play an important role in sonic coagulation of aerosols. The maximal effect is attained at t_p equal to $4\,\tau$–$8\,\tau$. It should be noted that these calculations hold only if the period of oscillation, t_p, is much greater than the time necessary for the flow disturbance caused by one particle to reach the other particle, or in other words, if the amplitude of oscillation of the medium is much larger than the mean distance between the particles. On the other hand, an appreciable effect is possible only at $\tau/t_p > 0{\cdot}1$, when the entrainment of the particles by sonic waves is incomplete (see Fig. 20). The values of τ given in Table 13 show that at $U_0 = 400$–1000 cm/sec the range of t_p values over which both inequalities hold good (assuming $\varrho = 100\,r$) is rather narrow at $r = 1\,\mu$, but widens considerably at $r = 10\,\mu$.

Dörr [664] measured the interaction between two hollow glass spheres of $r = 2{\cdot}8$–$4{\cdot}4$ mm suspended on glass filaments in the antinode of a standing sonic wave of the frequency of 525 c/s. The frequency of the free oscillations of the suspended spheres was about 1 c/s and there were practically no forced oscillations. The amplitude of the oscillation velocity U_0 was varied from 0 to 35 cm/sec and the interaction force was found to be strictly proportional to U_0^2, in agreement with theory. With the angle θ between the direction of oscillations and the line of centres equal to $\pi/2$, the absolute value of the force at $\varrho/r \gg 3$ agreed with the theoretical formula (23.6), but as the spheres approached one another, it increased much faster than it should have according to the formula. In the case of $\theta = 0$ the agreement with theory appeared to be more complete, but there are no definite indications in Dörr's paper.

§ 24. ELECTROSTATIC SCATTERING OF AEROSOLS

In conclusion, electrostatic scattering of aerosols [164, 165, 166, 335] will be discussed. This occurs if the algebraic sum of the charges on the particles differs from zero. The simplest case of an isodisperse aerosol, all the particles of which have the same positive charge q, will be taken first. The total charge σ on unit volume of aerosol is then nq. A fundamental equation of electrostatics states

$$\mathrm{Div}\,\mathbf{E} = 4\pi\sigma = 4\pi nq, \qquad (24.1)$$

where **E** is the field intensity vector which is related to the velocity vector of the particles by the equation

$$\mathbf{V} = B\mathbf{E}q. \tag{24.2}$$

Hence

$$\text{Div}\,\mathbf{V} = 4\pi n q^2 B. \tag{24.3}$$

The divergence of the velocity is defined by

$$\text{Div}\,\mathbf{V} = -\frac{1}{n}\frac{dn}{dt}, \tag{24.4}$$

giving

$$-\frac{1}{n}\frac{dn}{dt} = 4\pi n q^2 B, \tag{24.5}$$

whence

$$\frac{1}{n} - \frac{1}{n_0} = 4\pi q^2 B t, \tag{24.6}$$

where n_0 is the initial particle concentration.

The same equations will be obtained for negatively-charged particles.

It follows from these equations that an aerosol consisting of particles which all bear a charge of the same sign is uniformly dispersed under the action of its own bulk charge.† The rate of scattering of the aerosol, i.e. the rate at which the concentration decreases, is the same in all parts and depends on neither the size nor the shape of the space occupied by the aerosol: it is an internal property of the aerosol. A cloud increases uniformly in volume while retaining its shape, and the rate of relative increase in volume is equal to the rate of relative decrease in concentration, $-(1/n)(dn/dt)$, which is given by the right hand side of equation (24.5). If the aerosol is in a confined space the particles are deposited on the walls and the same quantity, $-(1/n)(dn/dt)$, shows what fraction of the particles existing at a given instant is deposited in unit time. Since the concentration of the aerosol is the same in all places at any instant during electrostatic scattering, motion of the medium has no effect on the scattering, so that the equations given above are valid for an aerosol experiencing either laminar or turbulent flow.

Suppose next that 1 cm³ of an aerosol contains n_+ positive and n_- negative particles, each bearing a charge of the same absolute value q. Let $n_+ > n_-$ so that the total charge per unit volume of cloud is $(n_+ - n_-)q$. Repeating the reasoning given above, the scattering of the positive particles is expressed by the following equation, equivalent to (24.5),

$$-\frac{1}{n_+}\frac{dn_+}{dt} = 4\pi q^2 B(n_+ - n_-). \tag{24.7}$$

For negative particles

$$-\frac{1}{n_-}\frac{dn_-}{dt} = 4\pi q^2 B(n_+ - n_-). \tag{24.8}$$

This means that the negative particles (the particles in the minority) are driven to

† In German literature the phenomen under discussion has been called "electrical diffusion" [165] but, since it consists of an ordered motion of particles, such a term can hardly be accepted as correct.

the middle of the aerosol where their concentration consequently increases. The charge density $(n_+ - n_-)q$ in this central nucleus of the aerosol decays asymptotically until the nucleus becomes practically neutral and consequently stable. At the same time scattering of the purely unipolar part of the aerosol outside the nucleus continues (see below).

The equations can easily be extended to polydisperse aerosols and to particles carrying charges of different magnitudes (see below).

Electrostatic scattering has been investigated experimentally by Wolodkewitch [165] and by Fuchs and Petryanov [166]. Wolodkewitch worked with highly dispersed amicroscopic aerosols ($r \approx 10^{-6}$ cm) obtained by heating magnesium oxide to 1000–1200°. Some of the particles so produced are charged and most of these probably have one elementary charge each (see page 114). The mobility of the particles was determined by passing the aerosol through a condenser connected to an electrometer (see § 27). By passing the aerosol through a charged grid which removed particles of the opposite sign it was possible, on applying a sufficiently high potential to the grid, to obtain purely unipolar aerosols. The rate of scattering of these was determined from the reduction in bulk charge of the aerosol as measured by means of a condenser. Satisfactory agreement between the experimental data and theory was obtained.

Fuchs and Petryanov studied the scattering of an oil mist ($\bar{r} = 0{\cdot}55\,\mu$) charged by means of a unipolar corona discharge. From 70 to 100 per cent of the particles were charged negatively, the rest remaining uncharged, but the charges on the particles differed quite widely in absolute value; the polydispersity of the mist also increased markedly on charging. The rate of scattering was measured by the particle count method. In agreement with equation (24.6) the reciprocal of particle concentration increased linearly with time at a rate which rose with the average charge on the particles (from 0 to 48 elementary charges) but in absolute value the rate of scattering was several times less than theoretical. No explanation could be found for this discrepancy.

In aerosols having a charge which is not completely unipolar it has been possible to produce a central nucleus experimentally [167].

It is characteristic of electrostatic scattering that, unlike precipitation under gravity, particles are precipitated with almost the same velocity at the bottom as at the side and top walls of the chamber containing the aerosol. This could be used in the disinfection and disinfestation of buildings if a simple enough portable apparatus for unipolar charging could be devised. Whereas the usual insecticidal aerosols obtained from cannisters, cartridges and special generators† fall largely on the floor and have a very weak effect, unipolarly charged aerosols would allow ceilings and walls, and even cracks in them, to be treated. Similar considerations apply to inhaled medicinal aerosols; by means of unipolar charging, their deposition in the respiratory passages might be increased (see page 237). On the other hand the opinion expressed by Dessauer [168] and Chizhevskii [169] on the therapeutic properties of unipolarly charged aerosols cannot be considered conclusively proved.

For polydisperse aerosols carrying charges of both signs

$$-\frac{dn_i}{dt} = 4\pi B_i n_i q_i \sum_i n_i q_i = 4\pi B_i q_i n_i n\bar{q} \tag{24.9}$$

† Mechanical atomizers, freon bombs and other apparatus of this sort may produce a stream of very coarse mist or dust which is deposited on walls by inertial forces.

for each size fraction, where $\bar{q}$ is the mean charge per particle. The solution of such a set of non-linear differential equations (24.9) is very complicated and has never been attempted. The change in the total particle concentration $n = \sum_i n_i$ is

$$-\frac{dn}{dt} = 4\pi n\bar{q} \sum_i B_i n_i q_i = 4\pi n^2 \bar{q}\, \overline{Bq}. \tag{24.10}$$

It is impossible to integrate this equation since $\bar{q}$ and $\overline{Bq}$ are time functions which can be determined only by integrating the whole set of equations (24.9). If the electrical mobilities of all particles, $B_i q_i$, are more or less identical and $\bar{q}$ and $\overline{Bq}$ therefore undergo no change during dispersion, (24.6) can be applied to polydisperse aerosols.

This equation can be written

$$n = \frac{n_0}{1 + \beta t} \tag{24.11}$$

where

$$\beta = 4\pi \overline{q^2}\, \overline{B} n_0. \tag{24.12}$$

When an isodisperse unipolarly charged aerosol, of n_0 particles per cm³ and equal charges on the particles, is drawn at a mean velocity $\overline{U}$ through an earthed tube of radius R and length x, the concentration of the aerosol emerging from the tube (according to 24.11) for either laminar or turbulent flow is equal to

$$n = \frac{n_0}{1 + \beta x/\overline{U}}. \tag{24.13}$$

The total mass of the particles deposited per second on the walls of the tube, owing to electrostatic dispersion, is

$$\Phi_x = \pi R^2 \overline{U}(n_0 - n)m = \frac{\pi R^2 \beta x n_0 m}{1 + \beta x/\overline{U}} = \frac{\pi R^2 \beta x c_0}{1 + \beta x/\overline{U}} \tag{24.14}$$

where m is the mass of the particle and c_0 the initial weight concentration of the aerosol. Flowing through the tube is an electric current

$$I_x = q\Phi_x/m. \tag{24.15}$$

Equation (24.14) can be written as

$$x = \pi R^2 c_0 \overline{U} x/\Phi_x - \overline{U}/\beta. \tag{24.16}$$

Foster [665] experimented with a mist, produced by the thermal decomposition of wood, which was charged by a corona discharge and drawn through a tube made up of a number of insulated sections. The mass of deposit in each section and the current passing through it were measured, and Φ_x and I_x in the function of x were calculated. A linear relationship between x and x/Φ_x was obtained, in agreement with (24.16), from which c_0 and β were estimated. The density of the droplets was also measured.

As c_0, β and I_x are known functions of the three unknown quantities n_0, m and q, the latter could be calculated. The values of r thus obtained varied between 0·075 and 0·144 μ; q ranged from 6 ε to 32 ε (ε is the elementary charge), depending on the intensity of charging, and the value of q had little effect upon r. The satisfactory agreement between experimental data and theory indicates that the mist was relatively isodisperse and that the charges on the particles were similar.

In the paper of Drozin and La Mer [666] nearly isodisperse aerosols of stearic acid having $\bar{r} = 0{\cdot}3\text{–}1{\cdot}0\,\mu$ were charged unipolarly, by a corona discharge, and the mean charge per particle was determined as a function of radius. The electric mobility of the particles increased with $\bar{r}$ in agreement with theory.

The charged aerosol was drawn into a wide flat horizontal condenser and deposited by the field upon the lower plate of the condenser. The charge imparted to the plate in unit time was measured continuously until the aerosol was completely deposited and the particle size distribution was calculated from the curve obtained. The values of $\bar{r}$ thus determined agreed with those measured by means of high order Tyndall spectra.

According to the authors, the mobility of particles of the sizes indicated could be calculated from the simple formula of Stokes without correction. To justify this surprising point of view, which is in contradiction both with theory and much experimental evidence, the authors quote experimental results obtained at their laboratory, which are not available to the author of this book.

In their experiments the authors made no allowance for electrostatic dispersion. From Fig. 4 of their paper it is possible to estimate that the initial volume charge of the aerosol in one of the experiments was 2×10^{-3} electrostatic units per cm^3. Hence it follows that at the beginning of this experiment the strength of the electric field at the upper and lower boundaries of the aerosol, caused by the volume charge, was equal to $\sim 0{\cdot}03$ electrostatic units, which is of the same order of magnitude as the external field (0·055 electrostatic units). The charge given up to the condenser plate in unit time, at the beginning of the experiment, was thus much increased by the volume charge, but the time required for complete deposition of the aerosol must have been longer, as was actually observed by the authors. It is possible that the compensation of this error necessitated neglect of the correction to Stokes' law. Another point is that it is hardly likely that convection could have been completely eliminated in the relatively large condenser used in the experiments.

Apart from medicine, unipolarly charged aerosols are of use in the treatment of rooms, barns, greenhouses, gardens etc. with insecticides since, in contrast with uncharged aerosols, they are deposited in considerable quantities on the walls and ceilings [667] and on the under surfaces of leaves [668], thus yielding much larger deposits when blown through the foliage of trees. According to Göhlich [668] unipolarly charged dusts give more uniform and less aggregated deposits, apparently due to repulsive forces between the particles. In addition to this, the deposits stick to the surface of the leaves better (see also page 348).

Electrostatic dispersion is spectacularly manifested in free aerosol jets which widen considerably [668] if the charge is unipolar.

CHAPTER IV

CURVILINEAR MOTION OF AEROSOL PARTICLES

§ 25. GENERAL THEORY OF CURVILINEAR MOTION OF PARTICLES. SETTLING OF AEROSOLS IN A VARIABLE HORIZONTAL ELECTRIC FIELD

The theory of the curvilinear motion of aerosol particles is comparatively simple at low values of Re where the resistance of the medium is proportional to the velocity of the particle. Assuming that the non-inertial character of the resistance of the medium, valid as shown above for rectilinear motion of the particles, is preserved for a curved trajectory, the equation of motion in vector form is

$$m\frac{d\mathbf{V}}{dt} = -6\pi\eta r(\mathbf{V} - \mathbf{U}) + \mathbf{F}, \tag{25.1}$$

where **V** and **U** are the velocity vectors of the particle and the medium, and **F** is the external force vector.

In Cartesian form (25.1) becomes

$$m\frac{dV_x}{dt} = -6\pi\eta r(V_x - U_x) + F_x; \quad m\frac{dV_y}{dt} = -6\pi\eta r(V_y - U_y) + F_y, \tag{25.2}$$

where V_x is the velocity resolute of the particle along the x-axis etc.

As equations (25.2) show, the component of particle motion along any axis obeys the same equation as in rectilinear motion, since the motions resolved along different axes are independent of one another. This greatly simplifies the analysis of the curvilinear motion of particles.

The situation is different at large Re. In this case, assuming that

$$\mathrm{Re} = \frac{2\gamma_g r\,|(\mathbf{V} - \mathbf{U})|}{\eta}, \tag{25.3}$$

the vector equation of particle motion is

$$m\frac{d\mathbf{V}}{dt} = -\psi\left(\frac{2\gamma_g r\,|(\mathbf{V} - \mathbf{U})|}{\eta}\right)\gamma_g\frac{\pi r^2}{2}(\mathbf{V} - \mathbf{U})\,|(\mathbf{V} - \mathbf{U})| + \mathbf{F}, \tag{25.4}$$

where $|\mathbf{V} - \mathbf{U}|$ is the length of the vector $\mathbf{V} - \mathbf{U}$.

Resolving equation (25.4) along the co-ordinate axes we obtain

$$m\frac{dV_x}{dt} = -\psi\left(\frac{2\gamma_g r\,|(\mathbf{V} - \mathbf{U})|}{\eta}\right)\gamma_g\frac{\pi r^2}{2}(V_x - U_x)\,|(\mathbf{V} - \mathbf{U})| + F_x \text{ etc.} \tag{25.5}$$

The first term on the right hand side of this equation involves the product of the resolute of relative velocity along the x axis and the absolute value of the relative velocity. The motions resolved along the different axes are therefore not independent. The system of equations (25.5) cannot be solved in general, so that analytical investigation of the curvilinear motion of particles at larger Re is possibly only in special cases. Curvilinear motion will therefore be discussed only at low Re.

Problems of the curvilinear motion of aerosol particles fall into two groups, one having the medium stationary and the other a moving medium. Of the first group only the fall of particles vibrating horizontally under the action of an external force such as an electric field will be studied. This is one of the few cases in which curvilinear motion of aerosol particles has been studied experimentally. Using the derivation above, the equations for steady motion of this kind give

$$V_z = mgB = \tau g, \tag{25.6}$$

$$m\frac{dV_x}{dt} = F - 6\pi\eta r V_x \tag{25.7}$$

where x is the horizontal axis and the z-axis points vertically downwards.

If the force F varies sinusoidally with time, $F = F_0 \sin \omega t$, and the horizontal velocity of the particle is given by formulae (19.6) and (19.11), the horizontal displacement being

$$x = -\frac{F_0 B \cos(\omega t - \varphi)}{\omega\sqrt{1 + \omega^2\tau^2}} \tag{25.8}$$

where

$$\tan\varphi = \tau\omega = V_z\omega/g. \tag{25.9}$$

Since the vertical displacement of the particle in the time t is

$$z = \tau g t, \tag{25.10}$$

the trajectory of the particle is sinusoidal

$$x = -\frac{F_0 B \cos\left(\dfrac{\omega z}{\tau g} - \varphi\right)}{\omega\sqrt{1 + \omega^2\tau^2}}. \tag{25.11}$$

Experimental verification of these results has been performed by Abbott [170] who photographed water droplets of radius 30–40 μ falling in a sinusoidal horizontal electric field of frequency 60 c/s. The particle trajectories proved to be accurately sinusoidal. By means of a special device instants of time were marked on the photographs corresponding to values of $F = 0$, and thus the phase shift φ of the oscillations was determined. The values found for φ were in satisfactory agreement with formula (25.9). Small discrepancies (1–2°) can evidently be explained by deviations from Stokes' law for the horizontal and vertical motions.

This method can, of course, be used for the simultaneous determination of the size and charge of aerosol particles. It has proved more convenient to replace the sinusoidal field by one which is constant in magnitude but changes in direction. Zigzag lines composed of straight sections are then obtained on the photographs. From the distance between the turning points the vertical displacement of a particle in one oscillation period is determined, and hence, from (25.10), τ and also the particle radius. The horizontal velocity of the particle is $V_x = BEq$ (E is the field strength and q the charge on the particle). The tangent of the angle of inclination of the sections of the trajectory to the horizontal is equal to $V_z/V_x = gm/Eq$ and hence, with a knowledge of the particle radius the charge can be determined. This "oscillation" method, developed by Fuchs and Petryanov [171], has been very useful in investigating the distribution of charges on mist droplets with radii $> 0{\cdot}5\,\mu$. When studying smokes it is necessary to remember that the apparent density of the particles is appreciably less than the true density. The oscillation method was first proposed by Wells and Gerke [172], who did not allow the particles to settle. While slowly drawing the aerosol through a chamber, they measured only the amplitude of the oscillations and calculated the radius assuming that the particles had one elementary charge each. However, the particles to which the oscillation method is applicable ($r > 0{\cdot}5\,\mu$) carry on an average considerably more than one elementary charge (see page 114); thus the technique of Wells and Gerke can find application only in exceptional cases, for example with aerosols obtained by seeding with a microscopic nuclei [173].

Tauzin [886] used a vertical alternating field and horizontal illumination. Due to photophoresis of the particles (largely excluded in the work of Fuchs and Petryanov by use of bilateral illumination) zigzag trajectories more or less inclined to the vertical were obtained is this case.

A variant of the oscillation method, developed by Rosenblum [174], has proved convenient for investigating charges on coarsely dispersed aerosols. The particles fall in a horizontal sinusoidal field, produced by alternating voltage. The particles then appear as luminous horizontal dashes, the length of which is twice the amplitude of the oscillations. By means of an eyepiece grid and a timer the rate of settling and the length of the dashes are determined simultaneously. The calculations are performed as shown § 19.

The oscillation method in its original form is not applicable to size determinations of particles with $r < 0{\cdot}5\,\mu$, because their sedimentation velocity is small and the Brownian motion vigorous. By using a strong electric field, however, it is possible to determine the amplitude of oscillation of particles quite accurately for r down to $0{\cdot}15\,\mu$; by recharging them their charge, and hence their size, can be measured (Gladkova and Natanson [671]). This method is very helpful when working with iso-disperse aerosols, since in this case recharging is unnecessary.

Robinson [669] considered the motion of a system of uniform particles, allowing for their inertia, as the flow of a hypothetical compressible fluid of a density equal to the concentration of the particles. He confined himself to potential flow of the medium and proportionality of the drag to relative velocity, and proved that if the flow of the aerosol fluid was at some time potential, for instance at a very large distance from an obstacle where the velocities of the particles and of the medium coincided, it would always remain so. It follows from this, incidentally, that the trajectories of particles cannot intersect. The theorem holds good only when any external force acting upon the particles also has a potential.

By applying the operator div to the fundamental equation of motion of the aerosol fluid

$$\frac{DV}{Dt} = \frac{1}{\tau}(V - U) + F. \tag{25.12}$$

DV being the total differential of fluid velocity, Robinson proved that div $V \leqslant 0$, so that the concentration of the flowing aerosol can only increase. The correctness of this conclusion becomes apparent on consideration of a curved tube of flow of constant cross-section. In the region of a bend the radius of curvature is smaller on the inside, and the velocity, and hence the centrifugal force, due to the properties of the potential flow (see page 127), are greater than on the outside. The tube of flow of the aerosol fluid will therefore be shifted relative to the tube of flow of the medium and at the same time will become narrower, so that the concentration in it increases. It should be noted that rotation of an aerosol as a whole, with a non-potential, vortex flow, causes the concentration of the aerosol to decrease.

Levin [670] proved that when the field of external forces is solenoidal (div $\boldsymbol{F} = 0$), as in the case of gravity, and in the absence of inertial forces, the aerosol concentration along the trajectory of the particles is constant. At $\tau \neq 0$ and constant $\boldsymbol{F}$ the concentration of the flowing aerosol may remain constant only if $\boldsymbol{U} =$ const., that is in a uniform field of flow. Otherwise, the concentration must increase as seen above. Since in the absence of inertia, and at $\boldsymbol{F} =$ const., the concentration does not change, assuming the medium to be incompressible, it follows that the increase in the concentration must become greater with increasing τ.† The particle size distribution in a flowing polydisperse aerosol therefore changes and a segregation of the different sizes must occur.

§ 26. PRECIPITATION OF AEROSOLS FROM LAMINAR FLOW UNDER THE ACTION OF GRAVITY

The motion of particles in a gas flow is the most important branch of the mechanics of aerosols. This chapter is devoted to motion in laminar flow. The much more complicated problem of turbulent flow is discussed in Chapter VI.

We will first examine the settling of particles of an aerosol in laminar flow along a horizontal tube of such small cross-section that vertical convection currents in it are either absent or at least have a velocity which is small compared with the settling rate of the particles. The calculations set out below are applicable to aerosols settling in laboratory instruments, sampling probes etc., but not to industrial settling chambers. Flow through a flat tube with a very large width-to-height ratio and through a circular tube will be dealt with.

The vertical motion of particles with respect to the medium is expressed by the formula [see (17.4)]

$$V = V_s\,(1 - e^{-t/\tau}). \tag{26.1}$$

The term $e^{-t/\tau}$ can be neglected if the average residence time of the particle in the tube is large compared with the relaxation time τ. In this case the vertical velocity of the particles can be considered constant and equal to V_s, while the horizontal velocity is identical with that of the medium at the same point. It will be seen from the values of τ and V_s, given in Table 13 (see page 184) for particles of various sizes, that these conditions are fulfilled in normal work with aerosols.

At the entrance to the tube the flow velocity is the same over the whole section but the flow near the walls is retarded until, at a distance of the order $0{\cdot}1\ R\,\mathrm{Re}_f$ from

† At very large τ particles in a tube of flow considered above will move rectilinearly and uniformly, and their concentration will not change.

the entrance [175] (where R is the radius or half-height of the tube and Re_f is the Reynolds number of flow in the tube), a constant velocity distribution is established which, in the case of a flat tube, is given by

$$U_x = \left(\frac{3z}{h} - \frac{3z^2}{2h^2}\right)\overline{U}, \tag{26.2}$$

where h is half the height of the tube, $\overline{U}$ is the mean velocity and z is the distance from the bottom of the tube.

The streamlines near the mouth of the tube are therefore inclined to the horizontal.

The velocity resolutes of laminar flow through a flat tube can be expressed in terms of a stream function ψ:

$$U_x = \partial\psi/\partial z, \quad U_z = -\partial\psi/\partial x. \tag{26.3}$$

If the inertia of the particles is neglected the velocity resolutes of the particles are equal to

$$\frac{dx}{dt} = U_x = \frac{\partial\psi}{\partial z}; \quad \frac{dz}{dt} = U_z - V_s = -\frac{\partial\psi}{\partial x} - V_s. \tag{26.4}$$

The differential equation of the particle trajectory is obtained by eliminating dt to give

$$-\frac{dx}{\partial\psi/\partial z} = \frac{dz}{(\partial\psi/\partial x) + V_s} \tag{26.5}$$

from which

$$-V_s dx = \frac{\partial\psi}{\partial x}dx + \frac{\partial\psi}{\partial z}dz = d\psi. \tag{26.6}$$

Integration along a trajectory over the entire length L of the tube gives

$$V_s L = \psi_0 - \psi_L, \tag{26.7}$$

where ψ_0 and ψ_L are the values of the stream function at the points occupied by the particle at the beginning and the end of the tube. The stream function is equal to the volume of gas flowing in unit time between the bottom of the tube and a given streamline per unit width of the tube. By putting $\psi_L = 0$, (26.7) gives the boundary trajectory separating the settling and non-settling particles in the tube. For the boundary trajectory at the entrance to the tube $\psi = \psi_0 = V_s L$; hence, if the total gas flow in the tube is equal to Ψ, the fraction of the aerosol settling in the tube, or the precipitation efficiency $\ni$ is equal to

$$\ni = \psi_0/\Psi = V_s L/\Psi. \tag{26.8}$$

Putting $\Psi = 2h\overline{U}$

$$\ni = \frac{V_s L}{2h\overline{U}}. \tag{26.9}$$

The important conclusion that the precipitation efficiency does not depend on the velocity distribution in the tube is thus reached. The length of the tube required for complete precipitation is

$$L_{cr} = \frac{2h\overline{U}}{V_s}. \tag{26.10}$$

For a circular tube with viscous flow the calculations are considerably more complicated. The velocity distribution is given by the formula

$$U = 2\bar{U}\left(1 - \frac{\varrho^2}{R^2}\right) \tag{26.11}$$

where ϱ is the distance from the axis of the tube. Assuming for simplicity that this distribution is established at the entrance to the tube, the length of tube, L, which is necessary for complete precipitation of an aerosol will be calculated. This requires the trajectory of a particle entering the tube at the uppermost point of its cross-section and moving in a vertical plane through the axis of the tube. The velocity distribution in this plane is given by $U = 2\bar{U}\left(1 - \frac{z^2}{R^2}\right)$, where z is the vertical distance from the axis.

The differential equation of the particle trajectory is

$$\frac{dx}{2\bar{U}\left(1 - \frac{z^2}{R^2}\right)} = -\frac{dz}{V_s}. \tag{26.12}$$

Integrating with respect to z from $-R$ to R, we find

$$L_{cr} = \frac{8R\bar{U}}{3V_s}. \tag{26.13}$$

This formula has been proposed by Chistov [176] for calculating the precipitation of aerosols in industrial equipment, but, as has already been pointed out, discrepancies may occur as a result of convection.

For the precipitation efficiency in a circular tube with viscous flow G. Natanson has derived the formula

$$\ni = \frac{2}{\pi}\left(2\mu\sqrt{1-\mu^{2/3}} + \arcsin\mu^{1/3} - \mu^{1/3}\sqrt{1-\mu^{2/3}}\right), \tag{26.14}$$

where

$$\mu = \frac{3V_s L}{8R\bar{U}}.$$

For $\mu = 1$, i.e. $L = 8R\bar{U}/3V_s$, formula (26.14) gives $\ni = 1$ in agreement with formula (26.13).

In calculating the sedimentation of particles in laminar flow, it must be remembered that the rectilinearity of laminar flow in straight tubes is disturbed long before the critical value of Re_f is attained. According to the observations of Prengle and Rotfus [672] in smooth round tubes, a coloured jet on the axis of the tube begins to deviate at $Re_f = 1220$. As Re_f increases, this break in rectilinearity extends towards the walls and reaches them at $Re_f = 2000$.

§ 27. PRECIPITATION OF AEROSOLS FROM LAMINAR FLOW BY AN ELECTRIC FIELD

The precipitation of charged particles from laminar gas flow through a parallel-plate condenser will be discussed, considering only fine particles for which inertia can be neglected, the motions, as in the preceding section, being quasi-stationary (see page 74).

Motion of particles in a direction perpendicular to the plane of the condenser is given by the equation

$$V_z = \frac{\Pi q B}{h}, \tag{27.1}$$

where Π is the voltage across the condenser plates, h is the distance between them, q is the charge on the particle and B is the mobility of the particles.

In the study of gaseous ions the velocity of an ion in a field of one volt per cm is called the mobility of the ion. This terminology is also carried over to heavy ions which are small, charged aerosol particles. The electrical mobility, u, is thus connected with the mechanical mobility B by the formula

$$u = qB/300 \tag{27.2}$$

(the number 300 is equal to the ratio of the e.s.u. of voltage to the volt). As pointed out in § 2, the particles in very fine aerosols are often studied by measuring electrically their mobility u, which is also the principal characteristic of heavy ions. The relationship between the radius of the particles and their mobility when they carry one electronic charge, $\varepsilon = 4{\cdot}802 \times 10^{-10}$ e.s.u., is presented in Table 3 and Fig. 10.

Brief reference must here be made to the magnitude of the charge on aerosol particles. The charge on a particle may originate in various ways. In crushing powders the particles receive triboelectric charges; the atomization of liquids produces droplets which are charged owing to fluctuations in the concentration of ions in the liquid; aerosols formed at high temperature are charged by thermionic emission; finally, a very important and common source of charges is the precipitation of gaseous ions and electrons on aerosol particles.

At the moment when aerosols are formed, their particles may be highly charged but, whatever the initial distribution of charges, a stationary state is gradually approached on account of precipitation on the particles of ions which are constantly being formed in the gas. Theoretical and experimental research in the author's laboratory has shown [177, 178], that the stationary distribution of charges produced by symmetrical bipolar ionization of the gas is given to a first approximation by the Boltzmann formula. The fraction of particles with ν elementary charges (ν being a positive or negative integer) is equal to:

$$f(\nu) = \frac{1}{\Sigma} \exp\left[\frac{-(\nu\varepsilon)^2}{2rkT}\right], \tag{27.3}$$

where

$$\Sigma = \sum_{-\infty}^{+\infty} \exp\left[\frac{-(\nu\varepsilon)^2}{2rkT}\right]. \tag{27.3a}$$

Thus the final stationary distribution of charge does not depend on the amount of ionization in the gas, although the time required to establish the steady state decreases as the degree of ionization rise.

The radii of most of the amicroscopic particles in atmospheric air, which are called condensation nuclei or heavy ions, lie between 1 and 5×10^{-6} cm. The charge distribution in this atmospheric aerosol is probably close to a stationary one, hence formula (27.3) has been employed to calculate the charge distribution shown in Table 11 in which n_0 denotes the percentage of uncharged particles, n_1 the particles with one elementary charge, and so on.

TABLE 11. DISTRIBUTION OF CHARGES ON PARTICLES IN HIGHLY DISPERSED ATMOSPHERIC AEROSOLS

$r \times 10^6$ cm	n_0	n_1	n_2	n_3
1	90	10	—	—
3	55	43	2	—
5	43	48	8·6	0·4

In aerosols with a stationary charge distribution the percentage of particles with several elementary charges is therefore small. The size of such particles can be determined approximately from their mobilities by means of Table 3. The classification of atmospheric particles according to mobility [179], as used in the physics of the atmosphere, is given in Table 12 where the particle size is also indicated.

TABLE 12. CLASSIFICATION OF ATMOSPHERIC AEROSOL PARTICLES ACCORDING TO THEIR MOBILITY u

Particles	u	$r \times 10^6$ cm
Ultraheavy ions	$<2{\cdot}5\times10^{-4}$	>5
Heavy (Langevin) ions	$2{\cdot}5 - 10\times10^{-4}$	2·5—5
Large "medium" ions	$10^{-3} - 10^{-2}$	0·7—2·5
Small "medium" ions	$>10^{-2}$	$<0{\cdot}7$

If the potential difference is expressed in volts, formula (27.1) becomes

$$V_z = \frac{dz}{dt} = \Pi u/h. \tag{27.4}$$

Motion of the particle parallel to the condenser plates is expressed by the equation

$$\frac{dx}{dt} = U(z), \tag{27.5}$$

where U is the gas velocity. From these equations it follows that a particle entering the condenser close to the plate bearing a charge of the same sign as its own charge will reach the other plate after travelling a distance x_0 (Fig. 24), between the plates where

$$x_0 = \frac{h}{\Pi u} \int_0^h U(z)\, dz = \frac{\overline{U} h^2}{\Pi u}. \tag{27.6}$$

Like (26.10), this formula is valid whatever the distribution of gas velocity between the plates may be.

The distance travelled by the particle through the condenser is therefore inversely proportional to its mobility u. This was the basis of a method proposed by Chapman [180] for determining the mobilities of fine particles, which deserves attention although it has not hitherto been much used. The aerosol is injected through a slot-shaped nozzle B into a rectangular tube (Fig. 25) through which clean air is blown. The aerosol stream is given a direction parallel to the walls of the tube and a velocity which is small in comparison with the air velocity. The flow in the tube must, of course, be laminar.

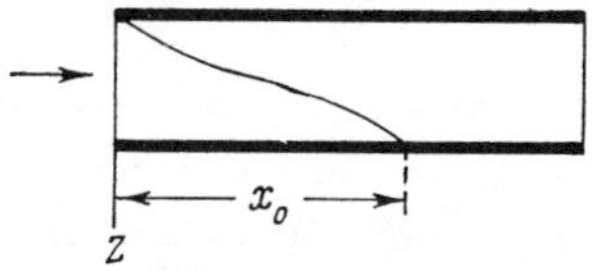

FIG. 24. Deposition of aerosol particles in a condenser.

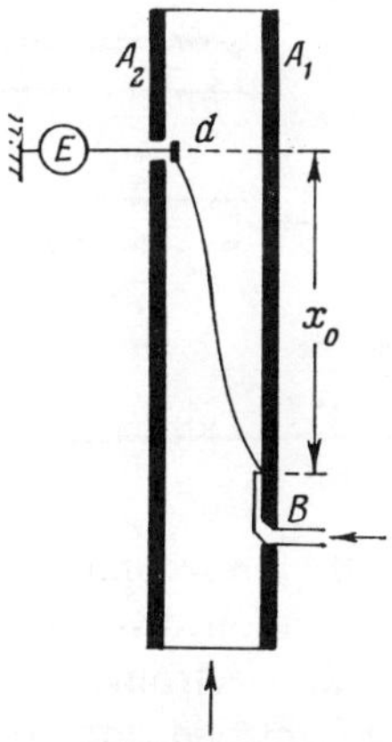

FIG. 25. Differential method of determining mobilities.

The walls A_1 and A_2 form the plates of a condenser, and particles having the same charge as A_1 move as described above and are precipitated on the wall A_2, after passing a distance x_0 along the condenser (see 27.6). It might be better to inject the aerosol stream midway between the walls with a velocity equal to the axial velocity of the air ($1{\cdot}5\,\overline{U}$), in which case x_0 will be reduced by a half.

The value of x_0 can be found either by microscopic examination of the deposit or by measuring the quantity of electricity given up as the particles deposit. In the first case a copper grid covered with a thin layer of collodion, on which the particles settle, is placed on the wall A_2. At the end of the experiment the deposit is examined under the electron microscope.

The distribution of mobilities, like the size distribution, is a continuous curve which can, however, have several maxima. The number of particles with mobilities lying within the limits u and $u + du$ is

$$dN = f(u)\, du. \tag{27.7}$$

The longitudinal distribution of particles in the deposit on A_2, obtained as above, reflects the distribution of mobility, the number of particles in the strip bounded by the co-ordinates x and $x + dx$ being given by

$$dN = \varphi(x)\, dx. \tag{27.8}$$

It follows from (27.6) that du is proportional to dx/x^2, so that $f(u)$ is proportional to $x^2\varphi(x)$. Thus if $\varphi(x)$ is determined experimentally the distribution of mobilities $f(u)$ of the particles can be found by equations (27.7) and (27.8). If the particle sizes are measured on the photomicrographs obtained and the mobilities u, which the particles would have for one elementary charge, are calculated by means of Table 3, the actual number of charges on the particles can be determined. Such a study, which has not yet been undertaken, would show how many elementary charges there are on very small particles.

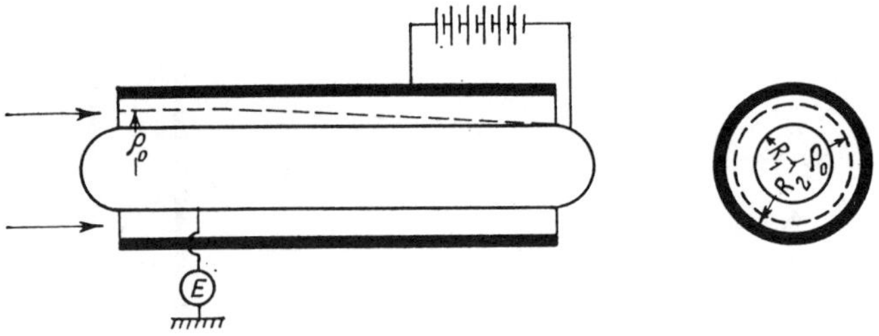

FIG. 26. Usual method of determining mobilities.

In working by the second method a narrow electrode, d (Fig. 25), insulated from the wall A_2 but positioned very close to it, is inserted into the condenser. The electrode is connected to an electrometer and can be moved along the condenser. The charge received by the electrode is the product of the number of particles settling on it and their mean charge and by (27.8) is equal to $\bar{q}\varphi(x)\,\Delta x$, where Δx is the width of the electrode. This method gives not the mobility distribution $f(u)$ but the function $\bar{q}(u) f(u)$, where $\bar{q}(u)$ is the mean charge of particles having mobility u. Only when the value of $\bar{q}(u)$ can be calculated theoretically, or when it can be assumed that the particles have one elementary charge each, is it possible to derive $f(u)$.

This differential method of determining mobilities has the advantage over the more common integral method of giving directly the percentage of particles with mobilities lying in a given range. The current flowing to the electrode is so small, however, that it is rather difficult to measure.

The integral method of determining mobilities involves passing the aerosol at a constant rate through a condenser, one plate of which is connected to an electrometer; the current, I, flowing to this plate is measured as a function of the potential, Π, across the condenser, and the distribution $\bar{q}(u) f(u)$ calculated [181, 182]. The condenser can be of any shape, but cylindrical ones have several advantages and are usually employed.

With laminar flow through such a condenser (Fig. 26) the gas velocity, and hence also the velocity of the particles in a direction parallel to the axis of the condenser, $V_x(\varrho)$, is a function of the distance ϱ from the axis. The field strength in a cylindrical condenser is equal to

$$E = \Pi/\varrho \ln\left(\frac{R_2}{R_1}\right), \tag{27.9}$$

where R_2 is the radius of the outer electrode and R_1 the radius of the inner.

The radial velocity of the particles is therefore

$$V_\rho = \Pi u/\varrho \ln\left(\frac{R_2}{R_1}\right). \tag{27.10}$$

Suppose the outside electrode is positively charged. A positively charged particle travels towards the inner electrode, which is connected to an electrometer, in time dt a distance

$$d\varrho = -\Pi u dt/\varrho \ln\left(\frac{R_2}{R_1}\right). \tag{27.10b}$$

In the same time it moves along the axis of the condenser through

$$dx = U(\varrho)\, dt. \tag{27.10c}$$

Elimination of dt from these equations leads to

$$dx = -\,U(\varrho)\varrho \ln\left(\frac{R_2}{R_1}\right) d\varrho/\Pi u. \tag{27.11}$$

If the particle enters the condenser at a distance ϱ_1 from the axis, it reaches the inner electrode after travelling along it to a point

$$x = -\frac{\ln(R_2/R_1)}{\Pi u}\int_{\varrho_1}^{R_1} U(\varrho)\varrho\, d\varrho = \frac{\ln(R_2/R_1)}{\Pi u}\int_{R_1}^{\varrho_1} U(\varrho)\varrho\, d\varrho. \tag{27.12}$$

If all the particles have the same mobility u and the length of the condenser is L, the inner electrode will be reached when $x \leqslant L$. Such particles have $\varrho_1 \leqslant \varrho_0$, where ϱ_0 is the limit of integration in

$$\frac{\ln(R_2/R_1)}{\Pi u}\int_{R_1}^{\varrho_0} U(\varrho)\varrho\, d\varrho = L. \tag{27.13}$$

Thus, for a given potential Π on the condenser, particles must enter it at a distance from the axis less than ϱ_0 in order to be precipitated. If 1 cm^3 of the aerosol contains n positively charged particles of charge q, the number of particles that will pass in unit time through an annular section bounded by circles of radius R_1 and ϱ_0 (Fig. 26) and settle on the inner electrode is

$$N = \int_{R_1}^{\varrho_0} n\, U(\varrho)\, 2\pi\varrho\, d\varrho. \tag{27.14}$$

Hence the current flowing across the inner electrode is equal to

$$I = qN = 2\pi qn \int_{R_1}^{\varrho_0} U(\varrho)\, d\varrho \tag{27.15}$$

or, using equation (27.13),

$$I = 2\pi qnuL\Pi/\ln\left(\frac{R_2}{R_1}\right). \qquad (27.16)$$

The current is proportional to the voltage across the condenser plates, and when the voltage is equal to

$$\Pi_s = \frac{\ln (R_2/R_1)}{uL}\int_{R_1}^{R_2} U(\varrho)\varrho\, d\varrho, \qquad (27.17)$$

ϱ_0 equals R_2 and all positive particles entering the condenser are trapped. The current I_s remains constant at saturation value if the voltage is further increased (Fig. 27a).

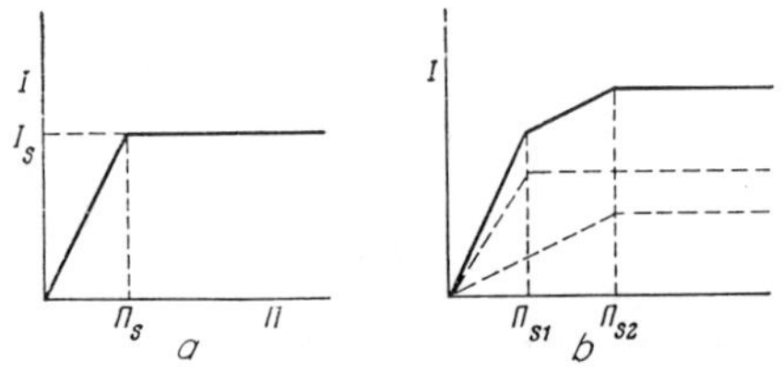

FIG. 27. Characteristics of aerosols with one and two values of mobility.

Since

$$\int_{R_1}^{R_2} 2\pi\varrho U(\varrho)\, d\varrho = \Phi \qquad (27.18)$$

where Φ is the volume rate of flow of the aerosol through the condenser, it follows from equation (27.17) that

$$u = \frac{\ln (R_2/R_1)\,\Phi}{2\pi L\Pi_s}. \qquad (27.19)$$

The particle mobility can thus be determined from the discontinuity in the slope of the current-voltage characteristic. As in previous examples, all the formulae hold for any distribution of flow velocities in the condenser as long as the streamlines are straight and parallel. The saturation current is

$$I_s = qn\Phi \qquad (27.20)$$

and from it can be determined the sum of the charges of positive particles contained in 1 cm³ of the aerosol. The same arguments apply also to negatively charged particles, when the condenser electrode connected to the electrometer is positively charged.

If the aerosol contains two groups of particles having mobilities u_1 and u_2, charges q_1 and q_2 and concentrations n_1 and n_2, equations (27.16) and (27.17) are applicable to each group. The broken lines in Fig. 27b are the characteristics for each group of particles, and the continuous line is the overall characteristic determined by experiment. The mobilities of the particles of each group are found from the discontinuities of the experimental characteristic.

The junctions between the branches of the characteristic are not sharply defined in practice; if only two groups of particles are present this does not matter, but for a large number of groups a smooth curve may be obtained instead of series of straight lines.

Conversely, a smooth curve will be produced by a continuous spectrum of particle mobilities, but random errors of measurement can easily provide an erroneous indication of the presence of discontinuities, thus suggesting the existence of discrete groups of mobilities. It is probable that such preferred mobilities, which have often been claimed, do not really exist but are merely the result of experimental errors.

When the characteristic is smooth the mobility distribution can be found in the following way. It will first be assumed that each particle has one elementary charge. At a potential Π_L let all the particles with mobility exceeding u_L be precipitated in the condenser. The relationship between Π_L and u_L is given by the equation

$$u_L = \frac{\ln(R_2/R_1)\,\Phi}{2\pi L \Pi_L}. \tag{27.21}$$

The current carried by these particles according to (27.20), is

$$I_1 = \Phi \varepsilon n \int_{u_L}^{\infty} f(u)\,du. \tag{27.22}$$

By (27.16), particles with mobility less than u_L contribute a current

$$I_2 = \frac{2\pi\varepsilon L \Pi_L n}{\ln(R_2/R_1)} \int_0^{u_L} u f(u)\,du. \tag{27.23}$$

Thus the total current is equal to

$$I = I_1 + I_2 = \Phi\varepsilon n \int_{u_L}^{\infty} f(u)\,du + \frac{2\pi\varepsilon L\Pi_L n}{\ln(R_2/R_1)} \int_0^{u_L} u f(u)\,du. \tag{27.24}$$

Differentiating (27.24) with respect to Π_L:

$$\frac{dI}{d\Pi_L} = -\Phi\varepsilon n f(u_L)\frac{du_L}{d\Pi_L} + \frac{2\pi\varepsilon L\Pi_L n u_L f(u_L)}{\ln(R_2/R_1)}\frac{du_L}{d\Pi_L} + \frac{2\pi\varepsilon L n}{\ln(R_2/R_1)}\int_0^{u_L} u f(u)\,du. \tag{27.25}$$

According to (27.21) the first two terms on the right hand side of this equation cancel, hence

$$\frac{dI}{d\Pi_L} = \frac{2\pi\varepsilon L n}{\ln(R_2/R_1)}\int_0^{u_L} u f(u)\,du \tag{27.26}$$

and it follows that

$$I - \Pi_L \frac{dI}{d\Pi_L} = h = \Phi\varepsilon n \int_{u_L}^{\infty} f(u)\,du, \tag{27.27}$$

where h is the length of the intercept on the ordinate axis made by the tangent to the characteristic at the point u_L (Fig. 28). This intercept is therefore proportional to the number of particles having mobilities between u_L and ∞, being determined from the appropriate value of Π_L by (27.21). The spectrum of particle mobilities or, if the charges on the particles are not all the same, the function $\bar{q}(u) f(u)$ is thus determined. This method has the disadvantage that small errors in I lead to considerable errors in the intercepts on the ordinate axis and hence in the value of the distribution function.

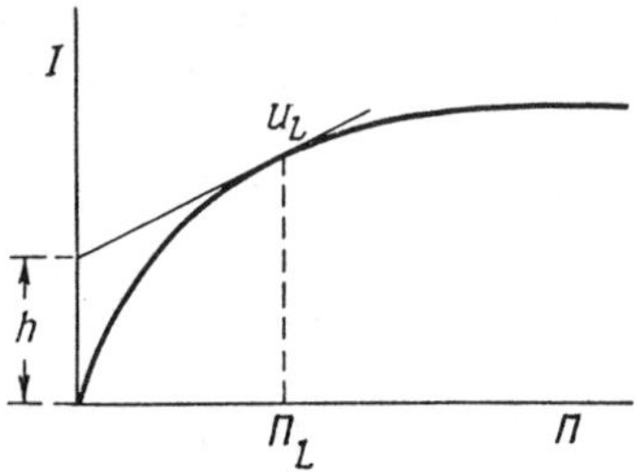

FIG. 28. Characteristic of aerosol with continuous distribution of mobilities.

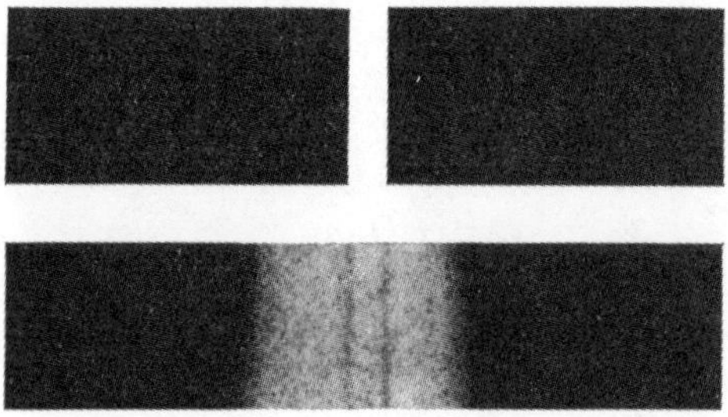

FIG. 28a. Widening of a jet of a charged aerosol in an electric field ($E = 5000$ V/cm). Above an uncharged aerosol.

In studying the mobility spectrum of particles in the atmosphere the method described above can lead to incorrect results due to change in the concentration of the atmospheric aerosol. This can be avoided by using two condensers [181]. Air is first drawn through a condenser at potential Π, the current in which (I_1, not measured) is expressed by formula (27.24), and then through a second condenser at such a potential that practically all the particles are deposited in it. The current in the second condenser I_2, measured by an electrometer, is then equal to $\Phi \varepsilon n - I_1$. If the potential of the first condenser is zero, a current $\Phi \varepsilon n = I_0$, proportional to the total number of charged particles in the air flows through the second. Measurements are made in the following way. After determining the current I_2, corresponding to a given potential Π, I_0 is measured and the ratio $(I_0 - I_2)/I_0 = I_1/I_0$ is calculated; Π is then increased a little, the corresponding current I_2 is determined, I_0 is again measured, and so on. In this way I_1 is standardized to the total number of charged particles in the air. Although the assumption underlying this method, that the mobility distribution of particles in the atmosphere does not depend on their concentration, can be true only in special cases, the two-condenser method nevertheless gives more reliable results than the single-condenser method. In either method it is desirable to eliminate gaseous ions from the air, which is done by passing the air through an auxiliary condenser charged to a small potential sufficient to remove all the gaseous ions with their high mobilities of order 1. The number of aerosol particles retained by this condenser is negligible.

It must be emphasized that everything above is only true for laminar flow through the condenser. In turbulent flow the nature of the phenomena is changed considerably and the current flowing across the condenser at a given potential becomes noticeably less than is indicated by formula (27.24) [181] (see page 265).

Until recently electrometric determination of the mobility of particles has been the principal method of investigating highly dispersed aerosols and, in particular, the highly dispersed part of atmospheric aerosols, because the particles are too small for ultra-microscopic examination. The accuracy of the electrometric method falls as particle size increases, and the equations show that it is necessary either to increase the field intensity in the condenser or to reduce the rate of flow of the aerosol. However, very intense fields result in leakages, local discharges and other complications, and at low flow rates serious difficulties are created by convection currents in the condenser. A value of r of approximately 10^{-5} cm can be considered the upper limit to the applicability of the electrometric method; this is a little above the lower limit of ultramicroscopy.

The electrometric method can be used for uncharged aerosols if they are first charged with the aid of a radioactive compound [183] or in some other way. The following observation is of practical interest [184]. While sampling atmospheric air through a rubber tube, a considerable loss of charged condensation nuclei was noticed, while uncharged nuclei were held back only in very small numbers. When the tube was bent the effect increased, but when it was smeared with conducting liquids the loss ceased. The explanation is that bending the tube gave rise to charges of opposite sign on the concave and convex walls, thus creating an electric field across the tube.

Cochet [185] has investigated precipitation of uncharged droplets from air at rest on a charged, infinitely long horizontal cylinder (for example, on an electric transmission line) under the simultaneous action of gravity and electric induction. Neglecting the inertia of the particles and assuming that the air resistance was proportional to the first power of the droplet velocity, Cochet found the collection efficiency of the droplets (see page 139) to be

$$\ni = \left(\frac{9\,\lambda E_0^2}{8\,\gamma g R}\right)^{1/3} \tag{27.28}$$

where $\lambda = (\varepsilon_k - 1)/(\varepsilon_k + 2)$.

ε_k is the dielectric constant of the droplets; R is the radius of the cylinder; and E_0 is the field strength at the surface of the cylinder, related to the charge on unit length of the cylinder by the formula $E_0 = 2\,q/R$.

Formula (27.28) is applicable only for $E_0 > \sqrt{4\pi\gamma g R/3\,\lambda}$; when this is not the case $\ni$ can only be determined by numerical methods. It does not depend on particle size because both gravity and the electric force are proportional to the volume of the particles.

Experimental determination of $\ni$ by photographing the trajectories of droplets falling on to a charged cylinder has shown that for $R = 0{\cdot}6$ cm, $E_0 = 47$ or 88 e.s.u. and $r < 25\,\mu$ the agreement between theory and experiment is very good, so that the inertia of the particles can in fact be neglected. For $r = 40\,\mu$ and $E_0 = 88$ e.s.u. the experimental value of $\ni$ is 14 per cent greater than that calculated from formula (27.28) because the inertia is becoming appreciable.

Electrical precipitation of particles is widely used in industry to free gases from dusts, mists etc. In electrostatic precipitators used for this purpose the negatively charged inner electrode of the condenser has a very small radius, of the order of 1 mm. A corona discharge forms upon it and creates in the precipitator a high concentration of gaseous ions and electrons which charge the particles. Charging of the particles and their precipitation thus proceed simultaneously.† The strong ion wind, accompanying the corona discharge, causes the gas flow in an electrostatic precipitator to have a disordered nature so that precipitation of particles must be calculated as shown in § 44.

Electrostatic gas cleaners with the charging and precipitation processes separate are now made. The first part of the apparatus is designed for charging the particles and the second for precipitation [186]. With such equipment it has proved possible to use a positive corona discharge electrode which generates fewer air oxidation products (nitrogen oxides etc., which are harmful to living organisms) than a negative electrode. This method of dust precipitation is useful for air conditioning.

The method of measuring the electric mobility of particles described on page 115 was used by Gillespie and Langstroth [673]. A laminar stream of air was drawn downwards through a flat vertical condenser and a thin jet of aerosol was introduced isokinetically into the middle of the stream.

Hinkle *et al.* [674] caused such a jet to pass between cylindrical electrodes and photographed it with illumination from one side. The jet of an uncharged aerosol retained its width, but that with charged particles widened considerably (Fig. 28a). From the optical density distribution in the cross-section of the jet it was possible to estimate roughly the distribution of the electric mobilities of the particles.

In experiments by Yoshikawa *et al.* [675] an aerosol of $r = 0{\cdot}5 - 10\,\mu$ was introduced into a wide metal tube through a plug made of thin plastic filaments parallel to the axis of the tube so that the aerosol stream flowed with a constant velocity over the whole cross-section of the tube. A wire on which a corona discharge occurred was fixed along the axis of the tube. A narrow metal cylinder serving as the inner electrode of a cylindrical condenser was placed downstream of this wire, also along the axis of the tube. Since the particles moving near the wire received a higher charge, but travelled a longer path to the external electrode, it was possible, by adjusting the potential and the dimensions of the device, to focus all particles of a given size on the same section of the external electrode. Charges received by particles in a corona discharge are proportional to r raised to a power greater than unity, hence the particles were deposited on the electrode in order of decreasing size.

In order to find the relationship between mean charge and particle mobility in microscopic aerosols, Nolan and O'Connor [676] along with the (I, Π) curves (Fig. 28), also determined by means of a counter of condensation nuclei the (n, Π) curves, where n is the number of the particles per cm^3 deposited in the condenser at the potential Π; the number of particles with mobilities between u_L and ∞ was then calculated and all the particles were divided into fractions according to their mobilities, the total charge and number of particles, and hence their mean charge, being calculated for every fraction. This work will be dealt with in § 39.

An interesting practical application of the motion of aerosol particles in a lateral electric field is the photographic process called xerography [348]. A metal plate is

† Coagulation of particles, due to the field (see § 52) which polarizes them, also takes place.

sensitized with a thin photoconducting layer of vitreous selenium charged to a uniform potential. During exposure the illuminated areas of the layer lose their charge. A charged coloured aerosol, such as carbon black, is then drawn in laminar flow through a parallel slit formed by the Se-covered plate and another metal plate charged with the opposite sign. The particles are deposited only on the charged regions of the selenium layer which have not been exposed to light. The effects of flow rate, size and charge of aerosol particles, width of slit etc., on the quality of the xerographic image involve the mechanics of aerosols.

§ 28. PRECIPITATION OF AEROSOLS IN A CENTRIFUGAL FORCE FIELD. THE AEROSOL CENTRIFUGE

The motion of an aerosol particle will be discussed in a medium rotating about a fixed axis with a velocity $U(\varrho)$, which is a function of the distance ϱ from the axis.

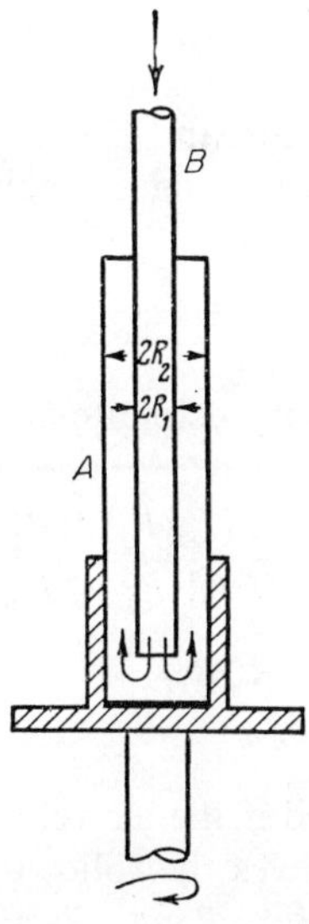

FIG. 29. Aerosol centrifuge.

Considering the motion of the particle to be quasi-stationary with its tangential velocity the same as the velocity of the medium, the radial velocity of the particle is equal to

$$V_\varrho = \frac{U^2(\varrho)\, mB}{\varrho} = \frac{U^2(\varrho)\, \tau}{\varrho}. \tag{28.1}$$

A simple example of this motion is seen in the aerosol (bacterial) centrifuge, for studying the bacterial microflora of the air [187]. A cylindrical glass vessel A (Fig. 29), rotating about its axis, is covered on the inside with a thin layer of culture medium. Air being sampled is admitted through a fixed tube B which reaches almost to the bottom of the vessel. At a certain distance from the bottom the air inside A attains the same angular velocity ω as that with which the vessel rotates so that particles contained in the air are precipitated by centrifugal force on to the walls of the vessel.

After a known volume of air has passed through the instrument the vessel is kept warm until the bacterial colonies have grown large enough for counting. In this device $U(\varrho) = \omega\varrho$ so that the radial velocity of a particle is

$$V_\varrho = \frac{d\varrho}{dt} = \omega^2 \varrho\tau. \tag{28.2}$$

The velocity distribution in the gas flowing along the vessel is given by the formula [188]

$$U(\varrho) = \frac{2\bar{U}\{R_1^2 \ln(R_2/\varrho) - R_2^2 \ln(R_1/\varrho) - \varrho^2 \ln(R_2/R_1)\}}{(R_1^2 + R_2^2)\ln(R_2/R_1) + R_1^2 - R_2^2}, \tag{28.3}$$

where $\bar{U}$ is the mean flow velocity; R_2 is the internal radius of the vessel and R_1 is the external radius of the tube *B*.

Taking the velocity of a particle parallel to the axis of the vessel as equal to the velocity of the air,

$$\frac{dx}{dt} = U(\varrho), \tag{28.4}$$

and eliminating dt from this equation and (28.2), we obtain the differential equation of the particle trajectory in a plane passing through the axis of the vessel and rotating with it

$$dx = \frac{U(\varrho)\,d\varrho}{\omega^2\tau\varrho}. \tag{28.5}$$

The height of vessel required to precipitate all particles of a given size

$$L = \frac{1}{\omega^2\tau}\int_{R_1}^{R_2} \frac{U(\varrho)}{\varrho}\,d\varrho = \frac{2\bar{U}\frac{1}{2}\ln(R_2/R_1)\{(R_1^2 + R_2^2)\ln(R_2/R_1) + R_1^2 - R_2^2\}}{\omega^2\tau\{(R_1^2 + R_2^2)\ln(R_2/R_1) + R_1^2 - R_2^2\}}$$

$$= \frac{\bar{U}\ln(R_2/R_1)}{\omega^2\tau}. \tag{28.6}$$

This expression is also obtained if the air velocity is constant over the space between the cylinders. Shafir [189] gives the following data on the centrifuge used by him: $\omega = 2\pi \cdot 50$, $L = 17$ cm, $R_2 = 2$ cm, $R_1 = R_2/3$, $\bar{U} = 36$ cm sec^{-1}. Taking the density of the bacteria as unity, equations (28.6) and (17.3) show that only particles of radius greater than $1{\cdot}4 \times 10^{-4}$ cm†, which is rather larger than many bacteria, can deposit completely in the centrifuge. The rate of settling of particles in the centrifuge is proportional to the square of their radius, hence it would be expected that the percentage of free bacteria escaping from the centrifuge must be appreciable. In fact Shafir's experiments show that 83–89 per cent of bacteria are retained in the centrifuge. Deposition is aided by the airborne bacteria being embedded in heavier material deriving from the culture medium. The air flow in the example just considered corresponds to an Re_f of approximately 300–400, so that the assumption of laminar flow is justified.

Ordinary centrifuges, rotating at the usual speeds, are only suitable for precipitating particles with sizes of the order 1–2 μ. For precipitating particles down to tenths

† Actually the radius is even larger, since no account has been taken of the lag of the air in acquiring the rotational velocity $\omega\varrho$.

of a micron in size a super-centrifuge with a speed of several hundred revolutions per second is required.

The "conifuge" (Fig. 30), proposed recently [190] for the size analysis of coarse aerosols, is of great interest. The rotor of the instrument consists of two rigidly connected coaxial cones A and A^1 with parallel generators. A flow of air is established by centrifugal action in the air gap between the cones, and as a result of the presence of the cover C this flow is in the nature of a circulation. The aerosol under examination is passed into the instrument through the tube B. Under the action of centrifugal force the particles move from the inner cone towards the outer one and are precipitated on the latter. The smaller the particles, the further will they manage to pass along the gap before being precipitated. A narrow glass strip, flush with the surface of the outer cone, is inserted along its generator, which is removed at the end of an experiment so

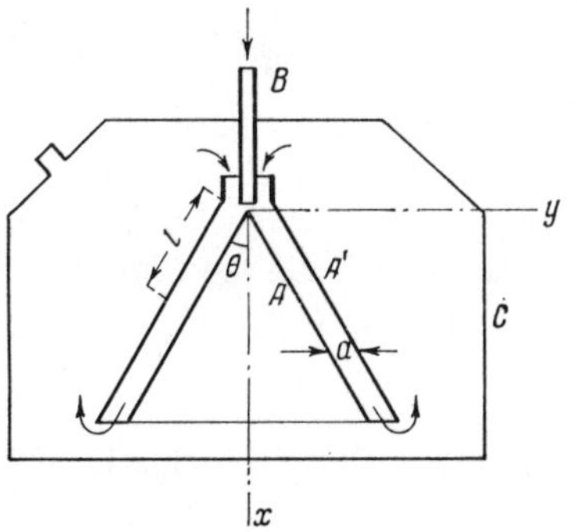

FIG. 30. Conifuge.

that the deposit formed on it can be examined under the microscope. Separation of the particles according to their size (in the cases of particles of irregular shape according to the value of the sedimentation radius) takes place very thoroughly, and from the preparations obtained it is easy to find the particle size distribution. The high resolving power of the instrument is due to the centrifugal force increasing with the distance from the apex of the cone while the air velocity decreases.

Fairly complicated mathematical calculation leads to the following equation for the trajectory of aerosol particles in the instrument on the assumption that the air is completely entrained by the rotor and therefore has the same rate of rotation

$$y = \pi^{1/2} \tan\theta \exp[A^2(x+\alpha)^2]\{\operatorname{erf} A(x+\alpha) - \operatorname{erf} A\alpha\}/2A, \qquad (28.7)$$

where

$$A = (\pi a \omega^2 V_s \tan\theta/\Phi g)^{1/2};$$

$$\alpha = \frac{1}{2} a \cot\theta; \qquad (28.7a)$$

and Φ is the volume rate of flow of the circulating air.

The meanings of the other symbols can be seen from the diagram. Since the generator of the outer cone is given by the equation

$$y = a + x\tan\theta, \qquad (28.8)$$

the distance l at which particles of a given size will settle can be determined from these two equations. Experiments with the conifuge have been conducted with

$$\theta = 30^\circ, \quad \Phi = 11{\cdot}6\ \text{cm}^3/\text{sec}, \quad \omega = 2\pi \cdot 50 \quad \text{and} \quad a = 0{\cdot}56\ \text{cm}$$

using particles with radii between 0·3 and 10 μ. The experimental values of l exceeded those calculated theoretically by 0·4 cm for particles of all sizes. This discrepancy is connected with the failure of the air to take up immediately the speed of rotation.

The chief advantage of using centrifuges for aerosols instead of cyclones (which are discussed in the next section) is that the flow rate in the rotor is low, which diminishes the risk of turbulence and of losing deposited particles by their being blown away.

Goetz's centrifuge [677] is in some respects similar to the conifuge, but the aerosol flows along a spiral channel of rectangular cross-section located on the periphery of the cylindrical rotor of the centrifuge. In contrast to the conifuge, the aerosol particles immediately experience the maximum centrifugal force, so that entrance effects must markedly distort the distribution of larger particles along the channel. Moreover, as seen from Goetz's paper, on entering the channel the aerosol fills up its whole cross-section, and the location of the settled particles depends not only on their size, but also on their inital position upon entering the channel; the separation of particles of different sizes must therefore be imperfect.

§ 29. THE CYCLONE [191, 192]

Of greater industrial importance are cyclones; these are centrifugal dust separators in which the apparatus itself is fixed and circulation of gas round the axis of the cyclone is produced by converting the forward motion. The gas enters an inlet tube A of rectangular cross-section (Fig. 31) and passes into the cylindrical part of the cyclone. Here it acquires a spiral motion, descending along the outer spiral (Fig. 32), and ascending by the inner spiral to pass out through the exhaust tube C (Fig. 31). Particles precipitated on the walls of the cyclone by centrifugal force slide down the wall and run out through the opening E into a collector. For cheapness, simplicity of construction and maintenance (associated with the absence of moving parts and filter surfaces needing periodical cleaning), comparatively low resistance and high output for given dimensions of the apparatus, cyclones are unequalled among equipment for cleaning gases of suspended particles.

However, they retain only coarse particles, so that their value is mainly for the removal of coarse dust from large volumes of gas where a small residue is unimportant, and also for preliminary cleaning before passing through more efficient equipment such as electrostatic precipitators. Cyclones have achieved most importance in the removal of ash and unburnt coal particles from flue gases, particularly in the combustion of pulverized fuel.

The motion of gas in a cyclone is very complex and has not been studied completely, with the result that the theory of cyclone operation is still inadequate and the efficiency cannot be calculated. In particular, the question of the most suitable form for a cyclone has hitherto been solved exclusively by empirical means.

In the elementary theory of the cyclone [193–197] a series of simplifying assumptions is made. Motion in the outer spiral is taken to be laminar, which is certainly untrue at high flow rates and the tangential velocity of the gas is expressed by the formula $U(\varrho) = b/\varrho^s$, where different authors take values from -1 to $+1$ for the index s. Radial motion of the gas from the periphery to the axis of the cyclone (discharge) is usually neglected, and the possibility of deposited particles being torn off by the gas flow is not taken into account; settling in the conical part of the cyclone is not considered, nor is re-entrainment of particles in the axial flow by the outer spiral. Many other phenomena occurring in cyclones are similarly neglected. Finally, the particle motion is considered quasi-stationary which, is justified by the fact that calculation of cyclones is always carried out for fine particles which are more difficult to precipitate.

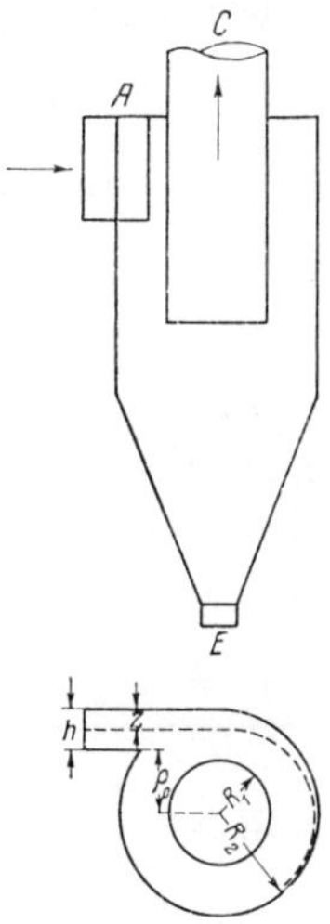

FIG. 31. Cyclone.

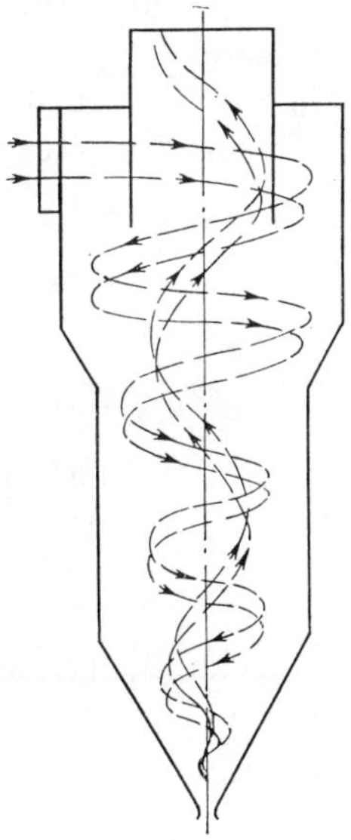

FIG. 32. Motion of gas in a cyclone.

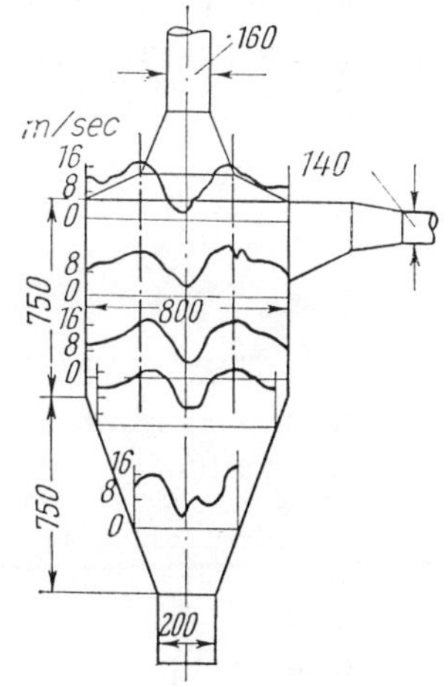

FIG. 33. Velocity distribution in a cyclone.

Experimental study of gas flow in a cyclone is difficult owing to the strong turbulence. Determination of the diameter of the inner spiral, which is roughly equal to the diameter of the exhaust tube, has also proved troublesome. The distribution of average tangential gas velocity in a cyclone is illustrated in Fig. 33 [198]. The velocity first increases from the periphery towards the axis of the cyclone, reaches a maximum, and then rapidly falls to zero. Near the axis a vortex core is formed which rotates as a rigid body. Commonly the width of the inlet tube, h, is equal to the width of the annular space, $R_2 - R_1$, between the exhaust tube and the walls of the cyclone (see Fig. 31) and the tangential velocity distribution below the exhaust tube in the outer spiral can then be expressed fairly accurately by the formula [197]

$$U(\varrho) = \frac{U_0}{2} \sqrt{\frac{R_2}{\varrho}}, \tag{29.1}$$

where U_0 is the velocity of the gas in the inlet tube; R_2 is the radius of the cylindrical part of the cyclone and ϱ is distance from the axis. For an ideal fluid the velocity distribution would be given by the formula $U(\varrho) = b/\varrho$ since potential (irrotational) flow of an ideal fluid in the inlet tube remains so inside the cyclone and non-vortex circulation of fluid must have the velocity distribution indicated. Actually, because of turbulent mixing in the cyclone, some equalization of the velocities occurs, which leads to a reduction in the exponent of $1/\varrho$ from 1 to 0·5.

The tangential velocity distribution in the annular space between the exhaust tube and the walls of the cyclone, when its width is equal to the width of the inlet tube h, is represented approximately by [197]

$$U(\varrho) = \frac{U_0 \varrho}{1{\cdot}2\, R_2}. \tag{29.2}$$

If the ratio $h/(R_2 - R_1)$ is not equal to unity then expressions (29.1) and (29.2) must be multiplied by the value of this ratio.

The motion of particles in a cyclone will be calculated from the gas velocity distribution (29·1). The radial velocity of the particles is then

$$\frac{d\varrho}{dt} = \frac{U^2(\varrho)\, mB}{\varrho} = \frac{U_0^2 R_2 \tau}{4\,\varrho^2}, \tag{29.3}$$

whence

$$dt = \frac{4\,\varrho^2\, d\varrho}{U_0^2 R_2 \tau}. \tag{29.4}$$

Let $\varrho_0 = R_2 - h$ be the least distance from the axis of the cyclone to the particles entering it (see Fig. 31). Then, according to (29.4) all particles of a given size will reach the wall in a time

$$t = \frac{4}{U_0^2 R_2 \tau} \int\limits_{\varrho_0}^{R_2} \varrho^2\, d\varrho = \frac{4\,(R_2^3 - \varrho_0^3)}{3\, U_0^2 R_2 \tau}. \tag{29.5}$$

The time of passage of the particles through the cylindrical part of the cyclone is†

$$t = \frac{2\pi\,\varrho_m}{U(\varrho_m)} = \frac{4\pi s \sqrt{\varrho_m^3}}{U_0 \sqrt{R_2}}, \tag{29.6}$$

where s is the number of turns of the outer spiral in the cylindrical part of the cyclone and ϱ_m is some average value of ϱ ($\varrho_0 < \varrho_m < R_2$). Replacing τ by its value $2\,r^2\gamma/9\eta$, the following is obtained for the minimum radius of the particles which are completely precipitated in the cyclone:

$$r_{\min} = \sqrt{\frac{3\,\eta\,(R_2^3 - \varrho_0^3)}{2\pi\,\gamma s U_0 R_2^{1/2} \varrho_m^{3/2}}}. \tag{29.7}$$

To simplify this formula write

$$R_2^3 - \varrho_0^3 = (R_2 - \varrho_0)(R_2^2 + R_2\varrho_0 + \varrho_0^2) = h\,(R_2^2 + R_2\varrho_0 + \varrho_0^2) \tag{29.8}$$

and replace ϱ_0 and ϱ_m by R_2 in the fraction $(R_2^2 + R_2\varrho_0 + \varrho_0^2)/\sqrt{\varrho_m^3}$. Both the numerator and denominator are thereby increased somewhat, but the value of the fraction does not change very much. Formula (29.7) then becomes

$$r_{\min} = 3\sqrt{\frac{\eta h}{2\pi\, U_0 \gamma s}}. \tag{29.9}$$

† The difference between the length of an arc of the spiral and its projection on a plane perpendicular to the axis of the cyclone has been neglected.

Starting from the distribution (29.2) in a completely analogous way the following formula is obtained.

$$r_{\min} = 3\sqrt{\frac{0{\cdot}3 \ln (R_2/R_1)\, \eta R_2}{\pi\, U_0\, \gamma s}}\,. \tag{29.10}$$

In practice both formulae give similar values of $r_{\min}$. In most cyclones h is between $R_2/3$ and $R_2/2$. For $h = R_2/2$, (29.9) becomes

$$r_{\min} = \frac{3}{2}\sqrt{\frac{\eta R_2}{\pi\, U_0 \gamma s}} \tag{29.11}$$

and (29.10) becomes

$$r_{\min} = 3\sqrt{0{\cdot}3 \ln 2}\sqrt{\frac{\eta R_2}{\pi U_0 \gamma s}}\,. \tag{29.12}$$

The numerical coefficients in (29.11) and (29.12) are equal to 1·5 and 1·37 respectively. The difference between them is of no significance, since the number of revolutions, s, in the cyclone is not known and is usually supposed to be between 1 and 3. For particles of irregular shape, and for aggregates, r denotes the sedimentation radius and γ the apparent density of the particles.

Formula (29.9) can also be used to determine the efficiency of a cyclone for particles of a given size by calculating the percentage retained. The distance h in this formula must then be replaced by z, which is the distance from the cylindrical wall of the cyclone of the limiting particle trajectory (see Fig. 31), and the resulting equation is solved for z. Then from (29.9)

$$z = \frac{2\pi\, U_0 \gamma s r^2}{9\eta} = \pi\, U_0 s \tau\,. \tag{29.13}$$

The value of z so determined is the maximum initial distance from the wall at which particles of radius r will deposit in the cyclone. If it is assumed that the aerosol particles are uniformly distributed in the inlet tube, then the efficiency of a cyclone for particles of a given size is z/h. Thus

$$\ni = z/h = \pi U_0 s\tau/h = \pi s l_i/h\,. \tag{29.14}$$

In polydisperse aerosols all particles with a radius greater than $r_{\min}$, determined from formulae (29.9) or (29.10), will be centrifuged out of the layer of thickness dz situated at a distance z from the wall of the cyclone. The mass of these particles is equal to $bG_a(r_{\min})\,dz$, where b is a proportionality factor and $G_a(r)$ is the cumulative weight distribution function for the particles showing what fraction of the particles, by weight, have radii $> r$. The total mass of particles retained in the cyclone is thus equal to

$$M = \int_0^h bG_a(r_{\min})\, dz\,, \tag{29.15}$$

and the total mass of particles passed through the cyclone

$$M_0 = \int_0^h b\,dz = bh\,. \tag{29.16}$$

Thus the efficiency of a cyclone for a polydisperse aerosol is

$$Э = M/M_0 = \frac{1}{h}\int_0^h G_a(r_{\min})\,dz = \frac{1}{h}\int_0^h G_a\left(\sqrt{\frac{9\,\eta z}{2\pi\, U_0 \gamma s}}\right) dz \qquad (29.17)$$

where $r_{\min}$ in the integral has been replaced by $(9\eta z/2\pi\, U_0\gamma s)^{1/2}$, in accordance with formula (29.13).

The efficiency so calculated is as a rule considerably higher than the actual efficiency, in the first place due to turbulence† which throws some particles approaching the wall towards the axis of the cyclone. Of fundamental importance also are redispersion of particles which have settled on the walls†† and the flow of gas to the axis of the cyclone. Particles of elongated or disk-like shape are oriented in the cyclone with their long axes parallel to the flow direction (see § 11) which reduces their radial velocity. The effect of turbulence, and possibly of redispersion, is apparent in the dependence of cyclone efficiency on the flow velocity U_0 of the gas. When U_0 increases, turbulent pulsations and redispersion must increase. Actually, experiment shows that the efficiency of the cyclone first increases considerably in agreement with the theory, but a maximum is reached, and finally it decreases [193]. The technical applications of cyclones, as of any gas-cleaning apparatus, demand the highest possible efficiency at the least hydraulic resistance, which determines the energy expended in passing the gas through the cyclone. Since the resistance is proportional to U_0^2 it is desirable to work at velocities rather less than the velocity corresponding to maximum efficiency. U_0 is usually of the order 10–20 msec^{-1}. It has been shown experimentally that cyclones which are different in size, but geometrically similar, have nearly the same resistance at the same velocity U_0 [194]. The effect of turbulence is evident also from the fact that internal smoothness of a cyclone reflects favourably on its performance. Any projections, screens etc., inside the cyclone noticeably reduce the efficiency and increase the resistance.

Formulae (29.9) and (29.17) suggest that the efficiency of a cyclone should depend not on its diameter but only on the width of the inlet tube h; this is not borne out in practice. Constriction of the inlet tube at constant U_0 leads, as shown above, to a reduction in the velocity of the gas inside the cyclone and therefore, as is well known from practice, does not have a large effect on efficiency. In this case, however, the capacity of the cyclone is obviously reduced. Cyclones are usually designed, on the basis of experimental data, with the width of the inlet tube from 1/3 to 1/2 the radius of the cyclone, in which case formulae (29.11) and (29.12) should be used.

It is therefore clear that the efficiency of the cyclone should increase as its diameter is decreased. In fact, the increase is even bigger than theory suggests, a fact explained primarily by the lower turbulence in small cyclones. This, in turn, may be due to the high pressure and gas density gradients, in small cyclones, from the axis to the periphery, which are proportional to U_0^2/ϱ. The density gradient, like a temperature inversion in the atmosphere, undoubtedly counteracts turbulent mixing. Small cyclones of diameter 5–15 cm have proved considerably more efficient than large types several metres in diameter. Connected in parallel, in a bank, they have become widespread in

† The effect of turbulence on cyclone efficiency is discussed in §46.

†† This is proved by the noticeable increase in efficiency when the walls of a cyclone are wetted [192].

industry under the name of "Multiclones" [196]. According to published reports they may have efficiencies up to 99 per cent for particles with $r \approx 3\,\mu$.

In the simple theory of the cyclone presented above no account was taken of change in direction of the particles along the axis of the cyclone at the end of the conical part, where they move from the external to the internal spiral [199]. That particles with high velocity fly out of the gas stream at this point is shown by the fact that small cyclones work excellently when inverted [196]; dust particles (but not the gas) in this case fly upwards out of the exhaust. The significance of this factor in the overall efficiency of a cyclone is difficult to assess.

When the concentration of dust passing through a cyclone is increased an increase in efficiency and a decrease in resistance is observed which is extremely interesting in connection with the mechanics of aerosols. In one experiment when the concentration of kaolin dust in air was increased from 0·1 to 100 g . mm^{-3} the efficiency of the cyclone increased from 70 to 90 per cent [193]. This phenomenon is undoubtedly caused by kinematic coagulation of the dust (see § 54) in the cyclone, fine particles being captured by coarser and more rapidly moving ones.

The effect of dust concentration on the resistance of a cyclone is illustrated by the following figures. A cyclone of diameter 22 cm, working with polydisperse dust of mean particle radius 8 μ and specific gravity 2·8, has resistance Δp related to dust concentration c (gm m^{-3}) according to the formula $1 - \Delta p/\Delta p_0 = 0{\cdot}013\,c$ (Briggs [200]) where Δp_0 is the resistance for clean air. Zaitsev and Shakhov observed an effect 2·5–3 times greater which, for a given weight concentration, increased with particle size.

The opinion is widespread amongst specialists in the field of gas cleaning that the main reason for the decrease in resistance is the high specific gravity, and consequently the high momentum, of the dust-laden gas entering the cyclone, compared with clean gas at the same velocity. This view overlooks the loss of the extra momentum by precipitation of particles on the walls of the cyclone which prevents its transmission to the gas.

In this connection it must be mentioned that when dust-laden gas is passed through a Venturi tube, if particles are not precipitated on the walls and are carried away by the stream, the resistance of the tube does not fall, but increases as the dust concentration increases. According to Farbar's experiments [201], also carried out with a highly polydisperse dust ($r = 1 - 100\,\mu$) of specific gravity 2·45, the following relationship is observed

$$\Delta p/\Delta p_0 - 1 = kc, \tag{29.18}$$

where c is the dust concentration expressed in grams of dust per gram of air, and k is a coefficient $\approx 0{\cdot}3$. Theoretical examination of the flow of an aerosol through a Venturi tube, based on Bernoulli's theorem, leads to the same equation (29.18) but with a coefficient unity. Here again the resistance is less than would be expected. In § 45 the effect of dust-laden gas on the resistance will be considered further.

In conclusion, an ingenious idea will be mentioned which has been tried in recent years, namely the passing through cyclones of a gas the dust content of which has already been raised to some extent. In the louvred dust collector [202] this concentration is achieved with the aid of a system of inclined baffles between which most of the gas escapes while the dust is deflected by the baffles, and leaves the apparatus as a concentrated gas-dust mixture which is led into a cyclone (Fig. 34a). In a tubular form of the apparatus [203] the gas on exit is given a rotatory motion (as in the cyclone) which forces the dust to the periphery of the tube (Fig. 34b). A decrease in the volume of gas passed through the cyclone and an increase in the dust concentration are thereby achieved. These two circumstances raise considerably the cleaning capacity and efficiency in cyclones.

In the theory of cyclone discussed above the radial component U_ϱ of the gas flow in the cyclone, which is directed towards the axis, was ignored.

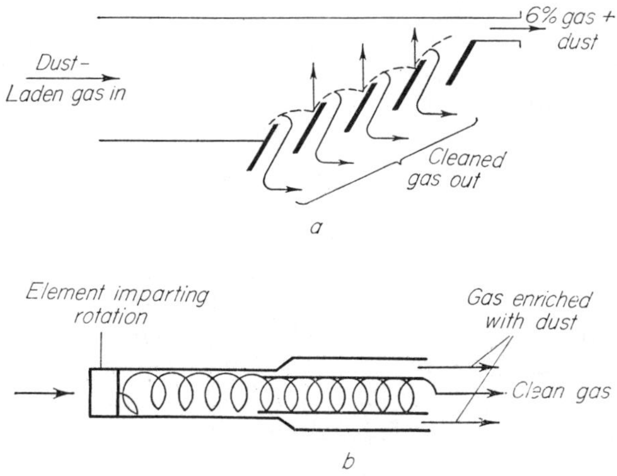

FIG. 34a and b. Inertial enrichment of aerosols.

According to Feifel [678] and Barth [679] the particles deposited in the cyclone must have radial velocities due to centrifugal force which are larger than U_ϱ at every point of their path to the walls. This determines the limiting size of the particles deposited in the cyclone. Thus, according to these authors, the cyclone functions as a classifier of particles. In practice, as is shown by experiments (see Fig. 34d), the classifying action of the cyclone is far from perfect.

Solbach [680] measured U_ϱ at various points in a straight-through cyclone and calculated the trajectories and the collection efficiency of particles of different sizes, making an allowance for the radial downflow. His measurements with clay dust having $r = 1 - 6\,\mu$ are in fair agreement with his calculations.

Maslov and Marshak [681] studied the deposition of $K_2Cr_2O_7$ powder fractions with $\bar{r} = 3{\cdot}5 - 12{\cdot}5\,\mu$ in a straight-through cyclone, having covered the walls with a thin layer of Vaseline. The radius of the cyclone, R, varied from 2·5 to 20 cm, the tangential inlet velocity ranged between 2 and 21 m/sec and the width of the feed pipe was $h = 0{\cdot}4\,R$. The flow in all the experiments could be regarded as geometrically similar. Experimental values of $\ni$ as a function of the Stk number (see § 31) fell on a common curve (Fig. 34c), the shape of which depended only on the ratio of the length L to the radius R of the cyclone.

Walter [682] used very narrow ($R = 0{\cdot}75$–$1{\cdot}4$ cm), long ($L = 16$–20 cm) cyclones having short outlet tubes. An initial increase of $\ni$ with flow rate was observed, after which it remained constant. The rise of $\ni$ occurring with an increase in the size and concentration of particles and a decrease in R was much greater than had been re-

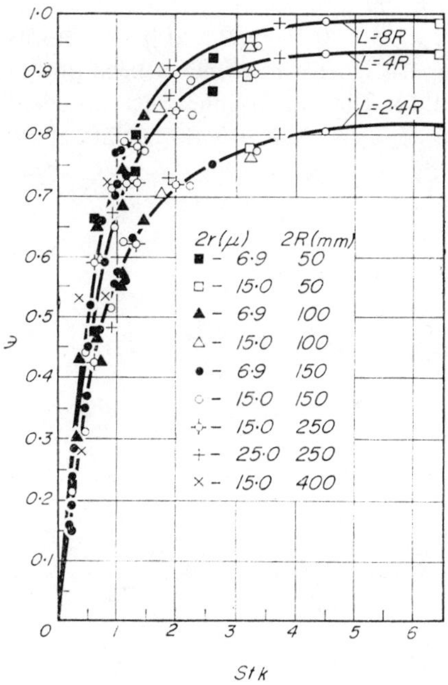

FIG. 34c. Deposition of dust in a straight-through cyclone.

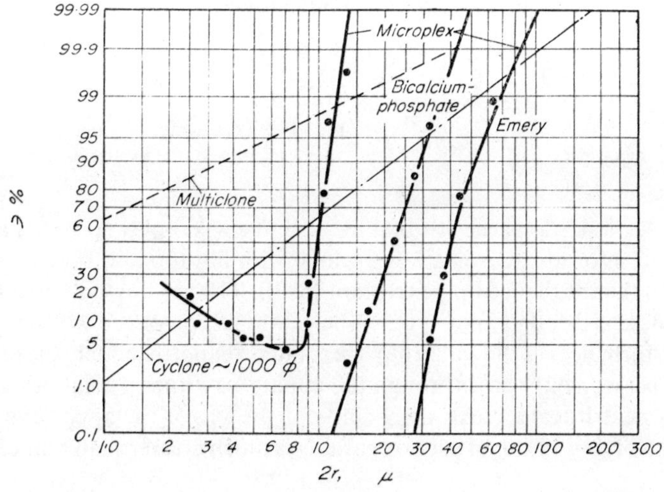

FIG. 34d. Efficiency of cyclones and of the Microplex apparatus as a function of particle size.

ported previously by other authors. Thus, in one experiment with $R = 0{\cdot}75$ cm at dust concentrations of 10, 100 and 500 mg/m^3 ∋ was equal to 0·8, 0·99 and 0·998 respectively; wiih $R = 0{\cdot}95$, r = 0·25, 0·30 and 0·35 μ, ∋ was, in turn, 0·6, 0·987 and 0·999. Such high cyclone efficiencies have never been obtained by anyone else. It is also difficult to account for the classifying effect observed in these experiments, which is at variance with theory as well as with the data of other authors (see Fig. 34d [683]).

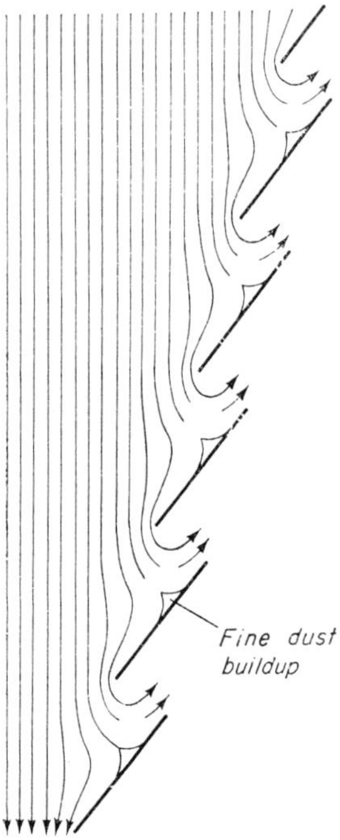

FIG. 34e. Field of flow in a louvred separator.

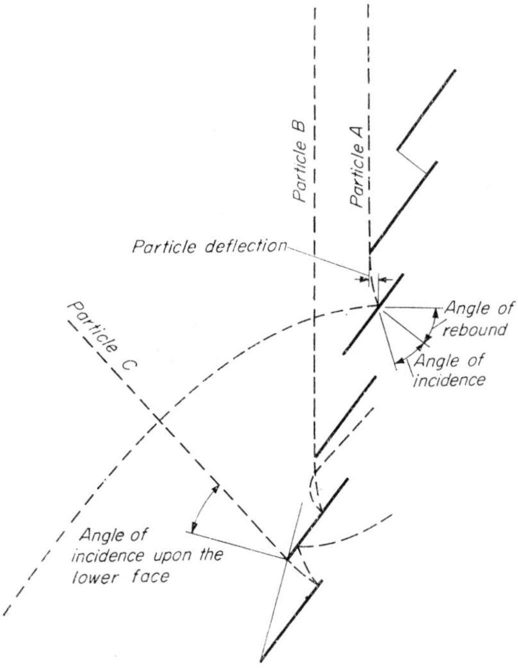

FIG. 34f. Trajectories of alundum particles in a louvred separator.

Sharp size selection is apparently obtained with the Microplex, a special apparatus for the classification of dusts. It is a low cyclone with flat top and bottom walls and a height equal to that of the feed pipe (Rumpf [683]). The gas flows along a flat spiral and leaves the device through an orifice in the centre of one of the walls. In order to eliminate the disturbances at the walls they are rotated at approximately the same speed as the gas. Particles can only pass through the apparatus if their radial velocity under the action of the centrifugal force in the vicinity of the axis of the apparatus is smaller than U_{ϱ}. The classifying action of such apparatus is much greater than that of cyclones (Fig. 34d).

Daniels [684] studied the efficiency of straight-through cyclones having $R \doteq 2{\cdot}5$ cm and $L = 25$ cm with fractionated quartz dust, and found that ∋ rose very slowly with r. At a flow rate of 20 l/sec ∋ ranged from 0·72 at $r = 10\,\mu$ to 0·83 at $r = 60\,\mu$. At higher rates of flow and $r \geqslant 25\,\mu$, ∋ began to decrease. Special experiments showed that large quartz particles bounced off the walls, carrying with them, also, some smaller particles. By wetting the walls with water ∋ was found to attain unity for $r \geqslant 12\,\mu$.

According to Stairmand [685], wetting the cyclone walls raises ∍ considerably for particles of all sizes investigated ($r = 1 - 25\,\mu$).

Yaffe *et al.* [686] developed a device of the cyclone type for the size measurement of aerosol particles. The device consists of a spiral channel of rectangular cross-section, the height of which is gradually reduced; the flow rate and the centrifugal force increase in the direction of the axis of the spiral, where there is an orifice to let the gas escape. The main drawback of this type of apparatus is the blowing off of the deposit by the air stream and the breaking up of aggregates in the boundary layer. Aerosol centrifuges, as has been already mentioned, are free of this drawback.

The mechanism of louvre dust separators was studied by Smith and Goglia [687] who worked with hard alundum dust. By blowing fine dust through the louvre and observing it with side illumination, the authors determined the flow pattern in the device (Fig. 34e) and then examined the paths of coarse particles. As shown in Fig. 34f, particles which bounce off as elastic bodies (*A*) do not pass through the louvre; those which are detached slowly (*B*), or those which move at a large angle to the louvre plane (*C*) are carried away by the air stream passing through the louvre. The efficiency of this type of separator is very low at $r = 10\,\mu$, but rises rapidly with r and attains $0{\cdot}95 - 0{\cdot}97$ at $r = 20\,\mu$. For soft particles, aggregates etc., the efficiency of this apparatus must be considerably lower (see page 378).

§ 30. DETERMINATION OF AEROSOL PARTICLE TRAJECTORIES IN CURVILINEAR FLOW

Very fine aerosol particles follow strictly the streamlines of a flowing gas. On this is based the use of aerosols in experimental aerodynamics for studying flow by means of thin filaments of smoke which trace out the streamlines. The thickness of the filaments determines the flow velocity at a given point because the product of velocity and filament cross section remains constant. Thus the velocity distribution in the gas is determined.

Coarser particles leave the streamlines to a greater or lesser extent and are displaced with respect to them because of their inertia; the order of magnitude of this displacement, as will be seen from the following paragraphs, is determined by the stop-distance, l_i, of the particles. This displacement leads to inertial precipitation of particles on walls when the flow direction changes at bends in pipes and channels, and in flow past various obstacles. Inertial precipitation of aerosols is widely used for separating the disperse phase of aerosols from the gas. It is responsible for the icing of objects in moving supercooled clouds and mists, and also governs collisions between coarse aerosol particles (see § 54), thus being of great importance in meteorology.

In § 28 and § 29 some examples of inertial deposition of aerosols in a rotating gas were discussed on the assumption that while the particles were completely entrained in the direction of rotation, they moved relatively to the gas, under the action of centrifugal force, in a radial direction. This simplification is permissible, generally speaking, when leading dimensions such as the thickness of the stream, radius of curvature of the streamlines etc., are large compared with l_i. This does not hold for the precipitation of aerosols on obstacles and in other cases where there is a sudden change in the direction of flow. The assumption of entrainment by the flow, as will be shown below (page 154), can thus lead to fundamental errors. Not only the normal but also the tangential component of the inertial force must therefore be taken into account, and the trajectory of a particle must be calculated from the differential equations of its motion. At low Re these equations have the form

$$\frac{dV_x}{dt} = \frac{1}{\tau}(U_x - V_x) + \frac{F_x}{m}; \quad \frac{dV_y}{dt} = \frac{1}{\tau}(U_y - V_y) + \frac{F_y}{m}, \tag{30.1}$$

where U_x and F_x are the resolutes of the velocity of the fluid and of the external force along the x-axis. The similar equation for the z direction has been omitted.

Solution of this set of equations in general presents great difficulties and has been achieved in comparatively few problems. It is usually necessary to resort to an approximate method for calculating particle trajectories which can be done as follows [204, 205, 206], restricting attention to cases when the main fluid flow is steady. It is possible to put **F** equal to zero without destroying the generality of the problem because this is equivalent to changing the flow field **U** (x, y, z) to the field $\mathbf{U} + (\tau/m)\,\mathbf{F}$. Let the time be divided into equal small intervals and the trajectory of the particle into corresponding segments using for the i-th interval the approximation

$$\frac{dV_x}{dt} = \frac{1}{\tau}(U_{xi} - V_x), \tag{30.2}$$

where U_{xi} is the value of U_x at the beginning, or better at the middle, of the interval. Integrating and assuming that at the beginning of the interval $t = 0$ and $V_x = V_{xi}$ gives

$$V_x = U_{xi} + (V_{xi} - U_{xi})\,e^{-t/\tau} = V_{xi} + (U_{xi} - V_{xi})\,(1 - e^{-t/\tau}), \tag{30.3}$$

and by a further integration the x-coordinate of the particle is obtained,

$$x = x_i + U_{xi}t + \tau\,(V_{xi} - U_{xi})\,(1 - e^{-t/\tau}), \tag{30.4}$$

where x_i is the position of the particle at the beginning of the i-th interval. Similar formulae are obtained for the other coordinates.

Calculation of a trajectory begins at a point at which the velocity of the particle is known. Usually in problems of this sort the velocity of the fluid in a definite region, for example at a sufficiently large distance from an obstacle, or a change in flow direction, can be considered rectilinear and constant, so that the velocities of particle and fluid, in the absence of external forces, are identical. The flow velocity must be known at all points through which the particle trajectory passes, either being given by theoretical hydrodynamical formulae or by experimental data. Beginning from the point where $V_x = U_x$ the path is calculated in steps by means of formulae (30.3) and (30.4) which give the position and velocity of the particle at the beginning of the successive stages.

When the steps are made shorter accuracy is improved but the time spent on the calculation increases. It is best to perform this work on a computing machine. To decide the required length of step it may be noted that in time Δt the velocity of the particle, according to (30.2), increases by $\Delta V_x \approx (U_{xi} - V_{xi})\dfrac{\Delta t}{\tau}$ and, according to (30.3), by

$$\Delta V_x = (U_{xi} - V_{xi})\,(1 - e^{-\Delta t/\tau}) = (U_{xi} - V_{xi})\left(\frac{\Delta t}{\tau} - \frac{1}{2}\left(\frac{\Delta t}{\tau}\right)^2 + \cdots\right). \tag{30.4a}$$

Both expressions must give a similar value of ΔV, hence it is clear that $\Delta t \ll \tau$; it follows from this that the following simpler formula may be used in place of (30.3) without loss of accuracy

$$V_x = V_{xi} + (U_{xi} - V_{xi}) \frac{t}{\tau}, \tag{30.5}$$

and, instead of (30.4),

$$x = x_i + V_{xi} t + (U_{xi} - V_{xi}) \frac{t^2}{2\tau} \approx x_i + V_{xi} t. \tag{30.6}$$

At large distances from an obstacle or a bend in the flow, longer steps can be taken in order to save work. Since $U_x - V_x$ is here very small the absolute error incurred in the calculation of V_x and x with greater lengths of step is negligible.

It is also possible to construct particle trajectories graphically which is particularly useful if the flow velocity is determined experimentally; the graphical method, however, is considerably less accurate than calculation.

§ 31. THE THEORY OF SIMILARITY IN THE MECHANICS OF AEROSOLS [207–209]

The method described in the preceding section for calculating the trajectories of aerosol particles provides a result for a given size of the system, a certain flow velocity and a particle with a known relaxation time. If the calculation had to be done afresh for every other value of each of these quantities the method would be too laborious to use but, thanks to the theory of similarity, results obtained can be generalized. The theory of similarity is frequently used when reducing and applying the results of experiments on the motion and precipitation of aerosol particles, and in the designs of various devices for separating airborne particles whether they depend on inertial precipitation, or operate by diffusion, electrostatic action etc. The value of similarity theory in the mechanics of aerosols is thus extremely great.

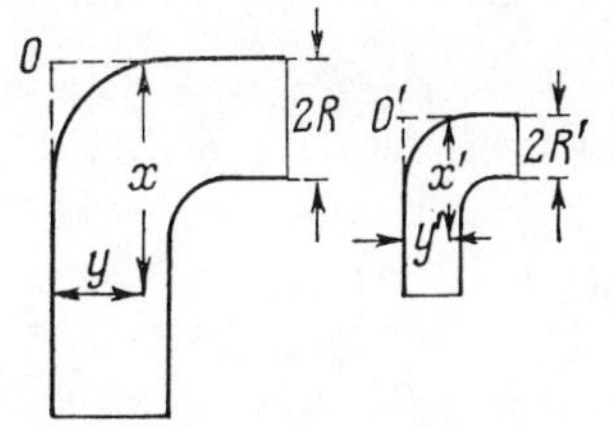

FIG. 35. Geometrical similarity.

For the motions of two aerosol systems to be similar, it is necessary to have

(1) geometrical similarity of the flow boundaries,
(2) similarity of the fluid flow in each system,
(3) similarity of the trajectories of the particles.

Geometrical similarity of the boundaries is expressed by the equation

$$x/x' = y/y' = R/R' = C_l, \tag{31.1}$$

where x and y are the co-ordinates of some point in the first system; x' and y' are the coordinates of the corresponding point in the second system; R and R' are quantities characterizing the linear dimensions of the two systems, such as the radii of the tube or of the objects past which the fluids are flowing.

For similarity of fluid motion in both systems it is first necessary that the kinematic condition should hold

$$U_x/U'_{x'} = U_y/U'_{y'} = U_0/U'_0 = C_u, \tag{31.2}$$

where U_x, U_y and $U'_{x'}$, $U'_{y'}$ are velocity resolutes at corresponding points in the systems; U_0 and U'_0 are constant characteristic flow velocities of each system, for example the mean flow velocity in a tube or the velocity at an infinitely large distance from an object in a flow.

Similarity of particle trajectories in the two systems necessitates an analogous equation

$$V_x/V'_{x'} = V_y/V'_{y'} = C_v. \tag{31.3}$$

In parts of the systems where the fluid moves steadily in straight lines and the external forces are zero the velocities of the fluid and of the particles are identical, i.e. $V_x/V'_{x'} = U_x/U'_{y'}$, hence it follows that $C_v = C_u$.

Corresponding instants of time in both systems are related by the equation

$$t/t' = C_t, \tag{31.4}$$

Since $U_x = dx/dt$

$$C_u = C_l/C_t. \tag{31.5}$$

Similarly, for the corresponding accelerations

$$\frac{dV_x}{dt}\bigg/\frac{dV'_{x'}}{dt'} = \frac{dV_y}{dt}\bigg/\frac{dV'_{y'}}{dt'} = C_a \tag{31.6}$$

and

$$C_a = C_u/C_t. \tag{31.7}$$

As is well known from hydrodynamics, the dynamical condition for similarity of the motion of the fluids in two geometrically similar systems is the equation

$$U_0/U'_0 = C_u = \frac{\nu/R}{\nu/R'g} \tag{31.8}$$

or

$$\frac{2U_0R}{\nu} = \frac{2U'_0R'}{\nu'} = \mathrm{Re}_f = \mathrm{const}. \tag{31.9}$$

All that has been said above does not, of course, depend on the law of resistance of the medium. If the resistance is proportional to the velocity the differential equations of motion of the particles have the form (30.1). If these relate to the first system, then to find the corresponding motions in the second system all the variable quantities in these equations must be replaced by expressions deduced from equations (31.2) and (31.6).

Denoting the acceleration F/m under the action of external forces by a we get from (30.1)

$$C_a\frac{dV'_{x'}}{dt'} = \frac{C_u}{\tau}(U'_{x'} - V'_{x'}) + C_a a'_{x'} \tag{31.10}$$

or

$$\frac{dV'_{x'}}{dt'} = \frac{C_u}{C_a \tau} (U'_{x'} - V'_{x'}) + a'_{x'}. \qquad (31.11)$$

Because the differential equation of motion of particles in the second system must have the form

$$\frac{dV'_{x'}}{dt'} = \frac{1}{\tau'} (U'_{x'} - V'_{x'}) + a'_{x'}, \qquad (31.12)$$

then for similarity of the particle motions in both systems it is necessary that the relaxation time τ of particles in the second system should be related to τ by the formula

$$\tau' = \frac{C_a}{C_u} \tau = \tau / C_t, \qquad (31.13)$$

which, incidentally, also follows directly from (31.4). Taking into consideration that $C_t = C_l/C_u = RU'_0/R'U_0$, formula (31.13) can be written

$$\frac{\tau U_0}{R} = \frac{\tau' U'_0}{R'} = \text{const} = \text{Stk} \qquad (31.14)$$

or

$$\frac{l_i}{R} = \frac{l'_i}{R'} = \text{Stk}. \qquad (31.15)$$

The dimensionless ratio of the stop distance of a particle to the characteristic dimension of the system is called the Stokes number (Stk). If from the conditions of the problem $a = a'$ (for example when F is the force of gravity and $a = g$) then $C_a = C_u^2/C_l = 1$, $U_0^2/R = U_0'^2/R'$, which can be written in the form

$$\frac{U_0^2}{2Rg} = \frac{U_0'^2}{2R'g} = \text{const} = \text{Fr} \quad \text{(Froude's number)}. \qquad (31.16)$$

If the Reynolds number characterizing the relative motion of a particle and the medium is large and the resistance of the medium is not proportional to the velocity, the use of similarity theory becomes complicated and offers no advantages except for steady rectilinear flow of the medium (see § 18).

Thus the condition for similarity of the motions of two aerosol systems is the equality of the dimensionless Reynolds' and Stokes' numbers, and when gravity is considered, Froude's number.

For laminar flow through pipes and channels similarity of the flow in geometrically similar systems is automatic for all values of Re_f less than the critical. Similarity of fluid flow is likewise maintained round spheres at low Re_f, and also in flow around spheres and cylinders at high Re_f, if the existence of a boundary layer can be neglected (see § 34).

A most important task in the mechanics of aerosols is to find the conditions for deposition of particles on walls, objects exposed to a flow etc. The collection efficiency (Э) of particles deposited on an object placed in a rectilinear flow is the name given to the ratio of the number of particles deposited on it to the number of particles, the centres of which would have passed through it if they had moved all the time in straight lines (Fig. 36). To find ∍ the outermost trajectories for which the centres of the particles still collide with the body must be determined. If h denotes the distance from the outermost trajectories to the centre line of the flow (directed towards the centre of a sphere, the axis of a cylinder etc.) at an infinitely large distance from the body, then for a cylinder $\ni = h/R$ and for a sphere $\ni = (h/R)^2$.

The arguments above show that in the case of purely inertial deposition ∍ depends only on the Stokes number.

From the appearance of the streamlines (see Fig. 45 on page 161) it follows that for the same value of Stk the collection efficiency for potential flow must exceed the values with viscous flow. In potential flow the streamlines bend round the body at a smaller distance and are more sharply curved than in viscous flow. It is not difficult to see that the presence of a laminar boundary layer on the surface of an obstacle leads to a reduction in ∍, because when the flow velocity is reduced the distance between the streamlines increases; the streamlines move away from the surface of the

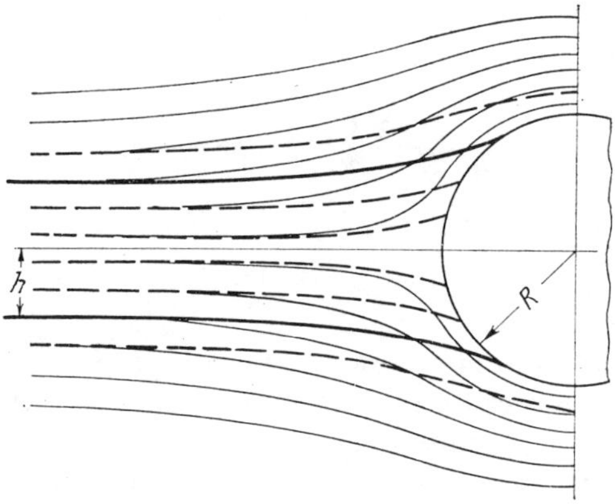

FIG. 36. Inertial deposition of particles and the collection efficiency.

body as a result of the comparatively small flow velocity in the boundary layer. With decrease in Re_f and the associated increase in boundary layer thickness the collection efficiency corresponding to a given value of Stk decreases continuously until it reaches a value corresponding to viscous flow conditions (this is discussed in more detail in § 34).

If gravity cannot be neglected, the condition (31.16) must also be taken into account. Combining it with the condition (31.15) we obtain

$$\frac{\mathrm{Stk}}{2\,\mathrm{Fr}} = \frac{2\gamma r^2 g}{9\eta U_0} = V_s/U_0 = \mathrm{const.} \tag{31.17}$$

If the inertia of the particles can be neglected only the condition (31.17) remains, according to which the fraction of particles deposited in a system of given geometrical shape depends only on the ratio of the settling rate of the particles to the flow velocity. In § 26, in particular, it was shown that the collection efficiency in a horizontal tube is a function of $L V_s/2\,h\overline{U}$, where L is the length and $2\,h$ is the height or diameter of the tube. Thus in geometrically similar tubes the efficiency depends only on the ratio $V_s/\overline{U}$.

By calculating, as shown above, or determining experimentally the value of ∍ as a function of Stk, when particle inertia predominates, or as a function of $V_s/\overline{U}$ when gravity governs deposition, we can find the value of the collection efficiency for particles on objects of a given shape for all flow velocities as long as the flows are similar.

If, unlike the cases just discussed, the velocity distribution in a flow depends on Re_f then only systems for which it is the same, for example when the ratio of boundary layer thickness to size of object is identical, can be compared. The use of similarity theory is therefore restricted considerably. For example, if only the inertia of the particles is taken into account, a combination of conditions (31.9) and (31.15) gives $\mathrm{Re}_f/9\,\mathrm{Stk} = R^2\gamma_g/r^2\gamma = \text{const.}$, hence, when the fluid and the particles are composed of the same substances in the two systems it is necessary to have

$$r/R = \text{const.} \tag{31.18}$$

For the motions of the particles to be similar in the two cases the particle sizes must be in the same ratio as the linear dimensions of the systems. The collection efficiency is then a function of the ratios l_i/R and r/R.

It is interesting that the same condition (31.18) is also reached when the effect of geometrical size (not the mass) of the particles on the value of ∍ is taken into account. It was hitherto assumed that a particle is deposited when the trajectory of its centre meets the surface of an obstacle. Deposition actually takes place however, when a particle reaches a distance from the surface equal to its own radius (the interception effect). This effect increases appreciably for precipitation of aerosol particles on thin cylindrical rods or on other aerosol particles. It will easily be understood that the condition (31.18) must be observed if similarity is to be preserved in the precipitation of particles when allowance is made for the interception effect.

Whatever the geometrical shape of the system and whatever the flow conditions, the inertial collection efficiency $∍_i$ increases with $l_i/R = \tau\, U_0/R$, and the collection efficiency for gravitational deposition, $∍_s$, increases with $V_s/U_0 = \tau g/U_0$. Thus the dependence of ∍ on the size and density of the particles and on the viscosity of the medium, which enter the expression for τ, is the same in both cases. The position is different for the dependence of ∍ on the linear dimensions of the system, R, and the flow velocity U_0 for $∍_i$ increases as U_0 increases and as R decreases, as long as the systems remain geometrically similar, while $∍_s$ does not depend on R and decreases as U_0 increases. The importance of this behaviour in everyday life and in technology is demonstrated by several examples.

It is well known that thin objects such as small twigs and wires quickly become covered with ice in supercooled water fogs (Fig. 37 shows diagramatically the cross-sections of an iced blade of grass and an iced tree trunk). Icing becomes more intense with increase in droplet size and wind speed, and becomes particularly severe on the leading edges of aircraft wings and propellers. Conversely, deposition of dust on horizontal surfaces (under gravity) occurs where there is little air flow.

Coarse dust inhaled from a dusty atmosphere is deposited in the upper respiratory passages, but fine dust penetrates to the lungs and is therefore particularly dangerous.

During sandstorms very fine grains of sand easily penetrate gaps in windows and doors so that it is impossible to find shelter from dust even in houses.

Rain to some extent clears the atmosphere of coarse dust but has little effect on the content of very fine particles, such as the heavy ions.

In conclusion we will mention the dynamical similarity of cyclones, which Soviet authors have studied in two ways. Syrkin [207] and Volkov [210] were guided by the consideration that, if Re_f is to be kept constant, the flow velocity and cyclone dimensions can be reduced simply by using water instead of a gas; for water η/γ_g is fifteen times smaller than for air and consequently $U_0 R$ can be reduced by the same factor. These authors therefore made a model cyclone using water.

Kouzov [192] started from the assumption that gas flows in cyclones are similar. At the very high values of Re_f which are characteristic of cyclones, the motion of the gas becomes more or less independent of Re_f. The dimensionless characteristics obtained by Kouzov (efficiency as a function of Stk and Fr) for cyclones of the particular type studied can therefore be used to predict the efficiency for any given values of the cyclone dimensions, flow velocity and particle properties. Unfortunately Kouzov determined not the collection efficiency for individual dust fractions, but the

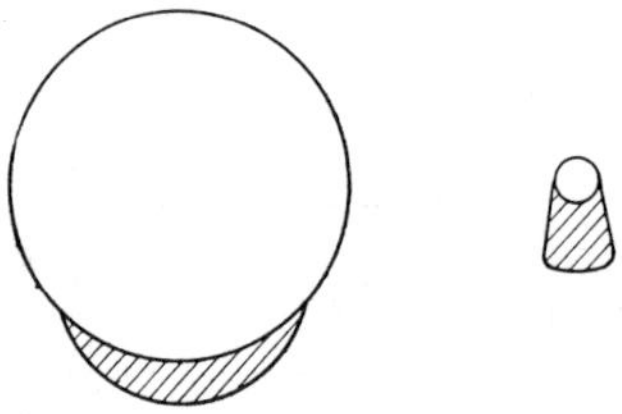

FIG. 37. Inertial deposition of ice.

overall efficiency, and the particle radius used in his expression for Stk was the mean value for an extremely polydisperse dust. The use of these data to compare cyclone theory with experiment is therefore limited. No reliable results suitable for this comparison exist because they would have to be obtained by experiments with more or less monodisperse dusts; this is rather difficult to arrange in view of the large amounts of dust required for cyclone operation.

§ 32. THE THEORY OF AEROSOL SAMPLING

Every aerosol study begins with taking a sample in some piece of apparatus such as a filter, condenser, tyndallometer or ultramicroscope cell. If correct results are to be obtained the difference in concentration and particle-size distribution between the original aerosol and the sample must be as small as possible. The difference is due partly to deposition of the aerosol in the instrument itself, particularly in the tube through which the aerosol enters, and partly to phenomena taking place at the inlet to the instrument (mouth of the sampling tube). Deposition of aerosol particles in the sampling tube and in other parts of the instruments is discussed in § 26, 39 and 46; here we shall be concerned only with the first stage of sampling, at the inlet to the instrument.

The following conditions must be fulfilled when a sample is taken from a flowing aerosol. If the sampling tube is at an angle to the flow direction (Fig. 38a) some particles, because of their inertia, will either be deposited on the inside wall of the tube, or else fail to enter it, so that the concentration of the aerosol in the sample will be less than the true concentration. If the sampling tube is parallel to the flow but the velocity in it is greater (Fig. 38b) than in the main flow, particles from the streamlines directed just outside the tube will not enter the tube; if the velocity in the tube is less than in the main flow (Figs. 38c, 39b) particles from the streamlines directed just inside the tube but passing outside it, will enter the tube. The aerosol concentration in the sample will be too low in the first case and too high in the second. In correct, or isokinetic, sampling the flow velocities in the tube and in the main flow must be equal (Fig. 39a).

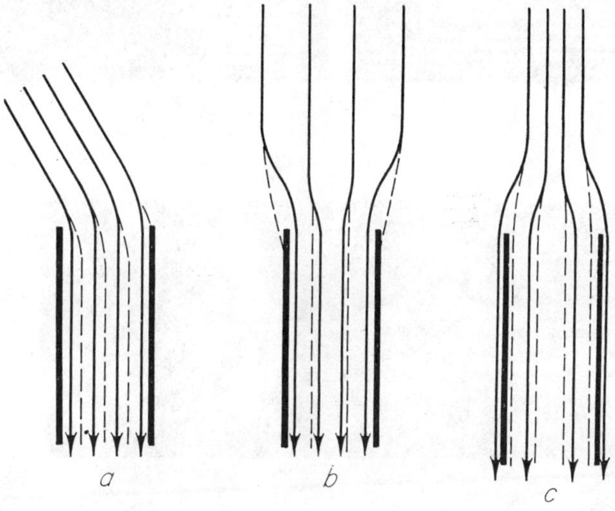

FIG. 38. Various cases of aerosol sampling.

In addition the wall at the mouth of the tube must be so thin that deposition of particles on its butt end can be neglected.

The following statements [211] are useful in connection with sampling an aerosol consisting of particles settling under the action of a constant force such as gravity. They are true only when the inertia of the particles can be neglected, that is for sampling at low flow velocities when the stop-distance is small compared with the size of the sampling head (page 141).

(1) No flow in a cloud of settling inertialess particles with a uniform initial concentration can change that concentration† (at a large enough distance from the boundary of the cloud). For non-settling particles this statement is obvious, because they behave like molecules of the medium. If the particles settle with a velocity V_s the result is the same as if the whole cloud was settling with the same velocity so that the concentration inside it does not vary from place to place. In a polydisperse cloud, settling only upsets the steadiness of the concentration at the upper and lower boundaries of the cloud.

(2) For a cloud of settling inertialess particles the number of particles passing each second through a given stationary area, the horizontal projection of which has an area S, is equal to the number which would have passed through in the absence of

† This statement is obviously untrue if inertia is present. Thus, in a vortex tube the concentration of particles will decrease.

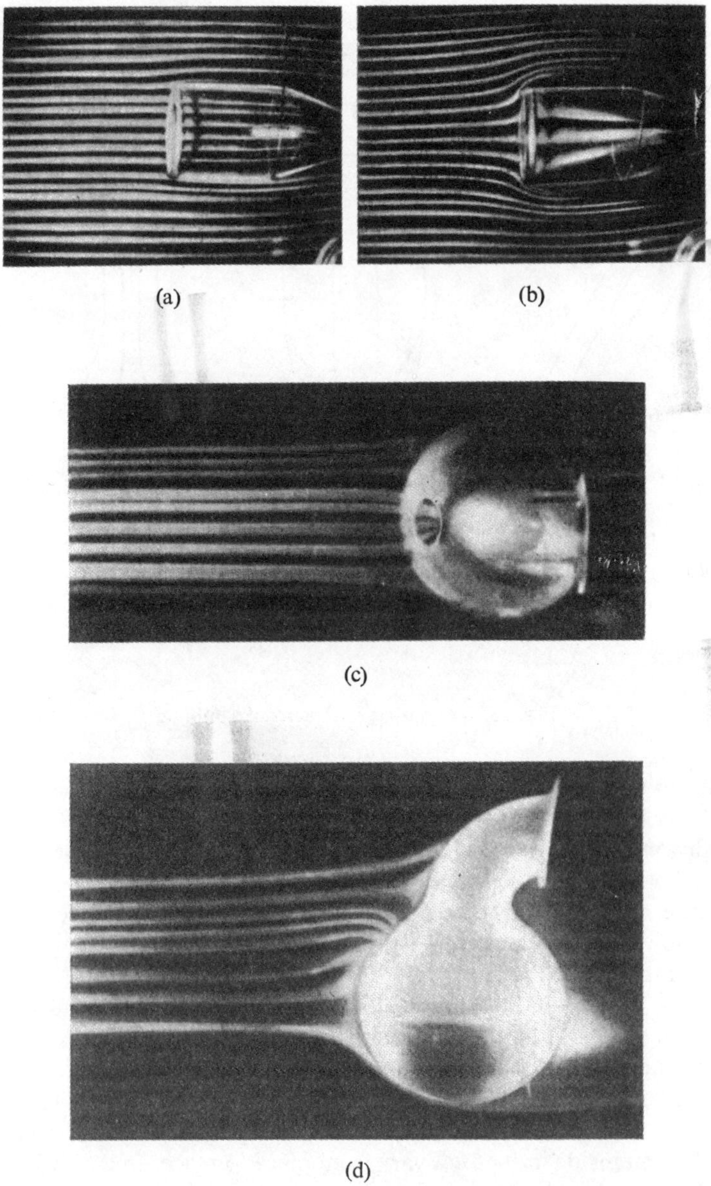

FIG. 39. Photographs of streamlines in sampling.

settling plus the number $V_s S$, which would have passed through if the medium were not moving. This statement follows from the fact that the total velocity of a particle in such a cloud is the vector sum of the velocity of the medium and the rate of settling in a stationary medium.

Hence it follows, in particular, that when an aerosol sample is taken through a horizontal tube from such places as rooms and buildings, correct results can be obtained if the inertia of the particles is negligible. When a sample is taken through a tube pointing vertically upwards the concentration of the aerosol in the sample will be greater than the true concentration by a factor $1 + (V_s/\overline{U})$, where $\overline{U}$ is the mean flow velocity in the tube.

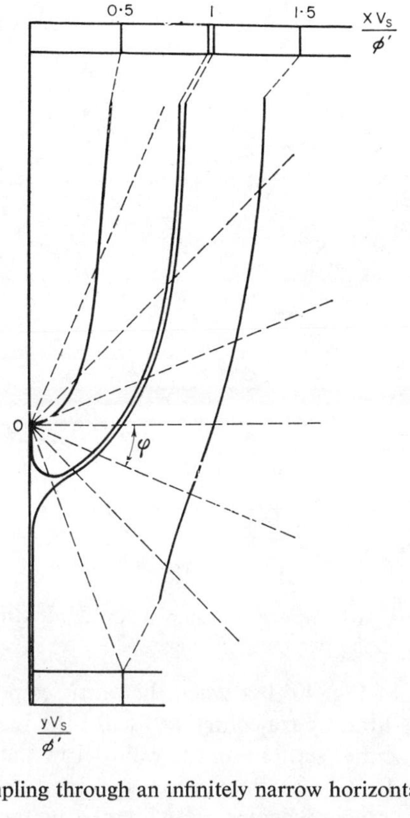

FIG. 40. Sampling through an infinitely narrow horizontal slit.

It is interesting to check the conclusion for horizontal sampling by a concrete example. Imagine that sampling proceeds through an infinitely narrow, infinitely long horizontal slit in an infinitely thin vertical wall.† The streamlines are straight lines (Fig. 40) and the flow velocity is equal to $\Phi'/\pi\varrho$, where Φ' is the volume of gas passing in 1 sec through 1 cm length of slit and ϱ is the distance from the slit. If the inertia of the particles can be neglected the velocity components are

$$V_x = -\frac{\Phi' x}{\pi \varrho^2}, \quad V_y = -\frac{\Phi' y}{\pi \varrho^2} + V_s. \tag{32.1}$$

† An analogous problem is that of sampling through a point opening, discussed by Davies [212].

Hence the equation of the particle trajectory can be found

$$(x_0 - x)\, V_s/\Phi' = \varphi/\pi, \tag{32.2}$$

where the meaning of the angle φ can be seen from Fig. 40 and x_0 denotes the initial coordinate of the particle at $y = -\infty$. If $x_0 V_s/\Phi' < 1$ the particles are drawn into the slit but if $x_0 V_s/\Phi' > 1$ they pass by. Thus, particles initially situated in a layer of thickness $\Delta = \Phi'/V_s$ adjacent to the wall enter the slit. At $y = -\infty$ in 1 sec $N = n V_s \Delta$ settling particles belonging to this layer (n is the aerosol concentration) pass across a strip 1 cm wide stretching along the x axis. The concentration in the sample is clearly $N/\Phi' = n$, i.e. the true concentration.

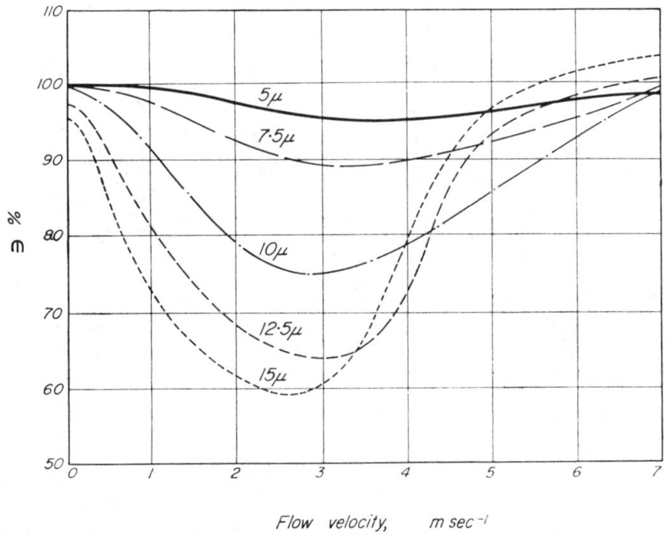

FIG. 41. Efficiency of sampling an aerosol from a flow.

It can be seen from Fig. 40 that when the particles possess inertia, those moving just to the left of the limiting trajectory will leave the inertialess trajectories and not pass through the slit, either settling on the wall below the slit or falling away from it. Thus the combined effect of sedimentation and inertia leads to a reduction of the concentration in the sample whereas, individually, neither sedimentation nor inertia gives this reduction.

The relationship between efficiency of sampling from a flow and the velocity of that flow at a constant sampling rate has been the subject of detailed experimental study by May [213] and the results of his experiments are shown in Fig. 41. Mists of different particle size were passed at various speeds through a horizontal wind tunnel and sucked out at a velocity of 6 m sec^{-1} through a 6 mm circular opening set at an angle of 45° to the horizontal (see Fig. 39d). These graphs have the same form when the aerosols are drawn off through a tube directed against the flow [214]. It can be seen from Fig. 41 that the sampling efficiency in still air is practically 100 per cent. When the air velocity is increased the efficiency first falls but then increases, again reaching 100 per cent under isokinetic conditions. The explanation of this is that, in sampling from still air, the streamlines are either straight or slightly convex towards the axis of the flow and consequently there are no inertia losses. When a sample is taken from moving air, the streamlines at a certain distance from the opening become concave towards the axis. The curvature first increases with the flow velocity, this leading to the loss of more particles, but on further increase in flow velocity the curvature of the streamlines begins to decrease, approaching zero for isokinetic sampling.

That particles of different size are not always arranged in Fig. 41 in the same order with respect to $\ni$ is probably explained by inaccuracies in the measurements. An efficiency of 100 per cent is, on the average, reached at a flow velocity greater than the sampling velocity, this being explained by deceleration of the flow near the instrument into which the sample is being drawn.

Exact coincidence of the sampling and flow velocities at 100 per cent efficiency can be observed only if a long narrow sampling tube with very thin walls is used.

Levin [670] calculated the efficiency of sampling from a uniform aerosol stream with a point sink, or very small orifice, allowing for inertia and particle sedimentation and obtained the formula:

$$\ni = \frac{n}{n_0} = 1 - 0{\cdot}8\,K + 0{\cdot}08\,K^2 + \cdots \qquad (32.3)$$

$$K = \tau \left(\frac{4\pi}{\Phi} \right)^{\frac{1}{2}} (|\,U_0 + V_s\,|)^{\frac{3}{2}} \qquad (32.4)$$

where n_0 is the initial aerosol concentration, n, that of the sample taken, τ, the particle relaxation time, Φ, the volume of the aerosol drawn in 1 sec, V_s, the particle settling velocity and U_0, the velocity of the undisturbed flow. The accuracy of (32.3) which is reduced with increasing K, is about 1 per cent at $K = 0{\cdot}25$ and $2{\cdot}5$ per cent at $K = 0{\cdot}5$. The formula is also valid for a finite orifice provided that the mean velocity in it $\overline{U} \gg 4U_0$.

An increase in concentration upon sampling was not allowed for in this theory since the flow rate in a point sink is always greater than U_0. In this case equations (32.3) and (32.4) show that for a decrease in the sample concentration, besides inertia of the particles, some motion in the undisturbed aerosol, due either to the flow of aerosol (U_0), or to sedimentation under the action of an external force (V_s). These formulae also show that greater aspiration rates should be used in order to obtain true samples at high aerosol flow rates with large particles and that the mean particle size in the sample is smaller than in the original aerosol.

The following formula, which is valid for $\overline{U} \gg \pi U_0$ was obtained for sampling with an infinitely narrow flat slit

$$\ni = 1 - 0{\cdot}451\,K + 0{\cdot}148\,K^2 + \cdots \qquad (32.5)$$

where

$$K = \frac{2\pi\tau}{\Phi'} (|\,U_0 + V_s\,|)^2 \qquad (32.6)$$

and Φ' is the aspiration rate per unit of the slit length.

Walton [688] considered sampling from an aerosol at rest through an inverted funnel with an aperture of radius R (Fig. 41a) fitted with a filter over the surface of which the flow rate U was uniform; the inertia of the particles was ignored. According to the principles outlined above (see page 143), the number of the particles drawn into the filter in 1 sec is equal to $\pi R^2 n_0 (U - V_s)$, and the sampling efficiency $\ni = 1 - V_s/U$. In this case the deposit on the filter must be uniform, which was confirmed by experiment. All this holds good only when the rim round the filter is very small.

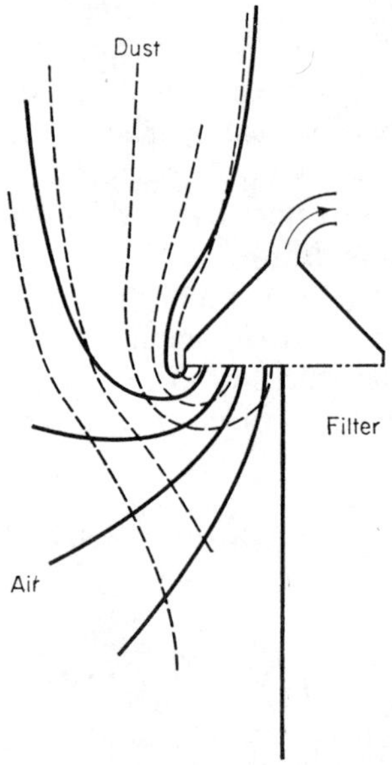

FIG. 41 a. Aerosol sampling with an inverted funnel and filter.

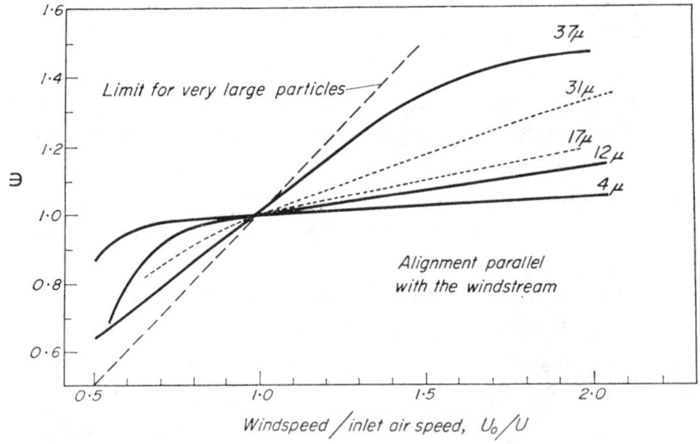

FIG. 41 b. Sampling efficiency and the ratio of flow rates inside and outside a sampling tube.

In Figs. 41 b and 41 c, taken from Watson's paper [689], the results of a number of sampling experiments are given. Fig. 41 b shows the relationship between the efficiency of sampling by a tube directed parallel to the flow and the ratio of the flow rates outside and inside of the tube; in Fig. 41 c the dependence of $\ni$ on the angle between the tube axis and the direction of the flow is plotted. The diameters of the particles are indicated on the graphs.

Imagine an aerosol of concentration n_0, flowing with a velocity U_0, drawn with a velocity U into a sampling tube of radius R, having infinitely thin walls, which is placed parallel to the flow. The gas streamlines entering the tube are confined, at a

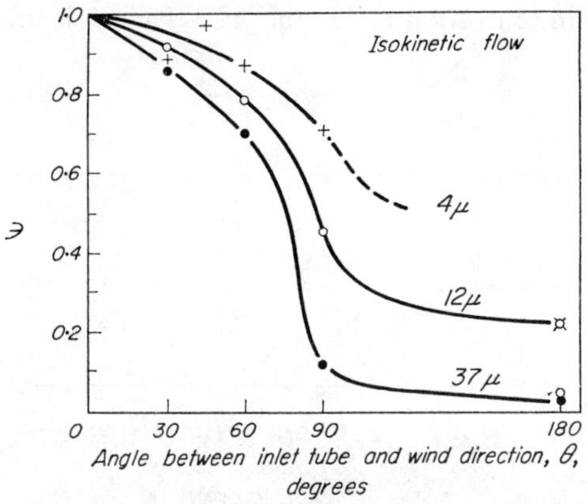

FIG. 41 c. Sampling efficiency and the angle between the axis of the sampling tube and the flow direction.

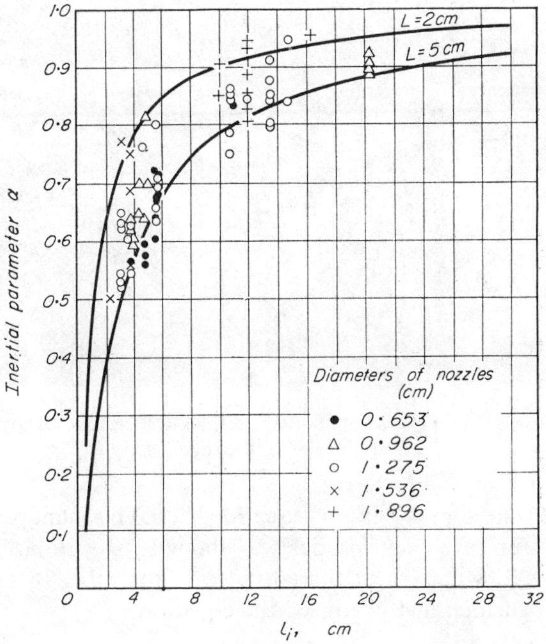

FIG. 41 d. Sampling parameter α and the stop distance of a particles.

distance from the tube, within a cylindrical surface of radius R_0 such that $\pi R^2 U = \pi R_0^2 U_0$. If $U < U_0$, $R > R_0$, the streamlines diverge when approaching the tube and all the particles moving within the cylindrical surface in question ($\pi R_0^2 n_0 U_0$ particles per second) enter the sampling tube. The streamlines issuing from the annular space confined within the cylinders of the radii R_0 and R pass outside the sampling tube but a fraction α of the particles located within this space will enter the tube due to their inertia. The number of these particles is equal to $\alpha\pi(R^2 - R_0^2)\, n_0 U_0$ per second. The sum of these two aerosol fluxes is equal to the flux entering the tube $\pi R^2 n U$, where n is the aerosol concentration inside the tube. Hence, it follows that

$$\frac{n}{n_0} = 1 - \alpha + \frac{\alpha U_0}{U}. \tag{32.7}$$

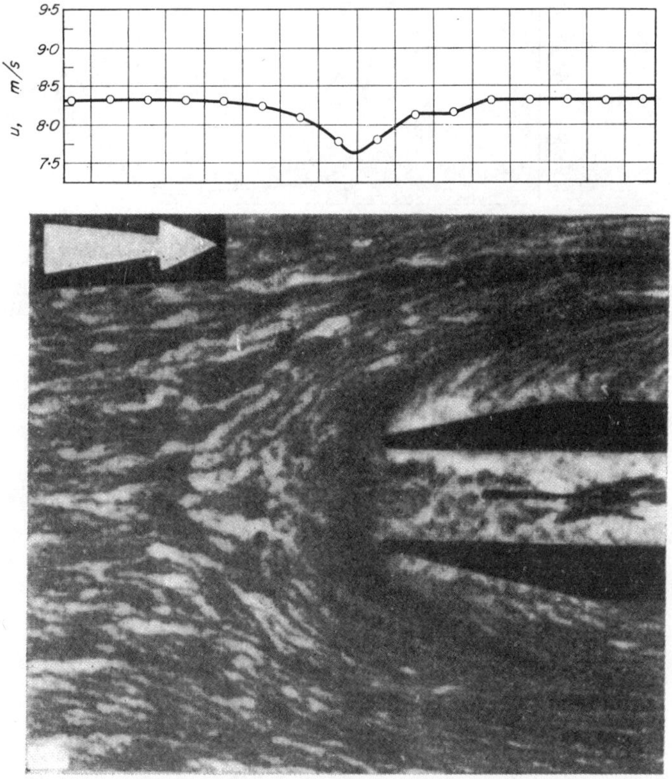

FIG. 41e. Field of flow in isokinetic sampling with a thick-walled tube. Above-flow velocity along the axis of the tube.

In the case of $U > U_0$ the sign of α in this equation should be changed. The value of α varies from about 0 for very small particles to about 1 for very large ones and certainly depends on the flow pattern at the entrance of the tube. By means of rather rough calculations Badzioch [690] deduced the equation

$$\alpha = (1 - e^{-L/l_i})\frac{l_i}{L} \tag{32.8}$$

where l_i is the stop distance of the particles and L the distance from the tube at which a marked divergence (or convergence) of the streamlines begins, a rather indefinite quantity. The author calculated l_i not from (18.11), but from Stokes' formula, which in this case (Re = 6–17) must result in appreciable error.

The results of Badzioch's wind-tunnel experiments, using spherical particles of zinc and rounded particles of a siliceous catalyst ($\gamma = 1{\cdot}4$) with mean sedimentation radii $\bar{r}_s = 10$–$15\,\mu$, are given in Fig. 41 d. As seen from the figure, the experimental points fall between two curves (32.8) having values of L equal to 2 cm and 5 cm. The experiments were made with widely varying values of R (from 0·33 to 0·95 cm), U_0 (from 8 to 24 m/sec) and U_0/U (from 0·25 to 4·1). The assumption of the author about the dependence of L on R is not well substantiated, nor is it confirmed by his experiments, and will not be discussed here. A semi-empirical method of the determination of the relationship between α and $l_i/R = \text{Stk}$ was suggested by Watson [689].

Walter [691] determined the flow pattern when sampling through a thick-walled tube (see Fig. 41 e). As seen from the picture, even for isokinetic sampling a stagnation zone with a lowered flow velocity is formed at the entrance to the tube where the streamlines are considerably distorted. When the aspiration rate was increased the streamlines became much straighter. In agreement with this finding Walter claims that very low (down to 50 per cent) and erratic values of the concentration are obtained by isokinetic dust sampling but when the aspiration rate is increased, the results are stable and more or less correct. This conclusion is evidently at variance with Figs. 41 a and 41 b.

Dennis *et al.* [692] investigated the efficiency of sampling tubes in which isokinetic flow was controlled by equality of the static pressures inside and outside the tube. This condition being observed, the dust concentration in the tube proved to be below the true one. To obtain correct results it was necessary that the external pressure should be greater than the internal, which means that the aspiration rate has to be increased, the increase being the greater the larger the flow rate outside the tube. However, according to Dennis, the effect is explained by the loss of pressure in the tube due to friction and formation of eddies rather than by the incorrectness of the principle of isokinetic sampling.

The problem of aerosol sampling has not yet been solved. The chief difficulty appears to be estimation of the true concentration in flowing, coarse aerosols.

§ 33. IMPINGEMENT INSTRUMENTS

Impingement instruments are widely used for sampling aerosol particles [215], particularly in hygiene investigations. In these instruments the aerosol is passed at a velocity of the order 100–200 m sec^{-1} through a flat slit or circular opening, behind which a glass plate perpendicular to the flow is located. The aerosol stream issuing from the slit impinges against the plate and flows out across it. Because of inertia the particles leave the streamlines and are deposited on the glass plate in a straight line or circular spot which is examined under the microscope. To improve the adherence of solid particles to the plate, the latter can be smeared with a material such as glycerin, poly-isobutylene or a mixture of resin and castor oil. Alternatively, a wet paper lining is used to create high humidity in the chamber of the instrument so that adiabatic expansion of the air leaving the slit causes moisture to condense on the particles and assist their adherence to the plate. If the glass plate is replaced by a slowly rotating Petri dish covered with a layer of culture medium the instrument can be used to determine the number of micro-organisms per cubic centimeter of air [216].

In another sampling instrument, called an Impinger, deposition occurs at the bottom of a test tube of water. The aerosol is sucked through a tube with a drawn-out end which is several millimetres from the bottom of the test tube. When suction commences the air jet displaces the water and uncovers the bottom of the test tube, the particles impinge against the wet glass surface and are washed off by the water.

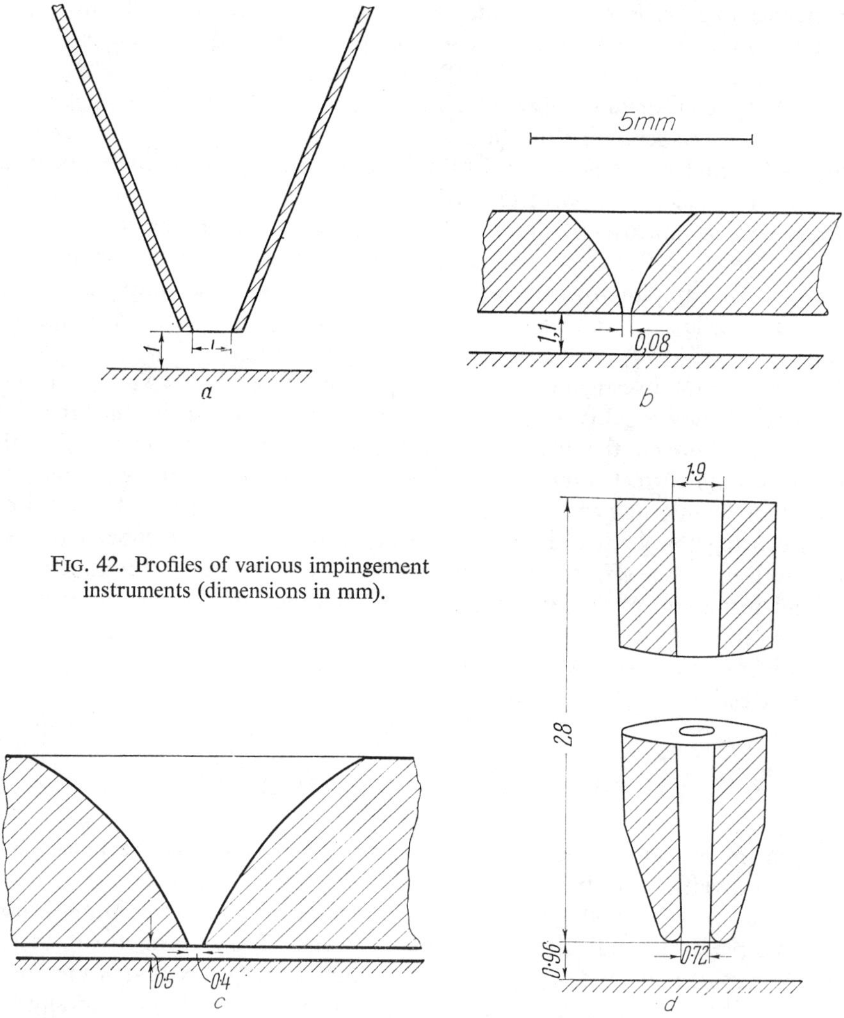

FIG. 42. Profiles of various impingement instruments (dimensions in mm).

Those particles which escape deposition at the bottom of the test tube may be caught as the air bubbles rise through the water. The particles are examined in the liquid suspension.

The walls of the impingement jets are made either flat (Fig. 42a), as in May's impactor [217], or rounded, as in the instruments [218] of Owens (Fig. 42b) or Bausch and Lomb (Fig. 42c). The walls of a circular opening may be made conical, as in the Kotze konimeter (Fig. 42d). As will be seen below, the slit profile in some cases has a substantial effect on the operation of the instrument. The ratio of the distance between the slit and the plate to the slit width, which is often between 1·0 and 2·0, may not have much effect on the collection efficiency.

Davies has attempted to calculate rigorously the motion of particles in impingement instruments [219]. He assumed that the length of the slit is infinitely large compared with the width, 2 h, that the flow is everywhere potential, and that the flow

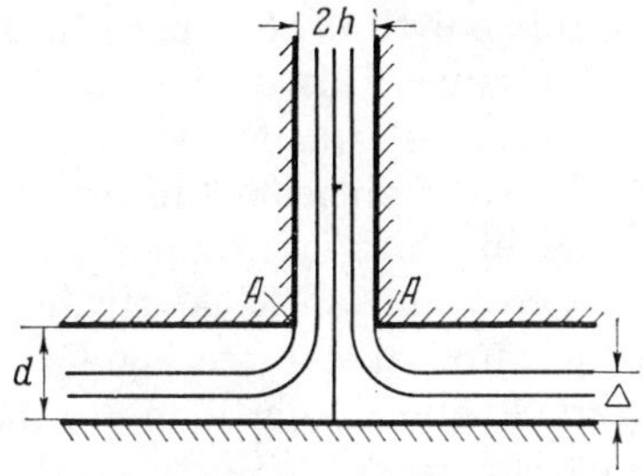

FIG. 43. Flow field in impingement instruments.

within the slit and over the plate at a sufficiently great distance from the slit is rectilinear and parallel to the walls, the velocity being constant throughout the cross-section of the stream (i.e. there is no boundary layer). The flow field was thus determined, and the following relationship between the quantities $\beta = d/2\,h$ and $\alpha = h/\Delta$, where d is the distance from the slit to the plate and Δ is the thickness of the stream flowing along the plate at an infinite distance from the slit (Fig. 43), was derived:

$$2\beta = \frac{(\alpha^2 + 1)\ln(\alpha + 1)}{\pi\alpha^2(\alpha - 1)} + \frac{1}{\alpha}. \tag{33.1}$$

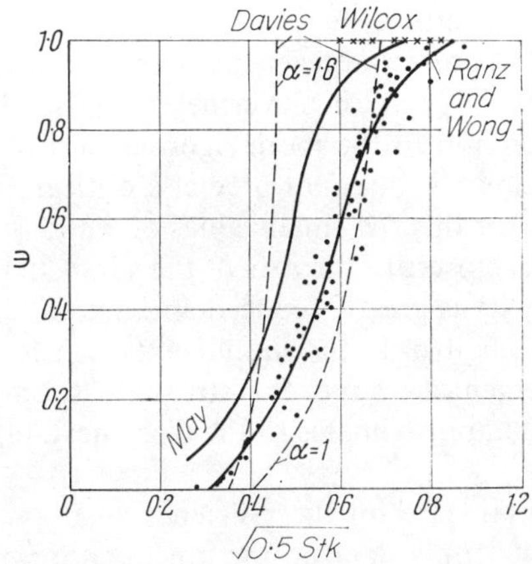

FIG. 44. Efficiency of impingement instruments.

In addition the approximate method given in § 30 was used to calculate particle trajectories so that the collection efficiency could be derived as a function of $\text{Stk} = l_i/h$ for $\alpha = 1{\cdot}0$, 1·6, 2·5 and 5·0. Figure 44 shows curves for $\alpha = 1{\cdot}0$ ($\beta = \infty$) and $\alpha = 1{\cdot}6$ ($\beta = 0{\cdot}8$) together with data obtained in other theoretical and experimental investigations which are discussed below. Values of $\sqrt{0{\cdot}5\,\text{Stk}} = \sqrt{l_i/2h}$, proportional to the radius of the particles, are plotted as absissae. It can be seen from Davies' curves that there exist critical values of the Stokes number (Stk_{cr}) below which there is no deposition in an impingement instrument. When $\alpha = 1$, $\mathrm{Stk}_{cr} = 0{\cdot}32$ and when $\alpha = 1{\cdot}6$, $\mathrm{Stk}_{cr} = 0{\cdot}24$. These values are close to those found experimentally (see below). However, in calculating the motion of particles along the central streamline passing through the axis of the slit, Davies arrived at the value $\mathrm{Stk}_{cr} = 0{\cdot}64$ which agrees with the value $2/\pi$ found analytically (see page 163) by Levin [220], but differs greatly from the experimental value. This discrepancy, which occurs similarly for other values of α, is rather difficult to explain.

The collection efficiency for an impingement instrument can be estimated by the following approximate method, in which it is supposed that $\Delta = h$, the flow velocity U_0 is constant over the entire cross-section of the stream and the streamlines at the bends are arcs of circles with centres at AA (see Fig. 43). It is also assumed that the tangential velocities of a particle and the medium are the same. The velocity of a particle normal to a streamline is $V_\varrho = U_0^2\tau/\varrho$ where ϱ is the radius of curvature of the streamline. In a time dt the particle is displaced relative to the medium by a distance $d\varrho = V_\varrho\, dt$ in the direction of ϱ or by $dz = V_\varrho \sin\varphi\, dt$ towards the plate. Considering the total displacement of the particle relative to the medium to be small for the time taken to traverse the curve, i.e., ϱ = constant, we get the following value for the component of this displacement towards the plate†

$$\Delta z = \int \frac{U_0^2 \tau \sin\varphi}{\varrho}\, dt = U_0 \tau \int_0^{\pi/2} \sin\varphi\, d\varphi = U_0\tau = l_i. \tag{33.2}$$

Thus the simple expression

$$\ni = l_i/h = \mathrm{Stk} \tag{33.3}$$

is obtained for the collection efficiency.

Although formula (33.3) predicts efficiencies of the the order of magnitude as those found by experiment at large $\ni$, it nevertheless possesses serious fundamental shortcomings. In contradiction to the more rigorous derivation given above and to experimental data (see below), it does not predict a critical Stokes' number. It is not difficult to see the reason for this. At small values of $\ni$ and Stk deposition only takes place from the streamlines closest to the axis of the slit, which pass near a stagnation point where the flow velocity is greatly reduced and inertial displacement of particles from the streamlines is negligible. In the simplified flow field the velocity along these lines and inertial displacement are large. For streamlines remote from the axis of the slit the simplified flow field approximates to the true one, and the crude calculation is more nearly correct.

Turning to experimental data on the efficiency of impingement instruments, it must be borne in mind that only data on the precipitation of mists can be used for comparison with theory because various complications, which will be discussed below, arise in experiments with solid particles. The most accurate data appear to be those of Ranz and Wong [222] for fairly monodisperse glycerin mists with mean particle radii ranging from 0·17 to 0·69 μ in various experiments. Flat slits 0·2 – 0·7 mm wide and circular openings 1·0 – 1·9 mm in diameter were used with velocities of

† The incorrect value $\frac{\pi}{2} l_i$ was obtained for Δz in Wilcox's work [221].

$10-180$ m sec^{-1} in the slit. The ratio $d/2h$ ranged from 1 to 3 and had no noticeable effect on the results. The results of the experiments with plane slits are shown in Fig. 44 (experimental points and the curve constructed through them.) A value of about 0·22 is obtained for Stk_{cr}. If the abscissae of all points of the curve for the plane slit are reduced by a factor of 1·5 the curve for a circular opening is reproduced fairly exactly; a circular orifice thus gives a greater efficiency than a slit of the same width.

May's experiments [217] were conducted with a cascade impactor in which the aerosol was passed successively through four jets, each with a smaller slit width and greater flow velocity than the preceding one, so that progressively finer fractions were retained. Efficiencies obtained at the third jet with a slit width of 1 mm and velocity of 50 m sec^{-1} are shown on Fig. 44. The curves given in May's article for the collection efficiency of each jet as a function of particle size terminate at about Stk = 0·12 (see Fig. 44). It may be conjectured that Stk_{cr} lies close to this value but May himself does not mention this and arbitrarily plots his (∍, Stk) curve through the origin of co-ordinates.

In most work on impingement instruments only the sizes of the smallest particles which are completely precipitated in the instrument at a given velocity, have been determined; these give the minimum values of Stk for ∍ = 1. The results of Wilcox's many experiments [221] on instruments of various designs with slit widths from 0·25 to 6 mm and velocities from 2·2 to 174 m sec^{-1} are widely scattered, as can be seen in Fig. 44. It is rather difficult to find any relationship between the operation conditions of the instrument and the values of Stk corresponding to complete precipitation.

It follows from Fig. 44 that the agreement between theoretical and experimental efficiency curves and also between the data of different authors is quite poor, and it is difficult at the present time to find any fundamental explanation for the discrepancies. A feature in which the theoretical curves differ from the experimental is that the latter bend appreciably to the right at large values of ∍. This means that particles moving along the streamlines farthest from the axis of the slit are especially poorly precipitated and the theoretical calculations presented above are not applicable to them. This is most probably due to vortices which are unavoidably formed at the free surface of the jet as it leaves the slit and which cause its surface to be ill-defined.

The efficiency may be reduced considerably for solid aerosol particles, which, after deposition, may be swept off the plate. Coarse particles are swept off more easily than fine ones, under the conditions of operation of impaction instruments, with the result that an upper limit exists as well as a lower limit to the size of particles collected. Using carborundum dust, Pik and Shurchilov [223] observed that the maximum radius of the particles collected in an Owens' instrument was 1·8 μ dry and 3·7 μ with moistening.

In experiments with an instrument shown in Fig. 42a Jordan [224] found that quartz particles of radius 2 μ and 1 μ are swept off a dry glass plate to a noticeable extent when the velocity in the slit reaches 60 and 150 m sec^{-1}, respectively. In instruments with circular openings this leads to a curious effect. Around a dense central deposit about 1 mm in diameter formed immediately opposite the opening it is often possible to observe a diffuse ring about 5 or 10 mm in diameter made up of particles swept by the flow from the central deposit and precipitated again when the velocity has fallen sufficiently [225].

This sweeping away of the particles is particularly pronounced in the Owens' instrument (see Fig. 42b), where the air velocity is very high (300 m sec^{-1}), and becomes noticeable even for particles of radius 0·5 μ. The effect is much smaller in the Bausch and Lomb instrument (Fig. 42b) in which the air velocity is only 150 m sec^{-1} [218].

The best means of prevention is to smear the plate with a viscous sticky composition which is a practically complete remedy for coal dust particles with radii up to 4 μ at flow velocities of 200 – 300 m sec^{-1} in the slit. Humidifying the air gives much poorer results, not only for coal dust but also for hydrophilic quartz dust [218].

Another secondary phenomenon inherent in impingement instruments is the disintegration of particles on impact with the glass plate. This is observed with such hard and brittle materials as quartz and orthoclase at particle diameters of 5 – 10 μ [226]. It is interesting that softer plastic particles such as plant spores remain whole. In work with dusts containing aggregated particles the aggregates are observed to be broken up by the air stream before hitting the plate. That the aggregates do actually break up before coming into contact with the plate was demonstrated by Davies [218] in the following way. Coal dust was passed through a 0·4 mm slit, the number concentrations of dust and the particle size distributions being determined before and after the slit with a thermal precipitator. At a velocity of 50 m sec^{-1} the dust was unchanged, but at 170 m sec^{-1} there was a considerable increase in the number of fine particles and a decrease in the coarse particles. When the dust was sampled with an impingement instrument using a well-greased plate the number of particles of radius 1 μ did not change; above 1 μ the number fell slightly and below 1 μ the number increased several times. In humid air the effect is smaller because moisture makes the aggregates stronger. It is strongly dependent on the slit profile, being much greater for rounded than for conical profiles. Davies attributes the disintegration of the aggregates to the enormous velocity of the particles relative to the medium at the slit entrance as a result of the inertia of the particles. In the slit illustrated in Fig. 42b the relative velocity is estimated to reach 150 m sec^{-1} for particles of radius 10 μ. Clearly, for a conical profile, which produces uniform acceleration of the flow, the relative velocity must be less than for a rounded profile. Another important factor in the disintegration of particles by an air stream is non-uniformity of the flow, which results in the existence of very large velocity gradients at the slit entrance (see page 374).

According to Davies, deposition of particles on the walls of the slit occurs when the profile is rounded. This increases with the linear air velocity in the slit and is therefore particularly pronounced in Owens' instrument.

Some equipment for removing dust from gases is based on this same principle of an aerosol jet impinging on a surface: the gas passes through a large number of holes 2 – 5 mm in diameter in a metal plate and impinges against another plate parallel to the first and 2 – 3 mm distant from it [227]. Unfortunately no data are available on the efficiency of these devices for particles of a specified size and so no comparison with theory is possible.

A comprehensive study of konimeters, which are impingement instruments with circular section jets, was made by Röber [693]. He used three kinds of nozzle, a cylindrical one, a conical and one having a curved profile, such that the flow velocity increased linearly with the distance from the inlet. The second and third types had the ratio of the diameters of the inlet and outlet apertures equal to 10.

Röber's experiments show that the flow in the nozzles was laminar although the Re_f numbers were as large as 4000–5000, a fact which was not allowed for in earlier work. There was not enough time for the parabolic velocity profile to become established in the nozzles, and by measuring the profile the author calculated the velocities attained by aerosol particles having various values of τ in nozzles of different shape. The relative velocity of the particles and the air was minimal in nozzles with a profile between cylindrical and curved. Such nozzles would cause the least possible disruption of aggregates.

Several factors contribute to displacing aerosol particles towards the axis of the nozzle. These include convergence of the streamlines due to the change in the profile of flow velocity, the effect described on p. 37 and inertial displacement of particles in nozzles with curvilinear profiles. The latter effect is the most important. Experiments

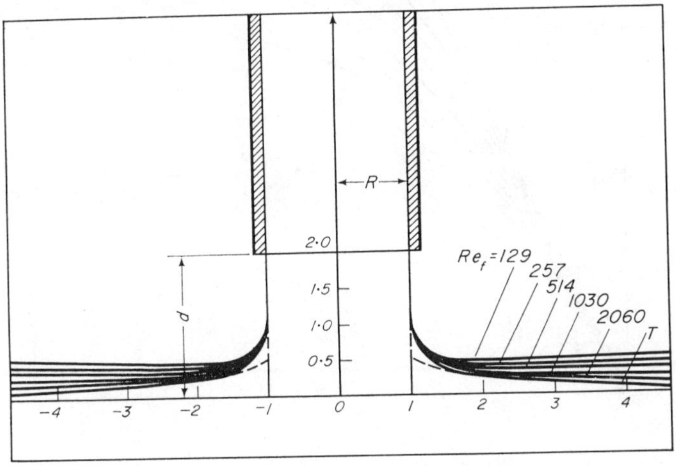

FIG. 44a. Contours of the air-jets issuing from a konimeter at different values of Re_f.

with polydisperse coal dust at $\bar{U}$ = 50–100 m/sec and a jet radius R = 0·2 mm indicated an appreciable narrowing of the dust jet (by about 15 per cent) compared to the width of the air jet in nozzles of curved profile. This could not be established with any degree of certainty for other types of nozzle. Particles moving near the walls of a nozzle issue from it at a relatively low velocity, so it is evident that the narrowing of the aerosol jet contributes to the efficiency of the konimeter, which in its turn shows the advantage of using nozzles with a curved profile.

Theoretical calculation of the flow field is more difficult for the jet issuing from a konimeter than for instruments which have a slit orifice. Röber measured the velocity contours of a smoke-jet issuing from a cylindrical tube of $2R$ = 10 mm with a parabolic profile of velocities; the distance d from the tube to the plate was equal to 10 mm and the equality of $2R$ and d was also observed in the experiments described below. His results are shown in Fig. 44a. The curve T corresponds to the ideal case when the spreading jet retains its mean velocity. Röber concluded that near the tip of a nozzle, where practically all the deposition of particles occurs, the flow pattern is similar at all speeds.

Rough calculation of the trajectories of aerosol particles, attempted by Röber is not in good agreement with the results of his experiments with *lycopodium* spores

The length of the cylindrical tubes, which had $R = 0{\cdot}45$ cm in these experiments, was sufficient to ensure that the parabolic profile of velocities was established, and that the spores had attained the velocity of the air. Re_f varied between 280 and 9500. The particle trajectories between the tube and the plate were photographed. The results of the experiments are shown in Fig. 44b, where $x_1 = \varrho_1/R$, $x_2 = \varrho_2/R$; ϱ_1 and ϱ_2 are the distances between the particles and the axis at the beginning and at the end of the trajectory; $\theta = R/\tau\bar{U}$. For *lycopodium* spores $\tau = 0{\cdot}0028$ sec. The straight line in Fig. 44b corresponds to $\tau = \infty$.

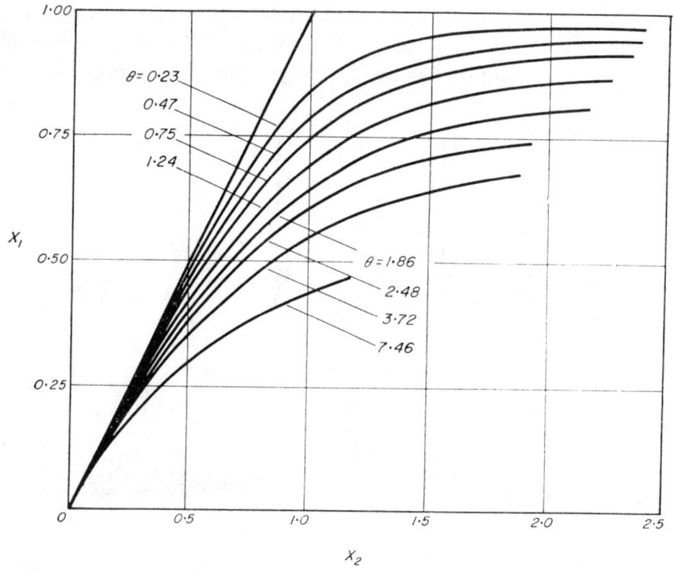

FIG. 44b. Particle trajectories in a konimeter.

The data on the mechanism of the formation of annular deposits in konimeters obtained by Röber are of great interest. These deposits are not formed when the plates are properly coated with adhesive, which shows that they are associated with blowing off or the rebounding of the particles after their initial contact with the plate. The shape of the deposit is due to eddies occurring beyond the point where the laminar boundary layer detaches itself from the plate and not to the decrease in flow velocity in the vicinity of the plate, as was stated above. The following facts support this statement. The deposits are formed only at a flow rate somewhat in excess of a certain critical value, the diameter of the deposit decreases with an increase in the flow rate and the inner edge of the deposit is very sharp. The picture would be quite different, if the formation of the deposit were due to a decrease in the flow velocity.

The data in the last part of Röber's paper connected with the adhesion of dust particles to the plate are considered in § 56.

On the basis of his investigations, Röber came to the conclusion that the optimal nozzle is that with a curved profile 15 mm in length having the diameters of the inlet and outlet apertures equal to 3 and 0·3 mm respectively; the mean air velocity at the outlet should be equal to 200 m sec^{-1}. His deduction that in such a konimeter all particles with $r \geqslant 0{\cdot}25\,\mu$ (at $\gamma = 1$) should be deposited is based on rather rough approximations and is unconvincing.

Cascade impactors have been widely used, particularly in the United States, for the determination of particle size in dusts, the following technique proving to be the most convenient. The impactor is first calibrated, the mean particle size of a given dust deposited on each stage being determined with the help of a microscope. The mass of the deposit on each stage is then measured by weighing on a microbalance, or by some analytical method, so that the fractional composition of the dust is evaluated.

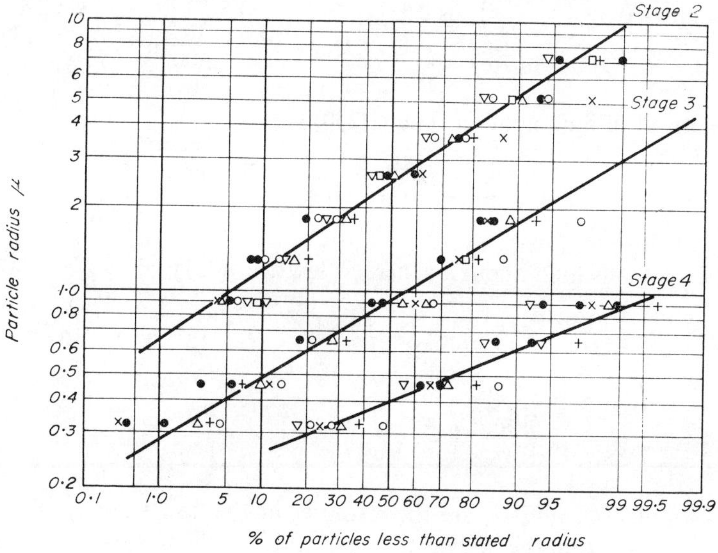

FIG. 44c. Deposition of U_3O_8 particles in a cascade impactor.

The completeness of the separation of different fractions in the impactor is of great importance. Some measurements made by Lippman [694] with U_3O_8 dust are shown in Fig. 44c, the dust concentrations being different in every case. The graphs show that a lognormal particle size distribution was observed in the deposits obtained in all three stages, the sizes of the smallest particles deposited on the second stage and the largest on the fourth stage being equal; the separation was thus far from perfect. The median radius and the mass of the deposit on each stage, however, are sufficient indications for practical purposes of the size range of the dust.

§ 34. INERTIAL AND ELECTROSTATIC DEPOSITION OF FLOWING AEROSOLS ON OBJECTS OF SIMPLE SHAPE

In spite of the great practical importance of inertial deposition from an aerosol stream upon objects of various shapes, little experimental work on this subject has been published. Much more has been done on calculating the collection efficiency theoretically (see page 135). These calculations have been carried out by the method given in § 30 for spheres and infinitely long cylinders and ribbons placed perpendicularly to the flow. The following hydrodynamic equations (given in dimensionless variables) provide the basis for such calculations.

1. Flow of an ideal fluid about an infinitely long circular cylinder (potential flow) [228]†

$$U_x = 1 - \frac{\cos 2\theta}{\varrho^2}, \quad U_y = -\frac{\sin 2\theta}{\varrho^2}. \tag{34.1}$$

2. Flow of a viscous fluid about an infinitely long circular cylinder at low Re_f [229]

$$U_x = \frac{\ln \varrho - 0{\cdot}5\left(1 - \dfrac{1}{\varrho^2}\right)\cos 2\theta}{2{\cdot}002 - \ln Re_f}, \quad U_y = -\frac{0{\cdot}5\left(1 - \dfrac{1}{\varrho^2}\right)\sin 2\theta}{2{\cdot}002 - \ln Re_f}. \tag{34.2}$$

3. Flow of an ideal fluid about a sphere [230]

$$U_x = 1 + \frac{1 - 3\cos^2\theta}{2\varrho^3}, \quad U_y = -\frac{3}{4}\,\frac{\sin 2\theta}{\varrho^3}. \tag{34.3}$$

4. Flow of a viscous fluid about a sphere at low Re_f [231]

$$U_x = 1 - \frac{3}{4\varrho}(1 + \cos^2\theta) - \frac{1}{4}\,\frac{1}{\varrho^3}(1 - 3\cos^2\theta),$$

$$U_y = -\frac{3}{8}\,\frac{\sin 2\theta}{\varrho}\left(1 - \frac{1}{\varrho^2}\right). \tag{34.4}$$

In these formulae U_x and U_y are the resolutes of the flow velocity; the velocity at an infinite distance from the body is taken as unity (in the positive direction of x); ϱ is the distance from the centre of the sphere or the axis of the cylinder which are taken to be of unit radius; θ is the angle between the radius vector and the x axis. The flow fields in cases 3 and 4 are illustrated in Figs. 45a and b.

The following points should be noted with regard to the above formulae. Equations (34.2) and (34.4) describe accurately the motion of a medium close to a cylinder or sphere, but become more and more inaccurate with increasing distance from the body. The motion of particles close to the body is of fundamental importance in the calculation of collection efficiency, and so calculations based on these formulae are probably close to reality. It is well known that Oseen's equations for the flow of a viscous fluid round a sphere are better than those of Stokes (34.4) for describing the motion of the medium a long way from the sphere, although they are less satisfactory close to the sphere [233]. It is therefore desirable to use Stokes' formulae for impaction calculations.

As Re_f increases, the envelope of flowing medium, which has been decelerated by the body and surrounds it under viscous flow conditions, gradually becomes thinner and turns into a boundary layer. When this happens the nature of the flow downstream of the body changes abruptly, vortices being formed and a turbulent wake developing. The flow upstream of the body remains laminar even at very large Re_f, unless the stream itself is turbulent, and formulae (34.1) and (34.3) are accurate

† See [232] for potential flow about an infinitely long ribbon.

right up to the boundary layer across which the velocity varies from zero to the value given by these equations. The boundary layer thickness decreases as Re_f increases. Since inertial deposition, calculated by the formulae given above, occurs on the upstream side of the body, vorticity and turbulence downstream of the body will have no effect. On the other hand the vortices sometimes lead to deposition of particles on the downstream side of the body (see page 271).

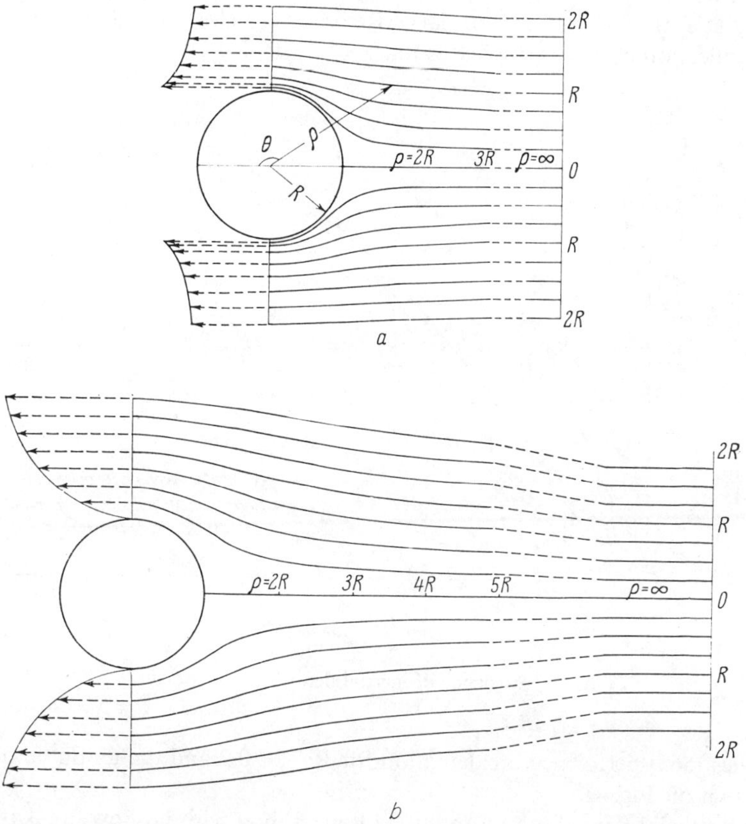

FIG. 45. Flow field for (a) potential and (b) viscous flow round a sphere.

Passing on to an account of published results for the calculation of collection efficiency on bodies of simple shapes, it must be emphasized that in most of these calculations no allowance was made for the interception effect and none took into account the effect of the boundary layer. They therefore apply only to the particles which are small compared with the body and, for potential flow, only at large Re_f. Errors introduced by these simplifications will be discussed below.

For potential flow about an infinitely long cylinder the first reliable data were those of Langmuir and Blodgett [234, 235] and Mazin [206] (Fig. 46), whose calculations are in very close agreement. In particular, Langmuir and Blodgett obtained a value of 0·125 for Stk_{cr}. The older work by Sell [204] and Albrecht [205] in this and other cases discussed below is sometimes obviously wrong and is not given here.

Landahl [236] performed a very rough calculation for flow about a cylinder at $\mathrm{Re}_f = 10$, using the flow velocity distribution that Thom [237] had calculated for this case, and arrived at the empirical formula

$$\ni = \frac{(\mathrm{Stk})^3}{(\mathrm{Stk})^3 + 1{\cdot}54(\mathrm{Stk})^2 + 1{\cdot}76}. \tag{34.5}$$

It can be seen from Fig. 46 that at large values of Stk the curves for collection efficiency at $\mathrm{Re}_f = 10$ and at very large Re_f are very close to one another; at small Stk the collection efficiency increases markedly with Re_f.

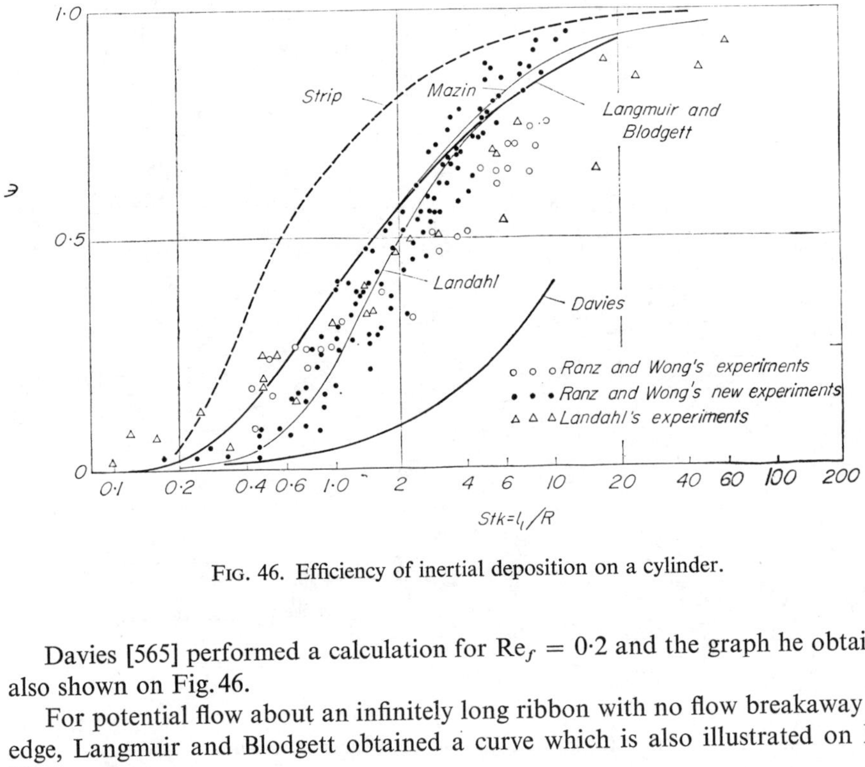

FIG. 46. Efficiency of inertial deposition on a cylinder.

Davies [565] performed a calculation for $\mathrm{Re}_f = 0{\cdot}2$ and the graph he obtained is also shown on Fig. 46.

For potential flow about an infinitely long ribbon with no flow breakaway at the edge, Langmuir and Blodgett obtained a curve which is also illustrated on Fig. 46 [238].

According to Langmuir and Blodgett [239] the collection efficiency for potential flow about a sphere can be expressed in the region $\mathrm{Stk} \geqslant 0{\cdot}2$ by the empirical formula (Fig. 47)

$$\ni = \frac{(\mathrm{Stk})^2}{(\mathrm{Stk} + 0{\cdot}25)^2}. \tag{34.6}$$

The value obtained for $\mathrm{Stk}_{\mathrm{cr}}$ is $0{\cdot}0834 \approx 1/12$. For viscous flow about a sphere Langmuir and Blodgett [239] obtained a curve which is represented by the empirical formula

$$\ni = \left[1 + \frac{0{\cdot}75 \ln(2\,\mathrm{Stk})}{\mathrm{Stk} - 1{\cdot}214}\right]^{-2}. \tag{34.7}$$

As would be expected, the collection efficiency is less than for potential flow and becomes zero at Stk = 1·214. For the intermediate hydrodynamic region Langmuir has suggested that ∍ for a sphere should be calculated by interpolating between the values found for potential and viscous flow [239]. However, this method lacks sufficient basis and the results obtained are unreliable; the results are therefore not quoted here.

It is worth mentioning that though ∍ decreases as the size of an obstacle increases, the total amount of deposit on it increases. This can be seen from the (∍, Stk) curves which are concave towards the axis of abscissae so that ∍/Stk, and hence ∍R decreases as Stk increases, that is as R decreases.

A general analytical method for determining Stk_{cr} for potential flow round obstacles of various shapes has been indicated by Levin [220]. Near a stagnation point

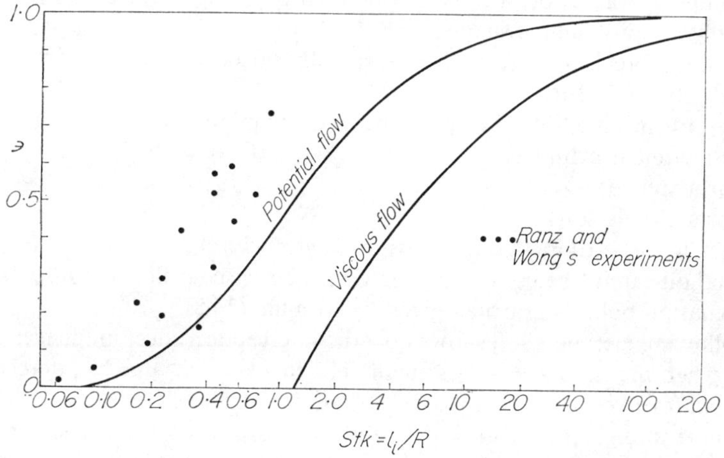

FIG. 47. Efficiency of inertial deposition on a sphere.

the flow velocity on the central streamline can be expressed approximately by

$$U_x = -ax, \tag{34.8}$$

where the stagnation point is taken as the origin. For flow round a circular cylinder the velocity on the central streamline ($\theta = \pi$), according to (34.1), is $1 - 1/\varrho^2$. With the stagnation point as origin and small values of x this reduces to

$$U_x = 1 - \frac{1}{(1-x)^2} \approx -2x, \tag{34.9}$$

hence $a = 2$. The differential equation of motion for particles along the central streamline [see (30.2)] is

$$\tau \frac{d^2x}{dt^2} + \frac{dx}{dt} - U_x \equiv \tau \frac{d^2x}{dt^2} + \frac{dx}{dt} + ax = 0. \tag{34.10}$$

The corresponding characteristic equation is $\tau\lambda^2 + \lambda + a = 0$ and for $1 - 4\tau a < 0$ the roots will be the complex conjugates $\lambda = \alpha \pm i\beta$. The general solution to equation (34.10) is therefore

$$x = A\cos\alpha t + B\sin\beta t. \tag{34.11}$$

Since the last equation has finite roots for any finite values of A and B (the points of intersection of the curves $A\cos\alpha t$ and $-B\sin\beta t$) then if $1 - 4\tau a < 0$ or $\tau > 1/4a$, the particles must reach the stagnation point in a finite time. If $\tau < 1/4\,a$ the particles reach the stagnation point only when $t = \infty$, as shown by Levin. Since in dimensionless co-ordinates, $\text{Stk} \equiv U_c\tau/R = \tau$, this gives†

$$\text{Stk}_{cr} = 1/4\,a. \tag{34.12}$$

The following values of Stk_{cr} were obtained by using this method for potential flow past obstacles of simple shape.

$1/4$ without breakaway and $4/(\pi + 4)$ with breakaway for an infinitely long ribbon of breadth $2\,R$;
$1/4\,(1 + \varkappa)$ for an elliptical cylinder with major/minor axis ratio $\varkappa$;
$1/8$ for a circular cylinder;
$1/12$ for a sphere;
$\pi/16$ for a circular disk;
$2/\pi$ for a flat jet striking a plane at right angles with $d/2\,h = \infty$ (see Fig. 43).

The method cannot be used to calculate Stk_{cr} for viscous flow because the velocity at the stagnation point is not then given by formula (34.8).

The following method [243] shows how the interception effect influences collection efficiency when the ratio $\varkappa = r/R$ is small. The increase in $\ni$ due to interception will be calculated for two extreme cases, first, when the inertia of a particle is so large that it moves in a straight line ($\text{Stk} = \infty$) and secondly when a particle possesses no inertia and moves along a streamline ($\text{Stk} = 0$). In the first case the increment in $\ni$ is

$$\Delta\ni = \frac{r + R}{R} - 1 = \frac{r}{R} = \varkappa \tag{34.13}$$

for a cylinder and

$$\Delta\ni = \left(\frac{r + R}{R}\right)^2 - 1 = \varkappa^2 + 2\varkappa \approx 2\varkappa \tag{34.14}$$

for a sphere. For potential flow round a cylinder of unit radius the equation of the streamlines is $\psi = (\varrho - 1/\varrho)\sin\theta = C$ (Fig. 48). For $\varrho \to \infty$, $C = \varrho\sin\theta$. Thus the constant C is the distance h from the corresponding streamline to the axis at an infinite distance from the cylinder. At $\theta = \pi/2$, in the equatorial section, $C = \varrho - 1/\varrho$. With the particle in this section, as illustrated in Fig. 48, $C = 1 + \varkappa - 1/(1 + \varkappa)$ and, since $\ni = 0$ for an inertialess particle with no interception effect,

$$\Delta\ni = 1 + \varkappa - \frac{1}{1 + \varkappa} \approx 2\varkappa. \tag{34.15}$$

† Taylor [240] derived formula (34.12) in 1940, but his proof that the particles cannot reach the stagnation point if $\tau < 1/4a$ is not rigorous because he assumed that the velocities of the particles and of the gas stream are equal not at $x = -\infty$ but at a point where equation (34·8) holds.

For potential flow round a sphere $\psi = (\varrho^2 - 1/\varrho)\sin^2\theta = C^2$ and

$$\Delta\ni = (1+\varkappa)^2 - \frac{1}{1+\varkappa} \approx 3\varkappa. \tag{34.16}$$

For viscous flow round a sphere $\psi = (\varrho^2 - 1{\cdot}5\,\varrho + 1/2\,\varrho)\sin^2\theta = C^2$ and

$$\Delta\ni = (1+\varkappa)^2 - \frac{3}{2}(1+\varkappa) + \frac{1}{2(1+\varkappa)} \approx \frac{3}{2}\varkappa^2. \tag{34.17}$$

For viscous flow round a cylinder it follows from (34.2) that

$$\Delta\ni = \left[(1+\varkappa)\ln(1+\varkappa) - \frac{\varkappa(2+\varkappa)}{2(1+\varkappa)}\right] \Big/ (2{\cdot}002 - \ln \mathrm{Re}_f), \tag{34.18}$$

$$\approx \frac{\varkappa^2}{2{\cdot}002 - \ln \mathrm{Re}_f}.$$

Thus $\Delta\ni$ lies between $\varkappa$ and $2\varkappa$ for potential flow round a cylinder and between $2\varkappa$ and $3\varkappa$ for a sphere.

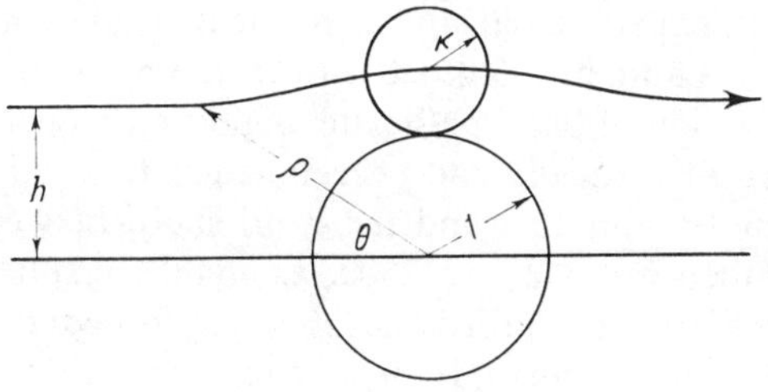

FIG. 48. The interception effect.

In practice the particle size may be of the same order as the transverse dimension of the obstacle. For example, atmospheric aerosols are sometimes sampled by deposition on very fine threads (of thickness 0·1 μ) stretched across the wind. The threads may be drawn out of solutions of high-polymer [244] or the thread spun by certain spiders may be used [245]. If the particle size is greater than the diameter of the thread, the deviation of the particle trajectory from a straight line can be neglected and the collection efficiency is practically equal to $1 + r/R$ [see (34.13)].

A boundary layer affects collection efficiency since the streamlines closest to the obstacle are moved slightly away from it. Using the "displacement thickness", δ, [246] of the boundary layer, the flow velocity will be taken as zero inside the boundary layer and given values according to equations (34.1) and (34.3) on the outside of the boundary layer.† In this case the streamline nearest the body moves away from it by a distance δ. As an approximation it will be supposed that the trajectories of particles which determine the value of $\ni$, namely those farthest from the axis which are still deposited on the obstacle, move away from it by the same amount. Hence the boundary layer effect has the same magnitude as the interception effect for particles of radius $r = \delta$, but the opposite sign.

† The boundary layer is taken as laminar up to the breakaway point ($\mathrm{Re}_f < 10^5$).

The displacement thickness increases slightly in passing from the stagnation point to the equatorial section, near which the boundary layer breaks away from the body. The ratio $\delta\ \mathrm{Re}_f^{\frac{1}{2}}/R$ for a cylinder is equal to 0·44 at the stagnation point ($\theta = 0°$), 0·57 at $\theta = 60°$ and 0·80 at $\theta = 90°$ [247]. Values of δ for a sphere are almost thes ame. Since the particles which determine the value of $\ni$ are deposited roughly at $\theta = 60$–$80°$, $\delta\ \mathrm{Re}_f^{\frac{1}{2}}/R$ will be given the value 0·7. For cylinders, therefore, the reduction in $\ni$ owing to the presence of a boundary layer lies, approximately, between 0·7 $\mathrm{Re}_f^{-\frac{1}{2}}$ and 1·4 $\mathrm{Re}_f^{-\frac{1}{2}}$; for spheres it is between 1·4 $\mathrm{Re}_f^{-\frac{1}{2}}$ and 2·1 $\mathrm{Re}_f^{-\frac{1}{2}}$.

It would be interesting to calculate the particle trajectories and collection efficiency, taking the boundary layer and interception effect into account. This would necessitate the determination of $\ni$ as a function of the two parameters Stk and Re_f or, what amounts to the same thing, of Stk and r/R.

Davies [565] performed such a calculation for a cylinder at $\mathrm{Re}_f = 0{\cdot}2$ and expressed his results by the empirical formula

$$\ni = 0{\cdot}16\,[\varkappa + (0{\cdot}25 + 0{\cdot}4\,\varkappa)\,\mathrm{Stk} - 0{\cdot}0263\,\varkappa\,\mathrm{Stk}^2]. \tag{34.19}$$

This formula has not been verified experimentally.

Let us turn now to experimental studies of the inertial deposition of aerosols on simply shaped objects. Landahl [236] determined the amount of mist, obtained by the mechanical atomization of butyl phthalate, deposited on cylinders with diameters from 0·1 to 9 mm placed vertically and perpendicular to a wind. The mean droplet radius varied between 1·6 and 13 μ and the wind speed between 0·5 and 4 m sec^{-1}, giving Reynolds numbers from 2·5 to 2500. Landahl's results, some of which are shown in Fig. 46, are scattered and it is not possible to deduce from them how Re_f affects the shape of the ($\ni$, Stk) curve. It must be remarked that the extremely polydisperse mists obtained by mechanical atomization are, in general, particularly unsuitable for this sort of investigation.

It is difficult to interpret the experiments of Yeomans [248], who studied the deposition of mist on circular disks, because they were conducted at very small values of Stk and $\ni$.

Ranz and Wong [222] experimented with more or less isodisperse sulphuric acid mists with mean droplet radii from 0·18 to 0·65 μ in a wind tunnel at air speeds from 12 to 97 m sec^{-1}. A wire of diameter 77 μ and a sphere of diameter 0·9 mm were used as collectors, Re_f varying from 55 to 450 in the first case and from 650 to 5000 in the second. Higher values of Re_f could not be obtained because liquid was blown off the surface at high velocities of air flow. It can be seen from Fig. 46 that the experimental data of Ranz and Wong for a cylinder agree with Landahl's theoretical curve at small Stk and fall somewhat below it at large Stk. The results of later measurements made by these authors [566] on wires, from 29 to 106 μ in diameter, at values of Re_f from 13 to 330 are also shown in Fig. 46. Gregory [371] worked with lycopodium spores ($r = 16\,\mu$) which were deposited on adhesive-smeared cylinders 0·2 to 20 mm in diameter in a wind tunnel at air speeds of 1 to 10 m sec^{-1}. His results (not shown in Fig. 46) are close to the new results of Ranz and Wong at Stk < 2 and to their old results for Stk > 2.

In view of the difficulty of obtaining accurate and reproducible results in these experiments, the agreement obtained between theory and experiment must be considered satisfactory. Conversely, the experimental values of ∍ for a sphere (see Fig. 47) are several times bigger than the theoretical values. The discrepancy obviously results from some error in setting up the experiments.

Khimach and Shishkin [249] used a microscope to study the growth of water droplets with radii of the order 0·1 mm suspended on glass threads in a wind tunnel through which an aqueous mist was driven at velocities from 0·3 to several metres per second. The mist was of condensation origin with droplet radii from 3 to 12 μ, and the value of Re_f was about 10. The authors assumed that the flow of air round the stationary droplets was viscous and used the empirical formula

$$\ni = \left(1 - \frac{1{\cdot}214}{\mathrm{Stk}}\right)^2 \tag{34.20}$$

to calculate the collection efficiency. This gives values that are fairly close to those calculated by formula (34.7) but it is more suitable for integration. Having found the droplet-size distribution from photomicrographs, the authors did not use this directly but determined from it, the mean droplet mass. Using Schumann's asymptotic distribution (49.39) for their calculations, they then calculated the theoretical rate of growth of the droplet due to inertial deposition, assuming that each collision between droplets resulted in their coalescence. The ratio of the calculated and measured growth rates varied widely in the individual experiments but the average over all the experiments (0·97) was very close to one. Similar results were obtained in Gunn's experiments (see page 299).

The values found by Boucher [250] for the collection efficiency of droplets on a plate of width 0·5 cm were very different from the calculated values. The deviations were in both directions and most probably arose from imperfections in the experimental technique. It must be acknowledged that the experimental study of inertial deposition is still far from perfect.

A problem of interest in connection with the icing of conductors in high-voltage transmission lines is that of the deposition of uncharged aqueous mist on a charged, infinitely long horizontal cylinder. Cochet [185] has investigated this, taking into account the settling of the droplets under gravity. Assuming that the flow of air round the cylinder is potential, that the motion of the particles obeys Stokes law, that the droplets are conducting so that the induction factor is $(\varepsilon_k - 1)/(\varepsilon_k + 2) = 1$, and neglecting the inertia of the droplets, Cochet calculated ∍ for values of E_0^2/R from 0 to 15,400 and U_0/r^2 from 0 to $1{\cdot}6 \times 10^8$ cm^{-1} sec^{-1}, where E_0 is the electric field strength (in e.s.u.) at the surface of the cylinder, R is the radius of the cylinder (cm) and U_0 is the flow velocity at a great distance from the cylinder (cm sec^{-1}). The curve of ∍ as a function of U_0 with allowance for the inertia of the droplets has also been calculated for $R = 0{\cdot}65$ cm, $r = 10^{-3}$ cm and $E_0 = 100$ e.s.u. (see also p.121).

A number of papers on inertial deposition of aerosols on objects of various shapes have appeared in the past few years. Comprehensive calculations of the collection of particles in potential flow were made at the Lewis Flight Propulsion Laboratory with a specially designed mechanical integrator. The equations of motion at large Re, for spherical particles (25·5) in the absence of external forces, were used as follows.

$$\tau \frac{dV_x}{dt} = \frac{\mathrm{Re}\,\psi}{24}(U_x - V_x) \quad \text{etc.} \tag{34.21}$$

Transforming to dimensionless variables

$$x' = x/R, \quad t' = tU_0/R, \quad U_x' = U_x/U_0, \quad V_x' = V_x/U_0 \quad \text{etc.,} \qquad (34.22)$$

$$\frac{dV_x'}{dt'} = \frac{\mathrm{Re}\,\psi}{24\,\mathrm{Stk}}(U_x' - V_x') \quad \text{etc.,}$$

where

$$\mathrm{Re} = 2r\,[(U_x - V_x)^2 + (U_y - V_y)^2]^{1/2}/\nu.$$

Only at small Re can these equations be simplified to

$$\frac{dV_x'}{dt'} = \frac{1}{\mathrm{Stk}}(U_x' - V_x'), \quad \frac{dV_y'}{dt'} = \frac{1}{\mathrm{Stk}}(U_y' - V_y'). \qquad (34.23)$$

so that it is possible to calculate separately the motion of the particles along the axes x and y, as shown in § 30. In general, it is necessary to known the value of Re at every point and to calculate the motion along both axes simultaneously.

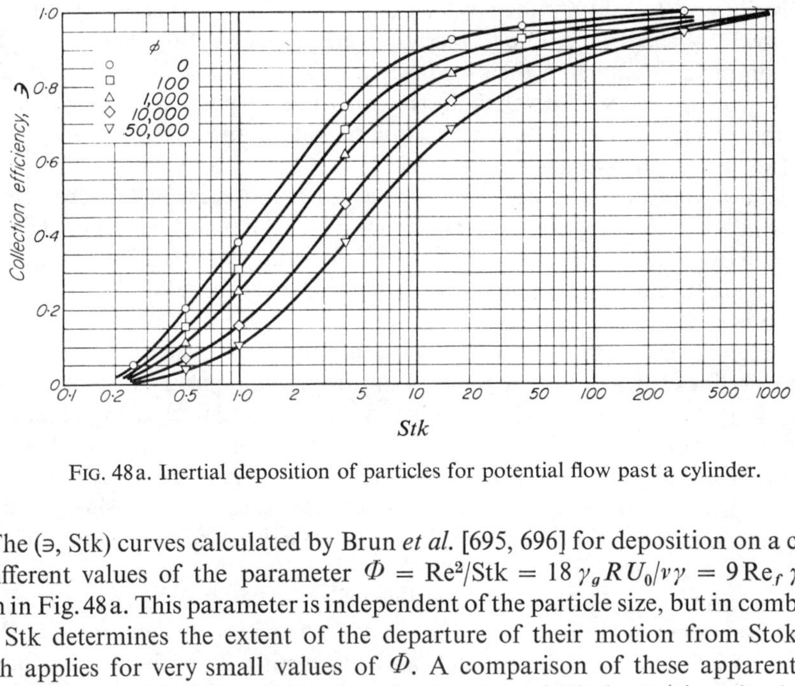

FIG. 48a. Inertial deposition of particles for potential flow past a cylinder.

The ($\ni$, Stk) curves calculated by Brun *et al.* [695, 696] for deposition on a cylinder at different values of the parameter $\Phi = \mathrm{Re}^2/\mathrm{Stk} = 18\,\gamma_g R U_0/\nu\gamma = 9\,\mathrm{Re}_f\,\gamma_g/\gamma$ are given in Fig. 48a. This parameter is independent of the particle size, but in combination with Stk determines the extent of the departure of their motion from Stokes' law which applies for very small values of Φ. A comparison of these apparently very accurate curves with those obtained by Langmuir and Blodgett (given in the paper of Tribus *et al.* [697]), suggests that the values of $\ni$ in the latter are too low (in some cases by as much as 10 per cent).

Curves for the local collection efficiency $\ni_{\mathrm{loc}}$ are shown in Fig. 48b. $\ni_{\mathrm{loc}}$ is the ratio of the number of particles deposited from a uniform flow of particles on a given surface element to that which would be deposited in the case of rectilinear particle trajectories upon an equal surface element placed normal to these trajectories. The maximum value of $\ni_{\mathrm{loc}}$ (at Stk = ∞) is obviously equal to cos θ. These curves also show the angular width of the deposit on the cylinder in radians, determined by the intersection points of the graphs with the axis of abscissa. Velocities of the particles along both axes at the moment of impingement upon the cylinder were also estimated.

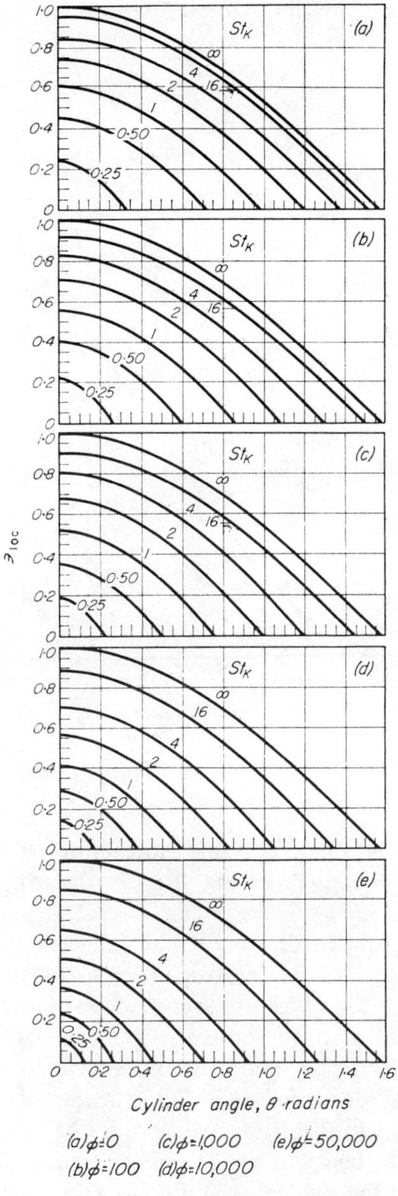

FIG. 48 b. Local collection efficiencies for potential flow past a cylinder.

Similar calculations were also made for deposition on ellipsoids of rotation [698], aeroplane wings [699, 700]†, two-dimensional channels with a 90° bend [702] etc. An estimate was made of the effect of air compressibility upon collection efficiency,

† Calculations of the deposition on aeroplane wings were also made by Levin [701].

which is important at flow velocities comparable to the velocity of sound and results in a small reduction of ∍ [703]. It is negligible at very large and very small Stk numbers; in the case of a cylinder, at $U_0 = 130$ m/sec it reaches its maximum value at Stk ≈ 5, where it produces a reduction in ∍ of 0·5 per cent at $\Phi = 0$, and 3 per cent at $\Phi = 50{,}000$.

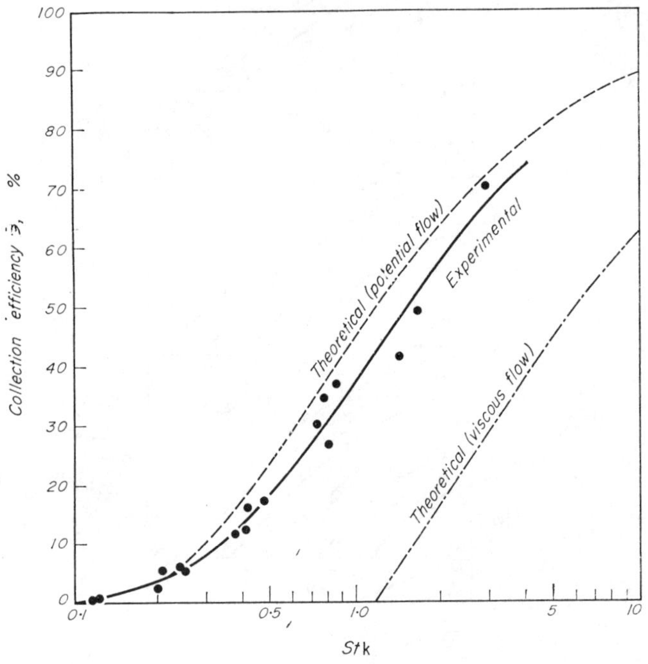

FIG. 48c. Inertitial deposition of particles on a sphere.

Fonda and Herne [704] made accurate calculations to find the deposition on a sphere in potential and viscous flow (Fig. 48c, for the experimental points see page 330).

Knowledge of the (∍, Stk) curves for cylinders makes it possible to assess the liquid water content and the mean droplet size in supercooled clouds [695] which are connected with the formation of ice on aircraft. Slowly rotating cylinders with different diameters are attached to an aeroplane for this purpose. Water drops which impinge upon the cylinder freeze, and it becomes covered with a thin, uniform layer of ice. The mass of ice Φ' deposited on a unit of length of each cylinder in unit time is measured and is equal to $2\,RU_0∍c$, where c is the weight concentration (water content) of the cloud, U_0 the speed of the aircraft and R the radius of the cylinder. The data are plotted as log $∍c$ against log $(1/R)$ and a curve is drawn through the points. If, for the sake of simplicity, the droplets are assumed to be all of the same size, since log $(∍c)$ = log ∍ + const. and log Stk = log $(1/R)$ + const., the experimental curve of [log $(∍c)$, log $(1/R)$] must be parallel to the theoretical (log ∍, log Stk) curve and the values of c and Stk, and hence of r, can be determined from their separation measured along the two axes. Since the parameter Φ changes with R, it is necessary to have resort to the (log ∍, log Stk) curves plotted at Φ Stk $= \mathrm{Re}^2 =$ const., rather than at Φ = constant, since the former quantity remains unchanged in the experiments, and to choose a value of the constant which brings the theoretical and the experimental curves into coincidence. For polydisperse clouds, as found in nature, the calculation is more complicated and the weight concentration and the weight median radius of the droplets cannot be established very accurately.

Tribus *et al.* [705] and Serafini [653], in great detail, considered the deposition of water drops on two-dimensional wedge-like profiles in supersonic flow (see Fig. 48d). The flow field is very different from those considered earlier. In front of the shock wave the flow (with the velocity $\mathbf{U}_1$) is undisturbed; in the wake of the wave the flow $\mathbf{U}_2$ is uniform, but directed parallel to the surface of the wedge. A drop moving initially with a velocity $\mathbf{V}_1 = \mathbf{U}_1$ has a relative velocity $\mathbf{V}_r = \mathbf{V}_1 - \mathbf{U}_2$ after crossing the shock wave and its absolute velocity is made up of the constant velocity U_2, and a gradually diminishing velocity $\mathbf{V}_r$. The drop reaches the surface of the wedge, if its stop distance l_i in the direction $\mathbf{V}_r$ is greater than the distance BD. The trajectories of all drops are identical in shape. By the use of (18·11), simple trigonometrical computations provide the critical value of the segment AE which is crossed by all drops reaching the surface of the wedge, the corresponding value of ED, the width of the deposit on the wedge, and the local collection efficiencies.

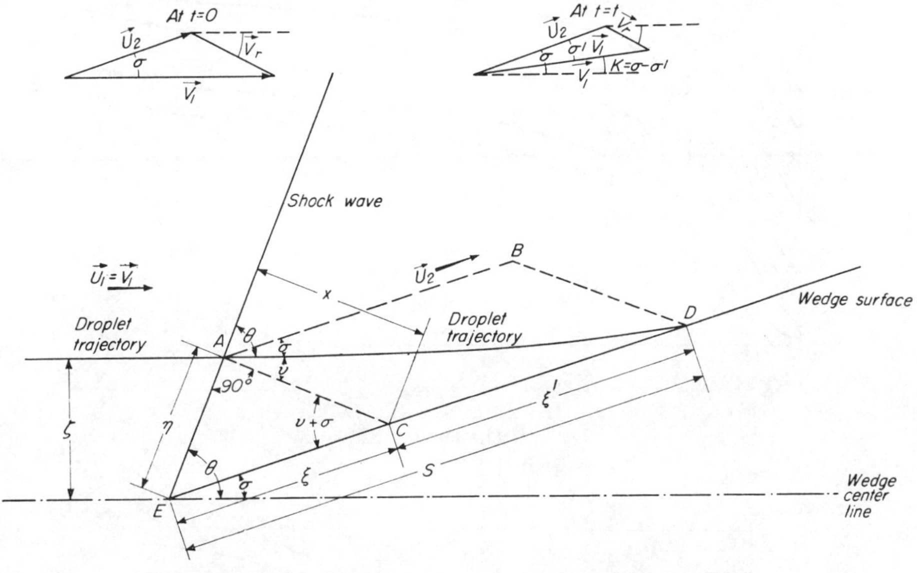

FIG. 48d. Droplet trajectories in supersonic flow past a wedge.

The very simple and popular method of sampling aerosol particles on a glass plate placed normal to the flow could give fairly good results, if the local collection efficiency at the centre of the plate were known as a function of the Stk number; in plotting the particle size distribution curve, the number of particles of a given size should be divided by the number $\ni_{loc}$ corresponding to this size.

The stream past a flat plate breaks away from its edges unless Re_f is very small and Levin calculated [706] $\ni_{loc}$ for the middle of the plate (see Table 12a). An important result was that it proved to be practically constant across more than a half the width of

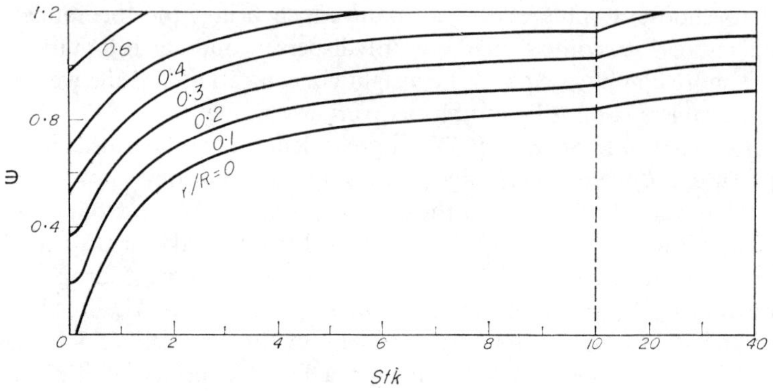

FIG. 48e. Inertial deposition of particles for potential flow past a cylinder allowing for interception effect.

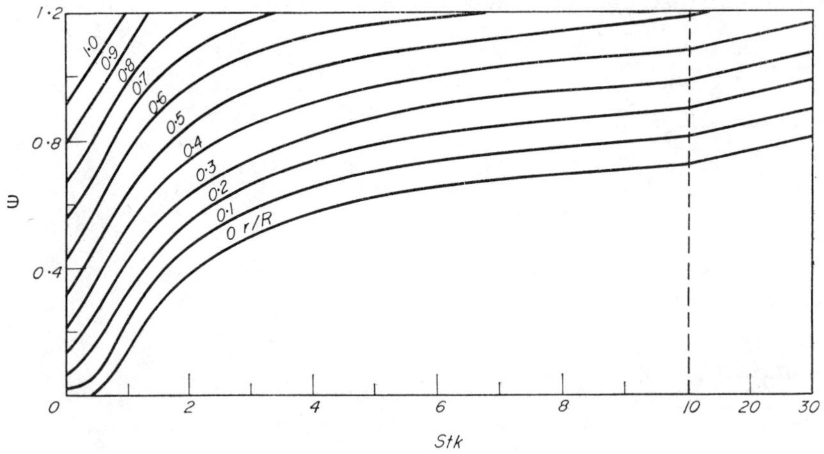

FIG. 48f. The same at $Re_f = 10$.

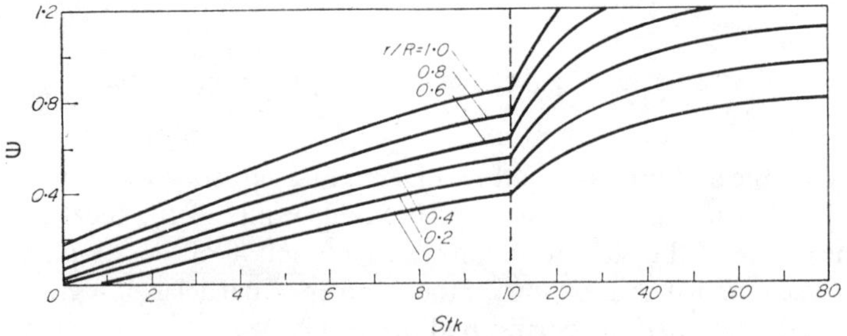

FIG. 48g. The same at $Re_f = 0·2$.

the plate. It should be noted, however, that no allowance was made in computing the table either for the effect of the boundary layer, or for that of interception; it can thus be used only at high values of Re_f and small values of r/h, where h is half the width of the plate ($\text{Stk} = 2\,r^2\gamma\; U_0/9\,\eta\, h$).

TABLE 12a. LOCAL COLLECTION EFFICIENCY AT THE CENTRE OF A FLAT PLATE IN POTENTIAL FLOW WITH THE STREAM BREAKING AWAY FROM THE EDGES

Stk	$\frac{4}{4+\pi} = 0{\cdot}560$	1	2	3	4	5	7·5	10	15	20	30
$\ni_{loc}$	0	0·15	0·35	0·44	0·49	0·56	0·63	0·67	0·72	0·77	0·81

Robinson [669], using his theory of the flow of aerosol fluid (see page 109), derived an approximate formula for collection efficiency in potential flow past a cylindrical surface of an arbitrary profile at small Re numbers. It was assumed in deducing it that $\mathbf{U} - \mathbf{V}$ was small everywhere, which is obviously incorrect in the vicinity of the surface of the cylinder. His formula, in fact, yields too high values of $\ni$ at all Stk values, except Stk $= \infty$.

Robinson also proved that when the settling velocity of particles is much smaller than the flow velocity, the collection efficiencies due to gravity and inertia are additive.

Using a numerical method and ignoring the interception effect by assuming $r \ll R$, Natanson [707] computed Stk_{cr} for particles obeying Stokes' law in viscous flow past a sphere, according to (34·4). He found $\text{Stk}_{cr} = 1{\cdot}21 \pm 0{\cdot}01$, in good agreement with Langmuir (see page 163). For flow past a sphere in Oseen's approximation, at $\text{Re}_f = 0{\cdot}1$, Stk_{cr} was $1{\cdot}15 \pm 0{\cdot}01$, and for the flow past a cylinder on Oseen's approximation, at $\text{Re}_f = 0{\cdot}1$, Stk_{cr} was $4{\cdot}3 \pm 0{\cdot}1$.

The computation of Stk_{cr} at small Re_f has been tackled in a general way by Levin [708]. The value of Stk_{cr} depends on the flow pattern in the vicinity of the forward stagnation point. The flow velocity along the x-axis can be expressed by an asymptotic formula $U_x = -ax^n$, if the origin of the coordinates is located at the stagnation point. Assuming $\text{Stk}_{cr} = A\,a^m$, where A and m are constants, it can be shown that $m = -1/n$. Thus, in potential flow, for which $n = 1$, $\text{Stk}_{cr} = 1/4a$ (see page 164). In viscous flow, equations (34.2) and (34.4) show that $n = 2$, so that $\text{Stk}_{cr} = A/a^{\frac{1}{2}}$. For Oseen flow past a sphere Levin showed that $a \approx (24 + 12\,\text{Re}_f)/(16 - 3\,\text{Re}_f)$; hence, in the range where Oseen's equations are valid, Stk_{cr} slowly decreases as Re_f rises.

Levin also considered the effect of a boundary layer on Stk_{cr} showing that it cannot exceed z determined by equation

$$z = \frac{1}{4a} + \left(\frac{0{\cdot}65}{z\beta a^{1/2}\text{Re}_f^{1/2}}\right)^{1/2}. \tag{34.24}$$

The constant β is equal to 3·8 for flow past a cylinder and to 1·8 for flow past a plate. In the case of a cylinder $z = 0{\cdot}17$ at $\text{Re}_f = 10^5$ and 0·25 at $\text{Re}_f = 1000$, and is thus greater than Stk in the absence of a boundary layer (0·125).

Davies and Peetz [709], allowing for interception, estimated the collection efficiency of Stokes particles on a cylinder for the following three cases: potential flow without a boundary layer and viscous flow at $Re_f = 10$ and 0·2. In addition, they calculated the particle velocities at the moment of impingement and the width of the deposit and the local collection efficiencies for $r/R \ll 1$. These results are shown on Fig. 48e, f, g. For $Re_f = 10$ the flow field was calculated by the method of Thom [237] and Stk_{cr} was found to be 0·42. Comparison with Fig. 46 shows that much lower values of ∋ were found than those of Landahl *et al.* who used a similar flow field. There is no doubt that Landahl's curve is erroneous since it meets the curve for potential flow at Stk = 6.

For $Re_f = 0{\cdot}2$ the authors used the field of flow past a cylinder calculated by Davies [710]. It has been suggested [707] that this field may be inaccurate in the vicinity of the stagnation point and that the value of $Stk_{cr} = 0{\cdot}899$ is low.

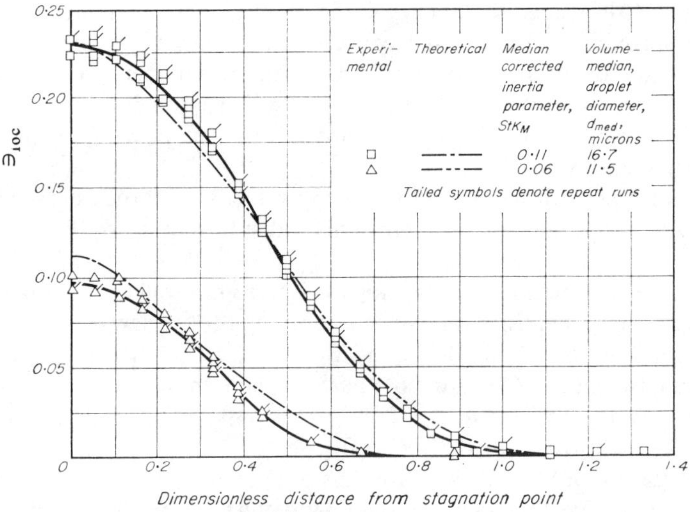

FIG. 48h. Inertial deposition of water droplets on a sphere.

Experimental work on collection efficiency will now be dealt with. Lewis and Ruggieri [711] sprayed coloured water in a wind tunnel with the cross-section 2 × 3 m and a wind speed of 76 m/sec. The quantity of water deposited on objects covered with filter paper was determined colorimetrically. Shapes studied included spheres of $R = 7{\cdot}5$ and 22·5 cm, ellipsoids and a cone. The water sprays ($\bar{r} = 5{\cdot}5$–$9{\cdot}5\,\mu$) emitted droplets over a range of sizes and the theoretical calculation of collection efficiency was made by allowing for the size distribution. Theoretical and experimental curves of local collection efficiency plotted against the dimensionless distance from the stagnation point s/R are shown on Fig. 48h for two values of the corrected Stokes number derived from the stop distance (see (18.11)). There is fairly good agreement between theory and experiment.

Ignatiev [712] measured the collection efficiencies of coal and iron dusts elutriated into fractions of $\bar{r}$ from 6 to 100 μ; cylinders of $R = 0{\cdot}6 - 2{\cdot}5$ cm were exposed in a descending vertical air stream with $\bar{U} = 2 - 16$ m/sec, and $Re_f = 1600 - 54{,}000$. A celluloid band covered with Vaseline was wound around the cylinder and the particles collected on it were counted under the microscope. The ratio r/R was always $< 0{\cdot}02$ so that the interception effect could be ignored. The dependence of ∋ on $\bar{r}$ at constant $\bar{U}$ and R was determined in each series of the experiments. The (∋, Stk) curve obtained at $\bar{U} = 2$ m/sec and $R = 2{\cdot}5$ cm was very close to the theoretical curve for large Re_f, but as $\bar{U}/R$ increased the curve shifted more and more to the right, and at $\bar{U} = 16$ m/sec and $R = 0{\cdot}6$ cm it nearly coincided with the Davies curve for $Re_f = 0{\cdot}2$ (Fig. 48g). The reason for this is not clear but it may be due to blowing off of the particles. As the factor Φ in these experiments was less than 1/100, according to Fig. 48a, the departures from Stokes resistance cannot be responsible.

The relationship between local collecting efficiency and polar angle θ, obtained by Ignatiev, is close to the theoretical one of Fig. 48b, but in the vicinity of the axis of the abscissa his curves flatten and approach the axis asymptotically; some deposition thus occurs at places where there should be none, according to theory. This is possibly due to displacement of the vaseline layer, and the particles in it, by the air flow, or to transverse turbulence.

Ignatiev also found a relationship between collection efficiency and direction of the flow. His ($∋_{loc}$, θ) curves, at $U = 2$ m/sec, $R = 2{\cdot}5$ cm and at different values of $\bar{r}$ for the descending (1) and the ascending (2) flows, are given in Fig. 48i. The graphs indicate that curves 1 and 2 are widely different at small and large values of $\bar{r}$ but almost coincide in between. This is accounted for by the rate of deposition due to sedimentation being determined by the value of V_s, which is proportional to Stk, while the (∋, Stk) curve has an inflection. Under the conditions of the experiment the (V_s, Stk) straight line intersects the (∋U, Stk) curve at two points between which $V_s < ∋U$; beyond these points $V_s > ∋U$. At Stk = 0·09, which is below Stk_{cr}, sedimentation deposition completely predominates, a small deposit on the lower side of the cylinder in the ascending flow being due to the range of sizes in the dust; the deposits on the upper side of the cylinder formed in ascending and descending flows are comparable. At $\bar{r} = 108\,\mu$ V_s attains 110 cm/sec, which is half the air velocity, so that sedimentation again has a great effect on ∋.

In experiments conducted by Amelin and Belyakov [713] with horizontal flow at $\bar{U} = 1 - 16$ m/sec and $R = 0{\cdot}5 - 1{\cdot}25$ cm a polydisperse ($r = 1 - 30\ \mu$) oil mist was employed, the different sizes of droplet in the deposit being counted separately; porcelain dust ($\bar{r} = 4{\cdot}4$ and $18\,\mu$) was also used. The (∋, Stk) curve obtained was characterized by extremely low ∋ values at small Stk (∋ = 0·01 at Stk = 1), and by a very sharp rise, the curve reaching the value ∋ = 1 at Stk = 18 at a finite angle rather than asymptotically. These surprising results appear to be due to a fundamentally erroneous method of measuring the concentration of a moving aerosol [714].

Jarman [715] drew an isodisperse kerosene mist with $\bar{r} = 8 - 24\,\mu$ through a large wind tunnel at a velocity $U_0 = 1$ m/sec and estimated the quantity of kerosene deposited on small pieces of metal gauze. The velocity through the gauze was not measured but calculated from the formula $U = U_0/[1 + (1 - \beta)/4\beta^2]$ [716], where β is the specific area of the apertures of the gauze (cm²/cm²). Since in these experiments $Re_f = 6 - 18$, a curve in Fig. 48f was chosen for comparison with theory. The experimental values of ∋ were up to 25 per cent greater than the theoretical. This was to be expected because the streamlines through a row of cylinders pass closer to their surfaces than in the case of an isolated cylinder. (For the deposition af aerosols on several gauzes in series, see page 209.)

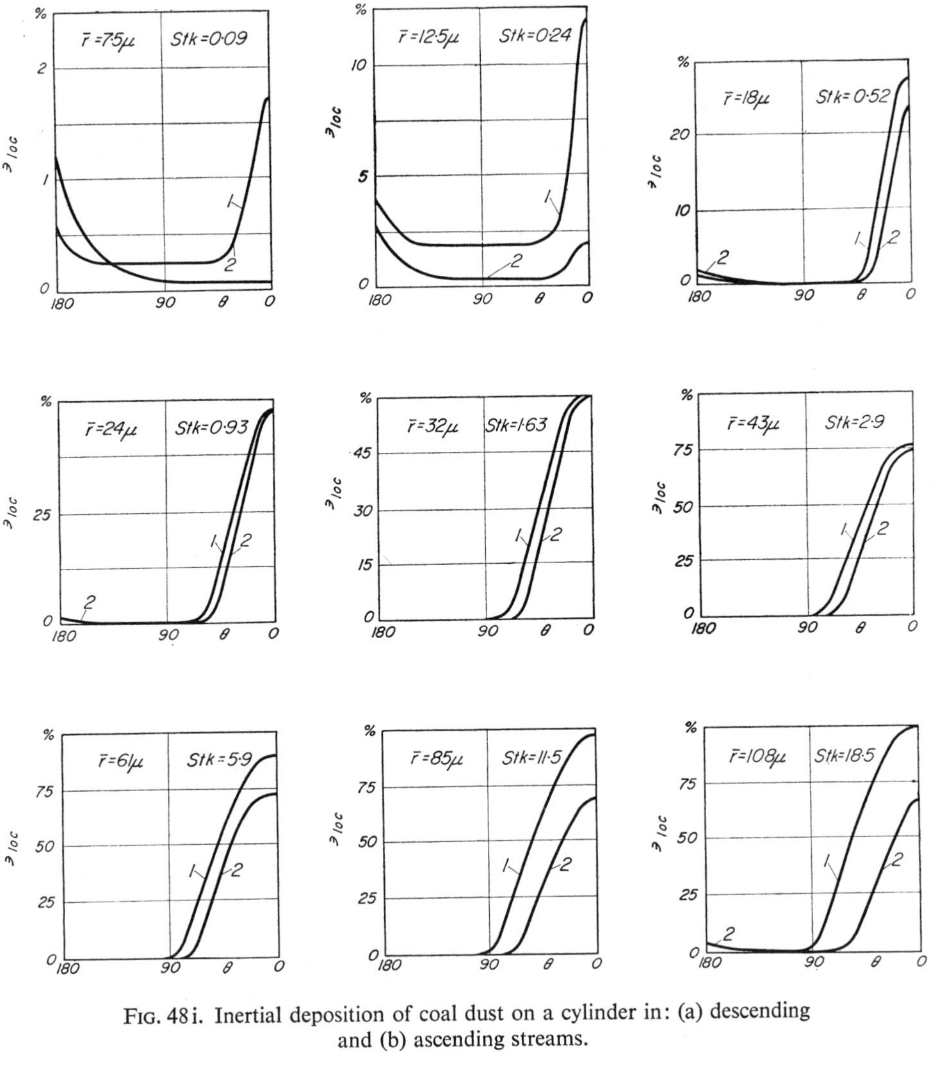

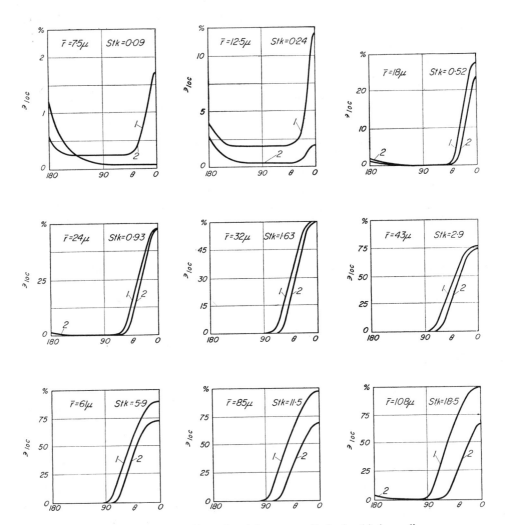

FIG. 48 i. Inertial deposition of coal dust on a cylinder in: (a) descending and (b) ascending streams.

Deposition of moving aerosols due to electric forces has been the subject of a number of papers. Kraemer and Johnstone [717] calculated the collection efficiency of aerosols for potential and viscous flows past a conducting sphere, due to electrostatic forces; allowance for interception was made, but the inertia of the particles was neglected. Different dimensionless parameters, K, expressed the ratio of the electric force acting upon a particle located near the collector surface to the quantity $6\pi\eta r U_0$, which characterizes the resistance of the medium to the motion of the

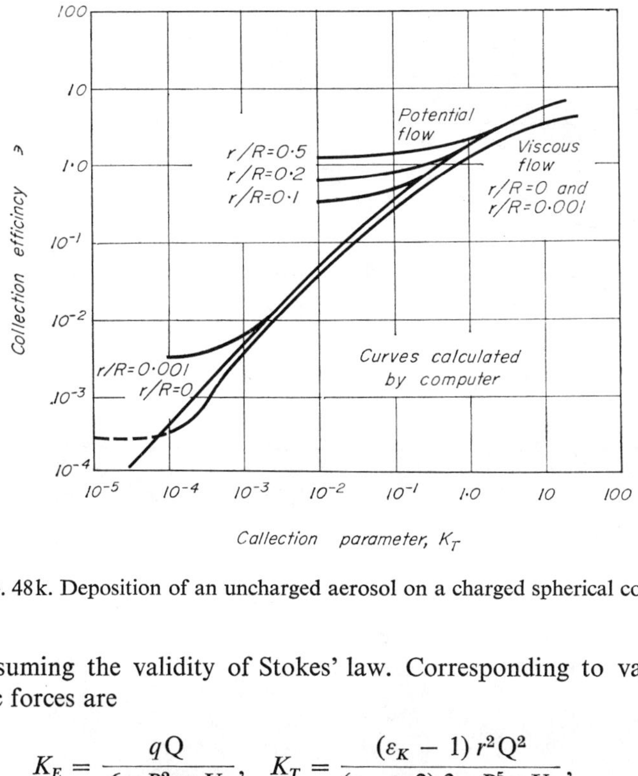

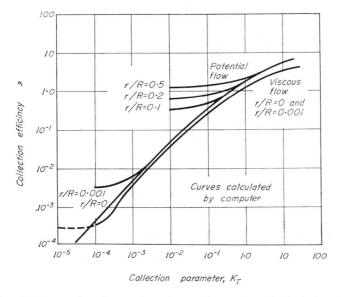

FIG. 48k. Deposition of an uncharged aerosol on a charged spherical collector.

particle, assuming the validity of Stokes' law. Corresponding to various cases of electrostatic forces are

$$K_E = \frac{qQ}{6\pi R^2 r\eta U_0}, \quad K_T = \frac{(\varepsilon_K - 1)\, r^2 Q^2}{(\varepsilon_K + 2)\, 3\pi R^5 \eta U_0},$$

$$K_M = \frac{q^2}{6\pi r\eta U_0 R^2}, \quad K_G = \frac{q^2 n R'^2}{3\eta r U_0 R}, \quad K_s = \frac{2Rnq^2}{9\eta r U_0} \qquad (34.25)$$

where q is the particle charge, Q the collector charge, R the collector radius, n the number of particles per cm^3, U_0 the flow velocity at a great distance from the collector and R' the radius of the aerosol cloud. K_E corresponds to the coulombic force between a charged collector and charged particles, K_T to the induction force between a charged collector and uncharged particles, K_M the same between charged particles and an insulated uncharged collector, while K_G and K_s refer to the case of a unipolarly charged aerosol. K_G corresponds to the force between the particles and the charge induced by them in an earthed collector, and K_S to the force exerted upon the particles by other particles. The inequalities $\frac{4}{3}\pi r^3 n = \varphi \ll 1$ and $\varphi R'^2/R^2 \ll 1$ are assumed to be observed here.

The results of the calculations for a charged collector and uncharged particles are given on Fig. 48k, those for an uncharged insulated collector and a bipolarly charged aerosol on Fig. 48l, and those for a charged collector and an aerosol with a

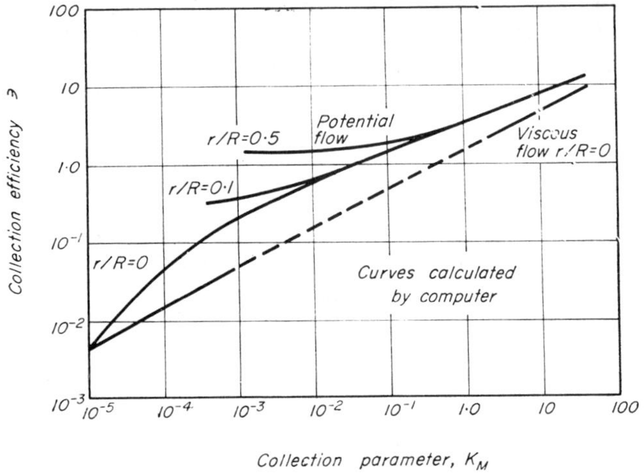

FIG. 48l. Deposition of a bipolarly charged aerosol on a uncharged spherical collector.

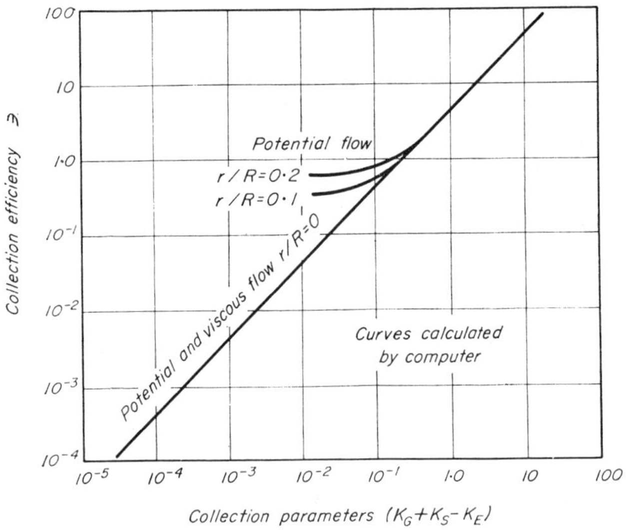

FIG. 48m. Deposition of a unipolarly charged aerosol on a spherical collector with the opposite charge.

unipolar charge of an opposite sign (the induced charges on the particles being neglected) on Fig. 48 m. The ratio r/R characterizes interception. It was demonstrated that simple addition of the separate collection efficiencies due to the various electric forces sometimes yielded values of ∋ which were much larger than those accurately calculated.

The experiments of Kraemer and Johnstone with dioctylphthalate aerosols having $\bar{r} = 0{\cdot}27 - 0{\cdot}59\,\mu$, unipolarly or bipolarly charged, at $R = 0{\cdot}3 - 0{\cdot}55$ cm, $U_0 = 1{\cdot}5 - 6{\cdot}9$ cm/sec and $\bar{q} = 0{\cdot}15 - 137$ elementary charges are shown on Fig. 48n. In all three series of experiments one of the parameters, K, prevailed over the others which could be ignored.

Under similar conditions to these, Natanson [718] investigated the deposition of a moving aerosol upon a cylinder due to electrostatic forces. The two basic cases to be

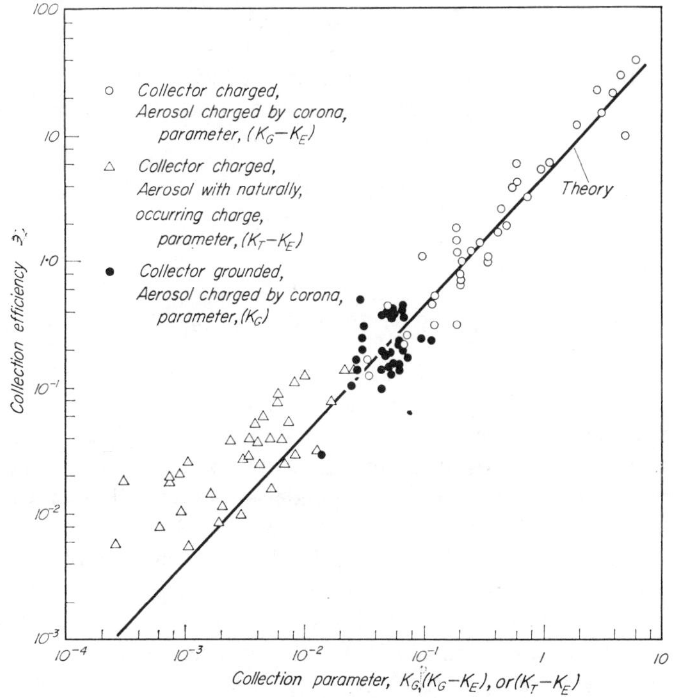

FIG. 48n. Deposition of a charged dioctylphthalate mist on a spherical collector.

distinguished are when the limiting trajectories of particles are either tangential to the cylindrical surface or else sweep past and curve back to meet it at the rear stagnation point. In the first case deposition occurs chiefly on the windward side of the cylinder and in the second particles are found over the entire surface. An example of the latter is when the radial component of the velocity of a particle near the surface, that is the vector sum of the radial flow velocity and the velocity of the particle due to the electric force, is everywhere directed towards to the cylinder; this occurs, for instance, at very small values of r/R. Only this case will be considered. With neglect of induction forces, Natanson found the following formula for the deposition of charged particles on an oppositely charged cylinder

$$\ni = \frac{2\pi BQ'q}{U_0 R} \tag{34.26}$$

where B is the mobility of the particles and Q' the charge of a unit of the cylinder length. For a charged cylinder and uncharged particles at large Q' and small U_0, i.e. at $\ni \gg 1$

$$\ni = \left[\frac{6(\varepsilon_K - 1) BQ'^2 r^3}{(\varepsilon_K + 2) U_0 R^3}\right]^{1/3}. \tag{34.27}$$

By substituting the sedimentation velocity of the particles V_s for U_0 in this formula, the formula of Cochet for the deposition of sedimenting particles on a charged cylinder is obtained (see page 167).

At $\ni \ll 1$

$$\ni = \frac{4\pi(\varepsilon_K - 1) BQ'^2 r^3}{2(\varepsilon_K + 2) U_0 R^3}. \tag{34.28}$$

In the cases considered $\ni$ does not depend on the flow pattern.

If the deposition of charged particles occurs on an uncharged cylinder and $\ni \ll 1$, as is usually the case under these conditions,

$$\ni = \left[\frac{3\pi(\varepsilon_K - 1) Bq^2}{2(\varepsilon_K + 2) U_0 R^2}\right]^{1/3} \tag{34.29}$$

in potential flow and

$$\ni = \left[\frac{(\varepsilon_K - 1) q^2 B}{(\varepsilon_K + 2) U_0 R^2 (2{\cdot}00 - \ln Re_f)}\right]^{1/2} \tag{34.30}$$

in the viscous flow (to Lamb's approximation).

Natanson also deduced the formulae for the first of the cases considered above, when $\ni$ depends on the location of the grazing point between the limiting trajectories and the cylinder surface. In this case $\ni$ always depends on the flow pattern.

Rough estimates of the deposition on a cylinder in potential flow under the action of the electrostatic forces were also given by Gillespie [719].

Since the electrical field around the charged object is solenoidal, the concentration n_0 in a flowing aerosol does not change in the absence of inertia (see page 110). If n_0 is constant at a great distance from the body, it retains its value also at the surface of the latter. The total flow of aerosol towards the surface, according to Dukhin and Derjaguin [720] is therefore equal to

$$\Phi = n_0 \oint (U_n + BF_n)\, dS \tag{34.31}$$

where U_n is the radial component of the flow velocity and F_n that of the electrostatic or some other central force. The integral is taken over the whole surface and there is no need to calculate the limiting trajectories of particles in this case.

In an electrostatic field $F_n = E_n q$, where E_n is the radial component of the field strength and $\oint E_n\, dS = -4\pi Q$; if the particles are very small, it can be assumed that $U_n = 0$ and

$$\ni = -\frac{4\pi q Q B}{U_0 S_M} \tag{34.32}$$

where S_M is the area of the middle cross-section of the object. Formula (34.32) obtained by Levin [721] is valid for objects of arbitrary shapes. In the case of a cylinder (34.26) can be obtained from it, and in the case of a sphere the following formula:

$$\ni = -\frac{4qQB}{U_0 R^2} \tag{34.33}$$

which yields (54.3).

CHAPTER V

BROWNIAN MOTION AND DIFFUSION IN AEROSOLS

§ 35. BROWNIAN MOTION IN AEROSOLS

Since the theory of Brownian motion has been treated adequately in text books [251, 253] this chapter is restricted to certain problems which are of direct interest in the physics of aerosols.

Thermal motion of particles suspended in a fluid is expressed by the following equations derived by Einstein [254]

$$\overline{x^2} = 2\,Dt, \tag{35.1}$$

$$D = kTB, \tag{35.2}$$

where x^2 is the mean square displacement of a particle, with respect to any coordinate axis, in time t and D is the diffusion coefficient of the particle, a quantity characterizing the intensity of Brownian motion and connected with the particle mobility by equation (35.2).

In the derivation of these equations the following assumptions are made:

(1) the particles move independently of one another;

(2) the average kinetic energy of a particle along each coordinate axis is equal to $\frac{1}{2}\,kT$, in agreement with statistical mechanics;

(3) the movements of a particle in consecutive time intervals $0 - t$, $t - 2\,t$, $2\,t - 3\,t$ and so on, are independent.

Let us examine these assumptions individually. The first implies that there are no forces of interaction between the particles. These can be hydrodynamic (§ 23), molecular and, if the particles are charged, electrostatic. The particles frequently change direction at random hence time-averaged hydrodynamic forces must reduce to zero. Molecular forces come into play only when the distance between particles is small compared with their size, while electrostatic forces can occur at distances several times greater than the particle size.

If the particles are so close together that the forces between them are appreciable, equation (35.1) is no longer applicable and must be replaced by a more complicated equation which takes into account the ordered motion of the particles resulting from these forces (see § 51).

There can, of course, be no doubt that in the absence of ordered motion due to external forces the mean thermal energy of a particle along each coordinate axis is

$\frac{1}{2}kT$. In order to explain the results of some experiments, which indicated that the horizontal and vertical displacements of aerosol particles were different, Frank [255] suggested that the energy of settling, $mV_s^2/2$, should be subtracted from the mean energy of vertical Brownian movement. An amount of energy less than $\frac{1}{2}kT$ would then remain for the thermal energy and the vertical Brownian displacement of the particles would therefore have to be decreased. This point of view clearly contradicts the principles of statistical mechanics.

The statement that particle movements in sucessive time intervals t are independent can also be expressed in the following way [253]: successive positions of a particle at times o, t, $2t$... form a "discrete Markov chain", so that the position of the particle at the instant $(n + 1)t$ depends only on its position at the preceding instant nt but not on the position at $(n - 1)t$ or any previous times. It can easily be verified that there is a lower limit to the value that can be chosen for t, and to show this we will examine the molecular-kinetic mechanism of Brownian motion. Figure 49 shows plane projections of the paths followed by a nitrogen molecule (a) and an aerosol particle (b), of radius $1\,\mu$ and unit density, in air at normal temperature and pressure. The scale of the diagram is $3 \times 10^5:1$.

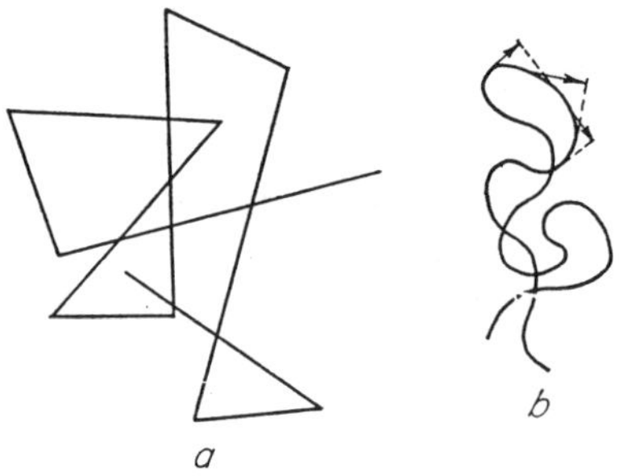

FIG. 49. Trajectories of gas molecules (a) and particles undergoing Brownian motion (b).

The trajectory of the gas molecule consists of straight segments each of which represents the path of the molecule between two collisions. At each collision the direction and speed of a molecule are changed abruptly. If persistence of velocity is neglected, all directions of a molecule after a collision are, on the average, equally probable. If $\overline{G_g}$ is the average velocity and l the mean free path of a molecule, $\tau = l/\overline{G_g}$ is the time taken to cover the mean free path. The velocities of a gas molecule in successive time intervals will therefore be independent if $t \gg \tau$.

Actually, even in collisions between gas molecules, there is some persistence of velocity, so that after a collision a molecule retains a definite velocity component in its former direction. With aerosol particles this effect is overwhelming because the mass of a particle is so much greater than the mass of the gas molecules with which it collides. The velocity of the particle therefore changes negligibly in a single collision. Appreciable changes in speed and direction occur only after a large number of collisions with molecules, with the result that the particle trajectory is almost a smooth curve.

The role of the mean free path is then assumed by an "apparent mean path", l_B, [256] defined as the distance covered by the particle before its direction is completely changed. It may be taken as the average length of a tangent to the trajectory from the point of contact to its intersection with the nearest perpendicular tangent (see Fig. 49 b) but a more exact definition can be obtained by the following argument. A particle always has a certain forward velocity with a mean square value of $\overline{G^2} = 3\,kT/m$ and the surrounding fluid constantly resists this motion according to the general equation (see 17.7).

$$G = G_0 e^{-t/\tau}. \tag{35.3}$$

On account of fluctuations in the impulses received from bombarding gas molecules, the particle acquires a velocity perpendicular to its original direction. The average absolute value of the velocity of the particle does not change, in spite of the resistance offered. According to equation (17.18) the path covered by the particle in the initial direction is

$$l_B = \overline{G}\tau \tag{35.4}$$

where τ is the relaxation time of the particle and $\overline{G}$ is its average thermal velocity. The direction in which the particle moves therefore changes substantially in a time of the order τ, and for the positions of the particle at times 0, t, $2\,t$.... to constitute a Markov chain the condition $t \gg \tau$ must hold.

If this argument is made more precise [258], the following expression is obtained for the mean square displacement of the particle along the x axis:

$$\overline{x^2} = 2\,\overline{G_x^2}\,\tau\,[t - \tau\,(1 - e^{-t/\tau})], \tag{35.5}$$

where $\overline{G_x^2}$ is the mean square particle velocity in the x direction. Since $\overline{G_x^2}\tau = \frac{kT\tau}{m} = kTB = D$, formula (35.5) can be rewritten in the form

$$\overline{x^2} = 2\,D\,[t - \tau\,(1 - e^{-t/\tau})]. \tag{35.6}$$

When $t \gg \tau$ (35.6) reduces to the fundamental equation (35.1). Expansion of $e^{-t/\tau}$ in (35.5) with $t \ll \tau$ gives

$$\overline{x^2} = \overline{G_x^2}t^2, \tag{35.7}$$

which is the equation of motion of a particle at constant velocity $\overline{G_x}$. Hence it is clear that $t \gg \tau$ is the condition for which equation (35.1) is applicable.

In the opinion of some authors [259] the persistence of Brownian velocity is greater downwards than horizontally owing to sedimentation of the particles, so that the downwards Brownian displacement should be greater than that given by equation (35.1). This complicated question will not be discussed beyond pointing out that this opinion is contradicted by the results of accurate measurements of Brownian motion in a gas (see § 36).

The important assumption that the particle mobility entering equation (35.2) is the same as the mobility under the influence of a constant force has been doubted by some authors in view of the nature of Brownian motion. That their doubts are without foundation is shown by Einstein's derivation of this equation.

Consider an isodisperse system in equilibrium in a gravitational field, in which the particles neither coagulate nor stick to the walls of the vessel. The particle concentration at a height z is given by the Boltzmann formula

$$n = n_0 e^{\frac{-mgz}{kT}}. \tag{35.8}$$

The number of particles passing downwards in unit time through a horizontal area of 1 cm² at the height z, due to sedimentation, is

$$I_1 = nV_s. \tag{35.9}$$

At the same time diffusion gives rise to a flux of particles in the opposite direction†

$$I_2 = -D\frac{dn}{dz} = \frac{nDmg}{kT}. \tag{35.10}$$

Since the system is in equilibrium $I_1 = I_2$ so that $D/kT = V_s/mg = B$, and equation (35.2) follows. It can be seen from this result that B is indeed the mobility under a constant force and that there there is no special need to prove this by a detailed study of the mechanism of Brownian motion [260].

Certain quantities associated with the Brownian motion and the mechanics of aerosol particles are shown for various particle size in Table 13. They comprise the diffusion coefficient D, the mean velocity $\overline{G} = \sqrt{\frac{8G^2}{3}}$, the relaxation time τ, the apparent mean path l_B and the mean displacement in 1 sec in a given direction $\overline{\Delta x_B} = \sqrt{4Dt/\pi}$, (see page 193). For comparison, the distance Δx_s fallen by a particle in 1 sec under gravity, for spherical particles of unit density in air at normal atmospheric pressure and 23 °C, is included. Expression (8.6) was used here for the mobility of the particles. The bottom row of the table shows the power of γ by which the figures given in each column must be multiplied if the particles are not of unit density.

TABLE 13. CHARACTERISTIC QUANTITIES IN THE MECHANICS OF AEROSOLS

r (cm)	D (cm² sec⁻¹)	$\overline{G}$ (cm sec⁻¹)	τ (sec)	l_B (cm)	$\overline{\Delta x_B}$ (cm)	$\overline{\Delta x_s}$ (cm)
10^{-7}	$1{\cdot}28\times10^{-2}$	4965	$1{\cdot}33\times10^{-9}$	$6{\cdot}59\times10^{-6}$	$1{\cdot}28\times10^{-1}$	$1{\cdot}31\times10^{-6}$
2×10^{-7}	$3{\cdot}23\times10^{-3}$	1760	$2{\cdot}67\times10^{-9}$	$4{\cdot}68\times10^{-6}$	$6{\cdot}40\times10^{-2}$	$2{\cdot}62\times10^{-6}$
5×10^{-7}	$5{\cdot}24\times10^{-4}$	444	$6{\cdot}76\times10^{-9}$	$3{\cdot}00\times10^{-6}$	$2{\cdot}58\times10^{-2}$	$6{\cdot}63\times10^{-6}$
10^{-6}	$1{\cdot}35\times10^{-4}$	157	$1{\cdot}40\times10^{-8}$	$2{\cdot}20\times10^{-6}$	$1{\cdot}31\times10^{-2}$	$1{\cdot}37\times10^{-5}$
2×10^{-6}	$3{\cdot}59\times10^{-5}$	55·5	$2{\cdot}97\times10^{-8}$	$1{\cdot}64\times10^{-6}$	$6{\cdot}75\times10^{-3}$	$2{\cdot}91\times10^{-5}$
5×10^{-6}	$6{\cdot}82\times10^{-6}$	14·0	$8{\cdot}81\times10^{-8}$	$1{\cdot}24\times10^{-6}$	$2{\cdot}95\times10^{-3}$	$8{\cdot}64\times10^{-5}$
10^{-5}	$2{\cdot}21\times10^{-6}$	4·96	$2{\cdot}28\times10^{-7}$	$1{\cdot}13\times10^{-6}$	$1{\cdot}68\times10^{-3}$	$2{\cdot}24\times10^{-4}$
2×10^{-5}	$8{\cdot}32\times10^{-7}$	1·76	$6{\cdot}87\times10^{-7}$	$1{\cdot}21\times10^{-6}$	$1{\cdot}03\times10^{-3}$	$6{\cdot}73\times10^{-4}$
5×10^{-5}	$2{\cdot}74\times10^{-7}$	0·444	$3{\cdot}54\times10^{-6}$	$1{\cdot}53\times10^{-6}$	$5{\cdot}90\times10^{-4}$	$3{\cdot}47\times10^{-3}$
10^{-4}	$1{\cdot}27\times10^{-7}$	0·157	$1{\cdot}31\times10^{-5}$	$2{\cdot}06\times10^{-6}$	$4{\cdot}02\times10^{-4}$	$1{\cdot}28\times10^{-2}$
2×10^{-4}	$6{\cdot}10\times10^{-8}$	$5{\cdot}55\times10^{-2}$	$5{\cdot}03\times10^{-5}$	$2{\cdot}80\times10^{-6}$	$2{\cdot}78\times10^{-4}$	$4{\cdot}93\times10^{-2}$
5×10^{-4}	$2{\cdot}38\times10^{-8}$	$1{\cdot}40\times10^{-2}$	$3{\cdot}08\times10^{-4}$	$4{\cdot}32\times10^{-6}$	$1{\cdot}74\times10^{-4}$	$3{\cdot}02\times10^{-1}$
10^{-3}	$1{\cdot}38\times10^{-8}$	$4{\cdot}96\times10^{-3}$	$1{\cdot}23\times10^{-3}$	$6{\cdot}08\times10^{-6}$	$1{\cdot}23\times10^{-4}$	1·21
	γ^0	$\gamma^{-1/2}$.	γ	$\gamma^{1/2}$	γ^0	γ

It will be seen that all the tabulated quantities depend very much on particle size, with the exception of the apparent mean path l_B which is of the same order right down to molecular sizes (in the atmosphere $l \approx 6\times10^{-6}$ cm). A further point shown clearly by this table is the practical impossibility of observing the actual trajectory of a particle

† See § 37 for the applicability of diffusion equations to Brownian particles.

with existing techniques; for example, in the case of particles of radius 10^{-4} cm it would be necessary to take no less than 10^6 photographs per second with a magnification of about 10^5. It is evident that the theoretical limitation imposed on the time interval between observations ($t \gg \tau$) has no practical significance because the minimum value of t allowed by theory cannot be achieved experimentally. The extremely rapid growth of the ratio $\overline{\Delta x_B}/\Delta x_s$, as r passes from 10^{-4} to 10^{-5} cm, should be noted. With a little experience visual estimation of particle size is possible in this important range purely from the nature of the particle motion, without the need of any measurements. Finally, it can be seen that Re for the thermal motion of particles is very small so that the resistance of the fluid is proportional to the velocity of the particle, as assumed in the derivation of equation (35.2).

§ 36. EXPERIMENTAL STUDY OF BROWNIAN MOTION IN AEROSOLS

The theory of Brownian motion was first confirmed experimentally in liquids but, with aerosols, considerable departures from the theory were often observed. In particular the mean square displacements in a horizontal direction were in some investigations less than and in others greater than those in the vertical [259, 261]. The departures were particularly noticeable at reduced pressures, when the displacements of coarse particles were too large and those of fine particles too small [262]. As already indicated, special hypotheses were put forward to explain these observations. It is very difficult to find the cause of discrepancies in individual investigations but one of the reasons is undoubtedly the insufficient number of observations made on a single particle. However, as will be seen below, where the number of observations was large enough and systematic errors were avoided, excellent agreement with theory was obtained.

Generally speaking, measurements of Brownian motion are more difficult in gases than in liquids as a result of the more intense convection in the fluid and the faster settling speed of the particles. On the other hand, aerosols differ from systems with liquid media in that the vertical electric field method can be used to determine particle mobilities very accurately (see § 15) and also for holding the particles over very long periods so that many measurements can be made on a single particle.

Brownian motion in aerosols can be studied by both the horizontal and vertical displacement of particles. In the first case a horizontal microscope with two vertical lines in the eyepiece is used and the time for the image of a particle to pass from one to the other is measured, the experiment being repeated many times. If t_ν is the time in the ν-th experiment and h is the actual distance corresponding to the distance between the two lines in the eyepiece, the diffusion coefficient of the particle is given by the formula (see page 197)

$$D = \frac{1}{2} h^2 \lim_{\nu \to \infty} \frac{1}{\nu} \sum \frac{1}{t_\nu} = \frac{1}{2} h^2 \overline{\left(\frac{1}{t}\right)}. \tag{36.1}$$

The mean relative error in D for ν observations is [263]

$$\overline{\left|\frac{\Delta D}{D}\right|} = \sqrt{\frac{2}{\nu}}. \tag{36.2}$$

For several reasons it is more convenient in aerosols to measure the vertical component of Brownian motion. In this case two horizontal lines in the eyepiece of a horizontal microscope are used and the time for a particle to settle under gravity from one to the other is measured. The particle is then raised by means of an electric field and allowed to fall again, this process being repeated many times. In order to prevent the particle from passing horizontally out of the field of view one of the condenser plates is made up of an inner and an outer part which are insulated from one another [264]. A particle which has strayed from the central part of the condenser can be pulled back by the application of a suitable potential difference between these two parts and an unlimited number of observations can then be made. By this means it has been possible to make as many as 6000 observations on a single particle.

The value t_s, obtained by averaging the values found for t_ν, determines the rate of settling $V_s = h/t_s$. The radius and mobility of a spherical non-aggregated particle are determined from this. There are two ways of determining the diffusion coefficient from the values found for t_ν. In the first of these the following formula [263], similar to (36.1), is used (see page 196):

$$D = \frac{1}{2} h^2 \lim_{\nu\to\infty} \left[\frac{1}{\nu} \sum_{\nu} \frac{1}{t_\nu} - \frac{1}{\frac{1}{\nu}\sum_{\nu} t_\nu} \right] = \frac{1}{2} h^2 \left[\overline{\left(\frac{1}{t}\right)} - \frac{1}{\bar{t}} \right]. \tag{36.3}$$

and the mean relative error is again given by formula (36.2).

In the second method [265] all the experimental values of t_ν are grouped into those greater than and those less than t_s. Suppose the mean value of the former is t_+ and of the latter t_-, and also $t_d = (t_+ - t_-)/2$. The diffusion coefficient is then found from the formula

$$\frac{1}{D} = \frac{4\, t_s^3}{\pi h^2 t_d^2} \left[1 - \frac{(\pi - 2)\, D t_s}{\pi h^2} \right]^2, \tag{36.4}$$

and the mean relative error is again approximately equal to $\sqrt{2/\nu}$ [266]. The experimental conditions for measurement of Brownian motion in gases are such that the second term inside the square brackets of equation (36.4) is usually of the order 0·01 and can be neglected.

TABLE 14. DIFFUSION COEFFICIENT OF OIL DROPLETS FROM MEASUREMENTS OF BROWNIAN MOTION

$r \times 10^5$, cm	3·12	3·21	3·39	3·41	3·99	3·68	3·31	4·11	3·64	2·79	3·93	4·04
ν	617	1713	1079	907	1854	1768	1354	1377	5950	380	557	1281
D_E/D_C	1·023	1·002	1·030	1·038	1·023	1·012	0·973	0·967	1·003	0·940	1·015	0·987
β_E	0·023	0·002	0·030	0·038	0·023	0·012	−0·027	−0·033	0·003	−0·060	0·015	−0·013
β_T	0·056	0·035	0·043	0·047	0·031	0·033	0·038	0·038	0·018	0·072	0·059	0·039

To check the theory of Brownian motion values of D found experimentally are compared with those calculated from equation (35.2); this is illustrated in Table 14 which shows the results of Fletcher's measurements [264] on oil droplets in air at pressures from 120 to 200 mm. Values shown in the table are the radius r of the droplet, the number of observations ν made on it, the ratio of observed (D_E) and calculated (D_C) values of the diffusion coefficient†, the relative error $(D_E - D_C)/D_C = \beta_E$ and the mean theoretical relative error β_T according to (36.2).

The weighted mean value of D_E/D_C from 18,837 observations is 1·003, the mean relative error is 0·003 and the mean theoretical error is 0·010. The relative errors in these experiments were thus systematically less than the theoretical value and it may be supposed that the author discarded those values which showed large deviations from the mean.

Schmid [267], working with selenium particles in an atmosphere of nitrogen, found from 9000 observations the mean values $D_E/D_C = 1{\cdot}015$, $\beta_E = 0{\cdot}015$ and $\beta_T = 0{\cdot}015$.

These works, executed with great care, prove that equations (35.1) and (35.2) are certainly applicable to Brownian motion in aerosols.

Brownian motion can be used to measure aerosol particle sizes, although this method is less accurate and convenient than the method described in preceding chapters in which the velocity of particles acted upon by a constant force is measured. Now suppose that the size of a particle is measured first by the mean time t_s required to traverse a fixed vertical distance h and then by the deviations from this mean, as shown above, ν observations being made in all. In the second case the mean relative error in D and B is given by formula (36.2) and to a first approximation the error in r will be the same. The error in determining t_s from ν observations will clearly be the same as in a single measurement of the time for the particle to traverse a vertical distance $\nu\, h$. In this time the particle will undergo a mean vertical Brownian displacement of $\sqrt{2\, D\nu t_s}$ and the relative error in determining the rate of settling is $\sqrt{2\, D\nu t_s}/h\nu$. Because V_s is very nearly proportional to r^2 the relative error in r will be half this.

The ratio of the two errors denoted by β is seen to be

$$\beta = \sqrt{2\, D\nu t_s}/2\, h\nu : \sqrt{2/\nu} = \frac{1}{2}\sqrt{\frac{D t_s}{h^2}} = \frac{1}{2}\sqrt{\frac{D}{h V_s}}. \tag{36.5}$$

As shown by Table 13, $D/V_s = 3{\cdot}6 \times 10^{-4}$ for particles of radius 0·3 μ. Taking into account the inconvenience of choosing h less than 10^{-2} cm shows that $\beta \leqslant 0{\cdot}1$, so that the error in particle size determination is ten times greater by Brownian motion than by the rate of settling. This difference becomes still greater if the velocity of particles in a sufficiently strong electric field is measured (see page 109). Only for particles with $r \leqslant 10^{-5}$ cm, which is near the limit of the ultramicroscope, can measurements of Brownian displacement give more accurate results than measurements of ordered velocity.

In determining the size of aerosol particles from Brownian motion it is best to use photography because this enables a large number of particles to be measured. A promising method was to photograph particles on a steadily moving horizontal ciné-film with intermittent illumination which served to mark time intervals [268]. Broken wavy lines were obtained on the photographs through which the best straight lines were drawn. The straight lines were used to determine the settling rate of the particles while the deviations from them gave the Brownian displacements. For particle size determination it was considered sufficient to take 100 points with the illumination interrupted 20 times/sec; the displacements of a particle from point to point on the photographs were then measured. The results indicated that particle

† In his calculations Fletcher used formula (36.4) with a different coefficient in the second term but, as mentioned above, this is not very important.

radii calculated from the rate of settling were systematically greater than those obtained from the Brownian displacements, which was attributed by the authors to downward convection currents in the centre of the containing vessel. There are, however, serious objections to this method. According to (36.2) the mean error in the particle radius determined on the basis of one hundred observations is about 15 per cent.; for particles of radius 0·2 to 0·6 μ, with which these experiments were conducted, the mean displacement from point to point (in 0·05 sec) is 1 to 2 μ which can be measured only very roughly. For particles of radius 0·2 μ the mean Brownian displacement and the distance settled in a time of 5 sec (0·05 × 100) are of the same order, and errors of 100 per cent are possible in the determination of settling rate from the photographs. Actually the authors found a settling rate of zero for some particles of this size. This work has been dealt with in detail because it shows clearly the misleading results which can be obtained from experiments on Brownian motion if the arrangements are insufficiently critical. The results could be made more reliable, and sufficiently accurate for practical purposes, by moving the ciné-film vertically to measure horizontal displacements, and by increasing several times the number of frames.

§ 37. PROBABILITY FUNCTIONS IN THE BROWNIAN MOTION OF AEROSOL PARTICLES

I. Certain probability functions are fundamental in the theory of Brownian motion. Consider the projection on the x-axis of a particle which, besides undergoing Brownian motion, is also moving regularly under the influence of an external force or the motion of the fluid. If the particle is at x_0 when $t = 0$ let $w(x_0, x, t)\, dx$ denote the probability of finding the particle between x and $x + dx$ after a time interval t. The function $w(x_0, x, t)$ obeys the integral equation [269]

$$w(x_0, x, t + t') = \int w(x_0, x', t)\, w(x', x, t')\, dx', \tag{37.1}$$

where the integration is performed over all possible values of x'. The physical meaning of equation (37·1) is that the transition from position x_0 to position x in a time $t + t'$ can be realized by a transition first from x_0 to any accessible x' in a time t and then from x' to x in a time t'. The probability of two such transitions is equal to the product of each, $w(x_0, x', t) \times w(x', x, t')$, provided they are independent, and the required probability $w(x_0, x, t + t')$ is obtained by summing all possible pairs of transitions leading from x_0 to x. Equation (37.1) applies only if $w(x_0, x', t)$ and $w(x', x, t')$ are independent or, in accordance with what was stated in § 35, if $t \gg \tau$.

From equation (37.1) it is possible to carry out a purely mathematical derivation of the Fokker–Planck equation† which for a uniform medium reduces to

$$\frac{\partial w}{\partial t} = -\frac{\partial (V_x w)}{\partial x} + D\frac{\partial^2 \omega}{\partial x^2}, \tag{37.2}$$

† Derivations of the most general form of the Fokker–Planck equation and a thorough analysis of the relevant mathematical problems have been given by Kolmogoroff [270] and Chandrasekhar [150]. A simplified derivation is given by Leontovich [253].

where D is diffusion coefficient of the particle and V_x is the resolute of the ordered velocity of the particle on the x-axis. Since the probability of transport to all accessible points is the same w must also obey the equation

$$\int w\,dx = 1, \tag{37.3}$$

with the initial conditions

$$w(x_0, x_0, 0) = 1,$$
$$w(x_0, x, 0) = 0 \quad \text{if} \quad x \neq x_0, \tag{37.4}$$

the meaning of which is apparent.

When there are no external forces and the medium is stationary, $V_x = 0$ and (37.2) becomes

$$\frac{\partial w}{\partial t} = D \frac{\partial^2 w}{\partial x^2}. \tag{37.5}$$

Equations (37.2) and (37.5) are the well-known equations of thermal conductivity and diffusion in moving and stationary media respectively. Note that when these equations are applied to Brownian motion they should be written in finite difference form and, as explained above, Δt must be much greater than τ. Because τ is so small this may be of practical significance only in some special problems (in the theory of coagulation, see § 49).

Since a gas is isotropic and diffusion of spherical particles is independent of direction the same equations are also obtained for the other coordinates. The probability, $w(x_0, x; y_0, y; z_0, z; t)\,dx\,dy\,dz$, of transition from the point (x_0, y_0, z_0) to the element of space $(x, x + dx;\ y, y + dy;\ z, z + dz)$ is equal to the product† $w_x\,dx.\ w_y\,dy.\ w_z\,dz$. Since w_x is independent of y and z etc., it follows that

$$\frac{\partial w}{\partial t} = w_y w_z \frac{\partial w_x}{\partial t} + w_x w_z \frac{\partial w_y}{\partial t} + w_x w_y \frac{\partial w_z}{\partial t}, \tag{37.6}$$

$$\operatorname{div}(\mathbf{V} w) = w_y w_z \frac{\partial (V_x w_x)}{\partial x} + w_x w_z \frac{\partial (V_y w_y)}{\partial y} + w_x w_y \frac{\partial (V_z w_z)}{\partial z}, \tag{37.7}$$

$$-D\Delta w = -D\left(\frac{\partial^2 w}{\partial x^2} + \frac{\partial^2 w}{\partial y^2} + \frac{\partial^2 w}{\partial z^2}\right)$$
$$= -D\left[w_y w_z \frac{\partial^2 w_x}{\partial x^2} + w_x w_z \frac{\partial^2 w_y}{\partial y^2} + w_x w_y \frac{\partial^2 w_z}{\partial z^2}\right], \tag{37.8}$$

where $\mathbf{V}$ is the velocity vector of the ordered motion of the particle and Δ is the Laplace operator. The addition of these equations, using equation (37.2) and the similar equations for the other coordinates, yields

$$\frac{\partial w}{\partial t} = -\operatorname{div}(\mathbf{V} w) + D\Delta w, \tag{37.9}$$

† $w_x = w(x_0, x, t)$ etc.

which is the well-known equation for diffusion or heat conduction in three dimensions. When $\mathbf{V} = 0$ it becomes

$$\frac{\partial w}{\partial t} = D \Delta w. \tag{37.10}$$

Normalization in this case is expressed by the equation

$$\int w dv = 1, \tag{37.11}$$

where the integration extends over the volume accessible to the particle and the initial conditions are

$$w(x_0, x_0;\ y_0, y_0;\ z_0, z_0;\ 0) = 1,$$
$$w(x_0, x;\ y_0, y;\ z_0, z;\ 0) = 0 \quad \text{if} \quad x \neq x_0, \quad y \neq y_0 \quad \text{or} \quad z \neq z_0. \tag{37.12}$$

II. $W(x, t)\,dx$ will be used to denote the probability that a particle is located between x and $x + dx$ at the instant t, the probability being given by the function $W(x_0, 0)\,dx$ at $t = 0$. This probability obeys the equation

$$W(x, t) = \int W(x_0, 0)\, w(x_0, x, t)\, dx_0. \tag{37.13}$$

Let us determine the value of the function

$$\Phi(x, t) = \frac{\partial W}{\partial t} + \frac{\partial (V_x W)}{\partial x} - D \frac{\partial^2 W}{\partial x^2}. \tag{37.14}$$

Differentiating equation (37.13) under the integral sign and using (37.2) we obtain

$$\Phi(x, t) = \int W(x_0, 0) \left[\frac{\partial w}{\partial t} + \frac{\partial (V_x w)}{\partial x} - D \frac{\partial^2 w}{\partial x^2} \right] dx_0 = 0. \tag{37.15}$$

Thus W satisfies the same equation (37.2) as w and the same normalization condition (37.3) but the initial conditions are determined by $W(x, 0)$. Similarly the three-dimensional location probability $W(x, y, z, t)$ satisfies equations (37.9) or (37.10) and (37.11).

If the equations are applied to a large number of particles, n_0, then $n(x, t) = n_0 W(x, t)$ denotes the number concentration of particles in the interval $(x, x + dx)$ at time t. Equations (37.5) and (37.2) then transform to the ordinary equations of one-dimensional diffusion in a stationary medium

$$\frac{\partial n}{\partial t} = D \frac{\partial^2 n}{\partial x^2} \tag{37.16}$$

and in a flowing medium

$$\frac{\partial n}{\partial t} = - \frac{\partial (V_x n)}{\partial x} + D \frac{\partial^2 n}{\partial x}. \tag{37.17}$$

Equations (37.10) and (37.9) become the equations of diffusion in three-dimensions

$$\frac{\partial n}{\partial t} = D \Delta n \tag{37.18}$$

and

$$\frac{\partial n}{\partial t} = - \operatorname{div} (\mathbf{V} n) + D \Delta n. \tag{37.19}$$

These equations are applicable either to the Brownian motion of an individual particle or to the diffusion of a system of particles viewed macroscopically.

III. A function describing the probability that a particle will reach a boundary was introduced into the theory of Brownian motion by Kolmogoroff and Leontowitsch [271] and is very important in some aerosol problems. Let x be the coordinate of a particle and $W^*(x, t)$ the probability that the particle will reach some boundary point in a time t. It is simplest to take x as the distance from this point. We will use $w^*(x, x', t)\, dx'$ to denote the probability that the particle will be between x' and $x' + dx'$ from the boundary at the instant t, never having touched the boundary. It is obvious that $w^*(x, x', t) \leqslant w(x, x', t)$.

The functions W^* and w^* are related by the equation

$$W^*(x, t) + \int_0^\infty w^*(x, x', t)\, dx' = 1. \tag{37.20}$$

The first term on the left hand side represents the probability that the particle will reach the boundary in a time t and the second term the probability that it will not reach the boundary. Moreover,

$$W^*(x, t + t') = W^*(x, t) + \int_0^\infty w^*(x, x', t)\, W^*(x', t')\, dx', \tag{37.21}$$

which means that a particle can reach the boundary in time $t + t'$ in two mutually exclusive cases: (1) if it has already reached the boundary in a time t; (2) if it reaches the boundary for the first time between t and $t + t'$. The first and second terms on the right hand side of the equation express these probabilities.

The differential equation

$$\frac{\partial W^*}{\partial t} = V_x \frac{\partial W^*}{\partial x} + D \frac{\partial^2 W^*}{\partial x^2} \tag{37.22}$$

can be derived from equations (37.20) and (37.21) while $W^*(x, t)$ must satisfy the initial conditions

$$W^*(x, 0) = 0 \quad \text{when} \quad x \neq 0, \quad W^*(x, 0) = 1 \quad \text{when} \quad x = 0 \tag{37.23}$$

and the boundary condition

$$W^*(0, t) = 1, \quad t \geqslant 0. \tag{37.24}$$

Equation (37.22) is identical with the Fokker–Planck equation only if V_x is independent of x. If there is a large number, n_0, of individual particles $n_0 W^*$ denotes the number of particles which were at the point x at $t = 0$ and have reached the boundary in a time t.

Similarly the probability $W^*(x, y, z, t)$ that a particle with coordinates (x, y, z) reaches a surface bounding a given space, or in general any surface, in a time t satisfies the equation

$$\frac{\partial W^*}{\partial t} = \mathbf{V} \operatorname{grad} W^* + D \Delta W^* \tag{37.25}$$

or, if $\mathbf{V} = 0$

$$\frac{\partial W^*}{\partial t} = D \Delta W^*. \tag{37.26}$$

It also satisfies the conditions

$$W^*(x, y, z, 0) = 1, \quad \text{for points} \tag{37.27}$$

on this surface and

$$W^*(x, y, z, 0) = 0, \quad \text{for points} \tag{37.28}$$

outside this surface.

Finally yet another function $W^{**} = \partial W^*/\partial t$ is needed in order to express the probability that a particle will reach the boundary between t and $t + dt$. By differentiating equation (37.25) with respect to t it can be verified that W^{**} also satisfies this equation provided that $\mathbf{V}$ does not depend on time.

The probability functions discussed above can be used to solve a series of important problems on Brownian motion in aerosols. Mathematically they amount to integrating the Fokker–Planck equation for various initial and boundary conditions; the well-developed mathematical theory of heat conduction can be employed for this purpose.

The expression obtained for the transition probability $w(x_0, x, t)$ is [272]

$$w(x_0, x, t)\, dx = \frac{1}{\sqrt{4\pi Dt}} e^{-(x-x_0)^2/4Dt}\, dx, \tag{37.29}$$

when the particle has no ordered motion, and

$$w(x_0, x, t)\, dx = \frac{1}{\sqrt{4\pi Dt}} e^{-(x-x_0-V_x t)^2/4Dt}\, dx \tag{37.30}$$

when the particle is moving in an ordered manner with a constant velocity V_x in addition to its random motion.

The transition probability in three dimensions with $\mathbf{V} = 0$ is

$$w(x_0, x;\ y_0, y;\ z_0, z;\ t)\, dx\, dy\, dz =$$

$$= \frac{1}{\sqrt{(4\pi Dt)^3}} e^{-[(x-x_0)^2+(y-y_0)^2+(z-z_0)^2]/4Dt}\, dx\, dy\, dz = \frac{1}{\sqrt{(4\pi Dt)^3}} e^{-\varrho^2/4Dt}\, dv, \tag{37.31}$$

where ϱ is the distance between the points (x, y, z) and (x_0, y_0, z_0) and dv is an element of volume.

The mean square displacement in one coordinate, $\overline{(x - x_0)^2}$, of a particle in time t can easily be determined by using formula (37.29). By means of the substitution $(x - x_0)/\sqrt{4Dt} = \xi$ we find

$$\overline{(x - x_0)^2} = \frac{1}{\sqrt{4\pi Dt}} \int_{-\infty}^{\infty} (x - x_0)^2 e^{-(x-x_0)^2/4Dt}\, dx = \frac{4Dt}{\sqrt{\pi}} \int_{-\infty}^{\infty} \xi^2 e^{-\xi^2}\, d\xi = 2Dt, \tag{37.32}$$

which is the basic equation (35.1). The mean square displacement in space $\overline{\varrho^2}$ can be determined from (37.31) but it can be found more simply by remembering that $\overline{(x - x_0)^2} = \overline{(y - y_0)^2} = \overline{(z - z_0)^2} = 2\,Dt$ so that

$$\overline{\varrho^2} = \overline{(x - x_0)^2 + (y - y_0)^2 + (z - z_0)^2}$$
$$= \overline{(x - x_0)^2} + \overline{(y - y_0)^2} + \overline{(z - z_0)^2} = 6\,Dt. \tag{37.33}$$

If a particle has a constant ordered velocity V_x equation (37.30) and the substitution $(x - x_0 - V_x t)/\sqrt{4\,Dt} = \xi$ can be used to obtain [273]

$$\overline{(x - x_0)^2} = \frac{1}{\sqrt{4\pi\,Dt}} \int_{-\infty}^{\infty} (x - x_0)^2\, e^{-(x-x_0-V_x t)^2/4Dt}\, dx$$
$$= \frac{1}{\sqrt{\pi}} \int_{-\infty}^{\infty} (V_x t + \xi\sqrt{4\,Dt})^2 e^{-\xi^2}\, d\xi$$
$$= \frac{1}{\sqrt{\pi}}\left(V_x^2 t^2 \sqrt{\pi} + 4\,Dt\,\frac{\sqrt{\pi}}{2}\right) = V_x^2 t^2 + 2\,Dt. \tag{37.34}$$

In this case, therefore, the mean square displacement is equal to the sum of the squares of the "ordered" displacement and of the Brownian displacement which would have occurred in the absence of the ordered motion.

The following expression for the average absolute value of the Brownian displacement when $V_x = 0$ is obtained in a similar way:

$$\overline{|x - x_0|} = \frac{1}{\sqrt{4\pi\,Dt}} \int_{-\infty}^{\infty} |(x - x_0)|\, e^{-(x-x_0)^2/4Dt}\, dx = \sqrt{\frac{4\,Dt}{\pi}}. \tag{37.35}$$

§ 38. DIFFUSIVE DEPOSITION OF AEROSOL PARTICLES FROM A STATIONARY GAS

Great practical importance attaches to the deposition of particles by Brownian motion on to the surfaces of solids or liquids which come into contact with the aerosol. From considerations presented in § 56 it may be seen that particles which touch a wall will stick to it; if they are not jarred too violently, and if there are no strong air currents, they do not return to a state of suspension. Determination of the rate at which particles are deposited on walls therefore amounts to calculating the probability that particles initially disposed in a definite way will reach known boundaries; this can be done by using the function W^* of the preceding section. In most cases, however, it is better to use a more physical approach regarding the deposition process as taking place in a system containing many discrete particles which diffuse towards an absorbing wall. Particles touching the wall are instantly removed from the gas and the particle concentration at the wall can therefore be taken as zero. In all the cases discussed below it is assumed, unless specifically stated, that the particles possess no ordered motion.

I. Imagine the simplest case of a plain vertical wall in contact with an infinitely large volume of aerosol which has the same initial concentration n_0 throughout. Construct the x-axis perpendicular to the wall and fix the origin at the wall. The particle concentration $n(x, t)$ must satisfy the equation

$$\frac{\partial n}{\partial t} = D \frac{\partial^2 n}{\partial x^2}, \tag{38.1}$$

the initial condition

$$n(x, 0) = n_0 \text{ if } x > 0 \tag{38.2}$$

and the boundary condition

$$n(0, t) = 0 \text{ if } t > 0. \tag{38.3}$$

The solution [274] is

$$n(x, t) = \frac{2 n_0}{\sqrt{4\pi D t}} \int\limits_0^x e^{-\xi^2/4Dt}\, d\xi = \frac{2 n_0}{\sqrt{\pi}} \int\limits_0^{x/\sqrt{4Dt}} e^{-\xi^2}\, d\xi = n_0 \operatorname{erf}\left(\frac{x}{\sqrt{4Dt}}\right), \tag{38.4}$$

where erf is the probability integral (error function). The particle concentration expressed by formula (38.4) is illustrated in Fig. 50. It follows from this expression that

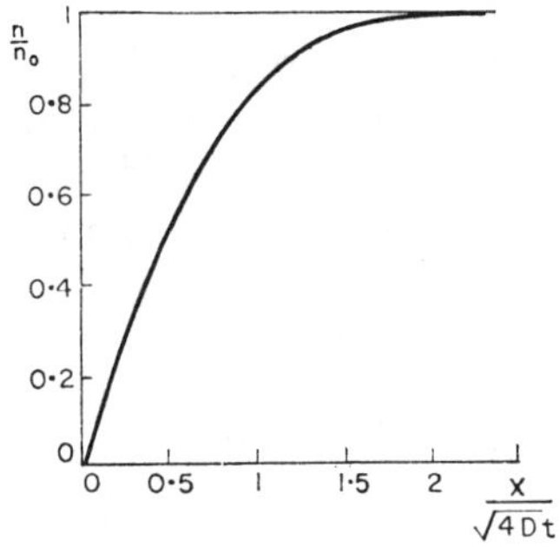

FIG. 50. Distribution of aerosol concentration near a vertical plane wall.

the concentration gradient at the wall is $\partial n/\partial x_{x=0} = n_0/\sqrt{\pi D t}$ so that the number of particles deposited on 1 cm² of the wall between t and $t + dt$ is

$$I\,dt = D \frac{\partial n}{\partial x_{x=0}}\, dt = n_0 \sqrt{\frac{D}{\pi t}}\, dt. \tag{38.5}$$

Integration with respect to time gives the number of particles deposited in a time t

$$N(t) = \int\limits_0^t I\,dt = 2 n_0 \sqrt{\frac{Dt}{\pi}}. \tag{38.6}$$

In aerosol work a simplified method is often used for determining the number of particles deposited on a wall by diffusion. At every instant half the particles are moving towards the wall and half away from it, from which it is argued that in a time t half the particles contained in a layer of thickness $\overline{\Delta x}$ adjacent to the wall will be deposited on it, $\overline{\Delta x}$ being the mean absolute displacement of the particles in a time t, given by expression (37.35). The number obtained in this way is half the correct number obtained from formula (38.6). It would seem, therefore, that all the particles in the layer must be deposited on the wall. The reason for this discrepancy becomes clear if the number of particles leaving the aerosol in a time t through a completely permeable wall, such as an opening in the vessel, is calculated.

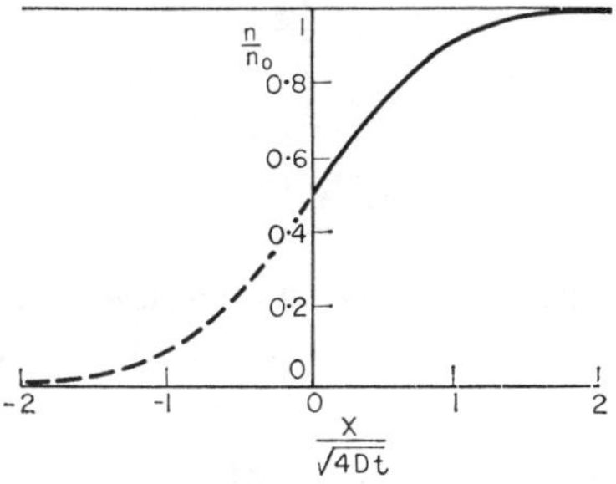

FIG. 51. Distribution of aerosol concentration near an opening.

In this case, as well as the initial condition $n(x, 0) = n_0$ for $x > 0$, there is also the condition $n(x, 0) = 0$ for $x < 0$, which means that there were initially no particles beyond the wall. The boundary condition (38.3) disappears because no particles leave the gas. The solution for these conditions is [272]

$$n(x, t) = \frac{n_0}{2}\left[1 + \operatorname{erf}\left(\frac{x}{\sqrt{4Dt}}\right)\right] \tag{38.7}$$

which shows that a constant particle concentration of $n_0/2$ is established at the wall (Fig. 51), while the concentration gradient and diffusion rate at the permeable wall will be half as much as at an absorbing wall. From the point of view of an individual particle undergoing Brownian motion the difference is explained by the fact that a particle passing through a permeable wall has a chance of returning to the space from which it came equal to the chance of its failing to return, whereas there is no chance of returning with an absorbing wall. Particles are therefore lost from the aerosol twice as rapidly in the second case.

The third possible case, that of a reflecting wall, is not discussed here because it is not applicable to aerosols in the particle-size range where Brownian motion and diffusion are of any significance at all.

II. Let us find the probability that a particle at a distance h from a plain wall and undergoing ordered motion towards it with a constant velocity V_0 reaches the wall between t and $t + dt$. This probability was previously called $W^{**}dt$ and was found by solving equation (37.22) with the appropriate initial and boundary conditions. It is, however, simpler to consider the equivalent problem of a large number N of identical particles which are a distance h from the wall at time $t = 0$ and are undergoing ordered motion towards the wall with a velocity V_0. It is required to find what fraction of them reaches the wall between t and $t + dt$. To do this, equation (37.17) must be solved for its former boundary condition (38.3) and for the initial conditions

$$n(x, 0) = 0 \quad \text{when;} \quad x \neq h; \quad \int_{h-\Delta}^{h+\Delta} n(x, 0)\, dx = N, \tag{38.8}$$

where the integral is taken over the infinitesimal range 2Δ enclosing the point h. The solution is [275]

$$n(x, t) = \frac{N}{\sqrt{4\pi D t}} \exp\left[-\frac{2V_0(x-h) + V_0^2 t}{4D}\right] \times$$
$$\times \left\{\exp\left[-\frac{(x-h)^2}{4Dt}\right] - \exp\left[-\frac{(x+h)^2}{4Dt}\right]\right\}. \tag{38.9}$$

It can easily be verified that this equation does indeed satisfy both equation (37.17) and the conditions stated. Hence the probability that a particle will touch the wall in the interval $(t, t + dt)$ is

$$W^{**}(x, t)\, dt = \frac{D}{N} \frac{\partial n}{\partial x_{x=0}} dt = \frac{h}{\sqrt{4\pi D t^3}} \exp\left[-\frac{(h - V_0 t)^2}{4Dt}\right] dt \tag{38.10}$$

and it follows that the average time required for a particle to reach the wall is†

$$\bar{t} = \int_0^\infty t W^{**} dt = \frac{h}{\sqrt{4\pi D}} \int_0^\infty t^{-1/2} e^{-(h-V_0 t)^2/4Dt} dt = h/V_0, \tag{38.11}$$

which is the same as if there were no Brownian motion. The mean of the reciprocal of this time is†

$$\overline{\left(\frac{1}{t}\right)} = \int_0^\infty \frac{1}{t} W^{**} dt = \frac{h}{\sqrt{4\pi D}} \int_0^\infty t^{-5/2} e^{-(h-V_0 t)^2/4Dt} dt = \frac{2D}{h^2} + \frac{V_0}{h} \tag{38.12}$$

and from the last two equations the important formula

$$D = \frac{h^2}{2}\left[\overline{\left(\frac{1}{t}\right)} - \frac{1}{\bar{t}}\right] \tag{38.13}$$

is obtained.

In the experiments mentioned in § 36 the time for the image of a particle to pass between two horizontal lines in a microscope eyepiece was measured. Formula (38.13) is clearly applicable to these experiments.

If $V_0 = 0$ the simpler formula

$$W^{**} dt = \frac{h}{\sqrt{4\pi D t^3}} e^{-h^2/4Dt} dt \tag{38.14}$$

† By the substitution $t = ht'/V$ these integrals are reduced to the forms

$$\int_0^\infty t^{-1/2} e^{-\frac{t^2+1}{qt}} dt \quad \text{and} \quad \int_0^\infty t^{-5/2} e^{-\frac{t^2+1}{qt}} dt,$$

values of which are given in books of tables.

is obtained instead of (38.10) and $\overline{\left(\frac{1}{t}\right)}$ is then given by

$$\overline{\left(\frac{1}{t}\right)} = \frac{2D}{h^2}. \tag{38.15}$$

This formula is used in the study of horizontal Brownian motion (see page 185).

Equation (38.10) can be reduced to a dimensionless form by referring times to the mean time $\bar{t} = h/V_0$ for a particle to reach the wall, i.e. putting $t' = t/\bar{t}$. If the dimensionless group $\sqrt{V_0 h/4D}$ is denoted by μ (38.10) can be brought to the form

$$W^{**}dt' = \frac{\mu}{\sqrt{\pi t'^3}}\, e^{-\mu^2(1-t'^2)/t'}\, dt'. \tag{38.16}$$

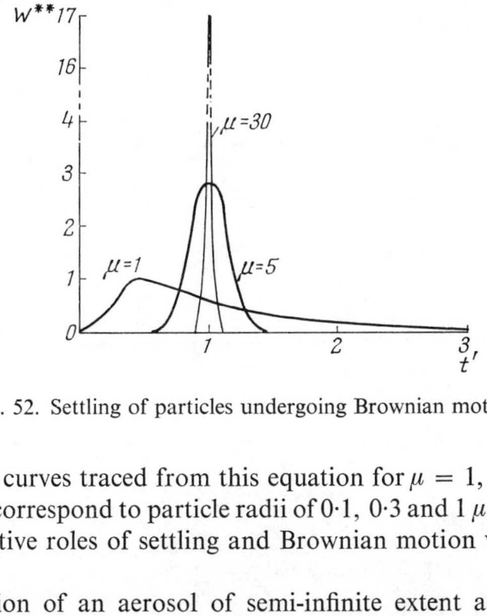

FIG. 52. Settling of particles undergoing Brownian motion.

Figure 52 depicts curves traced from this equation for $\mu = 1$, 5 and 30; when $h =$ 0·4 cm these values correspond to particle radii of 0·1, 0·3 and 1 μ. The diagram shows clearly how the relative roles of settling and Brownian motion vary in this range of particle sizes.

III. The deposition of an aerosol of semi-infinite extent and initially uniform concentration, n_0, on a horizontal surface will now be discussed, with allowance for settlement under gravity. This necessitates the integration of equation (38.10) for values of h from 0 to ∞ and yields the expression

$$I(t)\,dt = n_0\left\{\sqrt{\frac{D}{\pi t}}\, e^{-V_0^2 t/4D} + \frac{V_0}{2}\left(1 + \operatorname{erf}\sqrt{\frac{V_0^2 t}{4D}}\right)\right\} dt \tag{38.17}$$

for the number of particles deposited per cm² of horizontal surface between the times t and $t + dt$.

If $t \gg \frac{4D}{V_0^2}$ this formula becomes $I(t) = n_0 V_0$ and Brownian motion no longer has any effect on the rate of deposition, which is determined only by the rate of settling. If $t \ll \frac{4D}{V_0^2}$ the expression becomes $I(t) = n_0\left(\sqrt{\frac{D}{\pi t}} + \frac{V_0}{2}\right)$. In this case, therefore, the deposition consists of the diffusive deposition which would occur in the absence of settling [see (38.5)] and half the sedimentation which would occur in the absence of diffusion. This example shows clearly that serious errors may result if the rate of deposition of aerosols due to the simultaneous action of Brownian motion and external or inertial forces is calculated by simply summing the individual effects of each. Unfortunately one often has to resort to such summation because of the great mathematical difficulties.

The formulae given above hold if the medium is at rest, a condition which can only be realized in practice with comparatively small volumes of gas. Even in smoke chambers, let alone in the free atmosphere, convection completely changes the picture.

IV. In connection with the counting of particles with the ultramicroscope it is of interest to find the probability that a particle located at some point between two vertical parallel walls will be deposited on them by diffusion in a time t. By taking the origin of coordinates on one of the walls this problem reduces to solving equation (37.16) for the initial conditions

$$n(x, 0) = n_0 \quad \text{if} \quad 0 < x < h; \quad n(x, 0) = 0 \quad \text{if} \quad x < 0 \quad \text{and} \quad x > h, \tag{38.18}$$

where h is the distance between the walls, and the boundary conditions

$$n(0, t) = 0; \quad n(h, t) = 0 \quad \text{if} \quad t > 0. \tag{38.19}$$

The solution of this equation is the series [276]

$$n(x, t) = \frac{4\, n_0}{\pi} \sum_{\nu=1}^{\infty} \frac{1}{2\nu - 1} \sin\left[(2\nu - 1)\frac{\pi x}{h}\right] \exp\left[-\frac{(2\nu - 1)^2 \pi^2 D t}{h^2}\right] \tag{38.20}$$

and the total number of particles deposited on the two walls per cm^2 of each between times t and $t + dt$ is

$$I\, dt = 2\, D \frac{\partial n}{\partial x}_{x=0} dt = \frac{8\, D n_0}{h} \sum_{\nu=1}^{\infty} \exp\left[-\frac{(2\nu - 1)^2 \pi^2 D t}{h^2}\right] dt. \tag{38.21}$$

The number deposited similarly in a time t is†

$$N(t) = \int_0^t I\, dt = \frac{8\, n_0 h}{\pi^2} \sum_{\nu=1}^{\infty} \frac{1}{(2\nu - 1)^2} \left\{1 - \exp\left[-\frac{(2\nu - 1)^2 \pi^2 D t}{h^2}\right]\right\}$$

$$= n_0 h \left\{1 - \frac{8}{\pi^2} \sum_{\nu=1}^{\infty} \frac{1}{(2\nu - 1)^2} \exp\left[-\frac{(2\nu - 1)^2 \pi^2 D t}{h^2}\right]\right\}. \tag{38.22}$$

Dividing this equation by $n_0 h$, the initial number of particles between the walls per 1 cm^2 of their area, gives the required probability

$$W^*(t) = 1 - \frac{8}{\pi^2} \sum_{\nu=1}^{\infty} \frac{1}{(2\nu - 1)^2} \exp\left[-\frac{(2\nu - 1)^2 \pi^2 D t}{h^2}\right]. \tag{38.23}$$

† The formula

$$\sum_{\nu=1}^{\infty} \frac{1}{(2\nu - 1)^2} = \frac{\pi^2}{8}$$

is used here.

According to (38.21) the expression obtained for the mean life time of the particle between the walls is†

$$\bar{t} = \frac{1}{n_0 h}\int_0^\infty It\,dt = \frac{8\,D}{h^2}\int_0^\infty t\sum_{\nu=1}^{\infty}\exp\left[-\frac{(2\,\nu-1)^2\pi^2 D\,t}{h^2}\right]dt = \frac{h^2}{12\,D}. \tag{38.24}$$

In ultramicroscope cells where the volume is limited mechanically, particles are counted in a layer of thickness 0·1 mm between two parallel glass slides.†† The opinion has been expressed that, because of Brownian motion, some particles will diffuse to the walls before they have been counted. Suppose that a time of the order 0·1 sec is necessary for a particle to be counted after the aerosol is stopped. For particles of radius 0·1 μ, corresponding approximately to the lower limit of visibility in the ultramicroscope, $W^* = 0{\cdot}11$ according to equation (38.23) and for $r = 0{\cdot}3\,\mu$ $W^* = 0{\cdot}05$ so that the error will be comparatively small. The respective values of $\bar{t}$ are 3·8 and 16·7 sec. However, it can be seen from this that investigations requiring observations over longer periods, such as Brownian displacements, cannot be conducted in such narrow cells. The use of cardioid condensers and the like is thus excluded.

Radushkevich [278] investigated the rate of deposition of stearic acid particles in a cell with parallel walls 0·1 mm apart. He used an ultramicroscope to count the particles and determined their mean size by the number-weight method. In comparatively isodisperse freshly prepared smokes, fairly satisfactory agreement between the experimental results and formula (38.23) was obtained, but smokes which had coagulated, and were more polydisperse, deposited at a higher rate than the theoretical. Radushkevich correctly attributes this to the fact that, for a polydisperse smoke, the mean cube particle radius $\sqrt[3]{\overline{r^3}}$, determined by the number-weight method, is appreciably greater than the mean value r' given by the equation $\left(1 + A\frac{l}{r'}\right)\Big/r' = \overline{\left(1 + A\frac{l}{r}\right)\Big/r}$ on which the rate of diffusion of a polydisperse aerosol depends.

Figure 53 shows the decrease with time in the number of particles in the cell (the function $1 - W^*$), time being expressed in the dimensionless units Dt/h^2.

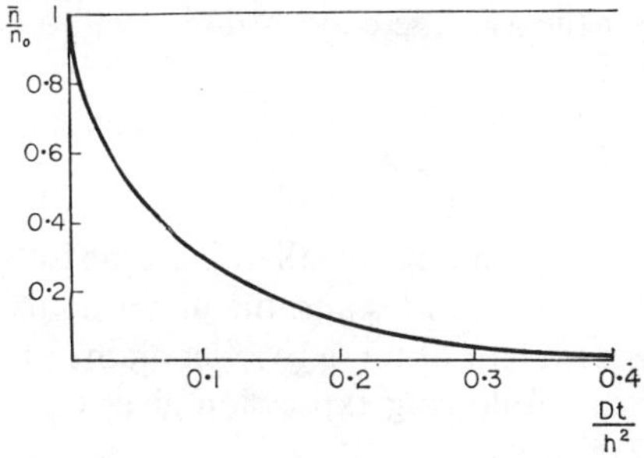

FIG. 53. Decrease in aerosol concentration between two vertical walls.

† The formula

$$\sum_{\nu=1}^{\infty}\frac{1}{(2\,\nu-1)^4} = \frac{\pi^4}{96}$$

is used here.

†† This method is now somewhat out of date. The method of counting flowing particles is much better [277].

The deposition of an aerosol which diffuses and settles between parallel horizontal walls is also of some interest [279]. If the dimensionless variables $x' = x/h$, $t' = t/t_0$ (where $t_0 = h/V_s$) are introduced and the quantity $\sqrt{V_s h/4D}$ is denoted by μ the expression obtained for the aerosol concentration distribution is

$$n = n_0 \sum_{\nu=1}^{\infty} \frac{2}{\pi\nu} \frac{1-(-1)^\nu e^{-2\mu^2}}{1+\dfrac{4\mu^4}{\pi^2\nu^2}} \sin \pi\nu x' \exp\left\{\frac{2x' - \left(1+\dfrac{\pi^2\nu^2}{4\mu^2}\right)t'}{\mu^2}\right\}. \tag{38.25}$$

Figure 54 shows the distributions for $\mu = 0$ (no settling), 1·2 and 5 and it can be seen that in highly dispersed aerosols it is not possible to obtain a sharp upper boundary, even under ideal conditions where there is no convection and the system is completely isodisperse.

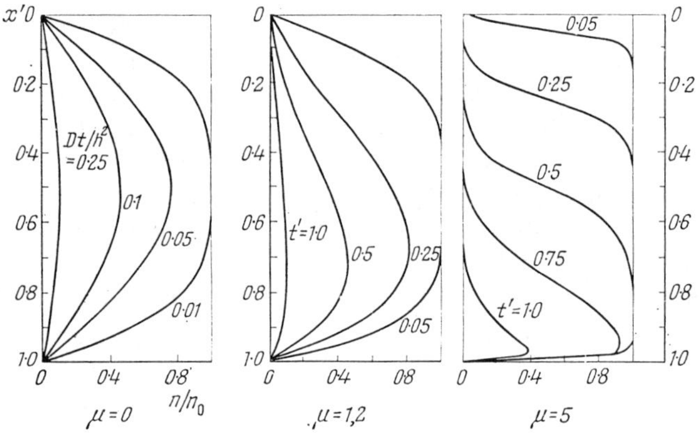

FIG. 54. Distribution of aerosol concentration between horizontal absorbing surfaces.

V. If an aerosol with an initial concentration n_0 is situated in a spherical vessel of radius R then in time t the average concentration will fall to the value [276]

$$\bar{n} = n_0 \frac{6}{\pi^2} \sum_{\nu=1}^{\infty} \frac{1}{\nu^2} e^{-D\pi^2\nu^2 t/R^2}. \tag{38.26}$$

The number of particles deposited on the wall in this time is $(4/3)[\pi R^3(n_0 - \bar{n})]$. Figure 55 shows the relationship between $\bar{n}/n_0$ and the dimensionless quantity Dt/R^2.

VI. In the analogous problem of an aerosol contained in an infinitely long cylindrical vessel of radius R the following expression gives the average concentration of particles at the time t [276]

$$\bar{n} = 4n_0 \sum_{\nu=1}^{\infty} \frac{1}{\beta_\nu^2} e^{-D\beta_\nu^2 t/R^2}. \tag{38.27}$$

Here $\beta_1^2, \beta_2^2 \ldots$ are the squares of the zeros of the zero-order Bessel functions of the first kind $I_0(x)$, which have the following values: $\beta_1^2 = 5{\cdot}784$; $\beta_2^2 = 30{\cdot}47$; $\beta_3^2 = 74{\cdot}89$; $\beta_4^2 = 132{\cdot}8$; $\beta_5^2 = 222{\cdot}9$.

Figure 56 shows the relationship between n/n_0 and Dt/R^2. If n is calculated by the simplified method mentioned earlier, considering that all the particles situated within a distance $2\sqrt{\dfrac{Dt}{\pi}}$ of the wall at $t = 0$ are deposited on the wall in a time t, then the broken curve, which is a useful approximation for $Dt/R^2 < 0{\cdot}1$, is obtained.

VII. A very important problem in the physics of aerosols is the diffusion towards the surface of a sphere situated in an infinitely large volume of aerosol with an initial concentration n_0. The initial condition for this problem is $n(\varrho, 0) = n_0$ if $\varrho > R$ and the boundary condition is $n(R, t) = 0$ if $t > 0$. Since the diffusion coefficient does

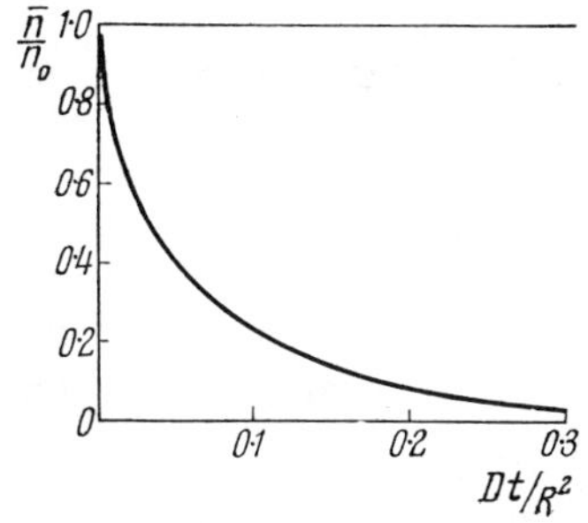

FIG. 55. Decrease in aerosol concentration in a spherical vessel.

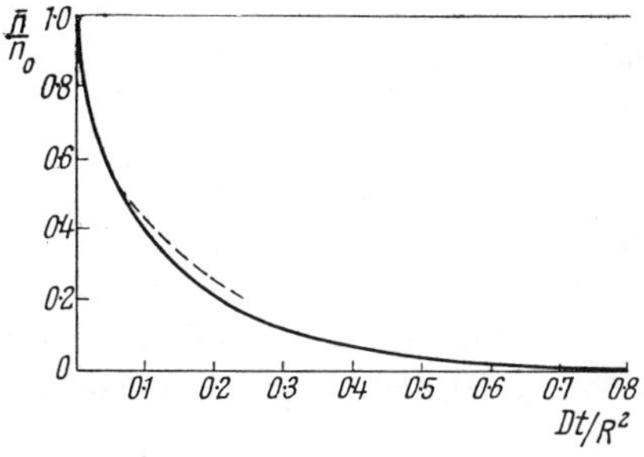

FIG. 56. Decrease in aerosol concentration in a cylindrical vessel.

not depend on direction, equation (37.18) becomes, in polar coordinates

$$\frac{\partial (n\varrho)}{\partial t} = D \frac{\partial^2 (n\varrho)}{\partial \varrho^2}. \tag{38.28}$$

Thus the function $\varrho n(\varrho, t)$ satisfies equation (38.28) and the conditions

$$\varrho n(\varrho, 0) = \varrho n_0 \quad \text{if} \quad \varrho > R, \tag{38.29}$$

$$R n(R, t) = 0 \quad \text{if} \quad t > 0. \tag{38.30}$$

Solution of this problem yields the expression [280]

$$n = n_0 \left[1 - \frac{R}{\varrho} + \frac{2R}{\varrho \sqrt{\pi}} \int\limits_0^{\frac{\varrho - R}{2\sqrt{Dt}}} e^{-\xi^2} d\xi \right]. \tag{38.31}$$

Hence it follows that the number of particles deposited on the sphere between times t and $t + dt$ is

$$\Phi \, dt = 4\pi R^2 D \frac{\partial n}{\partial \varrho_{\varrho = R}} dt = 4\pi D R n_0 \left(1 + \frac{R}{\sqrt{\pi D t}} \right) dt \tag{38.32}$$

and the number deposited in time t is

$$N = \int\limits_0^t \Phi \, dt = 4\pi R D n_0 \left(t + \frac{2R\sqrt{t}}{\sqrt{\pi D}} \right). \tag{38.33}$$

If $Dt/R^2 \ll 1$, particles are deposited on the surface of the sphere from a thin adjacent layer of aerosol at the same rate as onto a flat wall [see formula (38.6)]. If $Dt/R^2 \gg 1$, a practically constant concentration distribution is established around the sphere and is given by the equation

$$n = n_0\left(1 - \frac{R}{\varrho}\right). \tag{38.34}$$

The concentration gradient at the surface of the sphere is therefore given by

$$\frac{\partial n}{\partial \varrho_{\varrho = R}} = -\frac{n_0}{R} \tag{38.35}$$

and the rate of deposition (the number of particles deposited in 1 sec) on the sphere is constant and equal to

$$\Phi = 4\pi D R n_0. \tag{38.36}$$

VIII. The diffusive deposition of aerosols on the external surface of an infinitely long cylinder is more complicated [281] and only a table of values of $\Phi'/\pi R^2 n_0$ (Φ' is the number of particles deposited in time t on unit length of the cylinder) as a function of Dt/R^2 will be presented here.

TABLE 15. DIFFUSIVE DEPOSITION OF AN AEROSOL ON AN INFINITELY LONG CIRCULAR CYLINDER

Dt/R^2	0·001	0·005	0·01	0·02	0·05	0·1	0·2	0·4
$\Phi'/\pi R^2 n_0$	0·072	0·164	0·235	0·337	0·550	0·805	1·190	1·632
Dt/R^2	0·6	0·8	1·0	1·5	2·0	2·5	3·0	3·5
$\Phi'/\pi R^2 n_0$	2·279	2·721	3·125	4·04	4·87	5·67	6·38	7·10

All the results in this section must be corrected for interception, according to the size of the particles (see page 141). This is easily done by constructing an absorbing surface at a distance r, equal to the particle radius, from the actual surface under investigation. Since particles in contact with the actual surface have their centres in contact with the absorbing surface, it is then possible to consider the particles as points and to apply all the above equations to the system obtained.

When calculating diffusive deposition of aerosols it is usually assumed that the concentration at the wall is equal to zero. This assumption may not be accurate. Corresponding to the discontinuity in vapour concentration at the surface of an evaporating drop [722], there can be an analogous excess in particle concentration at an absorbing surface in the case of aerosols. It will be appreciated that this is of practical importance only when the apparent mean free path of the aerosol particles (see page 182) is comparable to the size of the object upon which the aerosol deposits. The problem is considered in more detail in § 49.

Todorov and Sheludko [723] studied the deposition of aerosols in a spherical vessel (see page 200), taking into consideration both diffusion and sedimentation of the particles; two kinds of boundary conditions at the walls of the vessel, making an allowance for the concentration discontinuity or ignoring it, were used. The sedimentation velocity of Stokes particles being proportional to r^2 and that of diffusion to

r^{-1}, it can be readily seen that the total deposition rate must have a minimum at a certain value of r, $r_m \approx (3\,kT/\pi R\gamma_g)^{1/3}$. In the pulmonary alveoli (see Table 20) $R \approx 0{\cdot}015$ cm so, when $\gamma = 1$, $r_m = 0{\cdot}14\,\mu$, in agreement with the estimate of Davies (see page 236).

A static diffusion method, to be distinguished from the dynamical diffusion method described in § 39, for determining the particle size in aerosols, was proposed by Pollak and O'Connor [724, 725] and by Fürth [726]; it is based on (38.27). An aerosol was aspirated from a large gasometer into the cylindrical tube of a photo-electric counter

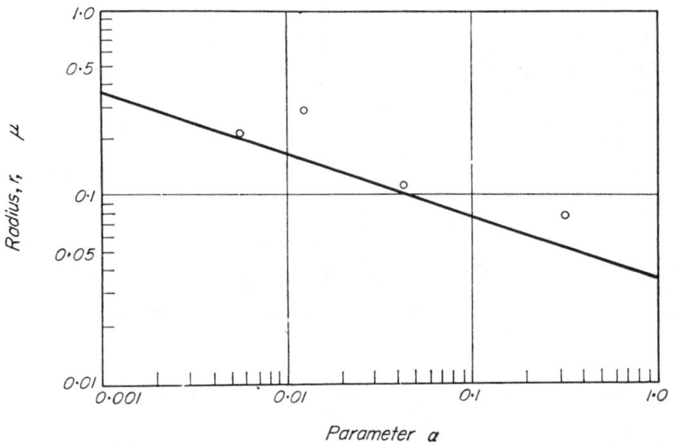

FIG. 56. Diffusion of an aerosol in a horizontal flat cell.

for condensation nuclei. The number Z of particles in the tube was determined immediately and a new sample of aerosol was then introduced and Z again determined after a time interval, the process being repeated several times. The value of the diffusion coefficient was calculated with the help of (38.27) and from it the mean particle size was found. Much lower values of D were always obtained than with the dynamical method for the reason that the aerosol was not drawn through the tube in these experiments, only enough to fill it being aspirated. A layer of uncertain thickness containing no particles must therefore have been retained at the walls of the tube, due to the parabolic profile of the flow, which resulted in a decrease in the rate of deposition of particles on the walls. The thickness of this layer was not determined by hydrodynamical calculations but was assumed to be constant and of a thickness chosen to eliminate the discrepancy mentioned above. Another point is that convection must inevitably have occurred in tubes with $2\,R = 2$ cm so that the employment of (38.17) was really inadmissible. It should be noted, however, that each diffusion method gives too low a value of $\bar{D}$, the mean diffusion coefficient of a polydisperse aerosol (see § 39).

Richardson and Wooding [727] studied experimentally the diffusion of an aerosol between parallel horizontal walls. Nearly isodisperse aerosols with $\bar{r} = 0{\cdot}05 - 2\,\mu$ were used and the distance between the walls was $h = 0{\cdot}2$ cm. Convection was carefully avoided. The aerosol concentration was measured as a function of time t and height z by tyndallometry and by ultramicroscopy. The ratios $\Delta n/\Delta t$, $\Delta n/\Delta z$ and

$\Delta(\Delta n/\Delta z)/\Delta z$ were calculated and substituted into the equation $\partial n/\partial t = -V_s \partial n/\partial z + D\partial^2 n/\partial z^2$ (see (37.17)), and hence the quantity $\alpha = D/hV_s = 3kT/4\pi r^3 \gamma g h$ was found. The theoretical graph of $(\log r, \log \alpha)$ is plotted on Fig. 56, as well as the experimental points. Taking into account the character of the measurements the agreement is reasonably satisfactory. The graphs of n/n_0 against z/h, obtained by these two methods, are given on Fig. 57; unfortunately, no comparison with theoretical curves (Fig. 54) was made by the authors.

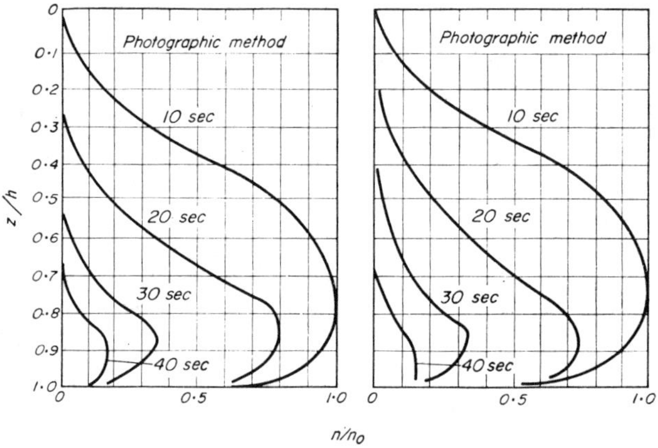

FIG. 57. Concentration distribution of an aerosol in a horizontal flat cell.

§ 39. DIFFUSION OF AEROSOLS IN LAMINAR FLOW

The considerably more difficult problem of the diffusion of aerosols in laminar flow has been solved exactly in only a few cases.

I. The problem of diffusion (or heat transfer) to the walls of a circular tube from a fluid passing through it in laminar flow has been handled by several authors who were unaware of the work of their predecessors [282–285]. The rate of longitudinal mass transfer, due to the flow, is assumed to be considerably greater than the diffusive transfer along the tube. In other words, if the x-axis is parallel to the streamlines, the term $D\partial^2 n/\partial x^2$ in the basic equation (37.17) is very small compared with $\partial(V_x n)/\partial x \approx U\delta n/\delta x$, U being the flow velocity, and can be neglected.

Let n_0 be the initial particle concentration in the aerosol, n the mean concentration in the aerosol that has passed through a tube of radius R and length x, and $\bar{U}$ the mean flow velocity in the tube. Gormley and Kennedy [285] obtained the following expression for $\bar{n}/n_0$:

$$\bar{n}/n_0 = 0{\cdot}819 \exp(-3{\cdot}657\,\mu) + 0{\cdot}097 \exp(-22{\cdot}3\,\mu) + 0{\cdot}032 \exp(-57\,\mu) + \cdots \quad (39.1)$$

where μ is the dimensionless group $Dx/R^2\bar{U}$.

Nusselt [283] derived a similar series with coefficients close to those of (39.1) and Townsend [284] found the value 0·78 for the first coefficient. It was assumed in the derivation of (39.1) that the concentration falls continuously along the axis of the tube and from the axis to periphery, corresponding to conditions of steady diffusion towards the walls. At small values of μ these conditions are not established and formula (39.1) is not applicable, $n = n_0$ being true over the complete cross-section except in the wall layer. For unsteady conditions with μ small Leveque [286] deduced the equation

$$\bar{n}/n_0 = 1 - 2{\cdot}56\,\mu^{2/3} \tag{39.2}$$

while Gormley and Kennedy [285] gave the more accurate expression

$$\bar{n}/n_0 = 1 - 2{\cdot}56\,\mu^{2/3} + 1{\cdot}2\,\mu + 0{\cdot}177\,\mu^{4/3}. \tag{39.3}$$

TABLE 16. FALL IN CONCENTRATION OF AN AEROSOL DUE TO DIFFUSION OF PARTICLES TO THE WALLS FOR LAMINAR FLOW THROUGH A CIRCULAR TUBE

μ	0	0·0002	0·0004	0·001	0·002
$\bar{n}/n_0$ (39·1)	0·948	—	—	—	—
$\bar{n}/n_0$ (39·3)	1·000	0·9915	0·9866	0·9756	0·962
μ	0·004	0·005	0·01	0·02	0·03
$\bar{n}/n_0$ (39·1)	—	0·916	0·885	0·833	0·789
$\bar{n}/n_0$ (39·3)	0·940	0·931	0·893	0·836	0·790
μ	0·04	0·05	0·10	0·20	0·30
$\bar{n}/n_0$ (39·1)	0·750	0·716	0·578	0·394	0·274
$\bar{n}/n_0$ (39·3)	0·750	0·715	0·576	0·383	0·245
μ	0·50	1·0	2·0		
$\bar{n}/n_0$ (39.1)	0·131	0·021	0·0005		
$\bar{n}/n_0$ (39·3)	0·056	−0·19	−0·25		

These equations are very important in the study of highly dispersed aerosols; besides providing a basis for one of the most important methods of determining particle size they permit estimation of the losses of particles in connecting tubes and channels which arise from diffusion. In Table 16 values of $\bar{n}/n_0$ calculated from these equations are given for a series of values of μ.

It can be seen from this table that for $0{\cdot}01 \leqslant \mu \leqslant 0{\cdot}1$ the values of $\bar{n}/n_0$ calculated from equations (39.1) and (39.3) differ by less than 1 per cent. In accurate work equation (39.2) is best avoided.

The above formulae were derived on the assumption that Poiseuille flow, given by equation (26.11), commences at the entrance of the tube but, in fact, the parabolic velocity distribution is not established until a distance of approximately $0{\cdot}1\, R\, \mathrm{Re}_f$ from the entrance has been travelled.

By comparing Table 16 with (38.27) or Fig. 56 and putting $x = t\bar{U}$, i.e. $\mu = Dt/R^2$ it is readily seen that the rate of aerosol deposition on the walls of the tube is much larger in a uniform flow than in the flow with parabolic profile. The overall deposition in the tube is therefore increased due to the entrance effect. It is obvious that the error thus committed increases with the diffusion coefficient of the particles. In fact this error reaches an appreciable value only for aerosols with particle size of the order of 10^{-7} cm.

Radushkevich [288] has used formula (39.1) to determine the mean size of particles in ammonium chloride smokes. The smoke was passed at a mean velocity of 1·84 cm sec^{-1} through a bank of vertical capillaries of length 20 cm and radius 0·012 cm, using the ultramicroscope to find the particle concentrations before and after passage through the tubes. In these experiments μ was about 0·5. The values found for $\bar{n}/n_0$ were used to calculate the mean particle radius which came out 20 to 30 per cent higher than the value determined by the number-weight method. The explanation of this discrepancy is probably that ammonium chloride smoke particles are more or less loose aggregates and their actual size is greater than that determined by the number-weight method.

II. The similar problem of a channel with parallel plane walls separated by a distance 2 h [289] leads at large values of $\mu = Dx/h^2\bar{U}$ to the formula:

$$\bar{n}/n_0 = 1{\cdot}066 \exp(-1{\cdot}83\,\mu) + 0{\cdot}0065 \exp(-22{\cdot}2\,\mu) \quad (39.4)$$

or at a more exact formula by Gormley [728]

$$\bar{n}/n_0 = 0{\cdot}9099 \exp(-1{\cdot}885\,\mu) + 0{\cdot}0531 \exp(-21{\cdot}43\,\mu). \quad (39.5)$$

Kennedy [729] has for small μ:

$$\bar{n}/n_0 = 1 - 1{\cdot}175\,\mu^{2/3} + 0{\cdot}1\,\mu + 0{\cdot}0175\,\mu^{4/3}. \quad (39.6)$$

DeMarcus [730], for any μ:

$$\bar{n}/n_0 = 0{\cdot}9149 \exp(-1{\cdot}885\,\mu) + 0{\cdot}0592 \exp(-22{\cdot}3\,\mu) + 0{\cdot}0258 \exp(-151{\cdot}8\,\mu). \quad (39.7)$$

Values of $\bar{n}/n_0$ calculated from these formulae are given in Table 16a.

As seen from the table, (39.7) covers all values of μ.

Nolan and Guerrini [289] passed atmospheric air through vertical parallel-sided channels and used an Aitken counter to determine the percentage of atmospheric condensation nuclei deposited on the walls of the channels. They calculated μ from (39.4) and obtained a value of $2{\cdot}85 \times 10^{-6}$ cm for the radii of the condensation nuclei. The experiment was then repeated with the channels horizontal and, by taking the difference between the results to be due to sedimentation, they calculated the mean rate of settling and hence the mass and density of the nuclei. The impossibly high density (1·7) which they obtained suggests that this simplified method of calculation may not be permissible.

Thomas [731] used in his experiments a diffusion battery consisting of 20 narrow rectangular channels 47 cm long with $h = 0{\cdot}05$ mm. Aerosols of dioctylphthalate with $\bar{r} = 0{\cdot}16$–$0{\cdot}5\,\mu$ were drawn through the battery at the rate of 4–20 cm/sec. To eliminate the entrance effects, in a parallel experiment the aerosol was drawn through another battery of the same construction but only 5 cm long, and the ratio of the concentrations of aerosols issuing from the two batteries was measured with a tyndallometer and used as the basis of calculations (which is undoubtedly wrong). The mean particle sizes were found from the light scattering at two different angles and by means of higher order Tyndall spectra.

The diffusion coefficients of the aerosols determined thus were found to increase considerably (as much as twice) with an increase in the flow rate. The same observation was made by Pollak *et al.* [725]. Proceeding from an obviously incorrect assumption that it was due to turbulence, Thomas extrapolated the values of D obtained to $\bar{U} = 0$.

TABLE 16A. FALL IN CONCENTRATION OF AN AEROSOL DUE TO DIFFUSION TOWARDS THE WALLS OF A FLAT RECTANGULAR CHANNEL IN LAMINAR FLOW

μ	0	0·001	0·002	0·005	0·01	0·02	0·05	0·01	0·15
$\bar{n}/n_0$ (39·5)	0·963	0·960	0·957	0·949	0·936	0·911	0·846	0·760	0·688
$\bar{n}/n_0$ (39·6)	1·00	0·988	0·982	0·966	0·946	0·915	0·846	0·758	0·685
$\bar{n}/n_0$ (39·7)	1·00	0·993	0·987	0·971	0·951	0·920	0·852	0·764	0·692

μ	0·20	0·30	0·40	0·60	0·80	1·0	1·5	2·0	4·0
$\bar{n}/n_0$ (39·5)	0·625	0·518	0·428	0·294	0·201	0·138	0·054	0·021	0·0005
$\bar{n}/n_0$ (39·6)	0·620	0·507	0·407	0·233	0·08	—	—	—	—
$\bar{n}/n_0$ (39·7)	0·629	0·520	0·430	0·295	0·202	0·139	0·054	0·021	0·0005

After this the values of $\bar{r}$ calculated from D coincided with those found by optical means at $r = 0{\cdot}16$–$0{\cdot}30\,\mu$ but were still too low for $\bar{r} > 0{\cdot}3\,\mu$. Regarding the use of the short battery to allow for entrance effects, Thomas wrote in one of his papers [732] that more accurate results could be obtained by this method, whereas in another paper [733] he claimed the opposite.

Pollak and Metnieks [734] pointed out that the diffusion coefficient of a polydisperse aerosol, determined from experiments, must appear to increase with the rate of flow in the channel. For the sake of simplicity, only the first exponential term in an equation of the type (39.5) was taken and it was assumed that the aerosol consisted of several isodisperse fractions. Let n_{i0} and $\bar{n}_i$ be the particle concentrations of the i-th fraction at the inlet to the channel and at its outlet, $n_0 = \sum_i n_{i0}$ and $\bar{n} = \sum_i \bar{n}_i$ the corresponding total concentrations and $p_{i0} = n_{i0}/n_0$, the relative concentration of the particles of each fraction. For the i-th fraction we have

$$\bar{n}_i = An_{i0}\exp(-KD_i/\bar{U}) = An_0p_{i0}\exp(-KD_i/\bar{U}). \tag{39.8}$$

The diffusion coefficient D' determined experimentally satisfies the equation

$$An_0\exp(-KD'/\bar{U}) = \bar{n} = \sum_i \bar{n}_i = \sum_i An_0p_{i0}\exp(-KD_i/\bar{U}). \tag{39.9}$$

Thus

$$\exp(-KD'/\bar{U}) = \sum_i p_{i0}\exp(-KD_i/\bar{U}). \tag{39.10}$$

This equation was differentiated with respect to $\bar{U}$, assuming D' to be a constant, which is incorrect since D' is a function of $\bar{U}$. The original deduction is therefore abandoned from this point.

Denoting $\exp(-KD_i/\bar{U})$ by z_i and putting $z = 1 - x$, $0 < x < 1$, we obtain from (39.10)

$$D' = -\frac{\bar{U}}{K} \ln \sum p_{i0} z_i;$$

$$\frac{dD'}{d\bar{U}} = \frac{1}{\bar{U}}\left(D' - \frac{\sum_i D_i p_{i0} z_i}{\sum_i p_{i0} z_i}\right) = \frac{1}{K}\left(-\ln \sum_i p_{i0} z_i + \frac{\sum_i p_{i0} z_i \ln z_i}{\sum_i p_{i0} z_i}\right)$$

$$= \frac{1}{K}\left(\frac{\overline{z \ln z}}{\bar{z}} - \ln \bar{z}\right)$$

$$= \frac{1}{K(1-\bar{x})}\left[\left(\frac{\overline{x^2}}{2} + \frac{\overline{x^3}}{6} + \cdots\right) - \left(\frac{\bar{x}^2}{2} + \frac{\bar{x}^3}{6} + \cdots\right)\right] > 0, \qquad (39.11)$$

as $\overline{x^n} > \bar{x}^n$. Thus, the value D' found experimentally increases with the flow rate.

At $\bar{U} \to \infty$, expansion of the exponential functions in (39.19) shows that

$$1 - \frac{KD'}{\bar{U}} = \sum_i p_{i0}\left(1 - \frac{KD_i}{\bar{U}}\right) \qquad (39.12)$$

or

$$D' = \sum p_{i0} D_i. \qquad (39.13)$$

In this case, therefore, the weighted mean value of D is found. It can readily be seen that this result is also obtainable from the general formula (39.7).

At $\bar{U} \to 0$ it is justifiable to consider only one term of the formula, and in the sum $\sum_i p_{i0} \exp(-KD_i/\bar{U})$ to retain one term corresponding to the fraction with the least value of D. Suppose this is the first fraction. In this case (39.10) becomes

$$\exp(-KD'/\bar{U}) = p_{i0} \exp(-KD_i/\bar{U}). \qquad (39.14)$$

As $\bar{U} \to 0$ it is evident that $D' \to D_1$, so that the experiments yield the lowest diffusion coefficient in the mixture. Whether this conclusion remains valid in the case of a continuous distribution of diffusion coefficients requires special consideration. However, in practice it is very difficult to estimate D' at very large as well as at very small flow rates, and the values of D' found experimentally always lie somewhere between the minimum and the weighted mean values, becoming closer to the latter as $\bar{U}$ increases.

In the light of these conclusions, it seems that Thomas deliberately used the lowest values of D found in his experiments in order to obtain better agreement with true particle sizes. It is rather difficult to explain his results, as the scattering of light by the larger particles, which pass through the channel with less loss, is greater than by smaller particles so that the tyndallometer method should have given D-values lower than the true ones. As regards Thomas's assertion that his aerosols were isodisperse because high order Tyndall's spectra were visible, reference can be made to the paper of Goyer and Pidgeon [735] who observed the spectra in a mist in which the radii of 80 per cent of the droplets were between 0·37–0·63 μ while the remaining 20 per cent were outside this range.

It is quite possible that the dependence of D' on the flow rate might be employed for finding the particle size distribution. For aerosols consisting of several isodisperse fractions such a method of computation has already been described in the paper of Pollak and Metnieks discussed above.

Chamberlain *et al.* [736] sucked atmospheric aerosols activated by thorium through narrow rectangular channels and determined the concentration of the effluent by counting the particles and by filtering the aerosol and determining the radioactivity on the filter. The first method, which is more correct, yielded values of D twice as large as the second. This is not surprising since the radioactivity of particles tends to increase with their size.

The charges carried by particles increase their rate of deposition, due to image forces, and this is an important, although often neglected factor, in diffusion experiments. Calculation of the deposition of particles, if both diffusion and image forces are taken into account, is very complicated but an estimate can be made of the effect of image forces. In a narrow rectangular channel with conducting walls the initial distance of the particle from the nearest wall, y, at the inlet to the channel, and the distance downwind at which it strikes the wall, x, are related as follows

$$\frac{Bqx^2}{6\,\bar{U}h^3} = \frac{1}{2}\left(\frac{y}{h}\right)^4 - \frac{1}{5}\left(\frac{y}{h}\right)^5 \tag{39.15}$$

where q is the charge and B the mobility of particles. The ratio of the number of particles issuing from the channel to the number entering it, allowing for the parabolic profile of the flow, is

$$\frac{\bar{n}}{n_0} = 1 - \frac{3\,y_0^2}{2\,h^2}\left(1 - \frac{y_0}{3\,h}\right) \tag{39.16}$$

where y_0 is the value of y at x equal to the length of the channel.

Since $Bq^2x/6\,\bar{U}h^3 = \mu q^2/6hkT$, the ratio of the rate of the electrostatic deposition to that due to diffusion depends only on the width of the channel and the charge of the particles. If $h = 0{\cdot}5$ mm and the values of μ lie within 0·05–0·8, the range in which measurements are usually made, the ratio is about 0·01 at 1, 0·1 at 10 and 1 at 100 elementary charges on the particles. At $h = 0{\cdot}125$ mm the same values of the ratio are obtained with charges half the size. It is thus permissible to neglect the effect of image forces only if q is around one elementary charge. In this deduction the image force from the other wall was neglected which can be shown to be justifiable.

Nolan and O'Connor [676] determined, by the method described on page 122, the mean charge and mobility in separate fractions of an aerosol prepared by bubbling air through a dilute solution of sodium chloride. Diffusion coefficients determined by parallel experiments with cylindrical capillaries were also used to find the mobilities of the fractions. The second method gave mobilities 7–11 times greater than the first, the ratio increasing with the mean charge on the particles. The latter varied from 9 to 106 electrons in the different fractions, whereas the mean radii of particles in the fractions, found by the diffusion method, decreased from $2{\cdot}7 \times 10^{-6}$ to $1{\cdot}1 \times 10^{-6}$ cm; the charge therefore diminished with increasing particle size, which is at variance both with theory and experiment. The calculations above suggest that the contradiction is due to neglect of image forces.

III. Diffusion from a flow towards the surface of a sphere has been investigated theoretically and experimentally in connection with the condensation of vapour on to the surfaces of droplets or their evaporation. The rate of diffusion onto a sphere, given by (38.36) for a stationary medium, increases in a flow for $\mathrm{Re}_f > 3$ [290] by the factor $1 + 0{\cdot}27\,\mathrm{Re}_f^{1/2}\,\mathrm{Sc}^{1/3}$ where $\mathrm{Sc} = \nu/D$ is the Schmidt number. However, D is several orders of magnitude less in aerosols than in vapours, while Sc, which is about unity for vapours, is very large in aerosols being about 10^5 for $r = 10^{-5}$ cm and about 10^3 for $r = 10^{-6}$ cm. For this reason the experimental data on diffusion of vapours cannot be extended to aerosols.

In problems of the type under discussion the Schmidt number indicates the ratio of convective and diffusive transfer rates (at constant Re). When $\mathrm{Sc} \approx 1$ the thickness of the diffusion layer and the hydrodynamic boundary layer round a body are of the same order of magnitude. When $\mathrm{Sc} \gg 1$ diffusive and convective transfer are comparable only at a very short distance from the surface of the body, since the diffusion layer is very thin in comparison with the friction layer.

The diffusion rates in these two cases therefore have different mathematical expressions. Since Sc is also very large for diffusion in solutions we can make use of the results of Levich [293] who found for the rate of diffusion towards a sphere in Stokes flow

$$\Phi = 7{\cdot}9\, n_0 D^{2/3} U^{1/3} R^{4/3}. \qquad (39.17)$$

This formula was derived on the assumption that the thickness of the diffusion layer is very small compared with the radius of the sphere. It is not applicable at very low flow velocities where the thickness of the diffusion layer increases.

The theory of diffusion towards a sphere during flow at small Re_f and large $\mathrm{Pe} = \mathrm{Sc} \times \mathrm{Re}_f$ (the Peclet diffusion number) has been worked out by several authors, who have obtained for the number of particles deposited on a sphere in 1 sec expression (39.17) with other values of the coefficient, such as 7·06 (Friedländer [737]) and 8·5 (Axelrud [738]). In dimensionless parameters Sh (the Sherwood number, or the diffusion number corresponding to the Nusselt number) and Pe are related by

$$\mathrm{Sh} = \beta\, \mathrm{Pe}^{1/3} \qquad (39.18)$$

where according to Levich $\beta = 1{\cdot}0$, to Friedländer $\beta = 0{\cdot}89$ and to Axelrud $\beta = 1{\cdot}07$.

There are no experimental data on the diffusion of flowing aerosols onto spheres but (39.18) can be verified by the data of Axelrud for the dissolution rate in oil of spheres of benzoic acid at $\mathrm{Re}_f = 0{\cdot}1$–$2{\cdot}5$ and $\mathrm{Sc} = \nu/D = 2{\cdot}3 \times 10^6$; these experiments give $\mathrm{Sh} = 1{\cdot}10\,\mathrm{Pe}^{1/3}$ in excellent agreement with theory.

For large Re_f (600–2600) and the same value of Sc Axelrud deduced

$$\mathrm{Sh} = 0{\cdot}8\,\mathrm{Re}_f^{1/2}\,\mathrm{Sc}^{1/3} = 0{\cdot}8\,\mathrm{Pe}^{1/3}\,\mathrm{Re}_f^{1/6} \qquad (39.19)$$

which was confirmed by experiment.

Garner *et al.* [739, 740] in similar experiments, using water as solvent at $\mathrm{Re}_f = 100$–700 and $\mathrm{Sc} \approx 10^3$, obtained a similar formula, but their coefficient was 0·95 instead of 0·8.

The corresponding case of diffusion to a cylinder at small Re_f and large Pe was solved by Natanson [741] using Lamb's equation for viscous flow. The number of particles deposited in 1 sec on unit length of the cylinder was found to be

$$\Phi' = \frac{4{\cdot}64\, n_0 D^{2/3} U^{1/3} R^{1/3}}{[2\,(2{\cdot}002 - \ln Re_f)]^{1/3}}. \tag{39.20}$$

Using a somewhat less accurate method Friedländer [737] obtained a similar expression with the coefficient 3·52. In dimensionless parameters (39.20) becomes

$$\mathrm{Sh} = \frac{\beta\, \mathrm{Pe}^{1/3}}{[2\,(2{\cdot}002 - \ln Re_f)]^{1/3}} \tag{39.21}$$

where $\beta = 1{\cdot}17$ according to Natanson and 1·035 according to Friedländer.

Friedländer considered his formula to be valid only at $Re_f < 0{\cdot}001$; for $Re_f = 0{\cdot}1$, by means of numerical integration, he obtained $\mathrm{Sh} = 0{\cdot}557\, \mathrm{Pe}^{1/3}$ which is in good agreement with measurements of diffusion from a cylindrical electrode at $Re_f = 0{\cdot}1$ and $\mathrm{Pe} = 1000$ [742].

The calculation of particle deposition by diffusion, allowing for interception, seems to be impossible analytically. Friedländer suggested the following approximate method. Proceeding from (34.17) and (34.18) for deposition due to interception alone, he added to the radius of the particle "a diffusion radius" equal to the radius of a hypothetical particle for which the collection efficiency due to interception is equal to that of the given particle due to diffusion, as calculated from (39.20). It is doubtful whether this method of calculation gives even approximately correct results.

Before the appearance of the above-mentioned papers, a simplified method of calculating deposition due to diffusion, suggested by Kaufman [304], was generally used. It was assumed that those particles would be deposited on the collector which moved along the streamlines and were located, on the average, at a distance x from the collector surface equal to the mean Brownian displacement $(4\,Dt/\pi)^{1/2}$, where t is the time of the flow past the collector. Langmuir [743] assumed x, in the case of a cylinder, to be equal to the root mean square distance of the streamline from the surface in the range of values of the polar angle θ from 30° to 150° while t was the time taken by the fluid to flow over this distance. Equation (39.20), with the coefficient 2·72, was obtained in this manner.

In an attempt to allow for the interception effect, Radushkevich [744] committed the same error as Müller (see page 324), assuming that the tangential component of the concentration gradient of particles at the surface of the cylinder was equal to zero; for the diffusion rate towards the cylinder at small Re_f he obtained a formula, which in the absence of interception reduces to $\Phi' = K n_0 D$, K being a constant, so that the deposition rate depends neither on the diameter of the cylinder nor on the flow rate, which is physically improbable.

For potential flow past a cylinder at large Pe the Boussinesq formula is valid

$$\Phi' = \frac{8}{\sqrt{\pi}}\, n_0 D^{1/2} U^{1/2} R^{1/2}. \tag{39.22}$$

IV. In conclusion the diffusion of an aerosol issuing from a narrow opening (point source) or from an infinitely long flat slit perpendicular to the flow (line source) will be discussed for a gas which is moving in straight lines and in which the velocity U is the same at all points. Settling of the particles will be neglected. Let the x-axis be taken parallel to the flow with the origin of coordinates at the source and, where the source is linear, take the y-axis along the source.

The case of an instantaneous source is the simplest [294] because the solution for a point source situated at the origin of a coordinate system which moves with the flow is given by equation (37.31) and can be adapted

$$n(x, y, z, t) = \frac{N}{\sqrt{(4\pi D t)^3}} e^{-(x^2+y^2+z^2)/4Dt}, \tag{39.23}$$

where N is the number of particles emitted by the source at the instant $t = 0$. In a fixed coordinate system (39.23) becomes

$$n(x, y, z, t) = \frac{N}{\sqrt{(4\pi D t)^3}} e^{-[(x-Ut)^2+y^2+z^2]/4Dt}. \tag{39.24}$$

Similarly, for an infinitely long line source,

$$n(x, y, z, t) = \frac{N'}{4\pi D t} e^{-[(x-Ut)^2+z^2]/4Dt} \tag{39.25}$$

where N' is the number of particles emitted per cm length of the source.

V. For a continuous source, a steady concentration distribution of aerosol in space must be established. The exact calculation of this is difficult, but, bearing in mind what was said above about the relative magnitudes of convective and diffusive transfer of particles, the term $D(\partial^2 n/\partial x^2)$ in equation (37.19) can be omitted and $(\partial n/\partial t)$ put equal to zero. For a line source this gives

$$\frac{\partial n(z,x)}{\partial x} = \frac{D}{U}\frac{\partial^2 n(z,x)}{\partial z^2} \tag{39.26}$$

with the boundary condition

$$n(z, 0) = 0 \quad \text{if} \quad z \neq 0 \tag{39.27}$$

and the normalizing condition

$$U\int_{-\infty}^{+\infty} n(z,x)\,dz = \Phi'. \tag{39.28}$$

This shows that any plane perpendicular to the x-axis is crossed in one second by the quantity of particles (Φ' per cm length of source) emitted by the source in one second. Equation (39.26) is mathematically equivalent to equation (37.16) and its solution is given by the formula obtained from (37.29) by replacing t by x, $x - x_0$ by z, D by D/U and multiplying by Φ'/U:

$$n = \frac{\Phi}{\sqrt{4\pi DUx}} e^{-Uz^2/4Dx}. \tag{39.29}$$

The formula for a constant point source is

$$n = \frac{\Phi}{4\pi D x} e^{-U(y^2+z^2)/4Dx}. \tag{39.30}$$

§ 40. TEXTILE FABRIC AND FIBROUS FILTERS

Two important types of mechanical filter for aerosols will be discussed, namely textile fabric and fibrous filters; the most important examples of the latter are filter papers and pads prepared from such fibres as cellulose, glass and asbestos.

Fabric filters are used for arresting comparatively coarse aerosol particles so that deposition of particles on the threads is mainly by the inertial process discussed in § 34. If the ratio of the space $2h$ between the threads to their diameter $2R$ is greater than 1 (a loose cloth) the flow of gas through the cloth is not much different from the flow round isolated threads and the filter efficiency at low flow velocities is very small. When h/R is decreased the streamlines move closer to the surface of the threads and the collection efficiency increases considerably. Unfortunately the velocity distribution in a gas flowing through a compact cloth ($h/r < 1$) is unknown and theoretical analysis of the filtration of aerosols in such filters must be left for the time being. Experiment shows that the filtering capacity of cloths increases rapidly as the compactness of the cloth increases (considerably more quickly than the number of threads per 1 cm^2 of cloth) [295].

Tests with multilayer fibrous filters have proved [296, 297] that the concentration n_ν of an aerosol after passing through ν layers can be represented approximately by the formula

$$n_\nu = n_0 e^{-\nu\alpha} \tag{40.1}$$

or

$$\log(n_\nu/n_0) = -0{\cdot}43\,\nu\alpha, \tag{40.2}$$

where n_0 is the initial concentration.

The logarithm of the fraction which penetrates is therefore directly proportional to the number of layers, but this is only true as long as each layer passes the same fraction $e^{-\alpha}$ of incoming particles, such as occurs with an isodisperse aerosol. In polydisperse aerosols α has a different value for each layer of cloth. For particles of radius greater than $0{\cdot}1\,\mu$, α increases as r increases, so that the coarser particles tend to be retained by the front layers of the filter. The fineness of the aerosol is thereby increased, with the result that α decreases and the curve of $\log(n_\nu/n_0)$ as a function of ν is more or less convex downwards depending on the polydispersity of the aerosol. The front layers of the filter quickly become clogged by the coarse particles and should be renewed from time to time. This, of course, applies to multilayer filters of cloth, gauze and filter paper [296, 297].

The efficiency of cloth filters increases as they become clogged with dust [298–300] because the spaces between the particles in the dust layer are considerably smaller than those between the threads. Asbestos dust is particularly effective in this respect [300, 301] and can, with advantage, be deposited on filter cloth before use. When cloth filters are periodically cleaned it is desirable to allow some dust to remain on them [301].

Thick fluffy fabrics, particularly woollen materials, are considerably more effective than cotton fabric and are notable in giving satisfactory results even without preliminary clogging with dust [298]. In addition, the accumulation of dust causes the resistance to rise much more slowly than for thin, smooth cloths [300]; because of their larger effective volume they possess a considerably greater dust-holding capacity.

In filters of metallic gauze the deposited dust is almost completely swept away by the gas flow unless they are impregnated with a viscous liquid [303]. They are used chiefly as air filters for internal combustion engines where only coarse particles need be eliminated.

Passing on to fibrous filters, filter papers and pads which have a comparatively high flow resistance, they are as a rule used at low flow velocities where the resistance is proportional to the rate of air flow, thus showing that the flow is laminar. In the capillary passages of fibrous filters Re_f is of the order 0·001–0·1 so that any attempt to calculate the magnitude of inertial deposition of particles by using the data given in § 34 for the deposition of particles on cylinders in potential flow [295] cannot lead to correct results.

The flow of gas inside a fibrous filter is very complex, because it continuously changes direction to pass round the randomly disposed fibres. To obtain even a rough idea of the extent to which aerosols are deposited in fibrous filters it is necessary to start with an idealized model of the filter, such as the one illustrated in Fig. 58. The

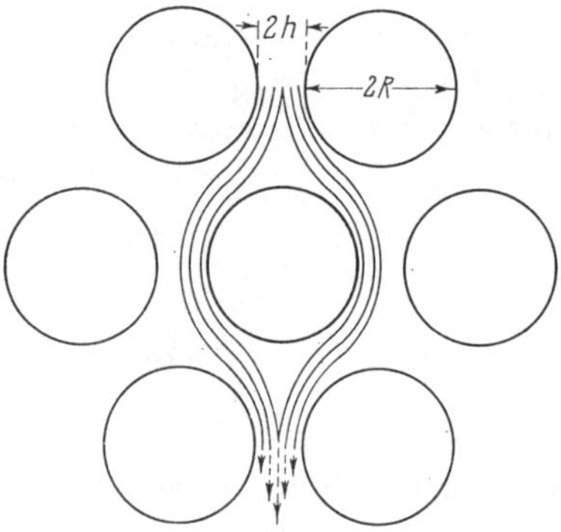

FIG. 58. Model of a fibrous filter.

drawing represents a cross-section of a system of individual parallel cylinders (fibres) in a staggered arrangement, the distance between the cylinders being $2h$ [304]. The porosity or volume of pore space per cm³ of the material of filter papers and pads varies from 0·70 to 0·98 so that the ratio h/R lies between 0·6 and 5·0, approximately.

Even in this much simplified model it is very difficult to determine the flow field and even more difficult, of course, to calculate the filter efficiency. The calculations given below should not be regarded as even roughly quantitative; they can only serve to give some initial guidance as to the way in which fibrous filters act.

Bearing in mind that Re_f is small, a start can be made with formula (34.2) for viscous flow past a cylinder. This method, to which we shall return later, is reliable for filters of high porosity (large value of h/R) since the fibres can then be treated as isolated cylinders. The author has chosen a different approach, correct for very dense filters with small values of h/R, by assuming that Poiseuille flow proceeds in the curved capillaries formed by the interstices of the filter. Several flow tubes of this type are shown in Fig. 58. The flow in each is successively decelerated, first from one side, then from the other. Because Re_f is so small it may be assumed that the Poiseuille flow profile is established at every section of the capillary.

Apart from electrostatic forces, the four mechanisms active in the deposition of aerosols in fibrous filters are inertia, interception, sedimentation and diffusion. The effect of inertia can be calculated as for impingement instruments (page 153) on the assumption that the streamlines are turned through 30° or 45° instead of 90°. The collection efficiency for one fibre, and hence for one row of fibres, is therefore given by (33.3) but with a numerical coefficient roughly half as large,

$$\ni_i = l_i/2\,h = \tau \bar{U}/2\,h, \tag{40.3}$$

where $\bar{U}$ is taken as the mean flow velocity in the interstices between the fibres. Clearly

$$\bar{U} = \left(1 + \frac{R}{h}\right) U_0, \tag{40.4}$$

where U_0 is the velocity of the gas entering the filter. A layer of thickness Δx parallel to the surface of the filter contains $\Delta x/\sqrt{3}\,(R + h)$ rows of fibres (more precisely their axes). When an aerosol of concentration n is passed through this filter, the concentration is reduced by an amount

$$\Delta n = \frac{l_i n \Delta x}{2\sqrt{3}\,h(R + h)}. \tag{40.5}$$

Hence it follows that the concentration of the aerosol after passing through a filter of thickness H is

$$n = n_0 e^{-\alpha_i}, \tag{40.6}$$

where n_0 is the initial concentration. The quantity

$$\alpha_i = \frac{\tau U_0 H (1 + R/h)}{2\sqrt{3}\,h(R + h)} = \frac{\tau U_0 H}{2\sqrt{3}\,h^2} \tag{40.7}$$

will be called the coefficient of inertial deposition of the aerosol by the filter. The efficiency of the filter (fraction of particles retained) is therefore†

$$\ni_i = 1 - e^{-\alpha_i}. \tag{40.8}$$

In order to calculate the interception effect it is necessary to start from the flow velocity at the actual surface of the fibre. Assuming that the velocity distribution in the spaces between the fibres is the same as in a parallel capillary channel of width $2\,h$ [see (26.2], the following approximate expression gives the velocity at a distance x from the surface of a fibre

$$U = 3\,x \bar{U}/h. \tag{40.9}$$

The flow of gas in 1 sec in the wall layer of thickness x is

$$\int_0^x U\,dx = 3\,x^2 \bar{U}/2h \tag{40.10}$$

per unit length of the fibre.

† Unlike § 22 and § 42 the coefficients α here do not relate to a 1 cm-thick layer of filtering medium, but to the entire filter.

The interception effect causes $3\,r^2 n \bar{U}/2\,h$ particles per second to be deposited from both sides on unit length of the fibre. Since the total flow of particles per second in the space between the fibres is $2\,hn\bar{U}$, the collection efficiency for the first row of fibres is

$$\ni_h = 2 \cdot 3\,r^3 n \bar{U}/2\,h : 2\,hn\bar{U} = \frac{3\,r^2}{2\,h^2} \tag{40.11}$$

and the deposition coefficient of the aerosol due to the interception effect is

$$\alpha_h = \frac{3\,r^2 H}{2\sqrt{3}\,h^2(R+h)}\,. \tag{40.12}$$

Sedimentation is usually neglected in the theory of fibre filters but it will be calculated for the above model. The number of particles deposited in unit time on unit length of a horizontal cylinder around which the aerosol is flowing is to a first approximation $n V_s \cdot 2\,R$. Since the volume of gas flowing round unit length of a fibre in one second is $2\,U_0(h+R)$, then for one row of fibres

$$\ni_s = V_s R/U_0(h+R) = g\tau/U_0(1+h/R) \tag{40.13}$$

and the deposition coefficient of the filter due to sedimentation is

$$\alpha_s = \frac{g\tau H}{\sqrt{3}\,U_0 R(1+h/R)^2}\,. \tag{40.14}$$

These equations show that α_i is directly proportional and α_s inversely proportional to the flow velocity. The two effects are comparable only at very small flow velocities of tenths of a centimetre per second and so, in practice, the sedimentation effect is indeed only of secondary importance.

Passing on to diffusive deposition of aerosols in filters it will be noted that the model is very similar to a tubular heat exchanger, thus suggesting that it might be possible to make use of the heat transfer data available for these in the filter problem. However, it has been pointed out (page 210) that such an analogy is not permissible, so that a simplified method of calculation as in § 39, I, is again resorted to. Using formula (40.9) the following expression for the thickness of the layer "absorbed" by the fibre is derived

$$\delta = \left(\frac{8\,LhD}{3\pi\bar{U}}\right)^{1/3}. \tag{40.15}$$

Taking the distance L along the surface of the fibre which is traversed by the diffusing particle to be half the circumference of the fibre, i.e. πR, gives

$$\delta = \left(\frac{8\,hDR}{3\bar{U}}\right)^{1/3}. \tag{40.16}$$

Use of the same procedure as in calculating the interception effect leads to

$$\ni_D = 3\,\delta^2/2\,h^2, \tag{40.17}$$

$$\alpha_D = \frac{2\,HD^{2/3}}{3^{1/6}U_0^{2/3}Rh^{2/3}(1+h/R)^{5/3}}\,. \tag{40.18}$$

It follows from (40.7), (40.12), (40.14) and (40.18) that inertial deposition increases with the size and density of the particles and with the flow velocity; sedimentation also increases with the size and density of the particles but decreases with flow velocity; deposition due to the interception effect increases with particle size but does not depend on their density or on the flow velocity. Diffusive deposition varies inversely with the particle size and flow velocity and does not depend on the density of the particles. All types of deposition increase rapidly as the distance apart of the fibres becomes smaller. The dependence on fibre thickness at constant separation is more complicated, but at constant filter porosity the deposition increases rapidly with decrease in fibre thickness. All these results are independent of the filter model taken and of the assumptions as to the nature of the flow: they are therefore of general significance.

TABLE 17. EFFICIENCY OF MODEL FILTER

r, cm	$U_0 = 1$ cm sec^{-1}				$U_0 = 20$ cm sec^{-1}			
	$\ni_i$	$\ni_h$	$\ni_s$	$\ni_D$	$\ni_i$	$\ni_h$	$\ni_s$	$\ni_D$
10^{-6}	0	0	0	1	0	0	0	1
3×10^{-6}	0	0	0	1	0·10	0	0	0·80
10^{-5}	0·03	0·02	0·01	0·95	0·43	0·02	0	0·33
3×10^{-5}	0·17	0·16	0·06	0·67	0·97	0·16	0	0·14
10^{-4}	0·80	0·86	0·41	0·35	1	0·86	0·03	0·06
3×10^{-4}	1	1	1	0·17	1	1	0·20	0·03

Further compression, or decrease in porosity, of an already fairly compact filter results in a rapid decrease in h so that the filter efficiency must increase markedly, as is evident from practical experience. The position is different for more porous filters which will be considered below.

As an illustration table 17 is presented which shows the filter efficiencies arising from the different mechanisms; they were calculated, by the formulae given above, for a filter of thickness 0·2 cm, a fibre radius of 10 μ and a ratio $h/R = 0{\cdot}7$ at two flow velocities.

All that can be said of the net efficiency by all mechanisms is that it is greater than each of the individual components but less than their arithmetical sum. It is clear that filter efficiency must tend to unity for very large and very small particle radii. The curve of $\ni$ as a function of r must therefore have a minimum at some value $r = r_{\min}$, where $r_{\min}$ decreases as the flow velocity increases (see below, page 222).

Since filter efficiency also tends to unity for very large and very small flow velocities, curves of $\ni$ as a function of U must have a minimum at some value $U_{\min}$ which depends on r and on the structure of the filter; this, in fact, is substantiated by experiment (Fig. 59).

A decrease in the space between the fibres, brought about by compressing the filter or reducing the diameter of the fibres, leads to increased filter efficiency but at the same time raises the resistance. The performance of a filter is determined by its efficiency at a given resistance. It is extremely difficult to calculate theoretically how the performance depends on the thickness and the arrangement of the fibres. Experimental data show that filtering efficiency increases if the fibre thickness is decreased [305, 308–310], while the performance may increase or decrease, depending on the conditions of filtration.

Davies [565], Langmuir and Chen [567] have used another method of calculating the efficiency of fibre filters. Starting from the equations of flow round a cylinder at low Re_f, they first calculate the efficiencies of collection on a single thread for the various mechanisms and then introduce theoretical or empirical corrections for the effect of neighbouring fibres. Clearly this method is applicable to filters of high porosity.

For the efficiency of inertial collection of very small particles on a cylindrical fibre Davies used his own calculated values, which are given in Fig. 46 (all Davies' calculations are for $\mathrm{Re}_f = 0{\cdot}2$). Langmuir's opinion is that inertial deposition can be completely neglected for highly dispersed aerosols. The American authors use formula (34.18) for the collection efficiency due to interception while Davies uses his own graphs. These give values of $\ni$ almost identical with those calculated by equation (34.18) for $\varkappa = 0{\cdot}75 - 1{\cdot}5$ but give higher values for $\varkappa < 0{\cdot}7$. Davies used formula (34.19) for the overall collection efficiency due to inertia and interception.

For calculation of the diffusive collection efficiency Langmuir applied formula (34.18), replacing $\varkappa = r/R$ by δ/R, where δ is the thickness of the layer, surrounding the cylinder, from which all the particles succeed in diffusing on to the cylinder surface during the time take to flow round the cylinder. He determined δ from the very approximate formula

$$\delta/R = [1{\cdot}12\,(2{\cdot}0 - \ln \mathrm{Re}_f)\, D/\bar{U}R]^{1/3}, \tag{40.19}$$

where $\bar{U}$ is the mean velocity of the air in the filter. Davies introduced the dimensionless parameter $D/\bar{U}R$, defining the intensity of diffusive deposition, and added it to Stk in formula (34.19) for reasons which are explained in his paper.

The effect of neighbouring fibres on $\ni$ was allowed for by Davies by means of the flow field calc ulated by Kovasznay [568] for a fluid flowing through a two-dimen sional lattice at low Re_f. This gave the expression

$$\ni_h = \varkappa\,(0{\cdot}16 + 10{\cdot}9\,\varphi - 17\,\varphi^2) \tag{40.20}$$

for the collection efficiency due to interception.

Combining this with (34.19), Davies derived the following expression for the overall collection efficiency

$$\ni = [\varkappa + (0{\cdot}25 + 0{\cdot}4\,\varkappa)\,\mathrm{Stk} - 0{\cdot}026\,n\,\mathrm{Stk}^2]\,(0{\cdot}16 + 10{\cdot}9\,\varphi - 17\,\varphi^2), \tag{40.21}$$

where the diffusion parameter is included in Stk. According to Davies R in the expression $\varkappa = r/R$ should be taken as the "effective" fibre radius R_{ef}, which differs from the actual mean radius because of the random, non-uniform distribution of fibres, the presence of aggregates and so on. R_{ef} can be determined from the resistance of the filter (see below).

The deposition coefficient of a filter can be obtained from the collection efficiencies of single fibres by the formula

$$\alpha = \frac{2\,\varphi \ni H}{\pi\,(1 - \varphi)\,R}, \tag{40.22}$$

which is derived on the basis of $2\,R \ni n\,\bar{U}$ particles per second being deposited on unit length of a fibre; $\bar{U} = U_0/(1 - \varphi)$, where U_0 is the air velocity in front of the filter, and the total length of fibre contained in 1 cm^3 of the filter is $\varphi/\pi R^2$. It is assumed that the fibres are arranged uniformly and perpendicularly to the flow.

Quite a large number of theoretical and empirical formulae have been proposed for calculating the resistance of fibrous filters. The well-known Kozeny–Carman formula, widely used for calculating the specific surface of powders from their permeability to liquids and gases, is applicable to very dense filters, but not to those of high porosity.

Experimental data on the resistance of pads of different fibres at various degrees of compression (φ = 0·006 to 0·3) have been expressed by Davies by the formula

$$\Delta p = \frac{16\,\eta U_0 H}{R_{ef}^2}\varphi^{1\cdot5}(1 + 56\,\varphi^3), \tag{40.23}$$

where R_{ef} is the effective fibre radius which is the same as the actual mean radius at small φ and exceeds it at large φ. According to Davies another effective radius R'_{ef} should be introduced into formula (40.21) when calculating filter efficiency. This is obtained from the equation

$$\Delta p = \frac{17\cdot5\,\eta U_0 H}{R'^2_{ef}}\varphi^{1\cdot5}(1 + 52\,\varphi^{1\cdot5}). \tag{40.24}$$

$R'_{ef} = R_{ef}$ only if $\varphi < 0\cdot02$ and at large φ $R'_{ef} > R_{ef}$. According to the experiments of Blasewitz and Judson [569] the resistance of glass fibre filters is proportional to $\varphi^{1\cdot5}$ over the range of φ from 0·004 to 0·04, which is in agreement with formula (40.23) because the term with φ^3 can in this case be neglected. Silverman and First [570] found for various filters that the resistance was proportional to $\varphi^{1\cdot4}$.

The resistance of a filter was calculated by Chen [567] starting from the drag force acting on a single fibre due to the flow. This is similar to the force discussed in § 14 acting on a system of particles settling in a limited space and depends on the distance between the fibres, that is on φ. The force on unit length of a fibre placed perpendicularly to the flow will be denoted by $F(\varphi)$. Using White's [571] empirical formula for the resistance experienced by a cylinder moving in a viscous liquid in a space bounded by walls, Chen showed that

$$F(\varphi) = -A\eta\bar{U}/(B + \log\varphi), \tag{40.25}$$

where A and B are constants.

Since the total length of fibre in 1 cm^3 of the filter is $\varphi/\pi R^2$, the expression for filter resistance obtained from (40.25) is

$$\Delta p = \frac{\varphi H F(\varphi)}{\pi R^2} = -\frac{\varphi H \eta U_0 A}{(1-\varphi)\pi R^2(B + \log\varphi)}. \tag{40.26}$$

Chen's experiments on glass fibre filters with fibre diameters from 0·5 to 14 μ at $\varphi \leqslant 0\cdot1$ showed, in agreement with (40.25) and (40.26), that Δp was proportional to the flow velocity up to the maximum velocity studied, which corresponded to $R_{ef} = 6$; it was found that $1/F(\varphi)$ varied linearly with log φ for the same material at different degrees of compaction. However, the constants A and B for filters of different fibre thickness and different construction deviated from the mean values $A = 5\cdot3$ and $B = 0\cdot388$. It is difficult to detect any relationship between these constants and the fibre diameter from Chen's data; the arrangement of fibres in the filter is obviously of great importance.

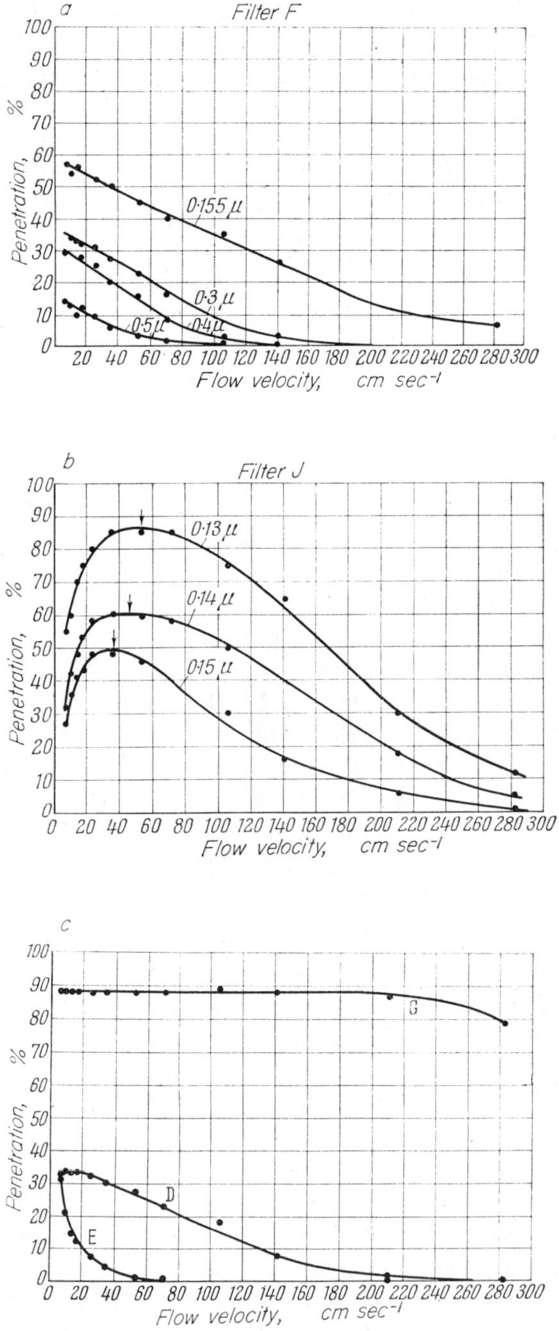

FIG. 59. Aerosol penetration through fibrous filters.

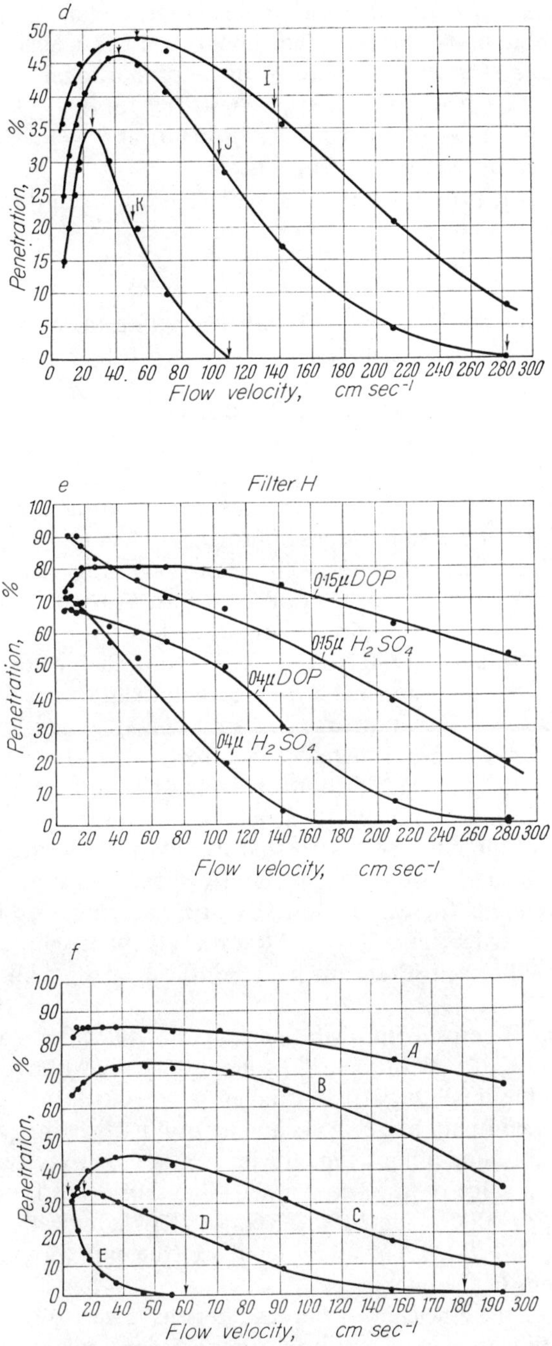

FIG. 59. Aerosol penetration through fibrous filters.

It can be seen from (40.25) that $F(\varphi)$ falls off as φ decreases. At a certain value of φ, depending on Re_f, $F(\varphi)$ attains the value corresponding to the resistance experienced by the fibre in the absence of neighbouring fibres, which is given by Lamb's formula (12.13); the resistance then ceases to be proportional to the flow velocity. This occurs for $\mathrm{Re}_f = 1$ at approximately $\varphi = 6 \times 10^{-2}$, for $\mathrm{Re}_f = 0{\cdot}1$ at $\varphi = 6 \times 10^{-3}$, for $\mathrm{Re}_f = 0{\cdot}01$ at $\varphi = 7 \times 10^{-4}$ and for $\mathrm{Re}_f = 0{\cdot}001$ at $\varphi = 7 \times 10^{-5}$. These calculations are confirmed by Wong's experiments.

It follows from the proportionality between resistance and flow velocity that the flow field in filters does not depend upon Re_f. Formulae (34.18) and (40.19), based on the assumption that the flow field in filters depends on Re_f, cannot therefore be used to calculate filter efficiencies.

From experiments on the variation of efficiency with the degree of compression, using glass fibre filters, Chen concluded that when $\varphi < 0{\cdot}1$, the efficiency of collection on a single fibre is given by

$$\ni = \ni_0 (1 + 4{\cdot}5\, \varphi). \tag{40.27}$$

Comparison of (40.27) and (40.21) shows that the latter predicts a much larger effect due to compression. It should be mentioned that in the experiments of Silverman and First [570] compression of loose fibre glass filters to increase φ from 0·0007 to 0·003 had no effect on the efficiency, while Wong [566] reached the same conclusion by compressing filters from $\varphi = 0{\cdot}045$ to 0·1. It must again be emphasized that this is not true when dense filters are further compressed.

Comparison of (40.27) and (40.26) or (40.23) shows that as loose filters are compressed their resistance increases much more rapidly than their collection efficiency; Langmuir's statement [567] that these two quantities may be considered proportional in calculating filter efficiencies is therefore incorrect.

Chen's experiments on the filtration of aerosols having different particle sizes ($r = 0{\cdot}07$ to $0{\cdot}4\,\mu$), by filters with fibre diameters of $2{\cdot}5\,\mu$ and $\varphi = 0{\cdot}02$ to 0·08, showed that the efficiency of the filters increased sharply with particle size at a flow velocity of 47 cm/sec, remained almost unchanged at $U_0 = 5{\cdot}2$ cm/sec, and decreased somewhat at $U_0 = 1{\cdot}7$ cm/sec. Increase in flow velocity is accompanied by a sharp reduction in efficiency for particles of radius $0{\cdot}075\,\mu$ and a very slight reduction for $r = 0{\cdot}36\,\mu$. Chen was able to find a particle size at which the efficiency was a minimum only when $U_0 < 4$ cm/sec.

In the filtration of aerosols through filter paper at velocities from 3 to 28 cm/sec LaMer [567] found that the efficiency falls monotonically with reduction in particle size right down to the smallest particles studied ($r = 0{\cdot}02\,\mu$).

Working with wood-pulp paper filters and aerosol particles up to $0{\cdot}1\,\mu$ Green and Thomas [572] failed to find a minimum efficiency at a flow velocity of 24 cm/sec, but for $U_0 = 6$ cm/sec a value of $r_{\min}$ between 0·03 and $0{\cdot}05\,\mu$ was found. For asbestos-cellulose filters $r_{\min} = 0{\cdot}15\,\mu$ for $U_0 = 2{\cdot}5$ cm/sec [307]. Indications existing in the literature that $r_{\min} = 0{\cdot}20$ [305] or $0{\cdot}17\,\mu$ [306] are true only with respect to specified operating conditions of the filters.

Detailed data on the efficiency of fibrous filters have been published by Ramskill and Anderson [310] who worked with fairly monodisperse sulphuric acid and dioctyl phthalate mists. Particulars of their filters are given in Table 18.

The results of their experiments are shown in Fig. 59† in which *a* and *b* illustrate

† On these graphs the penetration $1 - \ni$ is plotted on the ordinate axis instead of the efficiency $\ni$.

the dependence of efficiency on particle size, the latter being shown on the curves. It can be seen that, for the filter J at least, ϶ decreases appreciably in going from $r = 0{\cdot}15\mu$ to $r = 0{\cdot}13\mu$ over the whole range of velocities studied. Figure 59c shows the effect of compression on efficiency, filters G, D and E consisting of fibres of almost the same thickness. But, as seen from Table 18, when the filter is compressed, the increase in ϶ is accompanied by a very large increase in the resistance. In Fig. 59d it is shown (dioctyl phthalate mist, $r = 0{\cdot}15\,\mu$) how the filter efficiency rises when the fibre diameter is decreased. However, it can be seen by comparing the data in Table 18 that the performance of glass fibre filters increases with decrease in fibre diameter only at high flow velocities; at lower velocities it decreases. Figure 59e shows by how much ϶ increases when the particle density increases, this being characteristic of inertial deposition.

TABLE 18. CHARACTERISTICS OF FILTERS USED BY RAMSKILL AND ANDERSON

Filter	Material	Mean fibre diameter (μ)	Filter thickness (cm)	Resistance of filter (mm of H_2O at $U = 14$ cm/sec)	Resistance per 1 cm filter thickness	Method of preparation
A	Viscose	17	0·110	7	64	Wet, without calendering
B	Glass	3	0·05	10	200	Wet, without calendering
C	Glass	3	0·028	10	350	Dry, without calendering
D	Esparto	15	0·06	67	1100	Wet, moderate calendering
E	Wood pulp	15	0·015	219	15000	Wet, heavy calendering
F	Viscose	12	0·15	21	140	Wet, without calendering
G	Flock	16	0·075	4	53	Wet, without calendering
H	Viscose	17	0·22	14	64	Wet, without calendering
I	Glass	3	0·08	14	175	Wet, without calendering
J	Glass	2	0·07	40	570	Wet, without calendering
K	Glass	1	0·045	100	2200	Wet, without calendering

The (϶, U) curves for filters with fine fibres (B, C, I, J, K) exhibit clearly defined minima, while for thick fibres these minima are either poorly defined or completely absent, because they have been displaced to very low flow velocities where measurements have not been made. This can be explained as follows:

Equations (40·7) and (40·18) show that diffusive deposition increases rapidly with reduction in fibre radius, whereas inertial deposition does not depend directly on R. Hence, for small R diffusion predominates over inertia and varies inversely with U up to considerably higher flow velocities than when R is large.

Early smoke filters consisted of layers of cotton or similar materials possessing sufficient particle-removing ability only when employed as a fairly thick layer. The resistance of these filters is high. A big step forward was the introduction of pleated filters (Fig. 60) to give a larger surface [311]. Because of their large surface area (about 1000 cm^2) they operate normally at a very low flow velocity (about 1 cm/sec in gas masks) and so offer a very small resistance (less than 20 mm of water at a flow rate of 30 l./min of air). The resistance of filters used in gas masks and respirators is of the utmost importance, because if it is more than 20 or 25 mm of water difficulty is experienced in breathing.

The best modern filters are the asbestos-cellulose filter papers and pads [300, 307, 312] in which cellulose fibres about 15 μ thick act as a framework supporting asbestos fibres a fraction of a micron thick. For a resistance of approximately U cm of water, where U is the flow velocity in cm/sec, they have an efficiency of about 0·999 for highly penetrating aerosols with $r = 0{\cdot}15\,\mu$ when U is a few cm/sec. The efficiency and resistance of asbestos-cellulose and other fibrous filters both increase when dusts and smokes are passed through them and if aerosols contain comparatively coarse particles, which quickly clog the front layer of the filter, then a pre-filter which can be renewed as it becomes clogged must be fitted.

Rather less efficient than the asbestos-cellulose type are filters of glass fibres about 1 μ in diameter [308] covered with a special preparation to improve the adhesion of solid particles to the smooth surfaces of the glass fibres [305].

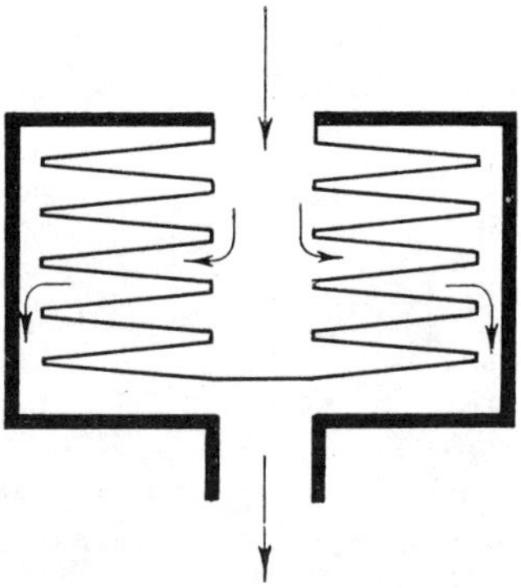

FIG. 60. Filter with folded surface.

Electrostatic filters made on carding machines from a mixture of wool and powdered resin are of extremely high quality [300, 305, 307]. Their efficiency undoubtedly arises from triboelectric charges produced on the resin particles during their preparation, particularly during combing. When highly insulating resins are used the charge leaks away from the particles very slowly, nevertheless these filters gradually deteriorate during prolonged storage [313]. Their range of application is somewhat limited by this but, when correctly prepared, they are more efficient for a given resistance than asbestos-cellulose filters. The efficiency of electrostatic filters is reduced appreciably by the passage of strongly ionized gases and by irradiation with X-rays, since the filter is discharged. The same effect is produced by the passage of oil and water mists, because liquid droplets settle on the resin particles, flow over their surfaces, and give them appreciable electrical conductivity. Filters which have lost their charge can be regenerated by recombing. Oil mists are particularly harmful to these filters as they probably dissolve the resin. When impregnated with polystyrene latex ordinary cellulose filters become very much more efficient because they are converted into electrostatic filters [314]. High efficiency has also been claimed for filters containing electrets or permanently polarized dielectrics, similar to permanent magnets [311].

It is likely that electric forces play a substantial part even in ordinary fibre filters since their efficiency falls considerably when moist gases pass through them [315, 305, 300]. Since moisture does not cause any appreciable clumping of the fibres, its effect is most probably to increase greatly their surface conductivity leading to a loss of charge. The marked increase in efficiency of fibre filters when used with aerosols of solid particles [316] is probably due not only to clogging of the capillary air passages but also to the generation of triboelectric charges. This is supported by the microscopic examination of dust in filters; large numbers of aggregates are seen to be present which were probably formed by the attraction of airborne particles to charged dust particles already deposited on the fibres. Ammonium chloride smoke is retained by a filter better when dry than moist [317].

The charge carried by the aerosol particles themselves is also of some importance. For a military type of filter the efficiency of 0·8 at $U = 2{\cdot}5$ cm/sec, with a dioctyl phthalate mist of $r = 0{\cdot}2\,\mu$, increased to 0·85 when the particles carried an average of about sixty elementary charges each [318].

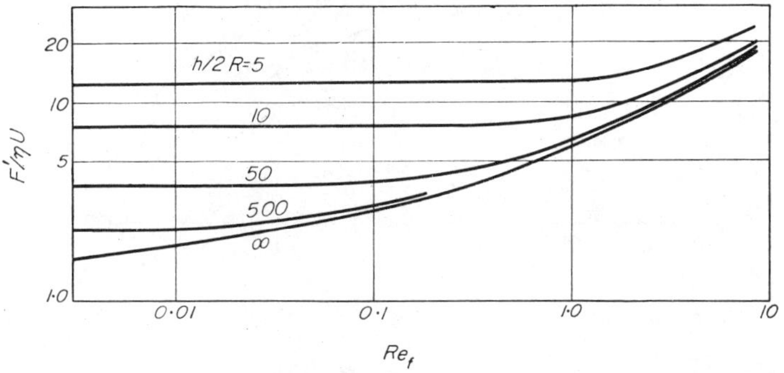

FIG. 60a. Hydrodynamical drag in a row of parallel cylinders.

In conclusion brief consideration will be given to the retention of aerosols by granular materials which, like fibrous filters, become more effective as the grain size decreases. When smoke formed by the sublimation of Sudan G dye, having particles a few tenths of a micron in radius, is passed at a velocity of 2 m/sec through a No. 2 porous glass filter with a mean grain radius of 23 μ the efficiency is 0·15, while for a No. 3 filter with grains 8·6 μ in radius it is 0·5 [295]. It is interesting that the efficiency of glass filters is increased by treatment with hydrofluoric acid, perhaps because particles adhere better to etched glass surfaces than to smooth ones. When heavy ions from the flame of a kerosene lamp ($r = 0{\cdot}8 - 2{\cdot}6 \times 10^{-6}$ cm) were passed through a layer of activated charcoal granules 1 cm thick at a velocity of 15 cm/sec, the efficiency of filtration was 0·18 with a granule radius of 0·125 to 0·275 cm, 0·50 at a radius of 0·05 to 0·125 cm, and 0·86 at a radius of 0·01 to 0·05 cm [179]. The internal structure of the charcoal did not influence the results, so that the particles evidently failed to penetrate the pores of the granules and were deposited on their outer surface.

Similar observations have been made on towers for the absorption of sulphuric acid mist. These towers contain a packing of coke wetted with sulphuric acid and the smaller the packing elements the greater the absorption [319]. Granular filters are less efficient, for the same resistance, than fibrous filters.

The development of the theory of fibrous filters has been rather slow, one of the main reasons being that it has been usual to proceed from the deposition of particles on isolated fibers using Lamb's viscous flow past a cylinder (34.2) which introduces the Reynolds number. However, it was pointed out on p. 222 that the flow pattern within filters is independent of Re_f, The same was shown by Tamada and Fujikawa [745] for flow through one row of parallel cylinders. The force F' per cm of cylinder length is proportional to the rate of flow for all small values of Re_f; this proportionality is maintained up to a higher Re_f the smaller the ratio of the distance h between the axes

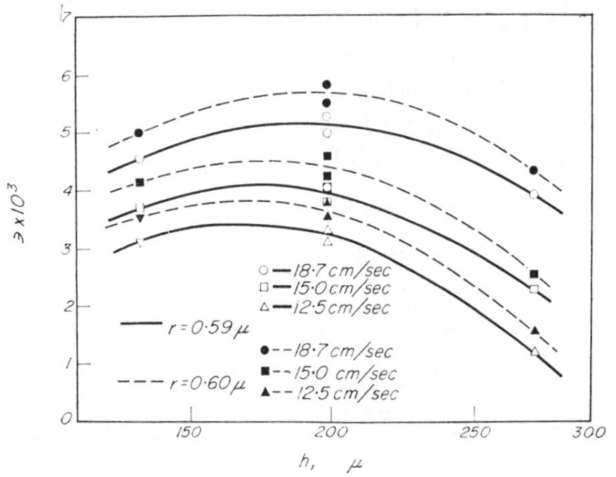

FIG. 60b. Particle collection efficiency on wires as a function of the distance between them.

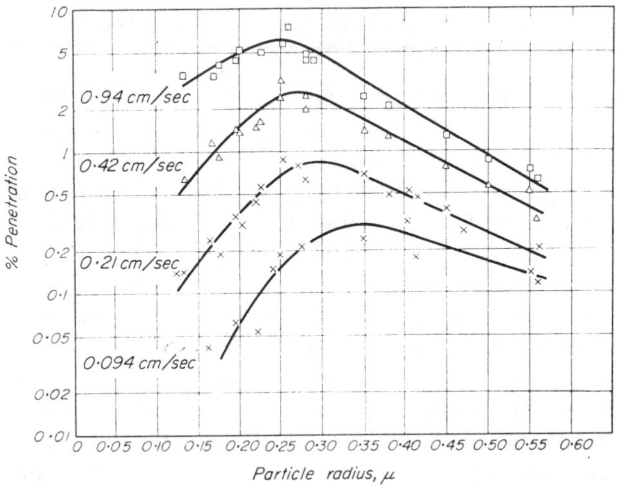

FIG. 60c. Efficiency of glass-fibre filters as a function of particle size and flow rate.

of the cylinders to their diameter (Fig. 60a). The theory of filtration should thus be based on viscous flow through a system of cylinders at small Re_f, either determined by model experiments, or calculated theoretically. Steps in the second direction have already been made by Japanese hydrodynamicists [746].

Great difficulty in calculating the efficiency of filters also arises because of the simultaneous action of several deposition mechanisms; simple addition of their separate effects, which is often resorted to [747, 748], cannot yield correct results. It is impossible to solve the problem by analytical methods. Numerical methods will, of course, involve a great deal of work, but there are no other means of solution.

Of the few papers on the theory of filtration which have been published recently, only that of Friedländer [749] will be discussed. It follows from (39·21) that the collec-

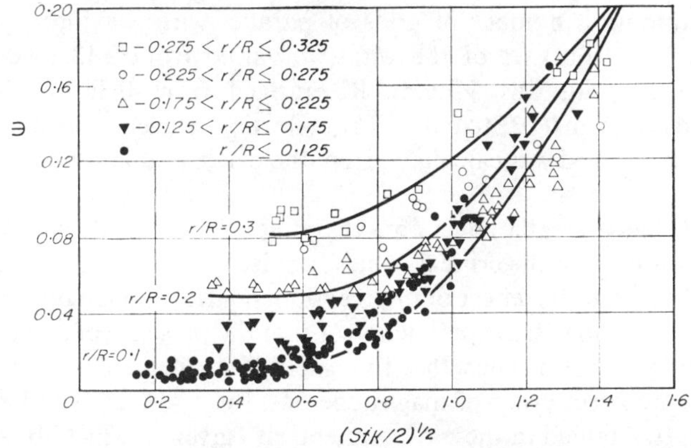

FIG. 60d. Efficiency of glass-fibre filters as a function of Stk and r/R.

tion efficiency $\ni_D$ due to diffusion on an isolated cylindrical fiber is proportional to $\mathrm{Pe}^{-2/3}(2{\cdot}002 - \ln \mathrm{Re}_f)^{-1/3}$. In the equation for viscous flow past a cylinder Friedländer replaces (without justification) $(2{\cdot}002 - \ln \mathrm{Re}_f)^{-1}$ by $K\mathrm{Re}_f^{1/2}$, K being a constant; it follows that $\ni_D$ is proportional to $\mathrm{Re}_f^{1/6}\,\mathrm{Pe}^{-2/3}$ and

$$\ni_D \,\mathrm{Pe}\, r/R = K'\,\mathrm{Pe}^{1/3}\,\mathrm{Re}_f^{1/6}\, r/R. \tag{40.28}$$

The interception collection efficiency is equal to $\ni_h \approx (r/R)^2 (2{\cdot}002 - \ln \mathrm{Re}_f)^{-1}$ (see (34.18)), whence

$$\ni_h \,\mathrm{Pe}\, r/R = K''\,\mathrm{Re}\,(r/R)^3\,\mathrm{Re}_f^{1/2} = K''\,[\mathrm{Pe}^{1/3}\,\mathrm{Re}_f^{1/6}\, r/R]^3 \tag{40.29}$$

At small values of Pe and of r/R, when the diffusion effect predominates, (40.28) should be used; at larger values (40.29) is appropriate. By plotting log ($\ni$ Pe r/R) against log ($\mathrm{Pe}^{1/3}\,\mathrm{Re}_f^{1/6}\, r/R$) from the many data for the filtration of aerosols through fibrous filters, Friedländer obtained a curve the slope of which was $\sim 1{\cdot}2$ at small values of $\mathrm{Pe}^{1/3}\,\mathrm{Re}_f^{1/6} r/R$, and ~ 3 at large values. He concluded that at Stk < 1 and $\mathrm{Re}_f < 1$, the inertial effect in fibrous filters could be neglected.

His argument is subject to the following objections: (1) the flow in the filters does not depend on Re_f, (2) over the wide range of values of Re_f (0·01–2) covered by the experimental data used by Friedländer the expression $(2{\cdot}002 - \ln \mathrm{Re}_f) - 1$ cannot be replaced by $K\mathrm{Re}_f^{1/2}$, (3) the upper part of Friedländer's curve is plotted from the data of Wong et al. (Fig. 60d), which show that from Stk $\sim 0{\cdot}7$ $\ni$ rises sharply with Stk. The conclusions derived from this work are therefore unconvincing.

In order to characterize the quality of a filter under standard conditions, at given values of r and U, the ratio of the logarithm of penetration to the resistance of the filter is useful,

$$\beta = \frac{\log(1 - \ni)}{\Delta p} \tag{40.30}$$

since this ratio, according to (40.2), does not change when several identical filters are connected in series to form one multilayer filter. It is a function of the quality of the filtering material, which is independent of the thickness of the filter.

The first of the more recent experimental work considered here is that of Gallili [750] who used a model filter. Isodisperse mist of uncharged dioctylphthalate particles with $\bar{r} = 0{\cdot}55 - 0{\cdot}70\mu$ was drawn at the rate of 12–18 cm/sec through a tube 5 cm in diameter containing a number of grids of parallel wires having $R = 0{\cdot}025$ mm. The distance between the axes of the wires, h, varied from 0·13 to 0·27 mm; the distance between the grids was 5·5 mm. Re_f ranged from 0·36 to 0·54, Stk from 0·018 to 0·043 and r/R from 0·022 to 0·028. The collection efficiency on separate wires, $\ni$, increased with r and U and was calculated from the penetration of the aerosol through the tube.

A change in U had a greater effect on $\ni$ than one in r^2, and values of Stk indicated above are much lower than the critical values for viscous flow past isolated cylinders. Hence it seems that either the effect of interception, contrary to the opinion of Gallili, played an essential role in these experiments, or that the presence of adjacent cylinders changed the flow pattern so much that the calculations made for isolated cylinders do not even give a correct order of magnitude of $\ni$ for a row of cylinders.

The values of $\ni$ obtained in these experiments are given in Fig. 60b. According to (34.19) $\ni = 4{\cdot}5$ to $5{\cdot}1 \times 10^{-3}$ at the values of Stk and r/R employed, which is of the correct order of magnitude. When the distance between the wires h was decreased, $\ni$ at first increased rapidly, as would be expected since the streamlines approached the wires. Further decrease in h caused little change in $\ni$, or even a slight decrease. Gallili accounted for this by the reflection of droplets from the surface of the wires but this is not convincing at the relatively low values of U in his experiments. Nevertheless, experiments with model filters are of great interest.

Thomas and Yoder [751] also used homogeneous mists of dioctylphthalate which were drawn through a filter 1·2 cm thick made of glass fibres with $R = 0{\cdot}75\,\mu$ and impregnated with resin. The results of these experiments (Fig. 60c) clearly show the existence of a particle radius $r_{\min}$, at which the efficiency of the filter is minimum. In agreement with theory, $r_{\min}$ decreased with an increase in the flow rate which accounts for the failure of experiments at large flow rates to reveal the existence of $r_{\min}$. When the results of Thomas and Yoder are compared with other experimental findings (Fig. 59 and 60d), it should be borne in mind that the former were obtained at much lower flow rates with very thin fibres, so that diffusion, as indicated by the decrease in $\ni$ with an increase in the flow rate, and interception were paramount. Over the left hand portion of the curves in Fig. 60c diffusion predominates over all other mechanisms, and $\ni$ diminishes with an increase in r. The right hand part of the curves corresponds to the left hand part in Fig. 59b and 59d. Here $\ni$ rises with r and decreases with an increase in U. Examination of Fig. 59 shows that this region is clearly defined only in the filters with very thin fibres; the diffusion effect in this region still predominates over that of inertia but the total effect of inertia and interception exceeds that of diffusion.

In the experiments of Wong *et al.* [752] homogeneous mists of H_2SO_4 with $\bar{r} = 0{\cdot}2$–$0{\cdot}65\,\mu$ were drawn through filters of glass fibres having $R = 3{\cdot}5$–$9{\cdot}6\,\mu$ and the volume fraction filled by the fibres $\varphi = 0{\cdot}045$–$0{\cdot}10$; with $U = 17$–260 cm/sec the value of Re_f was 0·04–1·4. From the experimental results the collection efficiencies on separate fibres were calculated as a function of Stk for different values of r/R and are shown in Fig. 60d. At Stk < 0·5 the inertia effect is practically zero; at Stk > 1 the results are quite close to the theoretical curves of Davies for $Re_f = 0{\cdot}2$ (Fig. 48g).

Humphrey and Gaden [753] aspirated spores of *B. Subtilis* ($\bar{r} = 0{\cdot}575\,\mu$) through a glass fibre filter impregnated with resin and found a very flat maximum of efficiency at $U = 30$ cm/sec. In experiments with multilayer filters proportionality between the logarithm of the penetration and the number of layers was strictly observed, which indicates the uniform size of the spores (see page 213).

The structure of filters seems to be important for their efficiency; loose spots have both smaller filtering efficiency and lower flow resistance, and with very fine aerosols the two effects should be additive, though experimental data are lacking.

Partial clogging of filters has an opposite effect and the structure of the deposit is of great significance. Coal dust which deposits on filter fibres aggregates in branched chains, jutting out on all sides, and greatly increases efficiency, whereas the spherical particles of methylene blue, prepared by spraying a solution and allowing the droplets to evaporate, rest on the surface of the fibres, do not form aggregates and have only a very slight effect on efficiency [754].

Leers [755] sucked aerosols of NaCl ($\bar{r} = 0{\cdot}2\,\mu$) through very thin (0·1 mm) filters so that the structure of the deposit could be examined. Particles tended to deposit on the ones deposted earlier and branched chains were gradually formed, thus making a second filter of very thin dust fibres inside the original filter. The efficiency of this filter increased much faster than its resistance. The growth of the aggregates was explained by mechanical capture, but electrical forces seem to play a decisive role (see page 314). Droplets of mist act the opposite way since they spread on the surface of the fibres, increase their diameter, and thus lower the efficacy.

A similar phenomenon was observed when a mist of sulphuric acid was passed through glass fibre filters by Fairs [756]; the preformed filter mats made by baking pressed glass wool had $R = 2{\cdot}5$–$7{\cdot}5\,\mu$. On prolonged testing these filters gave $\ni = 0{\cdot}95$–$0{\cdot}97$. Treatment of the fibres with silicone oil increased $\ni$ to 0·996–0·998. the mist droplets ran together on the hydrophobic fibres and formed large spherical drops, which easily ran off the fibres.

LaMer and Drozin [757] when filtering homogeneous aerosols with $\bar{r} = 0{\cdot}1$–$0{\cdot}6\,\mu$ through fibrous filters found that $\log(1 - \ni)$ linearly decreased with an increase in r; in these experiments $\ni$ varied from 0·70 to 0·93. There is no doubt that the relationship between $\ni$ and r is more complex than this result suggests. When aerosols with hard particles of synthetic wax were filtered, $\log(1 - \ni)$ decreased linearly as the weight of deposit rose. It is difficult to give a theoretical explanation of this relationship.

Filtration of hard particles does not always raise $\ni$. According to the observations of Hasenclever [758], who filtered polydisperse quartz dust with $r \leqslant 2{\cdot}5\,\mu$ through paper at a rate of 6 m/sec, $\ni$ at first increased from 0·85 to 0·95 and then gradually fell until the experiment was stopped at $\ni = 0{\cdot}3$. It is possible that under these conditions the particles were squeezed through the filter.

The effect of electric charges on particles and fibres has been described in a number of papers. The experiments of Rossano and Silverman [759] were performed with an aerosol of methylene blue having $\bar{r} = 1\,\mu$ and a glass fibre filter of $R = 25\,\mu$; the velocity was 16 cm/sec. With uncharged aerosols $\ni = 0{\cdot}65$; with unipolar charged aerosols, the mean charge being 1000 electrons per particle, $\ni = 0{\cdot}72$. The authors suggest that this is due to the deposited particles repelling the particles entering the filter and slowing down their motion in the front layer. In reality, the cause is most probably the mirror forces between the charged particles and the fibres.

Winkel [760] observed a marked increase in the efficiency of filters made of glass fibres when impregnated with a good electrical insulator such as polyethylene or polystyrene; a very thin layer of glycerol spread over the impregnated fibres annulled the effect. In order that wool-resin electrostatic filters may retain their efficiency it is necessary to use resins with a specific resistance of the order of 10^{21} ohm . cm [761]; such filters can be preserved for a period over 10 years but they lose their efficiency very quickly when wetting mists, such as tricresyl-phosphate, or radioactive aerosols are drawn through them, or they are exposed to X-rays etc. When electrostatic filters were compared with asbestos–cellulose ones, using an aerosol of methylene blue, the former proved to be more efficient at small flow rates, up to 30 cm/sec, and the latter at large ones [759]. This is not surprising, since the effect of electric forces decreases at high flow rate.

Gillespie [762] tested wool-resin filters with uncharged aerosols of stearic acid and charged aerosols prepared by spraying polystyrene solutions. As would be expected, the effect of the particle charges was pronounced and grew with a decrease in particle size. Irradiation of the filter with X-rays, which discharged it, somewhat diminished the efficiency for neutral aerosols but no experiments of this kind were made with charged aerosols.

Billings *et al.* [763] described an electrostatic filter with an external field: a filter (12 mm thick) of glass fibres of $R = 0{\cdot}5$–$1{\cdot}5\,\mu$ impregnated with a phenol resin was inserted between two metal grids. When an aerosol of $CuSO_4$, with $\bar{r} = 0{\cdot}9\,\mu$, was drawn through it, the efficiency of the filter was equal to 0·92, but the application of a potential difference of 10–15 kV to the grids caused $\ni$ to increase up to 0·98. The charge on the particles was not measured, so it is difficult to analyse the mechanism of this phenomenon, which might be due to a direct action of the field on charged particles, or to induction forces due to the inhomogeneity of the field between the fibres.

Membrane filters having very narrow capillary channels 0·1–1 μ in diameter are widely used for the investigation of aerosols. In addition to mechanisms of deposition already considered, mechanical retention of particles whose size exceeds the width of the channels plays a part. Diffusion in the channels appears to be an important factor in these filters.

Fitzgerald and Detweiler [764] found that when an aerosol of $KMnO_4$ is filtered through Millipore AA and HA the aerosol is completely retained if $r > 0{\cdot}05\,\mu$; with $r = 0{\cdot}01$–$0{\cdot}013\,\mu$, $\ni$ attains minimal values: 0·78 at $U = 10$ cm/sec, 0·62 at 20 cm/sec and 0·20 at 40 cm/sec; a further decrease in r causes $\ni$ to rise again. Walkenhorst [765], however, working with WO_3 aerosols at $U = 32$ cm/sec found that when fine German membrane filters are used, $\ni$ falls continuously with a decrease in r from 0·994 at $r = 0{\cdot}05\,\mu$ to 0·972 at $r = 0{\cdot}005$–$0{\cdot}01\,\mu$. It is difficult to be sure of the reason for this discrepancy, which may be due to different widths of the channels in the filters.

Membrane filters can be used to find weight and number concentration and for the microscopic examination of aerosols. Practically all the particles deposited lie on the front surface of the filter and are in one optical plane, which facilitates examination of the deposit with a microscope. Replicas of the deposits can be prepared for electron microscopy.

Snyder and Pring [766] studied the clogging of fabric filters by solid particles. The increase in resistance as they become clogged with coarse aerosols is given by Δp

$= kG$, where G is the mass of the deposit per cm^2 of the filter. With fine aerosols the coefficient k increases with G evidently due to increasing density of the deposit with growing G. Filters in which the specific surface area of the fibres is high have a low value of k. Fabrics with a nap, for example, have k smaller than similar ones without a nap. Using the same aerosol, a glass fabric had k 10 times, and an Orlon fabric with a small nap 4 times, larger than a similar fabric with a large nap. It is possible for k to vary a thousandfold in different fabrics and for various aerosols.

Eliseev [767] studied the formation of aerosol deposits using smokes of PbO and ZnO drawn through a metal gauze with the radius of the wires 25 μ and the space between the wires 80 μ (Fig. 60e). The deposits formed on the front of the wires grew until they exceeded the width of the wires and overlapped the spaces between them, finally making a continuous filtering layer which was impenetrable by the particles which were responsible for its formation. The solid phase in the layer had a volume fraction $\varphi \approx 0{\cdot}06$ and was very loosely packed, which accounts for the relatively small flow resistance.

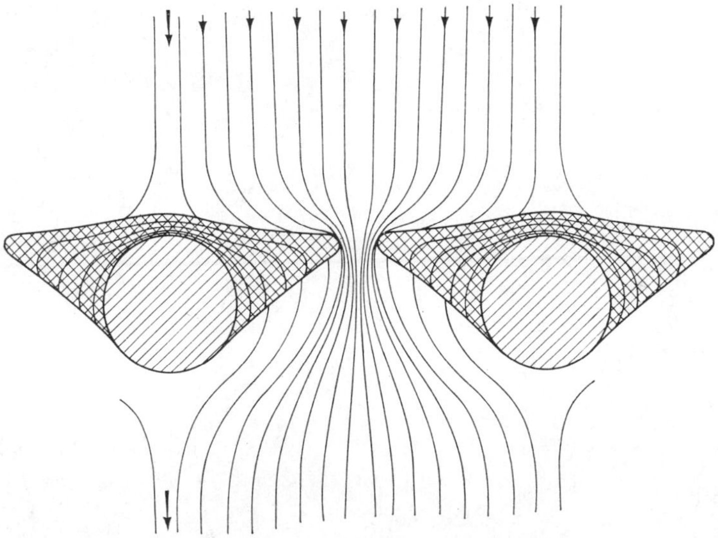

FIG. 60e. Formation of a dust layer on a wire-gauze.

Only a very small deposit, some 10 g/m^2, is necessary to form a filtering layer on clean fabric. Filtering fabrics are periodically blown out with air in the reverse direction and shaken to remove the deposit. Only a small part of the deposit can be removed thus and when work is protracted the density of the deposit may build up to 1000 g/m^2. Along with deposit which is active in filtering, a large quantity of a passive deposit seems to collect. After each reverse blowing channels are formed in the filtering layer, into which the aerosol penetrates. They become clogged very quickly, only 0·2–0·3 g/m^2 being necessary. It is notable that the resistance of the unbroken filtering layer is identical in each direction, while in the broken layer the resistance with reverse flow is one-half to one-third of the figure for direct flow, since aggregates on the front of the fabric act as valves. Eliseev also showed that reverse blowing is of very little effect, reducing the resistance only by 6–15 per cent, unless the fabric is de-

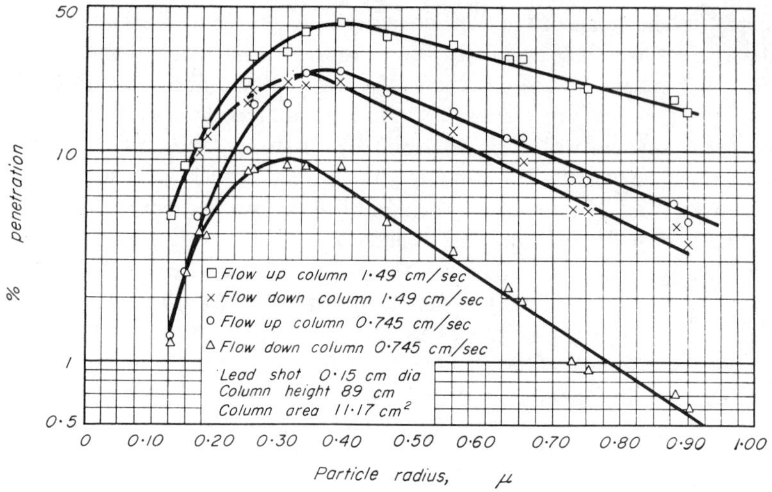

FIG. 60f. Filtration of aerosols through lead shot.

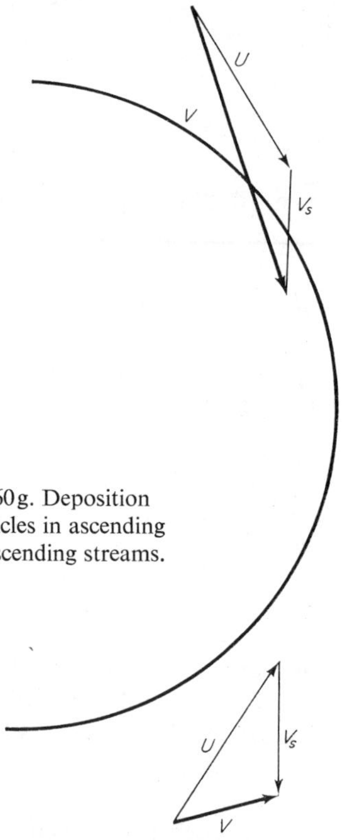

FIG. 60g. Deposition of particles in ascending and descending streams.

formed which breaks up the deposit and lowers the resistance by 70–94 per cent. In practice reverse blowing of bag filters is always accompanied by considerable deformation.

Silvermann *et al.* [578] showed that the efficiency of filtering fabrics is markedly increased when they are electrically charged, particularly at small flow rates, but if the relative humidity of the air exceeds 90 per cent the charges rapidly leak away and the effect disappears.

Thomas and Yoder [768] investigated the filtration of aerosols through granular packings in vertical columns. Curves for the filtration of dibutylphthalate aerosols through a column of lead shot are shown in Fig. 60f, all the data being indicated on the graph. The experiments were conducted at low flow rates and with very large R so that the effects of interception and inertia must have been negligible; the efficiency was very small. The key part was played by diffusion and sedimentation, the former prevailing in the ascending branches of the curves and the latter in the descending. Hence, $\ni$ decreased rapidly with an increase in flow rate. Owing to sedimentation $\ni$ differs for upwards and downwards flow as shown in Fig. 60g, which indicates the total velocity of a sedimenting particle in ascending and descending flows. Similar curves were obtained with columns of sand [751] in which $\ni$ diminished appreciably with increase of flow rate and grain size; grains of irregular shape gave better results than rounded ones.

Their experiments on filtration of a polystyrene aerosol, carrying large charges on the particles, through a lead shot column are of some interest. When the column contained a radioactive substance, which discharged the particles, the penetration increased as much as 7–15 times. Such a high effect of induction forces in the column is explained by the low efficiency of other mechanisms of deposition under the conditions of the experiments.

§ 41. DEPOSITION OF AEROSOLS IN THE RESPIRATORY SYSTEM

The deposition of aerosols in the breathing tract is of great concern in industrial hygiene. Not only is the total number of particles of a specified size which is deposited at a given frequency and depth of respiration important, but the distribution of the deposit over the various parts of the system must also be considered. Dust settling in the naso-pharynx, trachea and bronchi is comparatively harmless if insoluble and is removed by ciliary action, causing movement of the mucous lining, followed by expectoration and swallowing. Irritation of the upper breathing tract, hay fever etc. are associated with solubility. Dust entering the pulmonary alveoli is much more harmful and some may be permanently retained in the lungs so that it is often possible to relate a man's occupation to the appearance of his lungs, for example coal miners. Silica dust is particularly harmful and gives rise to a dangerous occupational disease called silicosis. On the other hand certain diseases are treated with medicinal aerosols and it is desirable that these should be deposited as deeply as possible, mainly in the lung alveoli.

Several experimental methods have been used for studying the deposition of aerosols in the respiratory system. One consists of microscopic examination of sections cut from the lungs, bronchi etc. and the use of a fluorescent dust, such as willemite, illuminated with ultra-violet light is particularly suitable [320]. This shows, for coarse dust, that the number and size of deposited particles fall off steadily with increasing depth in the respiratory system, and that deposits are heaviest where passages branch and turn, since inertial deposition plays a large part. There are many aggregates in the deposits and when the dust concentration is very large some narrow bronchioles may be completely blocked.

For quantitative measurements it is best to use the radioactive tracer method employing aerosols which have been given a known amount of radioactivity. In tests with animals individual parts are cut from the respiratory system, after exposure and sacrifice, and the amount of radioactive aerosol deposited in each is determined quantitatively by one of the usual methods. In human beings the amount of aerosol deposited in the peripheral parts of the lungs can be estimated by means of a Geiger counter in contact with the subject's chest [321].

The overall efficiency of deposition in the respiratory system can also be determined from the ratio of particle concentrations in the inhaled and exhaled air. To obtain an indication of the distribution over various parts of the system the aerosol concentration in successive fractions of each exhaled breath is measured [322], the first fraction containing air from the upper part of the system and the last fraction air from the alveoli.

TABLE 19. EFFICIENCY WITH WHICH OIL MISTS ARE DEPOSITED IN THE NASAL CAVITY OF A MAN

$\bar{r}$ (μ)	$\ni$ at rate of breathing of:	
	20 l. min^{-1}	29 l. min^{-1}
6	0·87	0·99
3·7	0·42	0·71
1·1	0·14	0·25
0·9	0·06	0·19

Two methods have been used specially to study deposition in the nasal passages. In one of them a person inhales an aerosol through his nose and immediately breathes out through his mouth. It is assumed that deposition occurs only in the nasal cavity, that in the deeper parts of the respiratory system being neglected [323]. In animal experiments the trachea of an anaesthetized animal is severed, the mouth is closed and the upper part of the trachea is connected to a pump which draws the aerosol through the nose [324].

Examination of experimental data on aerosol deposition in the respiratory system reveals large individual differences in the results of experiments conducted on both people and animals. According to Vigdorchik's observations [2] this is particularly true of coarse aerosols breathed through the nose. The efficiency of deposition can vary several times between individuals under the same test conditions, partly because of large differences in shape and width of the nasal passages [324]. The difference is not so great for breathing through the mouth and for more highly dispersed aerosols. The deposition of aerosols in the nasal cavity increases markedly when the rate of breathing is increased, thus indicating that inertial deposition plays the principal role [323, 324]. Averaged results of experiments on the deposition in the nasal cavity of oil mists of different particle size are presented in Table 19 [323]. The particles of pollen which cause hay fever are about 10 to 20 μ in radius [325] and are therefore deposited almost entirely in the nasal cavity as long as the subject breathes in through his nose.

When breathing through the mouth the overall deposition of aerosols in the respiratory system falls off as the frequency of breathing is raised, as shown in Fig. 61 where N is the number of breaths per minute. In this work fairly isodisperse glycerine mists containing radioactive ^{24}NaCl were used [321].

From Table 19 and Fig. 61 it can also be seen that deposition in the respiratory system increases with particle size over the entire range studied (from 0·2 μ), the relationship between ∋ and r being given by a straight line on a logarithmic plot when ∋ is not very large. The broken line on Fig. 61 is for breathing through the nose and shows that ∋ increases considerably more rapidly with particle size than it does for breathing through the mouth. The drop in ∋ with increase in the rate of breathing through the mouth and decrease in particle size shows that for $r > 0{\cdot}2\,\mu$ sedimentation plays the main part, much of the aerosol being deposited in the pauses between inhaling and exhaling. When the duration of these pauses is increased by holding the breath, the efficiency of deposition rises markedly [326] and this is the recommended procedure when treating lung diseases with aerosols.

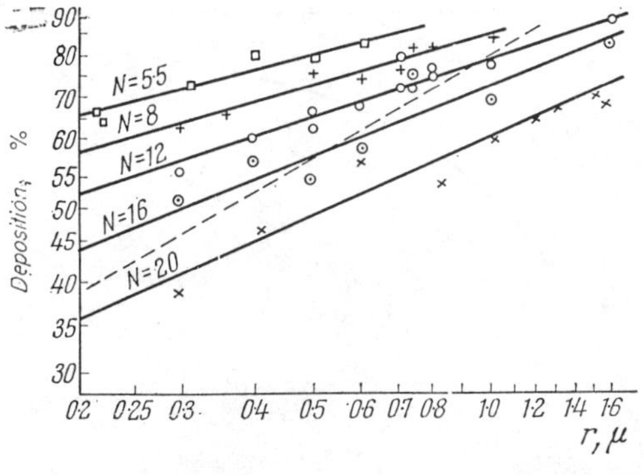

FIG. 61. Deposition of aerosols in the respiratory system.

The following data are available concerning the deposition of aerosols in various parts of the respiratory system. In experiments with rabbits [327] it has been found that with an aerosol having $\bar{r} = 0{\cdot}5\,\mu$ 29 per cent of inhaled particles retained in the system were deposited in the naso-pharynx, 13–19 per cent in the trachea, and 51–58 per cent in the lower part of the system. For $\bar{r} = 2\,\mu$ the proportion retained in the naso-pharynx rose to 65 per cent and for $\bar{r} = 4\,\mu$ to 98 per cent. The following figures (percentages) were found for the distribution of oil mist deposits in mice in the bronchi, bronchioles and alveolar passages respectively [328]: 26, 32 and 42 for r = 0·2–0·62 μ; 33, 33 and 34 for r = 0·62–1·05 μ; 35, 37 and 28 for r = 1·05–1·46 μ; 48, 37 and 15 for r = 1·46–1·88 μ; 46, 40 and 14 for r = 1·88–2·28 μ; 53, 37 and 10 for r = 2·28–2·71 μ; 62, 34 and 4 for r = 2·71–3·13 μ; 67, 33 and 0 for r = 3·13–3·55 μ; and 90, 10 and 0 for $r > 3{\cdot}55\,\mu$.

The overall efficiency of deposition thus increases with increase in particle size, mainly as a result of deposition in the bronchi. The amount of deposition in the alveoli, a most important consideration in industrial hygiene, is proportional both to the number of particles which penetrate the upper part of the system and to the efficiency with which these particles are subsequently deposited. When the particle size is increased the number of particles which penetrates to the alveoli decreases although the efficiency of deposition in the alveoli increases. The curve of deposition in the alveoli as a function of particle radius must therefore have a maximum, as verified by experiment. Figure 62 shows such curves obtained by several authors. Only curve 1, obtained by Van Wijk and Patterson [329] for an aerosol with γ = 2·6 at 15 breaths/min, relates to deposition in the whole respiratory system. Curve 2 was obtained by Wilson and LaMer [321] for a glycerine mist with $\gamma = 1{\cdot}2$ at 20 breaths/min; curve 3 by Brown and Hatch [330] for dust with $\gamma = 2{\cdot}6$ at 15 breaths/min; and curve 4 by Landahl [322] for triphenyl phosphate mist with $\gamma = 1{\cdot}17$ at 15 breaths/min. Since deposition in the upper respiratory passages can be neglected at small r, curves 1 to 3 agree fairly well with one another. The considerable difference of curve 4 is probably due to systematic errors in the determination of droplet size [331]. Particles about 1 μ in radius are mostly deposited in the alveoli and these therefore present the greatest danger to health, as far as insoluble particles are concerned. It must be remembered that, besides a particle size corresponding to maximum deposition in the lungs, there also exists, as in deposition in filters, a size corresponding to minimum deposition.

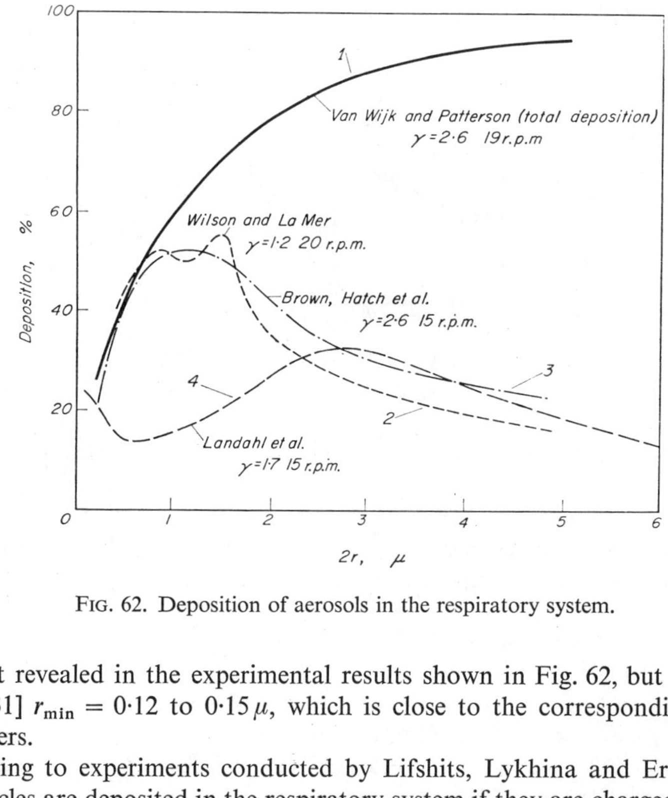

FIG. 62. Deposition of aerosols in the respiratory system.

This is not revealed in the experimental results shown in Fig. 62, but according to Davies [331] $r_{\min} = 0{\cdot}12$ to $0{\cdot}15\,\mu$, which is close to the corresponding value for fibrous filters.

According to experiments conducted by Lifshits, Lykhina and Erenburg [332] more particles are deposited in the respiratory system if they are charged than if they are uncharged. From aluminium dust containing 69 per cent of particles with radii less than $0{\cdot}5\,\mu$ and 27 per cent with radii between 0·5 and $1{\cdot}5\,\mu$, the amounts deposited when inhaled were 34 per cent uncharged and 66 per cent charged. This effect is due to induction or image forces which attract the charged particles to the walls of the respiratory passages.

Interesting theoretical work has been performed by Findeisen [333] who started from a much-simplified model of the respiratory system. He considered the respiratory passages to be straight tubes, the length, width and disposition of which corresponded roughly to anatomical data, and the alveoli of the lungs to be hollow spheres. Assuming that breathing in and out occur at the same constant rate (200 cm³/sec) and last for two seconds each with no pause between, Findeisen considered sedimentation and diffusion in the respiratory passages and inertial deposition at the junctions. Taking into account the deposition during breathing in and breathing out, he calculated the percentage of particles of a given size deposited in each part of the respiratory system. His assumption of constant air velocity over the cross-section of the tubes was bound to give too high values of ∍.

Table 20 shows Findeisen's calculated values for several values of r. The left-hand side of the table contains data on which the calculations are based. These are the number ν, length L, radius R and total cross-section S of the respiratory passages, the flow velocity U of the air and the time t that the particles took to pass through a given part of the system. The mean angle of inclination of the respiratory passages to the horizontal was taken as 50° in calculating the deposition due to sedimentation. Figures placed between the names of the respiratory passages refer to inertial deposition at the junctions. The angle turned through at the junctions was taken as 90° in going from the alveolar bronchioles to the alveolar passages and 30° for all the others.

It can be seen by comparing the data of Table 20 and Fig. 61 that the results of Findeisen's calculations differ rather widely from what is found experimentally. This is due to the crudeness of the calculation and to unavoidable simplification. Nevertheless his work is of interest because it provides a general picture of aerosol deposition in different parts of the respiratory system. Findeisen's calculations have been refined somewhat by Landahl [334].

As pointed out in § 24 a method of increasing the deposition of aerosols in the respiratory passages is to give the aerosol a unipolar charge [335]. The highly-dispersed magnesium oxide smokes, carrying a unipolar charge, which Dessauer [168] used for therapeutic purposes, contained about 10^7 particles per cm^3 and their mobility, u, [equation (27·2)] was 0·002 to 0·007. Assuming that each particle had one elementary charge, on the average, and using $u = Bq/300$, formula (24.6) suggests that 10–30 per cent of the particles were deposited by electrostatic expansion of the cloud in the time of about 2 sec during which the aerosol was in the respiratory passages. Dessauer's measurements gave 70–80 per cent for the total deposition, but it must be remembered that, according to Table 20, an aerosol spends considerably more time in the alveoli than in the upper respiratory passages, so that deposition by electrostatic expansion probably occurs mainly in the alveoli.

If unipolar charging of medicinal aerosols, such as penicillin, is to increase appreciably their deposition in the lungs, (24.6) shows that a high concentration of highly charged particles is required. It is also necessary for the particles to be small enough to avoid being deposited in the upper respiratory passages. The extreme difficulty of obtaining aerosols of high number concentration by mechanical atomization [336] renders the realization of this idea possible only with compounds which withstand the heating necessary to convert them into aerosols by evaporation and re-condensation.

In conclusion the effect of bipolar charging on the deposition of aerosols in the respiratory system will be determined, confining the analyses to deposition in the alveoli which are taken to be hollow spheres of radius $R = 0{\cdot}015$ cm. A particle with charge q situated at a distance x from the surface of an alveolus, x being small compared with R, is attracted to the surface by the force $q^2/4x^2$, due to the image charge, and moves towards it with a velocity

$$V = -\frac{dx}{dt} = \frac{q^2 B}{4x^2}. \qquad (41.1)$$

TABLE 20. EFFICIENCY WITH WHICH AEROSOLS ARE DEPOSITED IN VARIOUS PARTS OF THE HUMAN RESPIRATORY SYSTEM

Part of system	ν	R (cm)	L (cm)	S (cm²)	U (cm sec⁻¹)	t (sec)	Efficiency of deposition (per cent) at particle radius μ of:						
							0·03	0·1	0·3	1	3	10	30
Trachea	1	0·65	11	1·3	150	0·07	0·16	0·08	0·03	0·10	0·8	7·8	67
							—	—	0·02	0·16	1·2	11·0	33
Main bronchi	2	0·37	6·5	1·1	180	0·04	0·21	0·10	0·05	0·11	0·7	6·2	—
							—	—	0·03	0·27	2·5	20·0	—
1st order bronchi	12	0·20	3·0	1·5	130	0·02	0·28	0·13	0·07	0·07	0·4	2·5	—
							—	—	0·04	0·57	3·8	20·3	—
2nd order bronchi	100	0·10	1·5	3·1	65	0·02	0·55	0·26	0·13	0·14	0·8	2·9	—
							—	—	0·02	0·52	2·7	8·0	—
3rd order bronchi	770	0·075	0·5	14	14	0·04	1·03	0·51	0·29	0·35	2·0	5·3	—
							—	—	0·02	0·84	3·1	3·8	—
Terminal bronchioles	$5{\cdot}4 \times 10^4$	0·030	0·3	150	1·3	0·22	6·1	3·1	2·0	4·0	25·4	10·2	—
							—	—	—	0·79	1·5	—	—
Alveolar bronchioles	$1{\cdot}1 \times 10^5$	0·025	0·15	220	0·9	0·17	6·3	3·2	2·0	3·7	16·0	—	—
							—	—	—	1·8	2·5	—	—
Alveolar passages	$2{\cdot}6 \times 10^7$	0·010	0·02	8200	0·025	0·82	37·2	19·1	15·8	40·3	36·6	—	—
							—	—	—	1·1	—	—	—
Alveoles	$5{\cdot}2 \times 10^7$	0·015†	—	$1{\cdot}47 \times 10^5$††	0	1·2	14·1	8·6	12·7	41·6	—	—	—
Total	—	—	—	—	—	—	66	35·0	34·2	97·4	100	100	100

† Radius of sphere.
†† Total surface

A particle which is a distance x_0 from the wall at time $t = 0$ reaches the wall in a time

$$t = \frac{4}{3}\frac{x_0^3}{Bq^2}. \tag{41.2}$$

During the time of the order of one second in which the particles remain in the alveoli, those within a distance x_0 of the surface, where

$$x_0 = \left(\frac{3}{4} Bq^2\right)^{1/3}, \tag{41.3}$$

will reach the surface and the total number of particles deposited owing to image forces will be

$$N_1 = \frac{4}{3}\pi R^2 x_0 n = \frac{4}{3}\pi R^2 n \left(\frac{3}{4} Bq^2\right)^{1/3}. \tag{41.4}$$

The number of particles deposited in this time by gravity will be

$$N_2 = \pi R^2 n V_s = \pi R^2 n B m g. \tag{41.5}$$

According to the conclusions of Tunitskii, Tikhomirov and Petryanov [337] the aluminium particles about $0{\cdot}5\,\mu$ in radius used by Lifshits, Lykhina and Erenburg must have carried about 50 elementary charges per particle.† In this case formulae (41.4) and (41.5) show that $N_1/N_2 = 0{\cdot}67$ so that the image forces must have increased the deposition of charged particles appreciably. Such high charges arise as a result

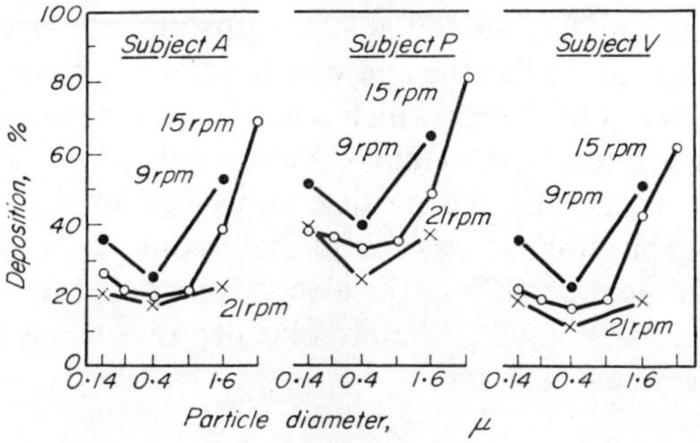

FIG. 62a. Aerosol deposition in the respiratory system of 3 subjects.

of the triboelectric effect. Bipolar charging of particles in the natural ionized atmosphere (see page 114) has only a negligible effect on the deposition of aerosols in the respiratory system.

The experiments of Morrow *et al.* [769] were conducted with NaCl particles with $r = 0{\cdot}02$–$0{\cdot}35\,\mu$; the deposition efficiency, in the whole respiratory system, varied from 50 to 80 per cent depending on the respiratory rate. It should be remembered that the air spaces of the respiratory system are almost saturated with water vapour so that particles of NaCl must grow appreciably in size.

Altshuler *et al.* [770] used homogeneous mists of triphenylphosphate ($\gamma = 1{\cdot}3$) which were inhaled through the mouth by three individuals, the total retention in

† This makes $x_0 = 1{\cdot}45 \times 10^{-3}$ cm so that the condition $x_0/R \ll 1$ is fulfilled.

the whole respiratory system being determined. Their results are shown in Fig. 62a, the respiratory rate being indicated on the graphs. The minimal $\ni$ was observed at $r = 0{\cdot}2\,\mu$. The great difference in the values of $\ni$ for different individuals is noteworthy.

In the experiments of Dautrebande *et al.* [771] with polydisperse dusts of coal and Fe_2O_3 and with particles obtained by spraying an Indian ink solution, the value of $\ni$ for particles of different sizes was determined by comparing the particle size

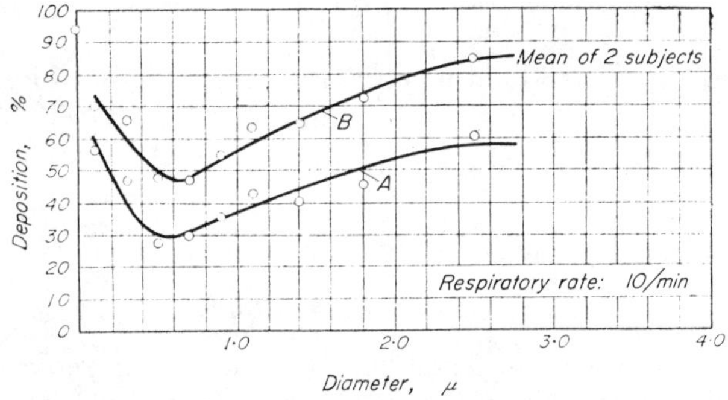

FIG. 62b. Aerosol deposition in the respiratory system.

distribution in the inhaled and exhaled air, a method which is not very reliable. The minimum of retention was found at $r \approx 0{\cdot}3\,\mu$ and, more recently, a feeble minimum at $r = 0{\cdot}12$–$0{\cdot}15\,\mu$. The results obtained at 10 breaths/min are given in Fig. 62b. The curve B refers to the air exhaled at the end of a normal expiration, and hence to the deposition of the aerosol in the lungs, which is much greater than the average deposition over the whole respiratory system (curve A). It is still greater in the alveoli proper and was explored by measuring the particle concentration in air retained in the lungs after a normal expiration. It was found that about 90 per cent of the particles of $r = 0{\cdot}1\,\mu$ which had penetrated into the alveoli were deposited there, independent of the respiratory rate; at $r = 0{\cdot}2\,\mu$ the alveolar deposition was 95 per cent and at $r \gg 0{\cdot}5\,\mu$ it was 100 per cent.

§ 42. ABSORPTION OF AEROSOLS BY BUBBLING

The absorption of aerosols when they are bubbled through liquids under normal conditions, although slight and of limited technical importance, has contributed to our knowledge of their properties. One of the earliest observations on the properties of aerosols was their poor absorption by bubbling, compared with gases, and the phenomenon was later encountered in the catalytic production of sulphuric acid. A large number of experimental investigations, many of them conducted by Remy, have been concerned with the absorption of aerosols by bubbling through water and aqueous solutions. These observations were extremely empirical in character and the authors did not attempt to give any theoretical explanation of their results. In addition the sizes of the particles in the hygroscopic mists with which most of the experiments were carried out were not determined exactly and it is therefore difficult to interpret the results. Nevertheless some conclusions about the mechanism of absorption during bubbling can be drawn from these studies.

The motion of gas bubbles through water and other liquids depends largely on the size of the bubbles [338, 339]. Very small ones with radii less than about 5×10^{-3} cm have Reynolds numbers less than one. Provided that there are no surface-active impurities in the water, circulation proceeds inside the bubble (Fig. 63) and the gas velocity at the surface of the bubble is given by the formula

$$U_\tau = 0{\cdot}5\, V_b \sin\theta, \tag{42.1}$$

where V_b is the rate of rise of the bubble and θ is the angle between the radius vector and the direction of motion of the bubble. When the bubble radius is between 5×10^{-3} and 10^{-1} cm the Reynolds number lies between 1 and 700. In this region the

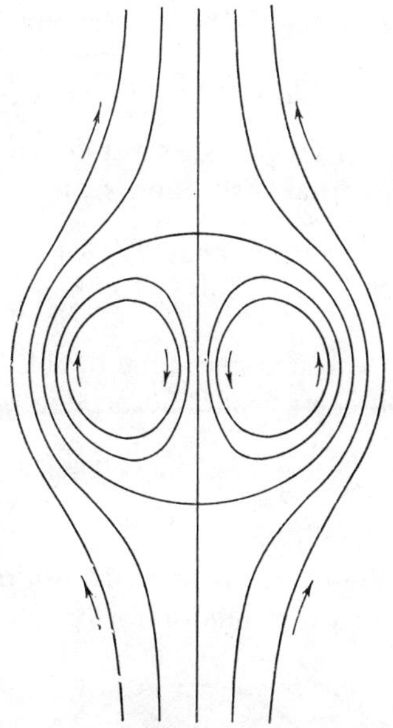

FIG. 63. Circulation in a spherical cloud and in a gas bubble.

bubbles retain their spherical shape but the flow of liquid round them resembles ideal flow round a sphere [338], so that the velocity of the circulating gas at the surface of the bubble is given approximately by

$$U_\tau = 1{\cdot}5\, V_b \sin\theta. \tag{42.2}$$

The velocity is therefore three times greater than in the previous case.

Finally when $R > 10^{-1}$ cm the bubble becomes flattened and pulsations of shape occur; the circulation of the gas becomes more complex and the rate of rise of the bubble reaches 20 or 30 cm/sec, varying only slightly with size.

This is the size which often has to be dealt with in practice. To simplify the calculations let us assume that the bubbles are spherical and that the rate of circulation at the surface is given by (42.2).

As in the cases discussed earlier, absorption of aerosols in bubbling is due to inertial deposition, sedimentation and diffusion. In this case the aerosol particles are considerably smaller than the bubbles and the interception effect can be neglected. The amount of inertial deposition in a spherical bubble can easily be calculated by formula (28.1). The velocity of the particles due to the centrifugal force is

$$V = \frac{U_\tau^2 \tau}{R} = \frac{9\, V_b^2 \tau \sin^2\theta}{4\, R}. \tag{42.3}$$

Hence the number of particles deposited per second on the bubble surface will be

$$\Phi = \frac{q V_b^2 \tau n}{4\, R} \int_0^\pi \sin^2\theta \cdot 2\pi R^2 \sin\theta \, d\theta = 6\pi V_b^2 \tau n R \tag{42.4}$$

and the number deposited while the bubble travels one centimetre is

$$\Phi_1 = 6\pi V_b \tau n R. \tag{42.5}$$

The ratio of the number of particles deposited per cm path to the total number in the bubble, or the coefficient of inertial absorption α_i, is

$$\alpha_i = \frac{6\pi V_b \tau n R}{(^4/_3)\pi R^3 n} = \frac{9\, V_b \tau}{2\, R^2}. \tag{42.6}$$

The number of particles deposited inside the bubble by sedimentation is $\pi R^2 n V_s = \pi R^2 n g \tau$ per second and the coefficient of absorption by sedimentation, α_s, is therefore

$$\alpha_s = \frac{3\, g\tau}{4\, R V_b}. \tag{42.7}$$

When $V_b = 25$ cm/sec this shows that inertial deposition is an order of magnitude greater than deposition due to sedimentation for bubble radii from 0·1 to 0·5 cm.

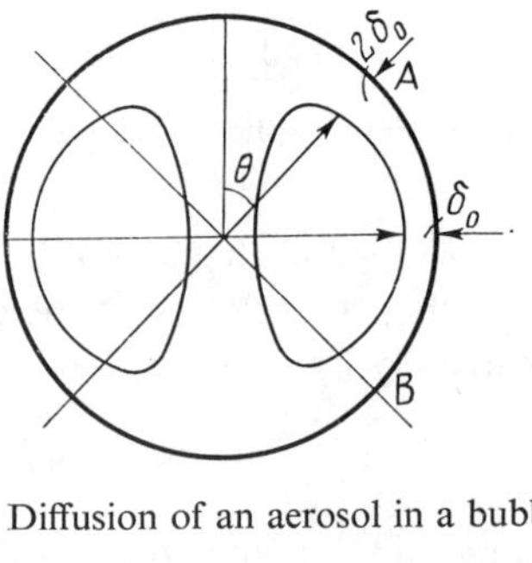

FIG. 64. Diffusion of an aerosol in a bubble.

It is difficult to calculate exactly the coefficient of diffusive deposition of aerosols during bubbling but an approximate value can be found in the following way. Consider a surface formed by the streamlines in a bubble which pass a small distance δ from its surface (Fig. 64). The volume of gas flowing per second between the two surfaces is

$$\Delta v = 2\pi R \sin\theta\, \delta\, U_\tau = 3\pi R V_b \delta \sin^2\theta = 3\pi R V_b \delta_0, \tag{42.8}$$

where δ_0 is the value of δ in the equatorial plane of the bubble. Diffusive deposition is small at the polar regions of the bubble because the flow velocity is low and the streamlines move away from the surface. Calculation will therefore be confined to the region from 45° above to 45° below the equator, i.e. $\pi/4 \leqslant \theta \leqslant 3\pi/4$. In this region δ varies from δ_0 to $2\,\delta_0$ and the average of $1{\cdot}5\,\delta_0$ will be used. The time, t, in which a particle travels from A to B (Fig. 64) is

$$t = \int\limits_{\pi/4}^{3\pi/4} \frac{R\,d\theta}{U_\tau} = \frac{2\,R}{3\,V_b} \int\limits_{\pi/4}^{3\pi/4} \frac{d\theta}{\sin\theta} \approx \frac{1{\cdot}2\,R}{V_b}. \tag{42.9}$$

By the simplified method of calculating diffusive deposition we find

$$\bar{\delta} = 2\sqrt{\frac{Dt}{\pi}} = 2\sqrt{\frac{1{\cdot}2\,DR}{\pi V_b}} \approx 1{\cdot}2\sqrt{\frac{DR}{V_b}}. \tag{42.10}$$

Particles situated between the two surfaces mentioned above diffuse towards the wall of the bubble and are absorbed by the liquid. The number of particles absorbed in unit time will be

$$\Phi = n\Delta v = 3\pi R V_b \delta_0 n = 2\pi R V_b \bar{\delta} n = 2{\cdot}4\pi n\sqrt{D V_b R^3}. \tag{42.11}$$

Hence the following expression is obtained for the coefficient of diffusive deposition of an aerosol during bubbling:

$$\alpha_D = 1{\cdot}8\sqrt{\frac{D}{V_b R^3}}. \tag{42.12}$$

It can be seen from these formulae that the absorption increases markedly with decrease in bubble radius. The effects upon absorption of particle size and rate of rise of the bubbles are therefore similar to those seen with fibrous filters.

An analysis of the experimental material obtained by Remy, who worked with ordinary gas-washing bottles and tall cylinders filled with aqueous solutions, will now be attempted. The bubbles were from 0·3 to 0·6 cm in radius and it was first of all noted that as the bubbles became smaller, the absorption increased considerably, as mentioned above [340]. A series of experiments conducted with highly dispersed "dry" aerosols of sulphuric acid and ammonium chloride, containing particles of 10^{-5} cm or less in radius, showed [341] that at a given temperature these aerosols were absorbed better when the vapour tension of the aqueous absorbing solutions was decreased. This can be explained as follows. Particles, of the size indicated, absorb moisture from the surrounding medium extremely rapidly and their increase in size depends on the pressure of water vapour inside the bubble. For particles of radius $\leqslant 10^{-5}$ cm the principal mechanism of deposition is diffusion. Under typical conditions of these experiments, with $R = 0{\cdot}5$ cm, $V_b = 25$ cm/sec and a particle radius of 10^{-5} cm, formulae (42.6), (42.7) and (42.12) give the following values for the coefficients of absorption by inertial deposition, sedimentation and diffusion respectively: $1{\cdot}0 \times 10^{-4}$, $0{\cdot}14 \times 10^{-4}$ and 15×10^{-4}. When the particle radius is increased to 2×10^{-5} cm the coefficients for inertial deposition and sedimentation rise to 3×10^{-4} and $0{\cdot}4 \times 10^{-4}$ respectively while that for diffusion falls to $9{\cdot}3 \times 10^{-4}$, so that the total absorption decreases.

Experiments with coarse mists ($r \approx 10^{-4}$ cm) of the same materials gave the opposite result, namely an increase in absorption with increase in vapour tension of the solution through which the aerosol was passed [341], that is when the particle size increases. This was to be expected because at $r = 10^{-4}$ cm $\alpha_i = 59 \times 10^{-4}$, $\alpha_s = 8 \times 10^{-4}$ and $\alpha_D = 3{\cdot}6 \times 10^{-4}$, i.e. absorption is mainly due to inertia and sedimentation, and increases with particle size.

The absolute values of the absorption coefficients found by Remy can be compared with those calculated theoretically. Only one of Remy's papers [342] is suitable for this; in it he presents values for the coefficient of absorption of a mist obtained by blowing air over aqueous solutions of sulphuric acid. The value $\alpha = 0{\cdot}005$ was obtained for $R = 0{\cdot}5$ cm and $V_b = 20$ cm/sec in a 5 per cent solution of sulphuric acid. The rate of settling of the upper boundary of the mist before bubbling was $3{\cdot}5 \times 10^{-3}$ cm/sec. Taking the concentration of acid in the droplets to be 80 per cent shows that the droplet radius before bubbling was 4×10^{-5} cm, giving $\alpha_i = 8{\cdot}3 \times 10^{-4}$, $\alpha_s = 1{\cdot}7 \times 10^{-4}$, $\alpha_D = 6{\cdot}8 \times 10^{-4}$ and an overall value of $\alpha \approx 0{\cdot}0015$. After the mist had come into equilibrium with the solution the acid concentration in the droplets fell to 5 per cent and, as shown by calculation, the droplet radius increased to $1{\cdot}2 \times 10^{-4}$ cm. Values corresponding to this radius are $\alpha_i = 67 \times 10^{-4}$, $\alpha_s = 13 \times 10^{-4}$, $\alpha_D = 4 \times 10^{-4}$ and an overall value of $\alpha \approx 0{\cdot}008$. The effective value of the absorption coefficient must therefore lie between 0·0015 and 0·008, which agrees in order of magnitude with Remy's findings.

In bubbling, as in the passage of aerosols through fibrous filters, a minimum occurs on the curve of absorption efficiency against particle size. A polydisperse aerosol therefore becomes more homogeneous during bubbling and the size-frequency distribution curve tends to become narrower with a peak at the size corresponding to the maximum transmission of particles. The effect increases in multi-stage bubbling and was used by Dautrebande [326] to obtain more or less monodisperse aerosols with $\bar{r} \approx 0{\cdot}2$–$0{\cdot}3\,\mu$ for physiological investigations.

Aerosols are absorbed very efficiently when they are passed through porous glass filters immersed in water [343]. A No. 1 filter which on its own retains only 50 per cent of a sulphuric acid mist retained about 99 per cent when immersed in water. This result is not surprising since the radii of bubbles formed in bubbling through glass filters are from 0·025 to 0·10 cm depending on the grade of filter.

It is interesting that the absorption efficiency in bubbling falls as the temperature of the water rises, becoming practically zero [344] at the boiling point. Two factors may act here: increase in size of the bubbles on account of the water vapour and repulsion of the particles from the walls of the bubble owing to Stefan flow (see page 67). Both effects are very large near the boiling point.

Mention must finally be made of the observation that some dusts, like coal dust, are poorly retained when passed through water because their particles are not wetted (see also page 345). This is partly due to the low apparent density of flocculent dust aggregates and to the plate-like shape of the particles which orient themselves perpendicularly to the direction of inertial forces or gravity (see § 11) with the result that their mobility in this direction is reduced markedly. The fact that absorption of dust in bubbling can sometimes be increased by adding surface-active materials to the water is partly explained by a reduction in bubble size and partly by the increased collision efficiency of dust particles with the walls of the bubbles (see page 351). An appreciable reduction of absorption is observed when gelatin is added to the water [344] which is probably due to a decrease in circulation.

The good results described by some authors [345–347] for the absorption of aerosols by foams, despite the absence of inertial deposition, are explained firstly by the small size of the foam cells and secondly by the comparatively large times of residence of the aerosol particles in these cells.

The foregoing does not apply to the foam-apparatus introduced for gas cleaning by M. Pozin [772]. In this apparatus the gases are drawn at a large velocity (1–3 m/sec over the whole cross-section of the apparatus) through a grating with large openings under the water, no addition of foaming agents being made. A thin layer of very coarse, quickly breaking and regenerating foam is formed. The principal mechanism of aerosol collection here is undoubtedly the turbulent-inertial type. For this reason the apparatus is very effective with coarse aerosols, but not so good, for instance, against H_2SO_4 mist having $r = 0{\cdot}1–1\,\mu$.

§ 43. BROWNIAN ROTATION. ORIENTATION AND DEFORMATION OF AEROSOL PARTICLES IN AN ELECTRIC FIELD

Brownian rotation is described by the equation [253]

$$\overline{\theta^2} = 2\,kTB_\omega t, \tag{43.1}$$

where $\overline{\theta^2}$ is the mean square angle of rotation of a particle about a given axis in a time t and B_ω is the angular mobility of the particle about this axis or the rate of rotation under unit torque. In other words

$$d\theta/dt = B_\omega P_\theta \tag{43.2}$$

where P_θ is the torque due to the external forces. For spherical particles

$$B_{\omega_0} = 1/8\,\pi\eta r^3. \tag{43.3}$$

Rotation of spherical particles is of little interest in the physics of aerosols because it does not manifest itself in any way and has no effect on the properties of aerosols. The rotation of non-spherical particles, however, gives rise to twinkling when they are illuminated from the side, a feature which enables them to be distinguished unmistakably from spherical particles. Particles which are elongated in one or two dimensions are oriented under the action of hydrodynamic and electric forces to an extent which is determined by the relative magnitudes of the orienting force and the Brownian rotation. Brownian rotation also affects the rate of coagulation of elongated particles.

For a prolate ellipsoid of revolution having a minor semi-axis a and an axial ratio β the angular mobility about the minor axis is [349]

$$B_\omega = \frac{3\left[\dfrac{2\beta^2-1}{\sqrt{\beta^2-1}}\ln\left(\beta+\sqrt{\beta^2-1}\right)-\beta\right]}{16\,\pi\eta a^3(\beta^4-1)}. \tag{43.4}$$

Table 21 shows the ratio of the angular mobility of a sphere of radius r (B_{ω_0}) to that of an ellipsoid with a minor semi-axis r (B_ω).

TABLE 21. ANGULAR MOBILITY OF PROLATE ELLIPSOIDS OF REVOLUTION ABOUT THE MINOR AXIS

β	2	3	4	5	6
B_{ω_0}/B_ω	3·0	7·0	13·5	23·1	36·4

The angular mobility of particles in the form of flat disks with radius r about their diameters is

$$B_\omega = 3/32\,\eta r^3. \tag{43.5}$$

The total angle θ through which the long axis of elongated particle turns in a time t can be derived from (43.1) and is given by [350]

$$\overline{\sin^2\theta} = \frac{2}{3}\left(1 - e^{-6kTB_\omega t}\right). \tag{43.6}$$

When $t \to \infty$, $\overline{\sin^2\theta} = \frac{2}{3}$, corresponding to equal probability for any direction of the axis.

The orientation of elongated particles in an electric field, due to polarization of the particles, is of great importance in the physics of aerosols. If an uncharged particle, shaped like an ellipsoid of revolution, is placed in a uniform electric field of intensity E so that the polar axis makes an angle θ with the field direction, the change in field energy caused by polarization of the particle is [351]

$$\Omega = -\frac{vE^2}{2}\left[\frac{\cos^2\theta}{\dfrac{1}{\varepsilon_k - 1} + \varkappa_1} + \frac{\sin^2\theta}{\dfrac{1}{\varepsilon_k - 1} + \varkappa_2}\right], \tag{43.7}$$

where v is the volume and ε_k the dielectric constant of the particle. The shape factors $\varkappa_1$ and $\varkappa_2$ are expressed in terms of the axial ratio β of the ellipsoid. For prolate ellipsoids they are

$$\varkappa_1 = \frac{1}{\beta^2 - 1}\left[\frac{\beta}{\sqrt{\beta^2 - 1}}\ln\left(\beta + \sqrt{\beta^2 - 1}\right) - 1\right], \tag{43.8}$$

$$\varkappa_2 = \frac{\beta}{2(\beta^2 - 1)}\left[\beta - \frac{1}{\sqrt{\beta^2 - 1}}\ln\left(\beta + \sqrt{\beta^2 - 1}\right)\right]; \tag{43.9}$$

and for oblate ellipsoids

$$\varkappa_1 = \frac{\beta^2}{\beta^2 - 1}\left[1 - \frac{1}{\sqrt{\beta^2 - 1}}\arcsin\frac{\sqrt{\beta^2 - 1}}{\beta}\right], \tag{43.10}$$

$$\varkappa_2 = \frac{1}{2(\beta^2 - 1)}\left[\frac{\beta^2}{\sqrt{\beta^2 - 1}}\arcsin\frac{\sqrt{\beta^2 - 1}}{\beta} - 1\right]. \tag{43.11}$$

For conducting particles (43.7) becomes

$$\Omega = -\frac{vE^2}{2}\left[\frac{\cos^2\theta}{\varkappa_1} + \frac{\sin^2\theta}{\varkappa_2}\right]. \tag{43.12}$$

Since, in practice, aerosol particles can be considered conducting (see page 55) only this last formula will be used.

Table 22 shows values of $\varkappa_1$ and $\varkappa_2$ for a series of values of B.

TABLE 22. SHAPE FACTORS FOR PARTICLE ORIENTATION IN AN ELECTRIC FIELD

PROLATE ELLIPSOIDS

β	1	1·1	1·5	2	3	5	10	∞
$\varkappa_1$	0·333	0·310	0·233	0·174	0·109	0·056	0·020	0
$\varkappa_2$	0·333	0·345	0·383	0·413	0·446	0·472	0·490	0·5

OBLATE ELLIPSOIDS

$\varkappa_1$	0·333	0·347	0·446	0·527	0·635	0·751	0·860	1
$\varkappa_2$	0·333	0·320	0·277	0·236	0·182	0·125	0·070	0

Particles try to arrange themselves so that the field energy is a minimum, hence it follows from these data that the stable position for prolate ellipsoids corresponds to $\theta = 0$ (the polar, or long, axis pointing along the field) and for oblate ellipsoids $\theta = \pi/2$ (the polar, or short axis, pointing across the field).

Orientation of particles in an electric field is therefore similar to that occurring in laminar flow (see page 35). The orientation of particles moving at large Re relative to the medium is the other way round, the long axes being arranged across the direction of motion.

In the absence of Brownian motion an elongated particle coming into an electric field will oscillate about the stable position and approach it asymptotically. The orientation of particles undergoing Brownian motion will be considered only for prolate ellipsoids. The probability that the angle between the polar axis of a particle and the field direction lies in the interval $(\theta, \theta + d\theta)$ is given by Boltzmann's law as

$$W(\theta)\, d\theta = b e^{-\Omega/kT} \sin\theta\, d\theta = b' e^{\lambda^2 \cos^2\theta} \sin\theta\, d\theta. \tag{43.13}$$

where

$$\lambda^2 = \frac{E^2 v\left(\dfrac{1}{\varkappa_1} - \dfrac{1}{\varkappa_2}\right)}{2kT} = \frac{2\pi E^2\left(\dfrac{1}{\varkappa_1} - \dfrac{1}{\varkappa_2}\right) a^3\beta}{3\,kT} \tag{43.14}$$

is the ratio of electrical energy to the energy of Brownian rotation of the particle, a is the minor semi-axis of the ellipsoid and b' is a constant whose magnitude is specified by the normalization condition

$$1 = \int\limits_0^{\pi/2} W(\theta)\, d\theta = \frac{b'}{\lambda} \int\limits_0^{\lambda} e^{x^2}\, dx. \tag{43.15}$$

The mean value of $\cos\theta$ is therefore

$$\overline{\cos\theta} = \int\limits_0^{\pi/2} b' e^{\lambda^2\cos^2\theta} \sin\theta \cos\theta\, d\theta = \frac{e^{\lambda^2} - 1}{2\lambda \int\limits_0^{\lambda} e^{x^2} dx}. \tag{43.16}$$

Values of $\overline{\cos\theta}$ for some values of λ^2 are given in Table 23.

Note that $\overline{\cos\theta} = 0{\cdot}5$ means there is no orientation while $\overline{\cos\theta} = 1$ denotes complete orientation, which is attained, for practical purposes, when λ^2 exceeds 10. Complete orientation of an ellipsoidal particle having an axial ratio of 3:1 and a minor semi-axis $a = 0{\cdot}1\,\mu$ necessitates a fairly strong field of about 1000 V cm^{-1} while for $a = 1\,\mu$ a field of 30 V cm^{-1} is sufficient.

TABLE 23. ORIENTATION OF ELLIPSOIDAL PARTICLES IN AN ELECTRIC FIELD

λ^2	0·01	0·1	1	2	4	6	9	16
$\overline{\cos\theta}$	0·500	0·508	0·587	0·676	0·817	0·864	0·946	0·993

The discrepancy noted in § 19 in measurements of the mobility of smoke particles under gravity and in an electric field is due to orientation; the mobility in the second case, when the particles were oriented in the direction of their motion, must have been greater than in the first case where there was no orientation. Particles oriented in an electric field are dipoles, which leads to the so-called "directed" coagulation of aerosols (see § 52). The intensity of light scattered by aerosols in an electric field has been measured for ammonium chloride smokes [352, 353]. Since the direction of observation, the illuminating rays and the field are each at right angles to one another the observer sees light scattered by the long sides of the oriented particles, and the intensity of the scattered light increases when the field is imposed. In ammonium chloride crystals the elongated axes coincide with the crystallographic axes and therefore the oriented smoke also exhibits the phenomenon of double refraction. The orientation of aerosols in an electric field has been studied very little.

In the work of Bourot *et al.* [866] photographs were taken of scale-like aluminium particles of an aerosol flowing laminarly across a sound field. Due to the changing particle orientation the brilliancy of their images varied and the sinusoidal tracks on the photographs were discontinuous. When an electric field was maintained normal to the plane containing the illuminating beam and the axis of the camera, the sinusoids became continuous and quite uniform because the particles were oriented by the field to give images of maximum brilliance.

Since the relaxation time for polarization of particles is negligibly small in comparison with the relaxation time for their rotation under the action of external forces, all that has been said above remains valid for the orientation of particles in a variable electric field.

Ferromagnetic materials exhibit similar orientation in a magnetic field [354].

Droplets can also be oriented by strong electric fields which produce deformation by induction forces. On the assumption that deformed drops are shaped like prolate ellipsoids of revolution, which is borne out by experiment, it can be shown [355] that the axial ratio, c/a, of the ellipsoids depends on the value of rE^2/σ, where r is the radius of the undeformed drop and σ is the surface tension of the liquid, and on the dielectric constant of the liquid.

In place of the complex formulae for the deformation of drops in an electric field, the table below is given which shows the deformation of water and dioctylphthalate drops at 25°, as calculated by O'Konski and Thacher [355]. e is the eccentricity of a drop deformed to a prolate ellipsoid and c/a is the ratio of its axes; E is expressed in electrostatic units and r in cm.

Calculation for e between 0·45 and 0·70 shows that for water drops, for which ε_k can be taken as infinite, the complex expression for e as a function of rE^2, upon which Table 24 is based, can be approximated by the simple formula $e = 0{\cdot}003\, rE^2 + 0{\cdot}223$. The experiments of O'Konski and Gunther [773], with water drops having $r = 0{\cdot}60$–$0{\cdot}94$ mm suspended on thin glass filaments in a field of $E = 38$ electrostatic units, demonstrated that when e was between 0·49 and 0·63, the formula was correct within 1–2%, but at $e < 0{\cdot}46$ the agreement was rapidly deteriorating.

TABLE 24. THE DEFORMATION OF DROPS IN AN ELECTRIC FIELD

e	c/a	rE^2 (water)	rE^2 (dioctylphthalate)
0	1	0	0
0·01	—	—	0·043
0·02	—	0·16	0·174
0·05	1·001	1·00	1·088
0·10	1·005	4·00	4·36
0·20	1·02	15·9	17·45
0·30	1·048	35·2	39·4
0·40	1·09	61·1	70·4
0·5	1·15	92·3	110·9
0·7	1·4	159	224·6

CHAPTER VI

CONVECTIVE AND TURBULENT DIFFUSION OF AEROSOLS

§ 44. DEPOSITION OF AEROSOLS WHICH ARE IN CONVECTIVE FLOW

Various examples of the motion of aerosol particles in a gas which is at rest or in laminar flow with a known velocity distribution have already been discussed. Owing to convection, however, these are rarely encountered in practice, even within the limits of experimental accuracy. Experiment shows [356] that if a gas is in contact with a wall which is warmer by an amount ΔT, a vertical flow develops with a maximum flow velocity at a height z from the base of the wall given by

$$U = 0{\cdot}55 \sqrt{gz\alpha\Delta T}, \tag{44.1}$$

where α is the coefficient of thermal expansion of the gas, equal to $1/T$†. Thus with $\Delta T = 0{\cdot}01°$ in a chamber one metre high the convective flow velocity reaches 1 cm sec^{-1}, which is the rate at which coarse particles, about 10 μ in radius, settle under gravity. In smoke chambers, therefore, convection of the aerosol is practically unavoidable and can be eliminated only in small, thick-walled cells made of metal by carefully absorbing the heat rays from the light entering the cell. For the same reason true laminar flow of a gas can be realized only in narrow tubes and channels; in a wide tube vertical convection currents are invariably superimposed. Convection can be eliminated by cooling the base of a vessel containing an aerosol, but this method is applicable only occasionally.

If one wall of a vessel is kept warmer than the other walls, as happens when the glass windows of a smoke chamber are heated by the illuminating beam, a steady circulation may be set up inside and the velocity distribution, and hence the particle trajectories, is amenable to calculation. Such a calculation can also be performed for heat transfer by free convection around a heated object of simple shape. In practical work with aerosols, however, convection is usually disordered and cannot be calculated so that the prediction of individual particle trajectories is not possible; this applies, also, for disordered forced convection produced by stirring an aerosol.

In such cases the motion of the medium, and of the particles suspended in it, can only be treated statistically, which calls for considerable idealization of the phenomenon of convection. It is assumed that convective transfer of the medium is realized through "convective" diffusion, the intensity of convection being defined by a "convective diffusion coefficient". The motion of a suspended particle is then made up of the

† Convection is less in liquids than in gases because of the small thermal expansion.

motion of the medium itself (ordered motion plus convective diffusion) and the motion of the particle relative to the medium which has already been discussed.

Consider the settling of an aerosol in a chamber, with and without convection, assuming that coagulation of the aerosol may be neglected. In the absence of convection the upper boundary of a monodisperse aerosol having particles too large to be affected by diffusion settles with a constant velocity V_s, while the concentration below the boundary, which is the same throughout the aerosol, remains constant. If the aerosol is slightly polydisperse the upper boundary gradually spreads but still remains clearly discernible for quite a long time. Such a picture is actually observed in the settling of more or less monodisperse mists having high weight concentrations which stabilize the upper boundary hydrostatically (see page 50).

If $n(r)dr$ is the concentration of particles with radii between r and $r + dr$, the number of particles deposited per second on 1 cm² of the bottom of the chamber is $V_s(r)n(r)dr$ and the number of particles deposited in a time t is

$$dN = V_s(r)n(r)t\,dr. \tag{44.2}$$

This formula is applicable only for $t \leqslant H/V_s(r)$ (H is the height of the chamber) because all particles of the indicated size will settle in a time $H/V_s(r)$. The total number of particles settling in a time t is obtained by integrating expression (44.2) with respect to r.

Very intense convection, with the mean velocity of the convection currents much greater than V_s, causes the aerosol concentration to be practically constant throughout the chamber, except near the walls, while it decreases steadily with time. The vertical component of the convection velocity, and hence of the convective diffusion coefficient, tends to zero on approaching the bottom of the chamber. At some small distance δ from the bottom convective and molecular diffusion are comparable in magnitude; within the wall layer of thickness δ molecular diffusion predominates.

It will be assumed that deposition on the bottom of the chamber is quasi-stationary, so that the aerosol concentration n_∞ a long way from the bottom remains constant while a particle traverses the wall layer, a condition that usually holds. The number of particles passing downwards in 1 sec through a horizontal area of 1 cm² is

$$I = V_s n + D_E \frac{dn}{dz} \tag{44.3}$$

where D_E is the effective diffusion coefficient which includes both mechanisms of diffusion and depends on z. At a distance much greater than δ from the bottom $n = n_\infty$ and

$$I = V_s n_\infty. \tag{44.4}$$

Taking deposition to be quasi-stationary, this formula also gives the number of particles settling on 1 cm² of the bottom in 1 sec. Putting equation (44.3) in the form

$$V_s(n_\infty - n) = D_E \frac{dn}{dz} \tag{44.5}$$

and solving it for the boundary conditions $n = n_\infty$ at $z = \infty$ and $n = 0$ at $z = 0$, yields the following expression for the aerosol concentration at a distance z from the bottom:

$$n = n_\infty \left[1 - \exp\left(-V_s \int_0^z \frac{dz}{D_E}\right)\right]. \tag{44.6}$$

Thus the total flow of particles to the bottom of the chamber consists of the sedimenttation part

$$I_1 = V_s n = V_s n_\infty \left[1 - \exp\left(- V_s \int_0^z \frac{dz}{D_E}\right)\right] \tag{44.7}$$

and the diffusion part

$$I_2 = D_E \frac{dn}{dz} = V_s n_\infty \exp\left(- V_s \int_0^z \frac{dz}{D_E}\right). \tag{44.8}$$

It can easily be verified that the latter becomes appreciable only at a very small distance from the bottom. Diffusion therefore changes the particle-concentration distribution near the bottom while leaving the rate of deposition unaffected.

As stated above, the rate of deposition of particles with radii between r and $r + dr$ on the bottom of the chamber is $V_s(r)n(r)dr$, as in a stationary medium, but $n(r)$ varies with time according to

$$- H\, dn(r) = V_s(r)n(r)\, dt \tag{44.9}$$

and hence it follows that

$$n(r) = n_0(r) \exp\left(- \frac{V_s(r)t}{H}\right), \tag{44.10}$$

where n_0 is the initial concentration. The concentration of particles of a given size therefore decays exponentially with time at a rate depending on the particle size. The mean particle size of the aerosol left in the chamber therefore decreases steadily with time, the total particle concentration at a given instant t being

$$n = \int_0^\infty n_0(r) \exp\left(- \frac{V_s(r)t}{H}\right) dr. \tag{44.11}$$

The number of particles of a mono-disperse aerosol settling in a time t on 1 cm^2 of the bottom of the chamber is

$$N = \int_0^t V_s n\, dt = V_s n_0 \int_0^\infty \exp\left(- \frac{V_s t}{H}\right) dt = n_0 H\left[1 - \exp\left(- \frac{V_s t}{H}\right)\right]. \tag{44.12}$$

The intermediate case for which the rates of settling and convection are comparable is more complicated and we shall not concern ourselves with it. It should be noted that for aerosols settling in smoke chambers convection has a very pronounced effect even in the absence of stirring, and the settling usually approximates to the second of the types discussed above. In the experiments of Gillespie and Langstroth [357], with ammonium chloride smoke having particles of radius from 0·3 to 2·0 μ, the concentration was the same at all points in a chamber 12 m^3 in volume while diminishing continually throughout the lifetime of the smoke. In Vigdorchik's experiments [358] with quartz dust in a chamber 1·2 m high, particles of radius 12·5 μ could be detected three hours after the dust had been injected into the chamber, whereas in still air they should have settled in 12 min.

It follows from formula (44.10) that

$$\ln n(r) = \ln n_0(r) - \frac{V_s(r)t}{H}. \tag{44.13}$$

Hence, for a monodisperse aerosol, the logarithm of the concentration decreases linearly with time at a rate which increases with particle size. Nearly all experiments on the kinetics of settling in chambers relate to polydisperse systems. The mean particle size of such an aerosol, and hence the slope of the (ln *n*, *t*) curve, decreases with time, so that curves for polydisperse aerosols are convex towards the axis of abscissae, their curvature increasing with the range of sizes present. Curves of this shape are usually observed in practice [358].

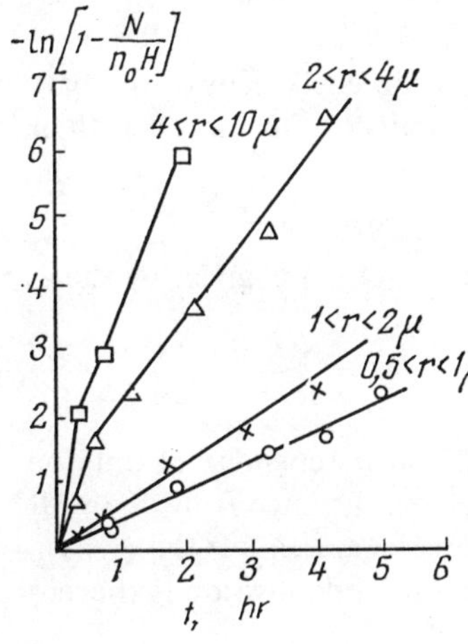

FIG. 65. Deposition of particles on the bottom of a chamber.

It can be seen from formula (44.12) that

$$-\frac{H}{V_s} \ln\left(1 - \frac{N}{n_0 H}\right) = t \tag{44.14}$$

for particles deposited under convective conditions. This formula was checked by Davies [218] on strongly aggregated coal dust and his curves for different ranges of particle radii in the individual dust fractions are shown in Fig. 65. For the fine fractions the linear form of the relationship (44.14) is obeyed, while the curvature observed for the coarse fractions is attributed by the author to the polydispersity within those fractions.

Deposition of aerosols on the vertical side walls of chambers has been studied much less. Whatever the mechanism may be the rate of deposition is proportional to the aerosol concentration n, so that $dn/dt = -\beta n$. Deposition on a vertical wall is thus expressed by

$$\ln(n_0/n) = \beta t, \tag{44.15}$$

which is similar to (44.13), and the problem reduces to deriving a theoretical expression for the coefficient β.

Inertial deposition of aerosols on the flat walls of a chamber must be very small when the cloud is not stirred and, in fact, it has been found that under these conditions coarse particles ($r \approx 7\,\mu$) are not deposited on vertical walls at all [248]. Deposition here must be purely diffusive with particles carried to the wall by convective diffusion and deposited upon it by molecular diffusion through the thin wall layer (see page 266). The way in which the convective diffusion coefficient behaves on approaching the wall is unknown, so it is necessary to simplify the problem drastically and assume that only molecular diffusion is effective within the wall layer of thickness δ while outside this layer convective diffusion is so intense that the aerosol concentration is constant.

The rate of deposition on a vertical wall is then given by

$$I = \frac{Dn}{\delta} \text{ particles cm}^2 \cdot \text{sec} \tag{44.16}$$

where D is the molecular diffusion coefficient of the particles. If v is the volume and S the surface area of the side walls of the chamber, then

$$-v\,dn = SI\,dt \tag{44.17}$$

and the coefficient β in formula (44.15) is given by the expression

$$\beta = \frac{SI}{vn} = \frac{SD}{v\delta}. \tag{44.18}$$

It must be borne in mind that δ depends not only on the intensity of convection but also on D; it is defined as the distance from the wall at which the coefficients of molecular and convective diffusion are equal: the greater the coefficient of molecular diffusion the greater is δ. The rate of deposition is therefore not directly proportional to D but to D^α, where $0 < \alpha < 1$.

It can be concluded from what has been said above that the smaller the particle size of an aerosol the greater is the proportion of the deposited particles to be found on the side walls.

By measuring the transmission of light through a smoke Shifrin and his co-workers [359] studied the kinetics of its deposition on the walls of a cylindrical chamber 9 m high and 4 m in diameter. The particle radius was about $0{\cdot}5\,\mu$. They found equation (44.15) to be applicable, the concentration being halved in two or three hours. The initial concentration of the smoke was only about 10^4 particles/cm^3 so that the reduction in concentration due to coagulation could be neglected; deposition on the bottom of the chamber was also negligible. It can be calculated from the data reported by Shifrin that the ratio of the number I of particles deposited in 1 sec on 1 cm^2 of the walls to the particle concentration was 5–7×10^{-3} cm sec^{-1}, which makes the thickness of the wall layer $\delta \approx 0{\cdot}5\,\mu$.

In the experiments of Gillespie and Langstroth [357], conducted in a chamber of volume 12 m^3 with ammonium chloride smoke having particle radii in the range $0{\cdot}3$ to $2{\cdot}0\,\mu$, the numbers of particles settling on the bottom, sides and top of the chamber were determined by direct counting on suitably placed slides. Deposition on the top was negligible, while in the first 100 min only a third as many particles were deposited on the sides as on the bottom, after which time deposition on the sides practically ceased. From the graph presented by the authors it can be roughly estimated that I/n in these experiments was about 10^{-4} cm/sec so that the thickness of the wall layer was about $20\,\mu$ or almost two orders greater than Shifrin's value. Since the rate of decrease of concentration observed by Shifrin in so large a chamber and in the absence of coagulation is in clear contradiction to many investigations on aerosols, while direct measurements of the deposition of aerosols on walls are more reliable than figures calculated from the fall in concentration, more confidence can clearly be placed upon the values of I/n and δ calculated by Gillespie and Langstroth until new experimental results are obtained.

Charging has a large effect on the rate of deposition of aerosol particles on walls. In some recent experiments by Gillespie [360], conducted with silica particles $0{\cdot}4\,\mu$ in radius, the total number of particles deposited in the first few minutes of the aerosol's lifetime on the walls and bottom of a chamber $0{\cdot}2$ m^3 in volume increased two to three times when the average charge per particle was increased from 7–14 elementary charges (the maximum charge changed from 50 to 100 electrons, approximately). Taking into account that the charge could have had no appreciable effect on the rate of deposition on the bottom of the chamber, it is evident that the deposition on the sides was in fact even greater. The effect of the electric charges gradually decreased and after a few tens of minutes, disappeared. Since the charging was bipolar and fairly symmetrical, and electrostatic dispersion (see § 24) was not appreciable, the effect is undoubtedly due to attraction of the charged particles to the walls by induction.

The magnitude of this effect can be calculated very roughly by assuming that a particle which has reached the wall layer is attracted towards the wall by the induced charge. According to formula (41.1) the resulting velocity at the edge of the wall layer is $q^2B/4\,\delta^2$ so that the number of particles deposited per second on 1 cm^2 of the wall is $I_q = q^2Bn_0/4\,\delta^2$. The number deposited by molecular diffusion is given by formula (44.16).

The ratio of these rates of deposition is

$$\frac{I_q}{I} = \frac{q^2Bn_0}{4\delta^2}\,\frac{\delta}{Dn_0} = \frac{q^2}{4\delta kT}\,. \tag{44.19}$$

Putting $q = \varepsilon\nu = 4{\cdot}8 \times 10^{-10}\,\nu$, ν being the number of elementary charges carried, $kT = 4 \times 10^{-14}$ and $\delta = 20 \times 10^{-4}$ cm gives

$$I_q/I = 0{\cdot}7 \times 10^{-3}\,\nu^2. \tag{44.20}$$

For particles bearing the maximum number ($\nu = 100$) of charges $I_q/I = 7$; these particles are thus deposited much more quickly than uncharged ones but the induction effect for particles carrying a small charge is negligible. After the more highly charged particles have been deposited by electrical attraction further deposition occurs by molecular diffusion.

It has been established in a series of investigations that the weight concentration of an aerosol in a chamber falls off more rapidly when it is kept stirred. This is partly due to more rapid growth of the particles as a result of the accelerated coagulation caused by the stirring, but the main cause is undoubtedly inertial deposition on the chamber walls and on the fans. This was proved directly by Gillespie and Langstroth [357] who measured the rate of deposition on the chamber walls and used the method described on page 332 to calculate the coagulation constant of ammonium chloride smoke at various rates of stirring. In a chamber of volume 1 m^3 I/n was found to increase linearly with the air flow velocity, being about five times greater at 50 m/min than in the absence of stirring; no significant effect of stirring on the value of the coagulation constant could be detected. According to Vigdorchik's experiments [358] on the stirring of air with a fan in a cubic chamber of side 1·2 m, the concentration of quartz dust drops three to four times more quickly at a mean velocity of 4 m $\sec^{-1}$ than in the absence of stirring. This is due to the decrease in thickness of the laminar wall layer and to inertial deposition of particles. Assuming that the chamber in Vigdorchik's experiments was cylindrical, the air velocity causes the centrifugal force at the walls to be $mV^2/R = m\,400^2/60$ or almost three times the force of gravity, mg.

In fact, inertial deposition due to stirring presents a very complicated picture since it occurs locally where eddies which approach the wall have their axes parallel to it. For an eddy of small diameter the inertial forces can be appreciably greater than that just calculated. The nature of the stirring is also very important. If it is produced by wide fan blades which almost reach the walls of the chamber [361] the rate of decrease in aerosol concentration (ammonium chloride smoke with $\bar{r} \approx 0{\cdot}5\,\mu$) is enhanced three or four times by a mean air velocity of 50 cm $\sec^{-1}$. This type of stirring undoubtedly encourages the formation near the walls of strong eddies with axes parallel to the walls.

It is extremely difficult to calculate the rate of inertial deposition on the walls due to stirring, but the number of particles deposited per second per square centimetre of wall area must in any case be proportional to $n\tau$, while the fall in aerosol concentration with time must obey the equation

$$v\frac{dn(r)}{dt} = -Sbn(r)\,\tau(r),$$

where b depends on the intensity and nature of the stirring. Hence it follows that

$$n(r) = n_0(r)\exp\left[-\frac{Sb}{v}\,\tau(r)\,t\right].$$

Since $V_s = g\tau$ the net variation of aerosol concentration with time (if diffusion is neglected) is given by the equation

$$n(r) = n_0(r)\exp\left[-\left(\frac{Sb}{v} + \frac{g}{h}\right)\tau(r)\,t\right], \tag{44.21}$$

which is like the equation derived in the absence of inertial deposition except that the factor g/h becomes $(Sb/v) + (g/h)$. Equation (44.11) is replaced by

$$n = \int_0^\infty n_0(r)\exp\left[-\left(\frac{Sb}{v} + \frac{g}{h}\right)\tau(r)\,t\right]dr. \tag{44.22}$$

As already stated a polydisperse aerosol gives a curve of ln n against time which is concave upwards. If, however, appreciable coagulation proceeds during deposition, the consequent growth in size of the particles tends to compensate for the decrease in mean particle size due to deposition so that the curvature is less marked. Graphs are frequently obtained [361] which are very nearly straight lines.

The weight concentration of a polydisperse aerosol varies with time as follows

$$c = \int_0^\infty n_0(r)\, m(r) \exp\left[-\left(\frac{Sb}{v} + \frac{g}{h} \right) \tau(r) t \right] dr, \tag{44.23}$$

where $m(r)$ is the mass of a particle of radius r. For an isodisperse aerosol $d \ln c/dt = d \ln n/dt$ but in polydisperse aerosols $|d \ln c/dt| > |d \ln n/dt|$, and the weight concentration falls faster than the number concentration. This is seen by writing

$$\frac{d \ln c}{dt} = \frac{1}{c} \frac{dc}{dt} = \frac{1}{n\bar{m}} \frac{\bar{\bar{m}}\, dn}{dt}, \tag{44.24}$$

where $\bar{m}$ is the mean mass of the particles contained in the aerosol and $\bar{\bar{m}}$ is the mean mass of the particles being deposited. Since $\bar{\bar{m}} > \bar{m}$ then $|d \ln c/dt| > |d \ln n/dt|$ and the difference $|d \ln c/dt| - |d \ln n/dt|$ increases with the polydispersity of the aerosol as well as with the absolute value of dn/dt, that is with the rate of stirring. Experiment confirms these conclusions [361].

Finally, a phenomenon caused by natural convection will be mentioned which is of interest in connection with the treatment of buildings with aerosols. If a closed vessel with holes in opposite vertical walls is placed in a chamber containing an aerosol some quite coarse particles ($r \approx 7\,\mu$) will pass through these openings as a result of horizontal convection or draught [248]. If there is only one opening there is no penetration. A similar phenomenon is observed at cracks in wooden walls in which an aerosol can penetrate fairly deeply if they go right through, although it will not enter a blind crack.

§ 45. MOTION OF AEROSOL PARTICLES IN TURBULENT FLOW

Theoretical and experimental study of the behaviour of aerosols is naturally much more difficult for turbulent than for laminar flow and, despite recent advances in the study of turbulence, little is known about the motion of particles suspended in a turbulent fluid. In particular, the extremely important problem of the extent to which particles follow the turbulent fluctuations has not been fully elucidated.

Turbulent flow is pictured as an average flow velocity with a continuous size spectrum of fluctuations superimposed. Fluctuations arising primarily from the breakaway of eddies have a scale comparable to the diameter of the pipe in which the fluid is flowing. Velocities due to these fluctuations depend on direction, being greater parallel to the walls than perpendicular to them.

The energy of large-scale eddies is gradually transferred to fluctuations on smaller and smaller scales, which become isotropic when they are small compared with the diameter of the pipe. On Kolmogoroff's theory [362] the total energy of fluctuations with the scale $\leqslant \lambda$ is proportional to $\lambda^{2/3}$. This dependence is valid as long as the transfer of energy to smaller-scale fluctuations is not accompanied by appreciable dissipation (conversion to heat) of the energy, the sizes must therefore be large compared with a certain critical value λ_0, the internal scale of the turbulence. At $\lambda < \lambda_0$ the decrease in fluctuation energy with diminishing λ proceeds much faster.

The spectrum of velocity fluctuations can be measured experimentally only at a fixed point which gives the function $F(\nu)$ showing what part of the turbulent energy belongs to fluctuations of frequency $\geqslant \nu$ sec^{-1}, as recorded by a stationary observer (this is termed the Eulerian frequency). Then, since $\lambda = U/\nu$ where U is the mean flow velocity, a "length scale spectrum" of fluctuations is obtained. This is expressed by the function $\Omega(\lambda)$ which shows what part of the turbulent energy must be attributed to fluctuations on a scale $\leqslant \lambda$. However, the spectrum of Lagrangian frequencies, referred to coordinates moving with velocity U, is still unknown.

Simmons and Salter [363] measured the spectrum of fluctuations in a 120 cm wind tunnel behind a screen having 75 mm openings. The degree of turbulence, u/U where u is the overall root mean square eddy velocity, was 0·03 and almost independent of Re_f. The turbulent energy was thus about 0·0009 of the total energy of the flow. Table 25 is based on results at $U = 7{\cdot}5$ m sec^{-1}; u_λ is the velocity corresponding to fluctuations on a scale $\leqslant \lambda$.

TABLE 25. SPECTRUM OF TURBULENCE IN A WIND TUNNEL

$U = 7{\cdot}5$ m sec^{-1}, $\mathrm{Re}_f = 600{,}000$

λ (cm)	$\Omega(\lambda)$	$u_\lambda/U \times 10^2$	u_λ (cm sec^{-1})	$u_\lambda/\lambda^{1/3}$	λ/u_λ
37	0·74	2·6	19·5	5·9	2·0
19	0·58	2·3	17·3	6·5	1·1
7·5	0·36	1·8	13·5	6·9	0·55
3·7	0·19	1·3	9·8	6·3	0·37
3·0	0·13	1·1	8·3	5·8	0·36
2·0	0·07	0·8	6·0	4·8	0·33
1·5	0·05	0·67	5·0	4·4	0·30
1·0	0·019	0·41	3·1	3·1	0·32
0·75	0·008	0·26	1·9	1·7	0·40
0·50	0·0025	0·14	1·1	0·9	0·45

It can be seen from Table 25 that the ratio $u_\lambda/\lambda^{1/3}$ which should be constant according to the relationship found by Kolmogoroff, decreases sharply when λ is below 2 cm. A similar result was observed when $U = 10{\cdot}5$ m sec^{-1}. The internal scale, λ_0, of the fluctuations in a wind tunnel with Re_f of the order 10^6 is therefore of the order 1 cm. Since substantial experimental errors are possible in the study of turbulence spectra it is desirable to check this result. From the formulae and experimental data presented by Obukhov and Yaglom [364] the following values can be found for λ_0: 1·1 cm when $U = 12{\cdot}2$ m sec^{-1}; 0·7 cm when $U = 24{\cdot}4$ m sec^{-1}; 0·6 cm when $U = 30{\cdot}5$ m sec^{-1}; these values are of the same order as above. Obukhov [365] gives $\lambda_0 \approx 0{\cdot}5$ cm at a height of 1·15 m in the atmosphere, while Taylor [366] gives $\lambda_0 = 2$ cm and 13 cm at heights of 2 m and 30 m respectively. Below λ_0 $\Omega(\lambda)$ falls off very rapidly with decreasing scale: it is proportional to λ^2 according to Obukhov and Yaglom, or λ^6 according to Heisenberg [367].

As already mentioned, the distribution of turbulent energy as a function of Lagrangian frequency or fluctuation period, which must be known to determine the extent to which particles are entrained by turbulent fluctuations, cannot be found directly from experimental data; it can be estimated roughly as follows. Suppose that fluctuations on a scale λ are generated by vortex streets moving with the flow at a velocity U, their axes being perpendicular to the flow. The mean eddy velocity u_λ will be taken as the circulation velocity at a distance $\lambda/4$ from the axis of the vortex. The Lagrangian period, t_L, of the corresponding fluctuations is then $0{\cdot}5\,\pi\,\lambda/u_\lambda$ and the Eulerian period, t_E, is $2\,\lambda/U$. The ratio $t_E/t_L \approx u/U$, and to a first approximation is thus equal to the degree of turbulence. According to Fig. 20 (page 82) complete (nominally 99 per cent) entrainment of particles by the fluctuations occurs if $\tau/t_L \leqslant 0{\cdot}02$; Table 13 shows that this means $\tau \leqslant 0{\cdot}01$ or $r \leqslant 30\,\mu$, for particles of unit density.

Using values of τ calculated in § 18 for particles beyond the Stokes region for a typical mean eddy velocity of 30 cm sec^{-1}, namely $\tau = 0{\cdot}1$ sec for $r = 0{\cdot}1$ mm and $\tau = 6{\cdot}3$ sec for $r = 1$ mm and unit density, shows that there is 70 per cent entrainment in the first case and 2 per cent in the second. Particles about 1 mm in size therefore take practically no part in fluctuations of the medium.

MacCready [368] investigated the spectrum of atmospheric turbulence 70 cm above the earth in mean wind of 2·3 m sec^{-1}. The root mean square eddy velocity in the vertical direction was found to be 27 cm sec^{-1}. It can be calculated from the author's data that the vertical eddy velocity on a scale $\leqslant \lambda$ is roughly three times less than in a wind tunnel under the conditions appropriate to Table 25. This means that the Lagrangian period is three times greater and consequently the degree of entrainment is greater still.

These calculations are, of course, very rough but they give the right orders of magnitude for the size of particles entrained and the size which is unaffected by the turbulent fluctuations. Exact determination of the degree of entrainment is possible either by studying the turbulence spectrum with an instrument moving at the velocity of the stream or by ultramicroscope observations in a flow containing both very fine and very coarse particles.

The topic under discussion is important for the solution of another fundamental problem in the mechanics of aerosols, the vertical distribution of particles in a turbulent stream, which was first investigated by Schmidt [369]. If an aerosol is swept in turbulent flow through a horizontal duct and deposited particles are being redispersed, a stationary distribution of particles with height becomes established. The number of particles passing downwards under the action of gravity through a horizontal area of 1 cm^2 in 1 sec is $V_s n$. The number of particles passing through this area in the reverse direction owing to eddy diffusion is $-D_t\,dn/dz$ where D_t is the eddy diffusion coefficient of the particles. Equating these expressions, we get

$$\frac{dn}{n} = -\frac{V_s}{D_t}\,dz, \tag{45.1}$$

whence

$$\ln\frac{n}{n_0} = -V_s\int_0^z \frac{dz}{D_t}, \tag{45.2}$$

where n_0 is the concentration at the bottom of the duct and z is the distance from the bottom. Since D_t changes rapidly with z near the bottom but comparatively slowly further away it can be considered constant in the central region of the duct, and (45.2) then takes the form

$$\ln\frac{n}{n_0} = -\frac{V_s z}{D_t}. \tag{45.3}$$

It should be stressed that the distributions represented by these formulae hold only if the deposited particles are returned by the flow to a state of suspension. If this is not the case it can easily be shown that most of the aerosol is deposited before the distribution is established.

The theory presented here for the vertical distribution of particles suspended in a turbulent flow has been criticized by certain hydrologists [370] who object to the assumption that the eddy diffusion coefficients of the particles and the medium are equal. This is dealt with below.

Fairly detailed experimental study of the vertical distribution of particles in turbulent flow has been carried out only in aqueous suspensions. In Kalinske's work [374] D_t was determined experimentally by the spreading of a stream of aqueous solution in the flow. The vertical distribution found for sand grains with radii up to 70 μ agreed more or less satisfactorily with formula (45.2).

Kalinushkin [375] studied the distribution of sawdust and other powders in a horizontal circular tube 25 cm in diameter at U = 10–17 m sec^{-1}. Turbulent flow in circular tubes is usually accompanied by rotation about the axis with the result that a vertical distribution tends to become radial and the particle concentration is maximal round the periphery of a horizontal tube and falls off towards the axis; the rotation must be eliminated by means of a screen to obtain the normal vertical concentration gradient. Kalinushkin's data are represented well by a straight line corresponding to formula (45.3).

Sherwood and Woertz [376] measured D_t in air flowing through a rectangular duct of height 5·3 cm having a large ratio of width to height. Except in the narrow zones adjoining the upper and lower walls, D_t was found to be practically constant at $\mathrm{Re}_f > 10{,}000$ and is given by the empirical formula

$$D_t = 0{\cdot}044\, \nu\, \mathrm{Re}_f^{0\cdot15} \tag{45.4}$$

in which ν is the kinematic viscosity of the gas. For particles of unit density and radius 5 μ the value V_s/D_t is equal to 0·016 when U = 13 m sec^{-1} (Re_f = 40,000) and the ratio of the concentrations at the upper and lower walls is 1·1, the distribution thus being almost uniform throughout the duct. However, at r = 20 μ the ratio is 3·7, showing that the bulk of the particles moves in the lower part of the duct.

Besides the effect of turbulent flow on particles suspended in it, the reverse problem of the influence that the disperse phase has on the turbulent flow carrying it is also of great interest. It is known from experiment that the critical value of Re_f is greater in clay suspensions than in pure water [377, 378]. In Vanoni's experiments [373] with sand suspensions in an open channel a marked reduction in the eddy viscosity coefficient, and hence in the hydraulic resistance, in comparison with pure water was noticed; the degree of turbulence of the flow and the eddy diffusion coefficient were also diminished.

The reason for this is that each particle of the disperse phase takes part, as we have seen, in the fluctuations of the medium and also moves continuously under the influence of gravity relative to the layer of medium adjacent to it. Additional mechanical energy due to the fluctuations is thereby dissipated as heat and the degree of turbulence must decrease.

Since the amount of energy dissipated by a particle in unit time is proportional to the product of the weight of the particle and its rate of settling, or to the mass m of the particle raised to the power 5/3, then at a constant weight concentration, $c = nm$, the decrease in turbulence must increase with particle size. The reduction in resistance of a cyclone when the dust concentration is increased, mentioned in § 29, is explained by the decrease in turbulence caused by the suspended dust, as has been suggested on several occasions [379].†

The theory of this effect has been worked out by Barenblatt [381]; its magnitude is determined by a dimensionless group which for aerosols has the form

$$K = -\frac{g\,dc/dz}{\gamma_g (dU/dz)^2} \; (> 0), \tag{45.5}$$

where z is distance from the bottom and U is the mean flow velocity.

When $K \ll 1$ the disperse phase has no influence on the degree of turbulence or on the velocity profile and formula (45.2) is applicable only under these circumstances. If K is comparable with unity the root mean square eddy velocity is $u = u_0(1 - K)^{\frac{1}{2}}$ where u_0 is the value of u for a pure gas. Formula (45.2) is not applicable in this case and calculation of the concentration distribution is more complicated.

Aerosols have been successfully used for recent investigations of turbulent diffusion, especially in the atmosphere. The motion of the particles and of the adjacent volume element of air are usually assumed to coincide, which is true only if the particles are small enough. In this case, turbulent diffusion theory applies to aerosol diffusion and the variables considered below can refer either to volume elements of air or to particles.

For simplicity, only stationary homogeneous turbulent flow will be considered in which the average flow rate and intensity of turbulence are independent of time and position, and a system of coordinates is chosen which moves at the average flow rate. Let U' be the instantaneous velocity and X the displacement of the particle, or of a volume element of air along one axis. With the moving system of coordinates $\overline{U'} = 0$ and $\overline{X} = 0$. In the theory of turbulent diffusion the principal role is played by the Lagrangian coefficient of correlation with respect to time

$$R(t) = \frac{\overline{U'(t)\, U'(t + \alpha)}}{\overline{U'^2(t)}}, \tag{45.6}$$

which gives a relationship between the fluctuation velocities of a particle at times t and $t + \alpha$, and the Lagrangian scale of turbulence

$$L = \int_0^\infty R(\alpha)\, d\alpha \tag{45.7}$$

which defines the interval of time which must elapse before the motion of the particle becomes independent of its initial movement.

Theoretical considerations and experimental data suggest that the probability of a particle transfer $w(X_0, X, t)$ (see § 37) which in the case of Brownian diffusion is

† In a note published in 1951 [380] the author of this book attributed the reduction in turbulence to the dissipation of energy caused by relative motion of the particles and medium. As can be seen from what has been written above, this effect must be very small.

given by the error function, is similarly expressed for turbulent diffusion. On account of the homogeneity of the flow, X_0 can be taken as zero so that

$$w(X, t) = \frac{1}{(2\pi \overline{X^2})^{\frac{1}{2}}} \exp\left(-\frac{X^2}{2\,\overline{X^2}}\right). \tag{45.8}$$

Comparison with (37.29) indicates that the coefficient of the turbulent diffusion must be

$$D_t = \frac{\overline{X^2}}{2t}. \tag{45.9}$$

Differentiating $\overline{X^2}$ with respect to t, gives

$$\frac{1}{2}\frac{d\overline{X^2}}{dt} = \overline{X(t)\,U'(t)} = \int_0^t \overline{U'(t)\,U'(t')}\,dt' = \overline{U'^2}\int_0^t R(\alpha)\,d\alpha \tag{45.10}$$

from which comes the principal formula of the theory of turbulent diffusion

$$\overline{X^2} = 2\,\overline{U'^2}\int_0^t (t-\alpha)\,R(\alpha)\,d\alpha. \tag{45.11}$$

For $t \ll L$, $R(\alpha) \approx 1$,

$$\overline{X^2} = \overline{U'^2}\,t^2 \tag{45.12}$$

and for $t \gg L$,

$$\overline{X^2} = 2\,\overline{U'^2}\,tL - 2\,\overline{U'^2}\int_0^\infty \alpha R(\alpha)\,d\alpha \approx 2\,\overline{U'^2}\,tL, \tag{45.13}$$

since the second term is of the order $\overline{U'^2}L^2$ and can be neglected. A complete analogy thus exists with Brownian diffusion [see (35.6) and (35.7)]. The concept of a coefficient of turbulent diffusion has a definite physical meaning only for $t \gg L$; the quantity L in the turbulent diffusion is similar to τ in Brownian diffusion. Besides (45.9), the following expressions for D_t can be deduced

$$D_t = \overline{U'^2}L = \frac{1}{2}\frac{d\overline{X^2}}{dt} = \overline{X(t)\,U'(t)} \quad (t \gg L). \tag{45.14}$$

In this case $(\overline{U'^2})^{\frac{1}{2}}\,L$ plays the role of the apparent mean free path of the particle (see page 182).

The theory of the relative diffusion of a system of particles is much less advanced. Batchelor [774] argues that when the distance between particles, ϱ, is greater than the internal scale of turbulence λ_0, but small compared with the scale of the largest eddies, then for high Re_f numbers

$$\frac{d\overline{\varrho^2}}{dt} \approx (\varepsilon\varrho_0)^{2/3}\,t, \quad \overline{\varrho^2} - \varrho_0^2 \approx \varepsilon^{2/3}\varrho_0^{2/3}t^2 \tag{45.15}$$

provided that t is small enough for the relative motion of particles to depend on the initial distance between them, ϱ_0. When t is large,

$$\frac{d\overline{\varrho^2}}{dt} \approx \varepsilon t^2 \approx \varepsilon^{1/3}\left(\overline{\varrho^2}\right)^{2/3} \tag{45.16}$$

where ε is the energy dissipation rate per unit mass of the medium. Equation (45.16) corresponds to (47.3) (see p. 275) since $^1/_2\, d\overline{\varrho^2}/dt$ can be regarded as the coefficient of the relative diffusion of particles. The rate of relative diffusion thus increases with the distance between the particles, ϱ. However, when ϱ exceeds the scale of the largest eddies, the motions of the particles become independent and the coefficient of relative diffusion no longer depends on ϱ (see page 275).

A fundamental problem in turbulent diffusion of particles which are so large that their motion cannot be considered identical to that of the medium was solved in Tchen's monograph [651]. From the general differential equation of motion for Stokes particles (17.23) Tchen proved by means of rather complicated calculations that the coefficient of turbulent diffusion of the particles is equal to that of the medium.

The physical meaning of Tchen's theorem is clear. Particles possess a smaller fluctuation velocity than the medium, but their velocities are more persistent; their time scale of turbulence and, consequently, the diffusion "steps" are larger than those of the elements of fluid. Soo [775] and Liu [776] came later to a similar conclusion, but in a less general form.

Tchen's theorem was confirmed recently by the experiments of Soo *et al.* [777]: glass spheres with $r = 57\,\mu$ and $115\,\mu$ were introduced into a wind tunnel and photographed at definite intervals of time. The positions of the spheres in the system of coordinates moving at the mean flow rate were calculated from the pictures and the values $\overline{V'^2}$, L and D_t deduced for the particles in longitudinal and transverse directions. Unfortunately, the corresponding values of the air were not measured. However, in agreement with theory, the values of D_t for the two sizes of sphere turned out to be very close. At $\overline{U} = 25$ m/sec, for $r = 57$ and $115\,\mu$, D_t in the longitudinal direction was equal to 2·54 and 2·60 cm²/sec, respectively, and, in the transverse direction, to 0·30 and 0·33 cm²/sec. At the same time $\overline{V'^2}$ for larger particles was less by 20 per cent than for smaller ones, thus the degree of entrainment by turbulent pulsations decreased with the rise of r.

The motion of aerosol particles in a free turbulent jet is of great interest, but little data are yet available. Kubynin [778] investigated the concentration distribution of a highly polydisperse ($r = 10$–$300\,\mu$) coal dust in a jet issuing from a tube 5 cm in diameter at the rate of $U_0 = 22$ and 38 m/sec. Velocity profiles of the jet were measured simultaneously and showed no appreciable change when the dust concentration was increased from 0 to 1·15 *g*/*g* of air.† The distribution of air velocity and dust concentrations in the jet proved to be similar. This can be readily understood from the closeness of the coefficients of turbulent diffusion D_t and turbulent viscosity ν_t. However, near the axis of the jet the proportion of coarse particles was greater. Unfortunately, no data for separate fractions were obtained.

Tchernov [780], working with fractionated dust of corundum, quartz and boric acid having $\bar{r} = 30$–$200\,\mu$ at $U_0 = 17$–37 m/sec and with a nozzle 2·5 cm in diameter, confirmed the independence of the velocity profile of the dust content. By making motion pictures of moving dust particles, he showed that their initial velocity V_0 on coming out of the nozzle hardly depends on r at all; at $U_0 = 28{\cdot}5$ m/sec $V_0 = 20$ to 22 m/sec (regarding this, see p. 346). The particles were rotating rapidly, at some

† Hoffman's observations [779] show that when the flow rate is measured with a Pitot tube, the dust content (quartz dust) of the air, up to 1 kg/m³, has no effect on the results, provided that no dust deposit is formed on the walls of the tube.

1,000 rev/sec, owing to their having struck the walls of the nozzle. At first the velocity of the particles V increased with the distance from the nozzle x, while the air velocity decreased; after some distance the velocities become equal and beyond this, owing to their inertia, the particles moved faster than the air and the ratio V/U increased more, with larger particles. Hence, it follows that the reported independence of the velocity profile on the dust content is not quite accurate; dust must lower the velocity of the air near the nozzle and increase it at larger values of x.

The vertical distribution of dust in horizontal turbulent flow was investigated by Dawes and Slack [781] who blew through a pipe of a square cross-section, 30 × 30 cm, fractions of coal dust with $r = 2$–$20\,\mu$ at a rate of 0·5–2 m/sec. Concentrations were measured at different levels above the bottom of the tube. For fractions with $r < 8\,\mu$ the concentration was practically constant over the whole cross section. The data for fractions with $r = 8$–11, 11–16, and 16–22 μ given by the authors show that in the second fractions, for instance, the concentration at the bottom of the pipe was 2–3 times as large as at the top. The coefficient of turbulent diffusion at $U = 1$ m/sec, calculated by the author of this book by means of the exponential formula (45.3) and allowing for the irregular shape of coal particles, varied for the fractions in question from 50 to 90 cm²/sec, whereas, according to the empirical formula (45.4), it should be 11 cm²/sec. Calculating from (46.15) and (46.29), assuming z to be half the radius of the tube, gives $D_t = 0{\cdot}035\,\mathrm{Re}_f^{7/8}\,\nu = 32$. Further measurements of this kind appear to be necessary.

§ 46. DEPOSITION OF AEROSOLS IN TURBULENT FLOW

Almost all that is written in § 44 about gravitational deposition in convection and artificial stirring is also true for turbulent flow. The root mean square eddy velocity transverse to flow is approximately 0·03 to 0·1 $\overline{U}$, $\overline{U}$ being the mean flow velocity [363, 382, 376, 383]. The rate of settling of particles with radii less than 10 μ in a horizontal tube, when $\overline{U}$ is several metres per second, is thus considerably less than the vertical component of the mean eddy velocity and such particles are distributed fairly uniformly over the whole section of the tube. The number of particles deposited in one second on unit length of a tube of radius R under the action of gravity is $2\,RnV_s = 2\,Rgn\tau$, when n is the particle concentration in the flowing gas. When 1 cm length of aerosol traverses a distance dx, in a time $dx/\overline{U}$, $2\,Rgn\tau\,dx/\overline{U}$ particles fall out of it. Since the volume of this length is πR^2

$$-\frac{dn}{dx} = \frac{2\,RnV_s}{\pi\,R^2\,\overline{U}} = \frac{2\,nV_s}{\pi\,\overline{U}R}, \tag{46.1}$$

which integrates to

$$n = n_0 \exp\left[-\frac{2V_s x}{\pi\,\overline{U}R}\right]. \tag{46.2}$$

This expression is similar to (44.21) and shows how the concentration of particles of a given size varies with distance x traversed by the aerosol. The rate of deposition of the aerosol, which is proportional to n, decays exponentially in the direction of the flow.

These considerations also apply to the theory of the cyclone. Suppose that continuous turbulent mixing of an aerosol is taking place in a cyclone and, in accordance with (29.1), the flow velocity at the outer wall is $U_0/2$. The radial velocity of the particles is then $U_0^2\tau/4\,R_2$ and the number of particles deposited per second on 1 cm² of the wall is

$$I = \frac{n\,U_0^2\tau}{4\,R_2}. \tag{46.3}$$

If L/s denotes the pitch of the spiral formed by the flow in a cyclone (L is the height of the cyclone and s the number of turns), IL/s particles per second will be deposited on unit length of this spiral. At the same time all the particles which entered the cyclone in one second, and have not been deposited on their way, pass along the spiral. These amount to $HhnU_0$ particles where H and h are the height and width of the inlet to the cyclone. We can therefore write

$$-\frac{1}{n}\frac{dn}{dx} = \frac{n\,U_0^2\,\tau L/4\,R_2\,s}{n\,HhU_0} = \frac{U_0\,\tau L}{4\,R_2\,sHh}, \tag{46.4}$$

where dx is the differential of length along the spiral. Since the total length of the latter is approximately $2\pi\,R_2 s$, equation (46.4) can be integrated to give

$$\ln\frac{n}{n_0} = -\frac{U_0\,\tau L\,.\,2\pi R_2 s}{4\,R_2 s\,Hh} = \frac{\pi\,U_0\,\tau L}{2\,Hh} \approx \frac{\pi\,U_0\,\tau s}{2h} \tag{46.5}$$

for the ratio of the number of particles leaving the cyclone to the number entering (the pitch of the spiral L/s has been taken as equal to the height of the inlet H). The efficiency of the cyclone is therefore

$$\vartheta = \frac{n_0 - n}{n_0} = 1 - \exp\left(-\frac{\pi\,U_0\,\tau s}{2h}\right), \tag{46.6}$$

which is considerably less than that calculated without taking account of turbulent mixing [formula (29.14)].

Although the absence of accurate experimental data on collection efficiency of cyclones as a function of particle size makes it impossible to check formula (46.6) quantitatively, there is no doubt that it is closer to the truth than (29.14). According to the latter, particles larger than a certain critical size must be deposited completely; in fact 100 per cent deposition in cyclones is not observed, even with comparatively coarse particles. There is only a steady increase in efficiency with particle size [384], as predicted by formula (46.6).

Turbulence has a similar effect on the deposition of highly dispersed aerosols between the charged plates of an electrical condenser (§ 27) where the transition from laminar to turbulent flow is accompanied by a considerable reduction in current below the figure calculated theoretically for laminar flow [181].

Before passing on to diffusive deposition of aerosols from turbulent flow some essential information on the hydrodynamics of turbulent flow will be communicated.

Experimental study of the distribution of mean velocities for turbulent flow in pipes and open channels has shown the existence of a logarithmic velocity profile $U = a \ln z + b$ where U is the mean velocity at a distance z from the wall. Starting from the concept of mixing length (a quantity playing the same role in the theory of turbulent viscosity and diffusion as mean free path of molecules plays in the theory of molecular viscosity and diffusion) Prandtl [385] derived the formula

$$U/U^* = \frac{1}{\varkappa} \ln \frac{zU^*}{\nu} + C \qquad z > \delta_L \text{ (see below)}, \tag{46·7}$$

in which $\varkappa$ is Karman's constant relating the mixing length l to the distance from the wall according to

$$l = \varkappa z \tag{46.8}$$

and C is a constant which cannot be determined theoretically. U^* is called the dynamic or friction velocity defined by

$$U^* = \sqrt{\tau/\gamma_g} \tag{46.9}$$

where τ is the shear stress or the momentum transferred in 1 sec through an area of 1 cm² parallel to the wall by the turbulent fluctuations. Near the wall τ can be considered constant (independent of z) and hence equal to the frictional force between the flowing gas and the wall per 1 cm² of the latter. It has been found from experiment that $\varkappa = 0{\cdot}4$ and $C = 5{\cdot}5$ so that formula (46.7) can be written

$$U/U^* = 2{\cdot}5 \ln \frac{zU^*}{\nu} + 5{\cdot}5 = 5{\cdot}75 \log \frac{zU^*}{\nu} + 5{\cdot}5 \quad (z > \delta_L). \tag{46.10}$$

U^* has the same order of magnitude as the root mean square eddy velocity u.

At very small distances from the wall this formula is not applicable because it leads to $U \to -\infty$ when $z \to 0$, whereas the flow velocity at the wall is zero. It must therefore be assumed that at the wall there is a thin laminar layer of thickness δ_L in which the gas moves according to the ordinary laws of viscous flow and momentum transfer occurs through molecular viscosity so that $\tau = \eta\, dU/dz$ with $z < \delta_L$. It follows that in the laminar layer

$$U/U^* = \frac{\tau z}{\eta U^*} = \frac{U^{*2} \gamma_g z}{U^* \eta} = \frac{U^* z}{\nu} \qquad (z < \delta_L). \tag{46.11}$$

Experiment confirms that the velocity profile is represented by formula (46.11) for $zU^*/\nu < 5$ and by formula (46.10) for $zU^*/\nu > 20$. A continuous transition occurs in the intermediate region [385]. The thickness of the laminar layer is thus of the order

$$\delta_L \approx 10\nu/U^*. \tag{46.12}$$

Using the concept of eddy viscosity, η_t

$$\tau = \eta_t dU/dz \qquad (z > \delta_L). \tag{46.13}$$

Substitution of the expression for dU/dz from (46.7) and for τ from (46.9) gives $\eta_t = \gamma_g U^* z\varkappa$ so that the kinematic eddy viscosity $\nu_t = \eta_t/\gamma_g$ is

$$\nu_t = \varkappa U^* z. \tag{46.14}$$

Since momentum and mass transfer occur by the same mechanism the coefficients of eddy diffusion, D_t, and eddy viscosity, ν_t, must be of similar magnitude. Prandtl [386] deduced that D_t/ν_t was between 1·4 and 2·0 and the experiments of Sherwood and Woertz [376] showed that the ratio was 1·6. These values for the eddy diffusion coefficient lead to the formula

$$D_t = (1{\cdot}5 \text{ to } 2{\cdot}0)\, U^* z\varkappa \approx (0{\cdot}6 \text{ to } 0{\cdot}8)\, U^* z \qquad (z > \delta_L), \tag{46.15}$$

which is, of course, only applicable outside the laminar layer.

Two points of view exist on the mechanism of diffusion within this layer. According to Prandtl and Taylor [387] there is no turbulence within the laminar layer so that both material and momentum transfer occur exclusively by molecular diffusion. Landau and Levich [388] consider that turbulent motion penetrates the laminar layer, dying out only at the actual surface of the wall; D_t is then proportional to the fourth power of z instead of the first

$$D_t \approx U^* z^4/\delta^3{}_L. \tag{46.16}$$

Immediately adjoining the wall a diffusion layer exists, of thickness δ_D, in which molecular diffusion predominates over turbulent diffusion ($\delta_L \gg \delta_D$).

Prandtl and Taylor took the boundary of the diffusion layer as coincident with the boundary of the velocity layer, within which the eddy viscosity is zero. Karman†, however, showed that it was necessary to define the diffusion layer as commencing where the coefficient of eddy diffusion vanished. When $\nu_t/\nu \approx D_t/D$ the velocity and diffusion layers actually coincide and the hypothesis of Prandtl and Taylor is correct; bearing in mind that $\nu_t/D_t \approx 1$ it is evident that the Schmidt number (Sc $= \nu/D$) must also be about unity as, for example, in the diffusion of gases. In aerosols, however, Sc $\gg 1$ so that $\nu_t/\nu \ll D_t/D$. Since the intensity of turbulence, the mixing length and, consequently, ν_t and D_t all decrease on approaching the wall, while ν and D remain constant, it follows that D_t/D in aerosols tends to unity much closer to the wall than ν_t/ν. In aerosols, therefore, δ_D is considerably less than δ_L. Furthermore, for given flow conditions, which define δ_L, the value of δ_D is not constant but must vary in the same sense as D. In particular, according to Landau and Levich it follows from the definition of δ_D that $D = D_{t\,(z\,=\,\delta_D)} \approx U^* \delta_D^4/\delta_L^3$. whence

$$\delta_D \approx D^{1/4} U^{*-1/4}\, \delta_L^{3/4} \approx \frac{0{\cdot}57\delta_L}{Sc^{1/4}}. \tag{46.17}$$

The rate of diffusive deposition on the walls of a tube, neglecting the change in concentration of the aerosol in the flow direction, will now be calculated. The concentration is taken to be independent of time and of the coordinate x along the flow direction, and is thus a function of the distance from the wall, z, alone. This simplification is justified because of the low rate of diffusive deposition and the comparatively large flow velocities under turbulent conditions. The same number of particles, I, passes in 1 sec through 1 cm² of every surface parallel to the wall, and in view of the large eddy diffusion coefficient at points far removed from the wall the aerosol concentration may be considered constant (n_0) everywhere except in the wall layer. Using D_E to denote the effective diffusion coefficient, including both molecular and eddy transfer,

$$I = D_E \frac{dn}{dz} = \text{const.} \tag{46.18}$$

† Th. von Karman. The analogy between heat transfer and fluid friction. *Trans Amer. Soc. Mech. Engrs.* **61**, 705, (1939).

To calculate I it is necessary to have some hypothesis as to how D_E varies with distance from the wall. The simplest is the Prandtl–Taylor one with only molecular diffusion considered as acting for $z < \delta_L$ and only eddy diffusion for $z > \delta_L$. In the laminar layer

$$I = D\frac{dn}{dz} \qquad (z < \delta_L), \tag{46.19}$$

so that $n = Iz/D + C_1$. Taking $n = 0$ at $z = 0$, this makes

$$n = Iz/D \qquad (z < \delta_L). \tag{46.20}$$

In the turbulent region D_E is given by equation (46·15) and, using α to denote the factor 0·6 to 0·8,

$$I = \alpha U^* z \frac{dn}{dz} \qquad (z > \delta_L), \tag{46.21}$$

which integrates to

$$n = \frac{I}{\alpha U^*} \ln z + C_2. \tag{46.22}$$

If h is the distance from the wall beyond which n becomes constant and equal to n_0,

$$n_0 = \frac{I}{\alpha U^*} \ln h + C_2. \tag{46.23}$$

The elimination of C_2 between (46.22) and (46.23) gives

$$n = n_0 + \frac{I}{\alpha U^*} \ln \frac{z}{h}. \tag{46.24}$$

Expressions (46.20) and (46.24) must be identical when $z = \delta_L$. Replacing z in them by δ_L and equating the two expressions leads to

$$I = \frac{n_0 D}{\delta_L - \dfrac{D}{\alpha U^*} \ln \dfrac{\delta_L}{h}}. \tag{46.25}$$

Since U^* equals $10\,\nu/\delta_L$, in agreement with (46.12),

$$I = \frac{n_0 D}{\delta_L - \dfrac{D\delta_L}{10\,\alpha\,\nu} \ln \dfrac{\delta_L}{h}} = \frac{n_0 D}{\delta_L \left(1 - \dfrac{1}{10\,\alpha\,\mathrm{Sc}} \ln \dfrac{\delta_L}{h}\right)}. \tag{46.26}$$

In aerosols $\mathrm{Sc} \gg 1$ and the second term in brackets can be neglected; hence, finally

$$I \approx \frac{n_0 D}{\delta_L}. \tag{46.27}$$

According to Prandtl and Taylor, therefore, the rate of diffusive deposition is proportional to the first power of the molecular diffusion coefficient of the particles. The concentration of the aerosol, as seen from formula (46.27), can be considered practically constant right up to the edge of the laminar layer, as on page 254. A similar calculation, due to Landau and Levich [388], allowing for a thinner diffusion layer, is more complicated and leads to

$$I = \frac{\beta n_0 D}{\delta_D}, \tag{46.28}$$

where β is a numerical coefficient (~ 1) which cannot be determined theoretically. Since δ_D is proportional to $D^{\frac{1}{4}}$ (46.17) this makes I proportional to $D^{\frac{3}{4}}$. The explanation of this relationship is that when the molecular diffusion coefficient is small, turbulent fluctuations carry the particles closer to the wall and in so doing partly compensate for the small value of D.

Although this description of diffusive deposition from turbulent flow is an improvement on the original approach of Prandtl and Taylor, experiment is necessary for the complete elucidation of the mechanism. A convenient way of doing this would be to use isodisperse sulphuric acid mists, produced by a LaMer generator [6] and to measure the amount of deposit on the walls of a tube. By using mists with different particle sizes it would be possible to find how I depends on D and thence to determine the law relating D_t to distance from the wall.

Formulae (46.27) and (46.28) can be put in forms which are convenient for practical use by using experimental data [389] obtained in smooth pipes at $\mathrm{Re}_f < 10^5$ which show that

$$U^* \approx \frac{0{\cdot}16\, U_M}{\mathrm{Re}_f^{1/8}} \approx \frac{0{\cdot}2\, \bar{U}}{\mathrm{Re}_f^{1/8}}, \tag{46.29}$$

where U_M is the flow velocity on the axis of the tube and $\bar{U}$ is the mean velocity. Hence

$$\delta_L = 10\, \nu/U^* \approx 50\, \nu\, \mathrm{Re}_f^{1/8}/\bar{U} = 100\, R/\mathrm{Re}_f^{7/8}, \tag{46.30}$$

R being the radius of the pipe, and

$$I = \frac{D n_0}{\delta_L} \approx \frac{D n_0\, \mathrm{Re}_f^{7/8}}{100\, R} \text{ (according to Prandtl and Taylor).} \tag{46.31}$$

According to (46.17)

$$\delta_D \approx 57\, R D^{1/4}/\mathrm{Re}_f^{7/8}\, \nu^{1/4} \tag{46.32}$$

and

$$I = \frac{\beta n_0 D}{\delta_D} = \frac{\beta D^{3/4} n_0\, \mathrm{Re}_f^{7/8}\, \nu^{1/4}}{57\,\mathrm{R}} \text{ (according to Landau and Levich).} \tag{46.33}$$

No experimental data has been available until lately for the diffusive deposition of aerosols on the walls of pipes. Alexander and Coldren [390] used an atomizing nozzle to spray water into a short horizontal duct and measured the droplet concentration at various points along the duct. They then calculated the rate of deposition of droplets on the walls. The fall in concentration from the axis to the periphery of the duct in these experiments, however, has nothing to do with diffusive deposition but depends on the concentration profile in a free jet.

Inertial deposition of aerosols from turbulent flow will now be considered.

Complete entrainment of the particles by turbulence does not mean that no deposition takes place. The calculations on page 258 relate to the central region of a turbulent stream; observations show that the eddy velocity increases on approaching the wall and begins to fall only at a very short distance from it [382, 391, 392]. On the other hand the mixing length, and hence the eddy diameter, decreases steadily on approaching the wall, and very small scale velocity fluctuations perpendicular to the wall can be observed with an ultramicroscope at a distance of a few microns from the wall [382]. It is thus possible for particles close to the wall to acquire considerable normal velocities which cause them to be deposited by inertia.

Unfortunately very few experimental data are available. The copious deposits of dust often seen on the walls of vertical pipes are possibly caused by deposition from turbulent flow. It must be borne in mind that when the flow velocity is high deposited solid particles can be re-entrained, so that the absence of a deposit does not mean that no deposition is taking place.

Experiments with mists or with the surface coated with adhesive would settle this point.

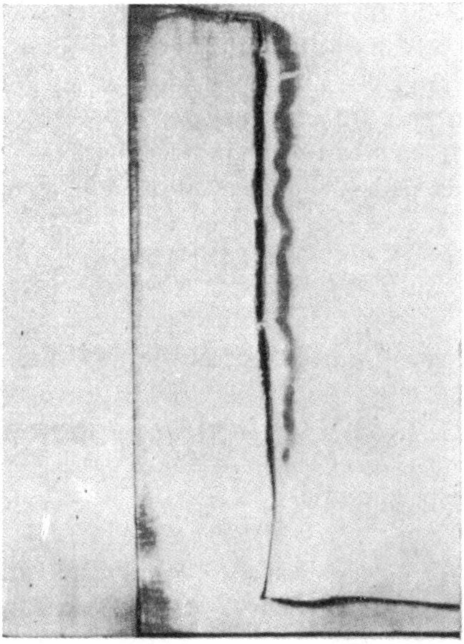

FIG. 66. Deposition of particles by eddies due to an obstacle.

Deposits caused by stationary vortices formed by flow round obstructions are well known. Figure 66 is a photograph taken by the author of a dust deposit on the wall of a ventilated building near a telephone cable. The deposit was formed in 1 year by a narrow current of air escaping from a vertical slit upstream from the cable. A narrow, sharply defined dust deposit in front of the cable and a broad, diffuse deposit behind it can be seen. An inertial deposit formed on the windward surface of the cable itself which was not due to vortices.

Flow round an obstacle gives rise to vortices in the way shown in Fig. 67. A very similar idea is found in Zhukovskii's theoretical study [393] of the formation of snowdrifts but the principal mechanism in this case may not be inertial deposition from vortices but sedimentation of decelerated snow from the air stream.

Eddy deposition under certain conditions of flow round isolated objects can cause deposits on both the upstream and the downstream sides. In Yeomans' experiments

[248] an airflow of 8 m sec^{-1} impinging on 7 cm diameter glass disks caused two and a half times as many droplets from a mist with $r = 5{\cdot}6\,\mu$ to be deposited on the downstream side than on the upstream side. Landahl [236] found that more particles were deposited on the downstream side of glass plates 2·5 cm wide at a velocity of 0·5 m sec^{-1} than at 1·5 m sec^{-1}. The experiments of Asset and Pury [394] showed that at a flow velocity of 2·3 m sec^{-1} droplets of an isodisperse mist having $r = 6{\cdot}5\,\mu$ were not deposited on the downstream side either of glass cylinders 7·5 cm in diameter or of a man's shaved forearm. The number of droplets deposited on the downstream side of a hairy forearm was about a quarter of that on the upstream.

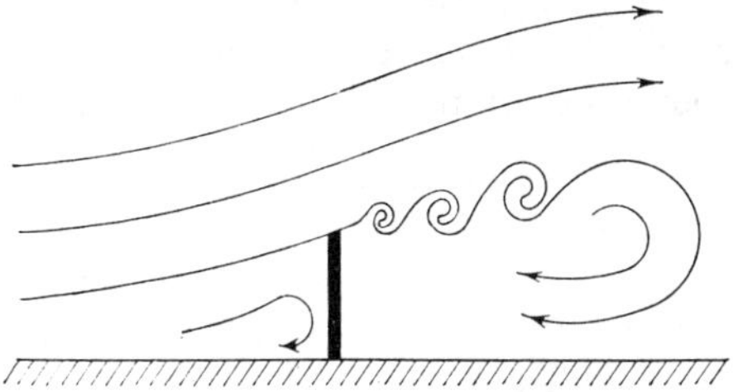

FIG. 67. Air eddies in front of and behind an obstacle.

Definite hydrodynamic conditions are probably essential for deposition to occur on the lee side: the flow velocity in the vortices must be big enough to ensure inertial deposition but the vortices must not be carried away from the object too quickly.

It is possibly appropriate to refer here to the thick dust deposit formed on the outlets of diffusers by the breakaway of vortices.

Two semi-empirical methods for calculating the turbulent diffusion flux in tubes with smooth absorbing walls should be mentioned. Lin *et al.* [782] assumed that in the laminar sublayer

$$D_t = \nu_t = (z^*/14{\cdot}5)^3\,\nu \tag{46.34}$$

where $z^* = zU^*/\nu$, U^* = the friction velocity. Hence it follows that in the intermediate sublayer

$$D_t = \nu\,(z^*/5 - 0{\cdot}959). \tag{46.35}$$

The expression for the diffusion flux obtained from these formulae, which is in good agreement with experiment, is not given here, as it is too cumbersome.

Deissler [783] proceeded from the velocity profile in a turbulent stream, deduced by him earlier, which is in excellent agreement with experiment, and arrived at the following expression valid at Sc > 200.

$$\mathrm{Sh} = \frac{4\sqrt{2} \times 0{\cdot}124\, U^* R\, \mathrm{Sc}^{1/4}}{\pi\,\nu} \tag{46.36}$$

which agrees well with the experimental data on diffusion in solutions. From (46.29) and (46.36) it follows that

$$I = \frac{n_0\, D^{3/4} R_f^{7/8} \nu^{1/4}}{90\, R} \tag{46.37}$$

as given by the Landau–Levich formula (46.33).

Friedlander and Johnstone [784] measured the deposition of aerosols in turbulent flow through vertical tubes. The experiments were done with almost spherical particles of Fe ($\bar{r} = 0{\cdot}4$–$0{\cdot}8\,\mu$) and Al ($\bar{r} = 2{\cdot}5\,\mu$) in tubes of radius 0·27–1·2 cm. The walls were covered with adhesive to prevent the particles deposited from being blown off. Since the deposition rate on the tubes was two orders of magnitude greater than that calculated from (46·37) and increased with the size of the particles, deposition in this case was undoubtedly due to inertia.

The results for Fe particles, having $\bar{r} \approx 0{\cdot}4\,\mu$, in a tube of $R = 0{\cdot}29$ cm are shown in Fig. 67a. The graphs show that the deposit begins at a distance x from the entrance of the tube, corresponding to $\mathrm{Re}_x = x\bar{U}/\nu = 1{\cdot}0 \times 10^5$ to $2{\cdot}5 \times 10^5$. As with flow along a sharp-edged plate the boundary layer becomes turbulent at $\mathrm{Re}_x \approx 1{\cdot}4 \times 10^5$ [785] so there is no doubt that deposition in these experiments was caused by the turbulence.

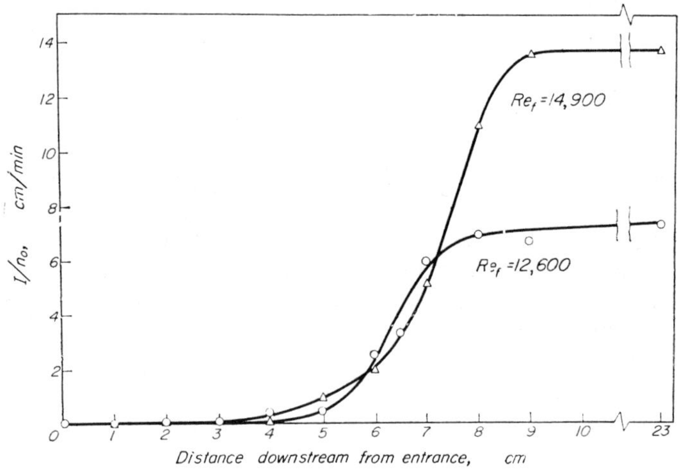

FIG. 67a. Deposition of iron particles on the walls of a tube during turbulent flow.

A theory of deposition was developed from the idea that those particles were deposited which approached within one stop-distance of the wall, using the r. m. s. velocity of the turbulent fluctuations v_n in the direction normal to the wall at the distance l_i from the latter. The experimental data of Laufer [786] were employed using the observation that as z^* increases from 0 to 80, v_n/U^* rises from 0 to about 0·9 and then remains steady. The calculations were simplified by assuming that $v_n = 0{\cdot}9\,U^*$. Had there been a purely laminar sublayer, there would have been no deposition in some of the experiments, as the thickness of the sublayer, $\delta_L \approx 5\,\nu/U^*$, was greater than l_i. By means of (46.34) and (46.35) the following expression for the deposition rate of particles per 1 cm² of the wall surface was deduced

$$I = \frac{n_0 U^{*2}}{\bar{U}\,[1 + U^*(14{\cdot}5^3/2l^{*3} - 50{\cdot}6)]} \tag{46.38}$$

$$\text{where } l^* = l_i U^*/\nu.$$

The theoretical values of I were between half and double the experimental ones; better agreement could hardly be expected in view of the present state of knowledge of turbulent flow near a surface.

The experiments of Dawes and Slack, discussed in § 45, included measurements of the deposition rate of coal dust on the bottom, sides and top of a channel. The following relationship was found for the bottom of the tube

$$I = Kr^2 n = Kr^2 n_0 \exp\left(-\frac{Kr^2 x}{\bar{U} h}\right) \qquad (46.39)$$

where n_0 is the initial concentration of the aerosol, h the height of the channel and K a constant which can be deduced, if the concentration is assumed constant over the cross section of the tube, from $Kr^2 = V_s$ [see (46.2)]. For $\bar{r} > 16\,\mu$ I diminished more rapidly than it should have done according to the formula, since the aerosol concentration, as was shown on p. 264, rises appreciably towards the bottom of the tube. The ratio of the deposition rate on side walls to that on the bottom of the channel, β, was found to be $2(2\,r)^{-1\cdot 35}$, where r is expressed in microns; β is only slightly dependent on the flow rate. The deposition rate of particles on the sides of the tube was thus found to be roughly proportional to $r^{2/3}$.

Pereles [787] showed that the deposition rate on the sides and top of the channel in these experiments was incompatible with the hypothesis of a purely laminar sublayer. Assuming the flow rate and the coefficient of turbulent diffusion to be constant over the cross-section of the tube, he neglected the inertia of the particles, which, as we have seen, is inadmissible.

The principal ideas of Owen [788] are near to those of Friedlander, but Owen used a more complex dependence of the coefficient of turbulent diffusion on the distance from the wall; he divided the boundary layer into three zones. His formulae for the deposition rate on the bottom, top and side walls of a rectangular horizontal channel are in quite good agreement with the experimental results of Dawes and Slack.

Koepe [789] made experiments with coal dust in a horizontal duct 1·8 m high at $\bar{U} = 1\cdot 4$ m/sec, and found that the number of particles deposited on 1 cm^2 of floor did not depend on the horizontal distance x. This result is probably accidental, being due to the large range of sizes of the dust particles. Madisetti [790], who worked under the same conditions but used fractionated dust ($\bar{r} = 0\cdot 5$–$3\,\mu$), found for the deposition of each fraction on the floor of the duct at $\bar{U} = 50$–60 cm/sec the relation $I = A \exp(-Kx)$, in agreement with (46.39). The deposition rate on the walls was very small compared with that found by Dawes and Slack, no particles of $r > 1\,\mu$ being found on the walls at all. This discrepancy is difficult to account for.

§ 47. THE SPREAD OF HIGHLY DISPERSED AEROSOLS IN THE ATMOSPHERE

The travel of aerosols such as factory smoke, military screening smoke and insecticidal mists is very important technically and from the point of view of industrial hygiene, including the protection of residential areas from pollution by industrial effluents. The problem is of special importance in connection with atomic bomb explosions which produce a huge cloud of radioactive aerosol which spreads in the atmosphere and may threaten human life several hundred kilometres from the place of detonation. In spite of the attention which has been devoted to this problem it has still been studied far from adequately, largely owing to the extreme complexity and variability of atmospheric currents which cause fluctuations in the measurements beyond the scope of mathematical analysis. It is therefore difficult to check the various theories experimentally.

The motion of aerosols in the atmosphere consists of motion of the air itself and motion of the particles relative to the air which, unless the particles are sufficiently coarse, amounts to their settling under gravity. To start with, settling will be neglected and the spreading of highly dispersed aerosols in the atmosphere discussed as if they acted like gaseous pollutants. This approach has been explored in detail by Sheleikhovskii [395], Andreev [396] and Sutton [397, 398] so that most attention will be devoted to the fundamentals of the problem without going into the various practical details. The mechanism of turbulent dispersion in the atmosphere must first be dealt with since it is covered rather sparsely, and sometimes wrongly, in existing literature.

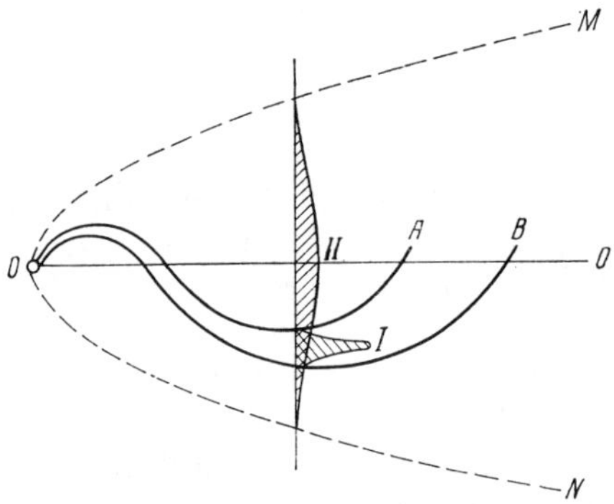

FIG. 68. Aerosol plume in a turbulent atmosphere.

The general nature of the spread of aerosols in the atmosphere is as follows. A continuous stream of aerosol, or a cloud emerging from some source, travels with the wind and is diffused simultaneously by atmospheric turbulence. Molecular diffusion plays no part in this process except in the very thin layer of air at the surface of objects in contact with the cloud.

Atmospheric turbulence possesses some special features which must be mentioned briefly. In turbulent flow in pipes, channels, rivers etc., there exists at every point a mean flow velocity $\overline{U}$, which is practically constant in magnitude and direction and can easily be determined from experiment; on this are superimposed random turbulent fluctuations. Variations in $\overline{U}$, if they occur at all (for instance daily or seasonal changes in velocity in rivers), have a period several orders of magnitude greater than that of the coarsest fluctuations. Furthermore, $\overline{U}$ is considerably greater than the root mean square eddy velocity and so in this case the line between mean flow and fluctuations can be drawn very sharply. In the atmosphere, on the other hand, wind speed, as measured by weather vanes, changes continually in magnitude and direction and the size of the variations increases with the time of observation. A continuous spectrum exists, beginning with the finest fluctuations having a period measured in hundredths of a second and finishing with daily and annual fluctuations in wind speed.

In thise case, therefore, no definite boundary can be established between the mean and the eddy velocities. The mean wind speed over a period of one minute measured by some meteorological instrument can, in observations made over a year, be regarded as a fluctuation velocity.

In order to examine the mechanism of diffusion in turbulent flow, picture an aerosol emerging continuously through a long narrow slit, perpendicular to the flow, at some point O on Fig. 68 where OO' represents a streamline of the mean flow and the curves OA and OB are instantaneously fixed contours of the aerosol plume. On these curves the aerosol concentration is equal say, to, 10 per cent of that on the axis of the plume. The mean square displacement of particles from the fixed axis OO' recorded by a stationary observer (absolute diffusion) is given by the usual equation

$$\overline{z^2} = 2\,D_t\,t, \tag{47.1}$$

assuming for simplicity that D_t, the eddy diffusion coefficient, is constant throughout the field of flow. The mean aerosol concentration distribution (curve *II*) for a continuous source is given by

$$n = \frac{\Phi'}{\sqrt{4\pi D_t U x}}\, e^{-\frac{U z^2}{4 D_t x}} \tag{47·2}$$

which was derived on page 212. This expression therefore gives the time-averaged concentration calculated from the usual laws of diffusion.

This is not the case for relative diffusion of particles in turbulent flow which comprises their displacement with respect to an observer moving with one of the particles. Whereas absolute diffusion is the result of all fluctuations, in relative diffusion only fluctuations having a scale, λ, the same as or less than the distance, ϱ, between the particles, contribute. These fluctuations constantly increase the distance between the particles, thus causing the plume to spread, but fluctuations with $\lambda \gg \varrho$, which obviously cannot do this, contort the plume and break it into individual clouds.

The instantaneous concentration distribution in the plume (curve *I*) is clearly determined by relative diffusion. This is illustrated by photographs (Fig. 69) of smoke leaving a factory chimney under differing degrees of atmospheric turbulence [399].

Expansion of the plume, as x increases, causes the scale of the fluctuations capable of changing the distance ϱ to increase also. In other words, the coefficient of relative diffusion D_L increases continuously during the motion and expansion of the aerosol cloud; this constitutes the characteristic property of turbulent diffusion. It must be emphasized again that this fact is relevant only to the instantaneous concentration distribution, not to the time-averaged distribution. Richardson [400] found the following relationship empirically:

$$D_L \approx 0{\cdot}2\,\varrho^{4/3}. \tag{47.3}$$

This formula, as shown by Obukhov [364], can be derived theoretically for the scale of fluctuations to which Kolmogoroff's theory applies.

It is not difficult to see that $D_L = 0$ for $\varrho = 0$ since particles packed closely together cannot move apart under the influence of turbulent fluctuations. On the other hand, sometimes, for example in a wide river, turbulent fluctuations at two points far enough apart may become independent. Formula (47.3) does not then apply and the relative diffusion coefficient becomes equal to the sum of the absolute diffusion coefficients at these points

$$D_L = D_{t_1} + D_{t_2} \tag{47.4}$$

in a similar way to what happens in molecular diffusion (see page 290). As a rule, $D_L \leqslant D_{t_1} + D_{t_2}$ and the averaged profile of the plume *MON* (see Fig. 68) is always wider than the instantaneous profile *AOB*.†

It is characteristic of diffusion in the atmosphere that there is a substantial difference between the vertical and horizontal fluctuations. The vertical component of the wind is usually small and can be neglected so that the mean vertical wind speed is taken to be zero and the problem of what to consider as fluctuations and what to consider as the mean velocity does not arise.

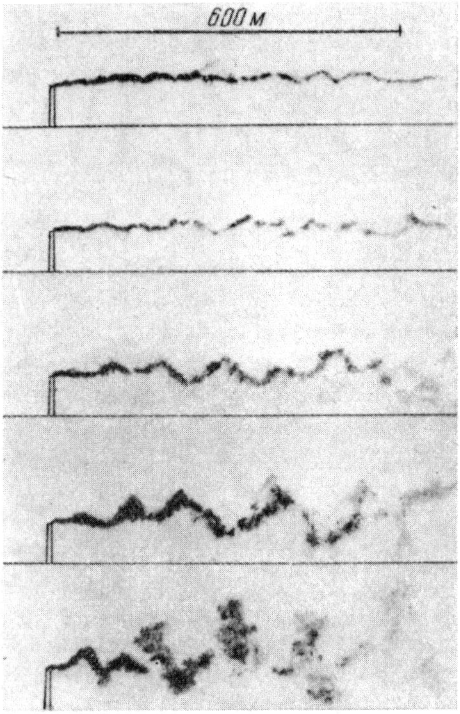

FIG. 69. Smoke plume at varying degrees of atmospheric turbulence.

The coefficient of absolute vertical eddy diffusion, D_z, which determines the kinetics of vertical exchange in the atmosphere and therefore plays a big part in meteorology, has been the object of numerous experimental and theoretical investigations and its value, which depends on the nature of the terrain, height above the ground, wind speed and other meteorological conditions, is known approximately.

Horizontal diffusion is more complicated because mean and eddy velocities cannot be distinguished, as explained on p. 274. A definite time of observation, t_{obs}, must therefore be adopted and the mean wind speed, U_{obs}, determined over this time; differences between U_{obs} and the instantaneous wind speed are regarded as the fluctuations.

† Yudin's article [401], in which these ideas on turbulent dispersion relative to a fixed axis and to the axis of a smoke plume were formulated more rigorously, became known to the author only after this book was finished.

The motion of an aerosol must likewise be considered as a steady motion with velocity U_{obs} plus dispersion by fluctuations. The mean square velocity and the scale of the fluctuations clearly increase with t_{obs}, since greater changes in wind speed are included over the longer period. The absolute diffusion coefficient D_{tx} also increases and the greater the value of t_{obs} the more does the time-averaged aerosol plume spread; the mean aerosol concentration thus decreases with time at a given point, either off the axis, or on it where the concentration is greatest.

Ryazanov's experiments [396] showed that the maximum concentrations of sulphur dioxide near a point source (factory chimney) were 4·5, 0·94, 0·75 and 0·3 mg m^{-3} respectively for values of t_{obs} of 5 min, 24 hours, 1 month and 1 year. Since the law of variation of the horizontal diffusion coefficient with time of observation is unknown, it is impossible to calculate theoretically the concentration in an aerosol emitted by a point source or a line source with a length comparable to the distance from the source.

Nothing precise is known about the magnitudes of the relative diffusion coefficients in the atmosphere, which are necessary for calculating the instantaneous concentrations, except formula (47.3), a very rough approximation not even involving the degree of turbulence of the wind.

The first attempts to establish a theory for the spread of gases in the atmosphere [294] made use of the usual diffusion equations with a turbulent diffusion coefficient, the wind speed being assumed constant and independent of height. This is the problem analysed at the end of § 39 and the concentration at a given point is given by (39.24), (39.25), (39.29) or (39.30), according to the type of source. This crude theory, however, disagrees with experiment, the concentration actually falling off much more quickly with distance from the source than is indicated by the formulae. According to the well-known theory of Sutton [403, 398], the reason for the difference lies in the increase in efffective diffusion coefficient with distance from the source. Since the mean concentration of the pollutant during a known time interval is always measured in experiments, Sutton's opinion might be correct for the horizontal spreading of a cloud from a point source but it could not hold for vertical diffusion from a line source at distances short compared with its length.

The differential equation of diffusion from a line source passing through the origin of coordinates and perpendicular to the plane xz is [(39.26)]†

$$U\frac{\partial n}{\partial x} = \frac{\partial}{\partial z}\left(D_{tz}\frac{\partial n}{\partial z}\right) \tag{47.5}$$

where U and D_{tz} are functions of height, z, and the wind blows along the x-axis. As usual, diffusion in the x direction is not considered. The boundary conditions are: $n = 0$ at $x = \infty$ or $z = \infty$; $n = \infty$ at $x = z = 0$; and $\int_{-\infty}^{\infty} Undz = \Phi'$, where Φ' is the strength per 1 cm of the source. The last condition follows from the stationary nature of the process and the law of conservation of matter. The function $D_{tz}(z)$ is determined from the wind profile, $U(z)$, by the method indicated on page 267, the coefficients of turbulent diffusion and viscosity being assumed equal.

† In this and the following section n denotes either number or weight concentration of an aerosol or the concentration of a gaseous pollutant.

Most experimental work relates to the dry adiabatic temperature gradient in the atmosphere. It has been shown under these conditions that the wind profile is logarithmic although it is somewhat different from the one discussed on page 266 for smooth pipes because the earth's surface, under normal meteorological conditions, is aerodynamically rough and the flow above it is given by the expression

$$U/U^* = \frac{1}{\varkappa} \ln \frac{z - d}{z_0}, \tag{47.6}$$

where z_0 and d are experimentally determined quantities depending on the degree of roughness of the surface (height and thickness of grass etc.) and on the wind speed [404].

The solution of equation (47.5) with a logarithmic wind profile presents insurmountable difficulties which can, however, be avoided by approximating expressions (46.7) and (47.6) by power functions

$$U = U_1 \left(\frac{z}{z_1}\right)^q \tag{47.7}$$

or

$$U/U^* = m \left(\frac{z - d}{z_0}\right)^q \tag{47.8}$$

and choosing the coefficients q and m so that the wind profile agrees as closely as possible with the logarithmic one over the required range of z [404].

Proceeding from his point of view, with the aid of some postulates from the statistical theory of turbulence, Sutton [403] reached the semi-empirical conclusion that the mean square of the relative particle displacement in a time t is

$$\overline{(\Delta z_1 - \Delta z_2)^2} = \frac{1}{2} C^2 (Ut)^s = \frac{1}{2} C^2 x^s, \tag{47.9}$$

where C is the generalized coefficient of turbulent diffusion and s is related to q, which occurs in the wind profile equations (47.7) or (47.8), by the expression

$$s = 2/(q + 1). \tag{47.10}$$

Taking $q = 1/7$, the experimental value for dry adiabatic conditions and a smooth surface, Sutton obtained $s = 1{\cdot}75$. It follows from (47.9) that the coefficient C is related to the relative eddy diffusion coefficient D_L, for which the formula $\overline{(\Delta z_1 - \Delta z_2)^2} = 2\, D_L t$ holds, as follows

$$D_L = \frac{C^2}{4} U^s t^{s-1} = \frac{C^2}{4} x^{s-1} U. \tag{47.11}$$

In solving the differential equation (47.5) Sutton ignores the difference between the relative and absolute eddy diffusion coefficients and the variation of both wind speed and diffusion coefficient with height. The wind speed at a given height (2 m) is employed for U and (39.29) and (39.30) are replaced by

$$n = \frac{\Phi'}{\sqrt{\pi}\, C_z^2 U^2 x^s} \exp\left(-\frac{z^2}{C_z^2 x^s}\right) \tag{47.12}$$

for a continuous line source and

$$n = \frac{\Phi}{\pi C_y C_z U x^s} \exp\left[-\frac{1}{x^s}\left(\frac{y^2}{C_y^2} + \frac{z^2}{C_z^2}\right)\right] \tag{47.13}$$

for a point source. Here y is the horizontal distance from the axis of the stream and C_y and C_z are values of C for horizontal and vertical diffusion, for which Sutton [397] derived expressions as functions of wind speed and gustiness, or degree of turbulence.

Apart from pointing out that C is proportional to U raised to the power $s/2 - 1$ these formulae are not given since, in his examples, Sutton [405] uses other values of C_y and C_z which are functions of the height of the source and were obtained by experiment over open downland with an adiabatic gradient and a wind speed of 5 m $\sec^{-1}$ at a height of 2 m (Table 26)

TABLE 26. SUTTON'S DIFFUSION COEFFICIENTS FOR A NORMAL GRADIENT (GRAM-METRE-SECOND UNITS)

H, *m*	0	10	25	50	75	100	200
C_y	0·21	0·21	0·12	0·10	0·09	0·07	0·05
C_z	0·12	0·12					

In order to obtain agreement with experiment Sutton therefore has to make C decrease with height, although, in fact, turbulent diffusion increases with height until z is several hundred metres [406]. Formulae (47.12) and (47.13) are therefore really empirical ones which give, as will be seen below, satisfactory agreement with experiment when the coefficients C_y and C_z are chosen suitably.

These formulae have been derived for the spread of a cloud in an unlimited space; it is now necessary to deal with the important question of boundary conditions at the earth's surface in the solution of the differential equations of turbulent diffusion. This is simple for substances, such as inert gases, which are not absorbed by the surface; the boundary condition at a non-absorbing wall is merely that the concentration gradient normal to the wall is zero, i.e.

$$\partial n/\partial z = 0 \quad \text{if} \quad n = 0. \tag{47.14}$$

Strictly speaking, the condition $n = 0$ at $z = 0$ should be taken for aerosols with particles which are not reflected from the walls but stick to them. In solving the diffusion equations for this case, however, variation of the effective diffusion coefficient with distance from the wall within the wall layer (page 267) would have to be taken into account and would lead to mathematical difficulties. Since the rate of diffusion of aerosols in the wall layer is very small, most of the particles carried into this layer by turbulent fluctuations are not deposited on the wall but are carried out again; it can therefore be assumed that the walls are reflecting for turbulent diffusion and that the condition (47.14) holds.

If the source of aerosol is at a height H above the earth, z, in formulae (47.12) and (47.13), should be replaced by $z - H$. The formulae so obtained are applicable only up to values of x smaller than that at which the cloud touches the ground. In order to obtain a solution satisfying the condition (47.14) the following procedure is used. An image source of the same strength as the actual one is placed, as if reflected in the earth's surface, with coordinates $z = -H$ and $x = 0$. The formulae

$$n = \frac{\Phi'}{\sqrt{\pi C_z^2 U^2 x^s}} \left\{ \exp\left[- \frac{(z-H)^2}{C_z^2 x^s} \right] + \exp\left[- \frac{(z+H)^2}{C_z^2 x^s} \right] \right\} \tag{47.15}$$

for a line source and

$$n = \frac{\Phi}{\pi C_y C_z U x^s} \exp\left(- \frac{y^2}{C_y^2 x^s} \right) \left\{ \exp\left[- \frac{(z-H)^2}{C_z^2 x^s} \right] + \exp\left[- \frac{(z+H)^2}{C_z^2 x^s} \right] \right\} \tag{47.16}$$

for a point source are then obtained instead of (47.12) and (47.13).

In checking these formulae experimentally Sutton [405] used an aerosol generated close to the earth's surface ($H = 0$). At various distances, x, downwind of the source the peak concentration was measured in the cloud at ground level ($z = 0$) for a line source and on the axis of the cloud ($z = y = 0$) for a point source. With the latter, as indicated above, the maximum concentration decreases during the time t required to take a sample of air for analysis but under Sutton's experimental conditions (dry adiabatic gradient) this decrease was very small, even for $t > 3$ min. Formulae (47.15) and (47.16) give the maximum concentration for a line source as

$$n_{max} = 2\Phi' / \sqrt{\pi C_z^2 U^2 x^s} \tag{47.17}$$

and for a point source,

$$n_{max} = 2\Phi / \pi C_y C_z U x^s. \tag{47.18}$$

Since the gradient in these experiments was close to dry adiabatic the value of s was 1·75, so that n_{max} should be proportional to $x^{-0·875}$ for a line source and $x^{-1·75}$ for a point source; the values actually found were $x^{-0·9}$ and $x^{-1·76}$.

It follows from (47.15), in view of the dependence of C on U, that $n_{max} \sim C_z^{-1} U^{-1} \sim U^{-0·875}$. For wind speeds in the ratio 1:2:3 the corresponding values of n_{max} are therefore in the ratio $1:2^{-0·875}:3^{-0·875}$ = 1:1/1·84:1/2·61; the figures actually found were 1:1/1·84:1/2·77.

At $x = 100$ m, $U = 5$ m sec^{-1} and $\Phi = 1$ gm sec^{-1}, with a point source ($C_y = 0·21$, $C_z = 0·12$), the absolute value obtained for n_{max} should have been 1·6 mg m^{-3} while 2·0 mg m^{-3} was actually found. For a line source with $\Phi' = 1$ gm sec^{-1} m^{-1} the theoretical value of n_{max} was 33 mg m^{-3} and the experimental value 31 mg m^{-3}.

Taking the concentration at the boundary of the cloud as 0·1 of the maximum concentration and putting $H = 0$ in formula (47.15), the "height" of the cloud z_0 from a line source is given by

$$0·1 = \exp\left(- \frac{z_0^2}{C_z^2 x^2} \right). \tag{47.19}$$

For $x = 100$ m this gives $z_0 = 10$ m, the value found experimentally. Similarly, the theoretical value of the width of a cloud from a point source on the earth at a distance of 100 m is 34 m and the experimental value is 35 m.

Teverovskii's experiments [407] suggest that (47.17) with $s = 2$ can be used over a fairly wide range of temperature gradient. His average value for C_z was 0·027 over smooth ground devoid of vegetation and 0·086 over tall grass; the values were practically indepent of wind speed. The figures in Table 26 thus relate only to the locality in which Sutton conducted his experiments.

A more satisfactory theory of the spread of a cloud from a line source at ground level, with a dry adiabatic gradient, has been given by Calder [408]. If the wind profile is given by (47.6) and the coefficients of eddy diffusion and viscosity in the vertical direction are assumed equal

$$D_{tz} = \varkappa U^*(z - d) \tag{47.20}$$

is obtained by the method described on p. 267. Since d, as a rule, is only a few centimetres it can be put equal to zero. As an illustration, Table 27 has been constructed from Deacon's experiments [409]. The height of the grass above which the measurements were made is denoted by h, and U_1 and U_2 are the wind speeds at heights of 1 and 2 m from which the parameters d, z_0 and U were calculated; $\tau_0 = \gamma_g U^{*2}$ is the tangential drag of the wind per cm² of the surface (see page 266). Direct measurements of τ_0 gave results close to those calculated from formula (47.6) [410], thus supporting the calculations.

The increase in turbulent diffusion with wind strength and roughness of the earth's surface can be seen clearly from Table 27.

If it is assumed that U is constant and D_{tz} proportional to z, equation (47.5) integrates to

$$n = \frac{\Phi'}{\varkappa U^* x} \exp\left(-\frac{Uz}{\varkappa U^* x}\right), \tag{47.21}$$

so that

$$n_{\max} = \frac{\Phi'}{\varkappa U^* x}. \tag{47.22}$$

TABLE 27. WIND PROFILE AND TURBULENCE FOR A NORMAL GRADIENT ABOVE GROUND COVERED BY GRASS

h (cm)	U_1 (cm sec⁻¹)	U_2/U_1	d (cm)	z_0 (cm)	U^* (cm sec⁻¹)	τ_0 (dyne cm⁻²)	D_{tz} (cm² sec⁻¹) at $z = 100$ cm
60 70	100	1·45	15	15·9	23·9	0·68	810
	200	1·35	16	8·8	35·5	1·51	1190
	300	1·32	21	5·6	45·4	2·47	1430
	450	1·28	32	3·0	57·8	4·00	1570
1·5	100–800	1·112	0	0·20	6·4–51	0·05–3·2	260–2100
3·0	100–800	1·140	0	0·71	8–65	0·08–5·0	320–2600
4·5	200	1·191	0	2·65	22·0	0·58	880
	450	1·170	0	1·74	44·5	2·38	1780

These expressions have been checked [408] under similar conditions to those used by Sutton. The coefficient z_0 was 3 cm, the wind speed 5 m sec⁻¹ at a height of 2 m, $d = 0$ and $U^* = 50$ cm sec⁻¹. For $\Phi' = 1$ gm sec⁻¹ m⁻¹ the theoretical height of the cloud at a distance of 100 m was 10·4 m and the experimental value was 10·0 m. The maximum concentration ($z = 0$) was 44 mg m⁻³ theoretical and 36 mg m⁻³ experimental. It was also found that $n_{\max}$ was proportional to U^{-1} and x^{-1} in agreement with formula (47.22).

The great advantage of (47.21) over Sutton's formulae lies in its fairly rigorous derivation, in which the only assumption is the constancy of U when integrating equation (47.5). It contains only quantities which can be determined experimentally and is free from coefficients which have to be adjusted to fit the experimental results.

Calder [404] has also used the power function (47.8) to approximate the logarithmic profile, obtaining D_{tz} as a function of z and then solving equation (47.5) quite rigorously without assuming U to be constant. The formula which this gives for the concentration distribution is too complex to be given here but agreement with experiment was even better than (47.21).

Deacon [409] obtained a similar result for non-adiabatic gradients, starting from the profile

$$U/U^* = \frac{1}{\varkappa(1-\beta)}\left[\left(\frac{z}{z_0}\right)^{1-\beta} - 1\right], \tag{47.23}$$

with $\beta > 1$ for a superadiabatic gradient and $\beta < 1$ for an inversion. Satisfactory agreement with experiment was found in both cases. Frost [411] has derived an expression for $n(x, z)$ using the profile $U = U_1(z/z_1)^q$. It is only permissible to calculate D_{tz} from the wind profile up to heights below which τ is constant and equal to the tangential drag of the wind upon the surface; this is true up to several tens of metres. Formulae for the concentration in a cloud, derived for a dry adiabatic gradient, are therefore true only for distances not exceeding several hundred metres.

The nature of horizontal diffusion in the atmosphere which is discussed on page 275 renders a point source on the ground more difficult to deal with than a line source. For a dry adiabatic gradient Calder [408] derived the following semi-empirical formula

$$n = \frac{\Phi U}{2\varkappa^2 \beta U^{*2} x^2} \exp\left[-\frac{U}{\varkappa U^* x}\left(\frac{y}{\beta} + z\right)\right], \tag{47.24}$$

whence

$$n_{\max} = \frac{\Phi U}{2\varkappa^2 \beta U^{*2} x^2}, \tag{47.25}$$

where Φ is the strength of the source and β is the ratio of the root mean square horizontal and vertical eddy velocities, which is readily measured. The following results were obtained when this formula was tested, under the meteorological conditions used to check formula (47.21), with $\beta = 2$. At $x = 100$ m the height of the cloud was 10·4 m theoretically and 10·0 m experimentally, the same as for a line source; the width of the cloud was 41 m theoretically and 35 m experimentally; $n_{\max}$ at $\Phi = 1$ gm sec^{-1} was 2·5 mg m^{-3} theoretically and 2·0 mg m^{-3} experimentally. Theoretically $n_{\max}$ was predicted to be proportional to x^{-1} and experiment gave $x^{-0\cdot 8}$, while $n_{\max}$ was found to be proportional to U^{-1} both theoretically and experimentally. The formula for a point source, as mentioned earlier, has to include the time required to make the concentration measurements at different points, which was three to five minutes in both Sutton's and Calder's experiments. That both expressions give more or less similar results is explained mainly by the fitting of Sutton's coefficients C_y and C_z to the results of the measurements.

A recent paper by Crozier and Seely [573] deals with the concentration distribution across wind in an aerosol emitted from a point source on the ground. Samples were taken from an aeroplane flying at various heights through the aerosol stream so that nearly instantaneous determinations of concentration were possible.

At distances up to several tens of kilometres from the source the concentration was approximately proportional to exp $(-By^n)$ where y is the distance from the instantaneous axis of the stream, B is a constant and n is about 1·7, fairly close to Sutton's index of 2 [see (47.13)].

The author considers that further progress in this field necessitates that:

(1) knowledge of vertical diffusion as a function of height, under various meteorological conditions and differing degrees of surface roughness, must be improved both theoretically and experimentally. Some advances have recently been made in this direction [412].

(2) The way in which the coefficient of horizontal absolute diffusion depends on the observation time t_{obs} (see page 276) must also be studied. Nothing has yet been done in this connection, so that a theory of diffusion from a point source cannot be constructed.

(3) The experimentally determined functions $U(z)$, $D_{tz}(z)$ and $D_{ty}(z, t)$ must be substituted in equation (47.5) or the corresponding equation for a point source and solved exactly with a computer for a series of typical wind and surface conditions.

The emission of smoke from factory chimneys is of special interest in public health, the important factor being the concentration of atmospheric pollutant at ground level. The expressions obtained for this from (47.15) and (47.16) are

$$n_{z=0} = \frac{2\Phi'}{\sqrt{\pi C_z^2 U^2 x^s}} \exp\left(-\frac{H^2}{C_z^2 x^s}\right) \tag{47.26}$$

for a line source of height H and

$$n_{z=0} = \frac{2\Phi}{\pi C_y C_z U x^s} \exp\left(-\frac{y^2}{C_y^2 x^s} - \frac{H^2}{C_z^2 x^s}\right) \tag{47.27}$$

for a point source. In this case the values given in Table 26 should be chosen for C_y and C_z.

Bosanquet and Pearson [413] obtained different expressions for the ground-level concentration. For a line source on the ground they derived

$$n = \frac{\Phi'}{b_z U x} e^{-z/b_z x}, \tag{47.28}$$

where b_z is an empirical coefficient. As in (47.21) they assumed that D_{tz} was proportional to z and that U was constant; their formula reduces to (47.21) if $b_z = \varkappa U^*/U$. Bosanquet and Pearson also proved a reciprocity law, the ground-level concentration of a cloud emitted from a line source of height H with D_{tz} proportional to z being the same as the concentration at a height H of a cloud emitted from a line source on the ground. Hence

$$n_{z=0} = \frac{\Phi'}{b_z U x} e^{-H/b_z x}. \tag{47.29}$$

For a point source they gave the semi-empirical formula

$$n_{z=0} = \frac{\Phi}{\sqrt{2\pi}\, b_y b_z U x^2} \exp\left(-\frac{H}{b_z x} - \frac{y^2}{2 b_y^2 x^2}\right), \tag{47.30}$$

where b_y is a coefficient, like b_z, characterizing horizontal diffusion.

Measurements of atmospheric pollution due to factory chimneys are very variable and the data obtained by averaging are inaccurate; it is therefore difficult to decide whether the formulae of Sutton or of Bosanquet and Pearson agree better with experiment. As a result of numerous experiments some authors [414, 415] conclude that the agreement is about the same for each, provided that the coefficients are chosen suitably.

It may be necessary to make allowances for gases leaving a chimney with a high velocity and being warmer than the surrounding atmosphere. The rise of a smoke plume above a chimney, due to these factors, has been calculated in the monographs mentioned at the beginning of this section and is not considered here.

The formulae given above show that the ground-level concentration of smoke from a factory chimney is small near the chimney, increases on going farther away, and then begins to drop. The maximum of the function $n(x)_{y=z=0}$, found in the usual way, indicates the distance downwind of the point of highest concentration.

$$x_{max} = H^{2/s}/C_z^{2/s}, \tag{47.31}$$

$$n_{max} = 2\Phi C_z/\pi C_y U H^2 e \tag{47.32}$$

are obtained from formula (47.27), and

$$x_{max} = H/2\,b_z, \tag{47.33}$$

$$n_{max} = 4\,b_z\Phi/\sqrt{2\pi}\,b_y U H^2 e^2 \tag{47.34}$$

from formula (47.30).

Since $2/s$ is nearly equal to 1, these equations show that the distance from the chimney downwind to the point of maximum air pollution is proportional to the height of the chimney, while the maximum concentration is inversely proportional to the square of the height of the chimney and to the wind speed.

A few words must be said about the appearance of a smoke plume leaving a chimney [416]. At very small values of D_{tz} which occur in an atmospheric inversion, a narrow plume, spreading very slowly in the vertical direction, is obtained. It remains sharply defined a long way from the chimney and is bounded by two practically straight lines embracing an angle of a few degrees. In a horizontal plane, however, the plume is strongly curved – a clear indication that vertical and horizontal diffusion differ.

With the normal adiabatic temperature gradient the spreading angle of the plume is much greater. Careful observation reveals that vortices with horizontal, crosswind, axes rotate as if they were rolling on the ground. In this case turbulence is actually due to friction of the wind against the ground. In a superadiabatic gradient the plume breaks up into large puffs of smoke which move rapidly up and down; the turbulence is then mainly due to local convection currents produced by uneven heating of the earth's surface. These appearances are illustrated in Fig. 69 showing smoke plumes for a wind speed of 10 m sec^{-1} at the chimney, and for various coefficients of vertical diffusion (in cm^2 sec^{-1}).

Frenkiel and Katz [791] studied the spread in the atmosphere of clouds generated by powder blasts from air balloons by taking photographs of them from the grou nd. The contour of a cloud on the photograph was determined by the extinction of light passing through its edges being imperceptible on the plate. An experiment lasted some 20 sec until the cloud disappeared. By measuring the photographs taken at successive intervals and constructing circles of the same area as the cloud it was found that their spreading could be described by (45.12). However, the moment of the explosion had to be taken as $t = 1$ sec, rather than as $t = 0$.

As Gifford [792] correctly observed, the spreading of clouds is a matter of relative diffusion (see p. 262) described by (45.15) and (45.16), rather than absolute diffusion (45.12). In fact, the data of Frenkiel and Katz were shown by Gifford to be described accurately by (45.15) for $t < 10$ sec and by (45.16) for $t > 10$ sec.

§ 48. DEPOSITION OF AEROSOLS FROM THE ATMOSPHERE

The motion in the atmosphere of coarse aerosols with an appreciable rate of settling possesses the following characteristics. Concentrated clouds, emerging from a source, at first move as a whole with the gas contained in them [§ 13] in a manner determined initially by their size and density.

Aerosols produced by thermal means have a higher temperature than the surrounding air and move upwards, at first, with considerable speed. Smoke leaving vertical chimneys has, in addition, the velocity acquired in the chimney. After a time, the temperatures of the cloud and the surrounding atmosphere become equal and, owing to turbulent mixing, any difference in density disappears; this sort of motion then stops and the effect of gravity is only manifest in the settling of the particles relative to the medium. The spread of an isodisperse cloud, discussed above, has superimposed on it uniform settling at a velocity, V_s, equal to the rate of settling of the individual particles in still air. A polydisperse clouu can be regarded as consisting of several isodisperse clouds settling at different rates.

The rate of deposition of aerosol particles, or the number deposited per sec per cm² on the earth's surface, from a moving cloud is of practical importance but, unfortunately, it is very difficult to calculate. At small distances x from the source ($x V_s/U \ll H$), the distributions derived in the preceding section for gaseous pollutants can be used without serious error for aerosols of fine particles to find the concentration at the earth's surface $n_{z=0}$. The rate of deposition at a point on the surface is then

$$I = n_{z=0} V_s. \tag{48.1}$$

If V_s is comparable with the rate of vertical spreading of the aerosol stream (for example under the conditions of experiment mentioned on page 281 with $V_s \approx 0{\cdot}1\ U \approx 50$ cm sec^{-1}, thus $r \approx 0{\cdot}1$ mm), this simple method of calculation is no longer applicable. This problem has been examined by Johnstone and his co-workers [417]. Since in a time t the aerosol particles descend a distance $V_s t$ relative to a gas, Johnstone found the concentration distribution in the stream from the expressions derived for gaseous streams with z replaced by $z + V_s t = z + V_s x/U$. This supposes that the concentration distribution is the same in settling as in non-settling clouds, which is possible only if the wind speed and coefficient D_{tz} are constant (independent of z). In fact, as an aerosol cloud settles its spreading will differ more and more from that of a non-settling cloud.

After changing the formulae of type (47.15) and (47.16) as indicated, Johnstone multiplied the second term in scroll brackets by α, the coefficient of reflection of the aerosol from the ground. For $\alpha = 1$ the formulae obtained correspond to a smoke stream reflected from the ground, and for $\alpha = 0$ to a stream which settled as if the earth were completely permeable to it. Johnstone assumed erroneously that $\alpha = 0$ corresponded to a stream absorbed by the earth's surface; the difference between these two cases has already been discussed in § 38, I. Starting from the expression containing α, for the aerosol concentration at the earth's surface, and formula (48.1), the equation of material balance in the settling stream was obtained. This gave an expression for α which made $\alpha = 1$ at $x = 0$; it then decreased with distance from the source, passed through zero and approached -1 when x tended to infinity. This shows clearly that α has no physical meaning and that the formulae proposed by Johnstone are essentially empirical. We have dwelt in detail on this work because little has appeared in the literature until recently apart from a formula given by Bosanquet [413] without any foundation.

Rigorous solution of the problem for a line source should start from the differential equation (47.5) with an extra term to take account of settling,

$$U\frac{\partial n}{\partial x} - V_s\frac{\partial n}{\partial z} = \frac{\partial}{\partial z}\left(D_{tz}\frac{\partial n}{\partial z}\right). \tag{48.2}$$

This equation should be solved by substituting theoretical or experimental expressions for U and D_{tz} as functions of z with the boundary condition (48.1).

The settling of a cloud of very coarse droplets from a height H has been examined by Davies [418] who was interested not in the place at which the cloud settled on the ground but in the concentration distribution in the deposit, due to relative diffusion in the cloud. Starting from Sutton's theory Davies obtained the formula

$$n(x, y) = \frac{Q\cos\theta}{\pi C^2 L^s}\exp\left(-\frac{y^2\cos^2\theta + x^2}{C^2 L^s}\right), \tag{48.3}$$

x and y are coordinates on the earth's surface with the origin at the centre of the cloud, Q is the mass of the cloud, θ the angle between the line of descent of the cloud and the vertical ($\theta = \tan^{-1} U/V_s$), L the length of the path ($= H/\cos\theta$) and C the averaged Sutton coefficient. Experiments with clouds of water droplets ($r \approx 0{\cdot}5$ mm) descending from heights of 300 to 1500 m in a normal gradient were claimed by Davies to be in good agreement with formula (48·3); C was determined from special experiments.

In conclusion the vertical distribution of aerosol particles in the atmosphere will be briefly mentioned. Numerous observations show that the concentration and mean size of particles decrease with height above the earth's surface. There is a continuous cycle in the atmosphere, some particles being deposited while new ones are formed and dispersed. A roughly stationary state, and hence a definite particle distribution with respect to height, can be regarded as existing and is given by the equation derived on page 258.

$$\ln(n/n_0) = -V_s\int_0^z \frac{dz}{D_{tz}}. \tag{48.4}$$

This relates the turbulence of the atmosphere to the distribution of particles of a given size, characterized by V_s, with respect to height.

From time to time theoretical papers on the deposition of aerosols from the atmosphere appear in which the sedimentation of the particles is either completely ignored [793], or taken account of by introducing into the theory of Sutton a coefficient of reflection from the ground, or some other artificial boundary condition [794]. A more correct approach to this problem, leading to a direct solution of (48.2) for a continuous line source and normal adiabatic temperature gradient, was made by Rounds [795]. By aproximating the logarithmic wind profile to the power function $U = U_1(z/z_1)^\alpha$ and assuming $D_{tz} = \varkappa U^* z$ (see (47.20)) Rounds deduced the following expression for the aerosol concentration at ground level

$$n_{z=0} = \frac{\Phi' (1 + \alpha) e^{-Ax} x^{p-1}}{h U_h \Gamma(1 - p) A^{p-1}} \tag{48.5}$$

in which $p = -V_s/\varkappa U^*(1 + \alpha)$, $A = h^2 U_h / D_h (1 + \alpha)^2$, h is the height of the source and U_h, D_h are the values of U and D_{tz} at the height h; Γ is the gamma function. The rate of deposition on the ground is $I = V_s n_{z=0}$.

Godson [796] suggested an approximate method of estimating $n_{z=0}$ from (48.5) when the temperature gradient was nôt adiabatic. Graphs of the rate of deposition on the ground (I, x) for different values of h, α and $V_s/\overline{U}$ ($\overline{U}$ is the mean velocity of the wind over the range of heights from 0 to h) are given in Godson's paper. Fortak [797] gave a general solution of (48.2) for $U =$ const. and $D_{tz} =$ const.

For a point source the term $\partial (D_{ty}\, \partial n/\partial y)/\partial y$ must be added to (48.2); the solution of the resulting equation with $U =$ const., $D_{tz} = k_1 Uz$ and $D_{ty} = k_2 Ux$ is given by Denissov [798] and for $U = U_1 z^m$, $D_{tz} = k_1 z$, $D_{ty} = k_2 z^m$ by Gandin and Soloveitchik [799].

CHAPTER VII

THE COAGULATION OF AEROSOLS

§ 49. THERMAL (BROWNIAN) COAGULATION OF AEROSOLS WITH SPHERICAL PARTICLES

When aerosol particles come into contact and coalesce or adhere to one another the process is called coagulation. Particles may come into proximity because of their Brownian motion, resulting in thermal coagulation, or superimposed on this there may be an orderly motion produced by hydrodynamic, electrical, gravitational or other forces. The velocity of approach due to such forces may be large enough to set up a rate of coagulation compared with which that due to Brownian motion is negligible. Thermal coagulation is spontaneous, as is the coagulation of aerosols containing charged particles. Other cases may be referred to as forced coagulation.

Theories of coagulation usually start with the assumption that particles adhere at every collision. Except for large particles this assumption has a firm theoretical and experimental basis and it will be accepted for the time being, postponing discussion to § 56. The theory developed by Smoluchowski [419], which has been applied to the thermal coagulation of aerosols with spherical particles, will be considered taking initially the simplest case when all the particles are the same size.

Imagining one of the particles to be stationary, while others collide with it due to their Brownian motion, the average time interval between collisions will be calculated on the assumption that every collision is effective. The problem is simplified by assuming that the shape and size of the stationary particle are preserved in spite of coagulation with the other particles; it will be shown that this does not lead to appreciable error in the early stages of coagulation.

Since spherical particles come into contact when the distance between their centres is equal to the sum of their radii, the stationary particle can be replaced by an absorbing sphere of radius $2r$ and the other particles by their centres. The problem is best solved by the method of Kolmogoroff and Leontowitsch (§ 37, III). Let $W^*(\varrho, t)$ denote the probability that the centre of a particle, which at $t = 0$ is a distance ϱ from the centre of the absorbing sphere, comes into contact with it in a time t. It was shown in § 37, III that the function W^* satisfies the Einstein–Fokker equation, which in the present case has the form [see equation (38.28)]

$$\frac{\partial(\varrho W^*)}{\partial t} = D \frac{\partial^2(\varrho W^*)}{\partial \varrho^2}. \tag{49.1}$$

The physical meaning of the function W requires that it should also satisfy the conditions

$$W^*(2r, t) = 1 \quad \text{and} \quad W^*(\varrho, 0) = 0 \quad \text{if} \quad \varrho > 2r. \tag{49.2}$$

The solution is then

$$W^*(\varrho, t) = \frac{2r}{\varrho}\left[1 - \operatorname{erf}\left(\frac{\varrho - 2r}{2\sqrt{Dt}}\right)\right]. \tag{49.3}$$

If the number concentration of the aerosol is n, then on average

$$d\Phi = n W^*(\varrho, t) \times 4\pi\varrho^2 d\varrho \tag{49.4}$$

particles, whose centres were initially at distances between ϱ and $\varrho + d\varrho$ from the centre of the stationary particle, will come into contact with it in a time t. The total number colliding with it in a time t is therefore

$$\Phi = 4\pi n \int_{2r}^{\infty} W^*(\varrho, t)\, \varrho^2\, d\varrho. \tag{49.5}$$

Since $W^*(\varrho, t)$ falls off very rapidly as ϱ increases, the upper limit to the integral can be taken as infinity instead of the radius of the vessel containing the aerosol. The number of particles coming into contact with the stationary particle between t and $t + dt$ is thus

$$\frac{d\Phi}{dt} dt = 4\pi n dt \int_{2r}^{\infty} \frac{\partial W^*}{\partial t} \varrho^2 d\varrho = 8\pi r D n \left(1 + \frac{2r}{\sqrt{\pi D t}}\right) dt \tag{49.6}$$

and between 0 and t

$$\Phi = 8\pi r D n \left(t + \frac{4r\sqrt{t}}{\sqrt{\pi D}}\right). \tag{49.7}$$

Equation (49.6) shows that if $2r/\sqrt{\pi D t} \ll 1$ the mean time interval between two collisions is $1/8\pi r D n$. The effect of the term $2r/\sqrt{\pi D t}$ is to shorten the intervals at the beginning of the process. The mean size of the first interval t_1 (from $t = 0$ to the first contact) can be found by putting $\Phi = 1$ in (49.7) and solving for $\sqrt{t}$. This gives

$$t_1 = \frac{1}{8\pi r D n}\left[1 + 64\, n r^3 - \sqrt{(64\, n r^3)^2 + 128\, n r^3}\right]$$

$$\approx \frac{1}{8\pi r D n}\left[1 + 15{\cdot}3\, \varphi - \sqrt{15{\cdot}3\, \varphi\, (2 + 15{\cdot}3\, \varphi)}\,\right], \tag{49.8}$$

where $\varphi = {}^4/_3\pi r^3 n$ is the total volume of the particles in 1 cm^3 of aerosol. For values of φ (about 10^{-8} to 10^{-5}) which are usual in experiments on thermal coagulation the first time interval is practically equal to $1/8\pi r D n$ so that the term containing $\sqrt{t}$ in (49.7) can be neglected. The physical meaning of this term is that at the very beginning another particle may happen to be in the immediate vicinity of the absorbing one; the first collision will then occur very quickly, although the probability of particles being near to one another is low when φ is small. After all the particles near at hand have coagulated or diffused away the rate of coagulation assumes a stationary value.

An advance will now be made from the idea of a stationary absorbing sphere in order to calculate the mean square relative displacement along the x-axis of two particles. If in a time Δt the x coordinate of the first particle has moved a distance

Δx_1 and that of the second particle Δx_2, the relative displacement is $\Delta x_1 - \Delta x_2$ and the mean square relative displacement is

$$\overline{(\Delta x_1 - \Delta x_2)^2} = \overline{(\Delta x_1)^2} + \overline{(\Delta x_2)^2} - 2\,\overline{\Delta x_1 \cdot \Delta x_2}.$$

Since Δx_1 and Δx_2 are independent, $\overline{\Delta x_1\, \Delta x_2} = 0$ and

$$\overline{(\Delta x_1 - \Delta x_2)^2} = \overline{(\Delta x_1)^2} + \overline{(\Delta x_2)^2} = 2\,(D_1 + D_2)\,\Delta t. \qquad (49.9)$$

The coefficient of relative diffusion of the two particles is thus equal to the sum of the diffusion coefficients of each particle or, if they are the same size, $2\,D$. A particle therefore collides with $16\pi r D n$ other particles in one second and in all there are $\frac{n}{2} \times n\, 16\pi r D n = 8\pi r D n^2$ collisions each second in 1 cm³ of the aerosol, the factor 1/2 being necessary since each collision has been included twice. Since every collision reduces the number of particles by one it follows that the basic equation of coagulation is

$$\frac{dn}{dt} = -\,K_0 n^2. \qquad (49.10)$$

This integrates to

$$\frac{1}{n} - \frac{1}{n_0} = K_0 t \quad \text{or} \quad n = \frac{n_0}{1 + K_0 n_0 t}, \qquad (49.11)$$

where n_0 is the initial particle concentration and the coagulation constant, K_0, is

$$K_0 = 8\pi r D. \qquad (49.12)$$

These equations were first obtained by Smoluchowski by the method given in § 38, VII, considering the concentration distribution in an aerosol diffusing towards an absorbing sphere.

Smoluchowski's arguments have occasionally aroused objections [420] but the derivation given above [253, 421] involving the function W^* seems to avoid controversial issues.

If D is replaced by kTB and B is given the values shown in Table 3 (page 28) it in possible to find values of the coagulation constant as a function of particle radius in air at 23°C and atmospheric pressure (Table 28).†

For highly dispersed aerosols the theory given above must be corrected for discontinuity in concentration at the surface of the absorbing sphere. This is analogous to the discontinuity in vapour concentration at the surface of a droplet that is either evaporating or else growing due to condensation of vapour, which was first pointed out by the author of this book [421]. The correction is appreciable when the apparent mean free path, l_B, of the aerosol particles (see § 35) is comparable with the radius of the absorbing sphere; according to Table 13 (page 184) this occurs when $r \leqslant 10^{-5}$ cm. The physical meaning of this can be explained as follows. It was mentioned in § 35 that diffusion equations can be applied to Brownian motion only for time intervals which are large compared with the relaxation time, τ, of the particles or for distances

† Cunningham's mobility formula is generally used for aerosol coagulation. This gives

$$K_0 = \frac{4kT}{3\eta}\left(1 + A\frac{l}{r}\right) \qquad (49.12\text{a})$$

which are large compared with l_B. Diffusion equations cannot describe the motion of particles inside a layer of thickness l_B adiacent to an absorbing wall; if the size of the absorbing sphere is comparable with l_B this layer has a substantial effect on the kinetics of diffusive deposition on the sphere.

TABLE 28. COAGULATION CONSTANT OF ISODISPERSE AEROSOLS IN AIR

r, cm	10^{-7}	2×10^{-7}	5×10^{-7}	10^{-6}	2×10^{-6}	5×10^{-6}
$K_0 \,.\, 10^{10}$, $cm^3 \,.\, sec^{-1}$	323	162	65·8	34·0	18·0	8·57
$K \,.\, 10^{10}$, $cm^3 \,.\, sec^{-1}$	4·5	2	9	12	11	7·2

r, cm	10^{-5}	2×10^{-5}	5×10^{-5}	10^{-4}	2×10^{-4}	5×10^{-4}	10^{-3}
$K_0 \,.\, 10^{10}$, $cm^3 \,.\, sec^{-1}$	5·56	4·18	3·44	3·19	3·06	2·99	2·98
$K \,.\, 10^{10}$, $cm^3 \,.\, sec^{-1}$	5·2	4·0	3·37	3·14	3·04	2·97	2·97

A rigorous derivation of the correction is difficult and only a simplified argument will be given which leads to a rough estimate of the correction. It is more convenient to use Smoluchowski's method and consider the diffusion of particles towards an absorbing sphere in an established concentration gradient (see § 38, VII).

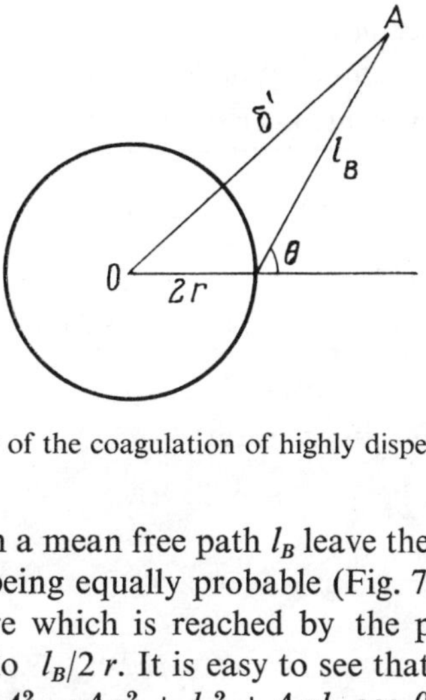

FIG. 70. The theory of the coagulation of highly dispersed aerosols.

Suppose that particles with a mean free path l_B leave the surface of the absorbing sphere in all directions, each being equally probable (Fig. 70). The mean distance δ' from the surface of the sphere which is reached by the particles after covering a distance l_B depends on the ratio $l_B/2\,r$. It is easy to see that $\delta' = l_B$ for $l_B \gg 2\,r$ and $\delta' = l_B/2$ for $l_B \ll 2\,r$. Since $OA^2 = 4\,r^2 + l_B^2 + 4\,r\,l_B \cos\theta$, averaging the value of AO over all directions θ shows that

$$\delta' = \frac{1}{6\,r\,l_B}\{(2\,r + l_B)^3 - (4\,r^2 + l_B^2)^{3/2}\} - 2\,r. \tag{49.13}$$

The corresponding quantity, δ, for two particles in relative motion is approximately $\sqrt{2}\,\delta'$.

Imagine a spherical surface of radius $2\,r + \delta$ concentric with the absorbing sphere; in the layer between them, the particles move as if in a vacuum, that is in straight lines with a mean velocity $\bar{G}$ (see Table 13 on page 184). If the particle concentration is n' at the outer surface and n_0 a long way from the absorbing sphere, the concentration outside the surface at a distance ϱ from the centre of the sphere, as in the case of (38.24), is given by

$$n = n_0 - (n_0 - n')\frac{(2r + \delta)}{\varrho}. \tag{49.14}$$

The number of particles diffusing inwards across the outer surface per unit time is

$$\Phi = 4\pi(2r + \delta)^2 \, 2D\left(\frac{dn}{d\varrho}\right)_{\varrho = 2r+\delta} = 8\pi D(2r + \delta)(n_0 - n'), \tag{49.15}$$

in which D has been replaced by $2D$ (see (49.9)). Under stationary conditions all particles which cross this surface come into contact with the absorbing sphere.

This number can also be determined by a method used in the kinetic theory of gases. If the inner sphere were a drop of liquid in equilibrium with its vapour, the number of molecules evaporating from it in unit time and reaching the outer sphere, according to the well-known kinetic theory formula, would be

$$\Phi_1 = 4\pi(2r)^2 n' \frac{\sqrt{2}\bar{G}}{4} \tag{49.16}$$

where n' is the number of molecules in 1 cm^3 of the saturated vapour and $\bar{G}$ is the mean velocity of the molecules.
During this time an equal number of molecules would also pass from the outer sphere to the inner. The factor $\sqrt{2}$ occurs in changing from the absolute to the relative particle velocity. Hence,

$$\Phi = \Phi_1 = 8\pi(2r + \delta)D(n_0 - n') = 4\sqrt{2}\,\pi r^2 \bar{G} n'. \tag{49.17}$$

Since according to (35.4)

$$l_B \bar{G} = \tau \bar{G}^2 = \frac{8kT\tau}{\pi m} = \frac{8kTB}{\pi} = \frac{8}{\pi}D,$$

equation (49.17) can be put in the form

$$8\pi(2r + \delta)D(n_0 - n') = 32\sqrt{2}\,r^2 Dn'/l_B.$$

This allows n' to be expressed as a function of n_0 so that substitution of the resulting expression in (49.17) gives

$$\Phi = 16\pi r D n_0 \beta, \tag{49.18}$$

where

$$\beta = \frac{1}{\dfrac{r}{r + \dfrac{\delta}{2}} + \dfrac{\pi l_B}{2\sqrt{2}\,r}} \approx \frac{1}{\dfrac{r}{r + \dfrac{\delta}{2}} + 1{\cdot}1\dfrac{l_B}{r}}, \tag{49.19}$$

or, if $l_B \ll r$,

$$\delta = \sqrt{2}\,\delta' = \frac{\sqrt{2}}{2}l_B,$$

$$\beta \approx 1\bigg/\left(1 + 0{\cdot}75\frac{l_B}{r}\right), \tag{49.20}$$

a formula similar to Cunningham's (see § 8). The actual value of the coagulation constant is therfore

$$K = K_0 \beta, \tag{49.21}$$

where K_0 is the limiting value (49.12) for coarse particles. Values of K for particles of unit density, calculated with the aid of Table 13 and formulae (49.13) and (49.19), are given in Table 28. Since these figures result from a much simplified argument great accuracy is not claimed for them. It can be seen from the table that the coagulation constant increases initially with a growth in the particle size of the aerosol but reaches a maximum at about $r = 10^{-6}$ cm, after which it decreases. For $l_B \gg r$

$$\beta \approx \frac{1}{\dfrac{\pi l_B}{2\sqrt{2}\, r}\left(1 + \dfrac{4 r^2}{\pi l_B^2}\right)} \approx \frac{2\sqrt{2}\, r}{\pi l_B}.$$

When D is replaced by $\frac{\pi}{8} l_B \bar{G}$, (49.21) becomes

$$K = \tfrac{1}{2}\pi (2r)^2 \sqrt{2}\, \bar{G} = 2\sqrt{2}\, \pi r^2 \bar{G}, \tag{49.22}$$

which is the expression for the coagulation constant under gas-kinetic conditions as calculated from the well-known formula for the number of collisions between gas molecules. The correction therefore expresses the transition from diffusive to gas-kinetic coagulation regime just as Cunningham's correction, given in § 8, expresses the transition from the hydrodynamic to the gas-kinetic regime of particle motion.

If the assumption that every collision results in coagulation is not made, a different value is obtained for the coagulation constant. Let α denote the collision efficiency, or the fraction of collisions which result in coagulation. In the discussion already given (49.15) remains unchanged but (49.16) has to be multiplied by α. As a result the expression for β becomes

$$\beta = 1 \Bigg/ \left(\frac{r}{r + \dfrac{\delta}{2}} + 1{\cdot}1 \frac{l_B}{r\alpha}\right) \tag{49.23}$$

in place of (49·19).

If $r \gg l_B$

$$\beta \approx 1 \Bigg/ \left(1 - 0{\cdot}35 \frac{l_B}{r} + 1{\cdot}1 \frac{l_B}{r\alpha}\right) \tag{49.24}$$

and when $l_B \gg r$ the coagulation constant is

$$K \approx 2\sqrt{2}\, \pi r^2 \bar{G} \alpha. \tag{49.25}$$

Proportionality between the coagulation constant and the collision efficiency, which Smoluchowski assumed in his theory of "slow" coagulation [422], only occurs if $l_B \gg r$. In the alternative case K is comparatively insensitive to changes in α. For example, if $r = 10^{-4}$ cm, K decreases by only 5 per cent as α goes from 1 to 0·25, while at $r = 10^{-5}$ cm a 5 per cent decrease in K results with $\alpha = 0{\cdot}60$. It is thus clear that approximate agreement between experimental and theoretical values of the coagulation constant does not provide evidence of perfect collision efficiency between the particles.

The physical meaning of formula (49.23) is that after an ineffective collision the particles are separated by a distance of the order of δ and may then collide again. The smaller the ratio δ/r the greater is the probability of repeated collision and hence of coagulation. The probability of coagulation in ν collisions is $1 - (1 - \alpha)^{\nu} > \alpha$. In Smoluchowski's theory a multiple collision is counted as a single one with the result that an incorrect conclusion is drawn regarding the proportionality of K and α.

In order to deal with polydisperse aerosols [423] it will first be supposed that two sizes of particle exist in the aerosol having radii r_1 and r_2. At a concentration of 1 cm^{-3} for each size half the number of collisions per second of particles of the first size with particles of the second size per cm^3 of the aerosol is written as $K(r_1, r_2)$, the coagulation constant of particles of the first size with particles of the second size. In this case the radius of the absorbing sphere is $r_1 + r_2$ and the coefficient of relative diffusion of the particles is $D_1 + D_2$ so that†

$$K(r_1, r_2) = 8\pi \frac{r_1 + r_2}{2} \frac{D_1 + D_2}{2} \beta, \tag{49.26}$$

TABLE 29. COAGULATION CONSTANTS OF UNEQUAL PARTICLES

$K(r_1, r_2) \cdot 10^{10}$ $(\text{cm}^3\ \text{sec}^{-1})$

r_1 (cm) \ r_2 (cm)	10^{-7}	2×10^{-7}	5×10^{-7}	10^{-6}	2×10^{-6}	5×10^{-6}	10^{-5}	2×10^{-5}	5×10^{-5}	10^{-4}	2×10^{-4}	5×10^{-4}	10^{-3}
10^{-7}	4·5												
2×10^{-7}	7·5	6											
5×10^{-7}	30	13	9										
10^{-6}	90	40	15	12									
2×10^{-6}	300	110	35	17	11								
5×10^{-6}	1600	550	120	40	15	7·2							
10^{-5}	5×10^3	1400	270	80	25	8	5·2						
2×10^{-5}	13×10^3	4100	600	170	47	11	5·3	4·0					
5×10^{-5}	37×10^3	9500	1600	420	115	24	9	4·7	3·4				
10^{-4}	77×10^3	20×10^3	3200	840	230	45	16	7·1	3·7	3·2			
2×10^{-4}	160×10^3	40×10^3	6600	1700	450	90	30	12	5·2	3·5	3·0		
5×10^{-4}	400×10^3	100×10^3	16×10^3	4300	1100	220	72	28	10·3	5·6	3·7	3·0	
10^{-3}	800×10^3	200×10^3	30×10^3	8500	2200	430	140	54	19	9·6	5·5	3·3	3·0

where β is the correction factor mentioned above. Repeating the previous argument for unequal particles leads to

$$\beta = 1 \Bigg/ \left(\frac{\bar{r}}{\bar{r} + \dfrac{\delta_r}{2}} + \frac{4\bar{D}}{G_r \bar{r}} \right) \tag{49.27}$$

where $\bar{r} = (r_1 + r_2)/2$; $\bar{D} = (D_1 + D_2)/2$; $\delta_r = \sqrt{\delta_1^2 + \delta_2^2}$; and $G_r = \sqrt{G_1^2 + G_2^2}$.

Values of $K(r_1, r_2)$ are given in Table 29, from which it is seen that the coagulation constant rises rapidly with increase in the ratio of the particle radii. Since the size

† In this case collision is taken into account only once, hence it is not necessary to divide the total number of collisions by two. In comparing the coagulation constants for equal and unequal particles, howerer, half the total number of collisions is used to evaluate $K(r_1, r_2)$, both here and below.

of an aggregate resulting from the coagulation of a fine and a coarse particle does not differ much from the size of the latter, the fine particles in polydisperse aerosols are rapidly collected by the coarse ones. As a rule, therefore, it is not possible to find aerosol particles appreciably smaller than the mean size, even if they are known to have existed in the early stages of the aerosol's existence, for example when condensation aerosols are formed from substances of very low vapour pressure; to some extent this counteracts the increase of polydispersity during coagulation.

The initial stages of coagulation of fairly homogeneous aerosols, particularly mists, are little affected by the range of particle size. A droplet of radius $r\sqrt[3]{2}$ results when two primary droplets of radius r coalesce, a droplet of radius $r\sqrt[3]{3}$ when three primary droplets coalesce, and so on. Using (49.26) and Table 3 the coagulation constants for primary droplets with double, triple etc. droplets have been found, taking the coagulation constant of primary droplets with one another to be unity. These figures are given in Table 30; the coefficient β can be considered constant for the particle sizes in the table.

TABLE 30. COMPARATIVE MAGNITUDE OF COAGULATION CONSTANTS FOR PRIMARY DROPLETS WITH PRIMARY (K_{11}), SECONDARY (K_{12}) DROPLETS ETC.

Radius of primary droplets (cm)	K_{11}	K_{12}	K_{13}	K_{14}	K_{15}
10^{-5}	1	0·96	0·97	0·98	0·99
3×10^{-5}	1	0·99	1·01	1·03	1·05
10^{-4}	1	1·00	1·02	1·04	1·06

Coagulation is usually studied with mists having particle sizes of this order and for them the differences between the constants are negligible and do not exceed the experimental error. The reason for the constancy of K is that for particles of these sizes the reduction in mobility of one of the droplets is almost exactly balanced by the increase in radius of the absorbing sphere. In highly dispersed aerosols this compensation no longer occurs. Note also that the coagulation constants of multiple particles with one another are appreciably lower than with primary particles. During the initial stages of coagulation of an isodisperse mist, when primary droplets still outnumber multiple ones, the coagulation constant thus remains practically constant, as was assumed in the derivation of the basic equation of coagulation above.

Two difficulties arise with smokes. Firstly, coagulation is complicated by the irregular shape of the primary particles and aggregates (see § 52), and secondly the number and volume of primary particles in an aggregate bears no simple relation to the size of the aggregates as with mist droplets.

In general the coagulation constant of a polydisperse aerosol is equal to

$$K = \int_0^\infty \int_0^\infty K(r_1, r_2) f(r_1, t) f(r_2, t)\, dr_1\, dr_2, \tag{49.28}$$

where $f(r, t)$ is the particle size distribution function (see page 5) at the instant t and $K(r_1, r_2)$ is the coagulation constant for particles of radius r_1 and r_2, which for coarse aerosols is

$$K(r_1, r_2) = 8\pi \frac{r_1 + r_2}{2} \frac{D_1 + D_2}{2} = \frac{kT}{3\eta}(r_1 + r_2)\left(\frac{1}{r_1} + \frac{1}{r_2}\right)$$

$$= \frac{kT}{3\eta}\left(\sqrt{\frac{r_1}{r_2}} + \sqrt{\frac{r_2}{r_1}}\right)^2. \tag{49.29}$$

As shown by Tikhomirov, Tunistkii and Petryanov [424], if the radii exceed 10^{-5} cm and Cunningham's formula (8.2), can be used for the mobility, calculation of K does not require a knowledge of the particle size distribution of the aerosol but only the mean values $\bar{r}$, $\overline{\left(\frac{1}{r}\right)}$ and $\overline{\left(\frac{1}{r^2}\right)}$. In this case

$$K(r_1, r_2) = 8\pi \frac{r_1 + r_2}{2} \frac{D_1 + D_2}{2} = \frac{kT}{3\eta}(r_1 + r_2)\left[\frac{1 + A\frac{l}{r_1}}{r_1} + \frac{1 + A\frac{l}{r_2}}{r_2}\right]. \tag{49.30}$$

Substituting this expression in (49.28) and integrating gives

$$K = \frac{2kT}{3\eta}\left[1 + \bar{r}\overline{\left(\frac{1}{r}\right)} + Al\overline{\left(\frac{1}{r}\right)} + Al\bar{r}\overline{\left(\frac{1}{r^2}\right)}\right]. \tag{49.31}$$

The quantity

$$\vartheta = \frac{1 + \bar{r}\overline{\left(\frac{1}{r}\right)} + Al\overline{\left(\frac{1}{r}\right)} + Al\bar{r}\overline{\left(\frac{1}{r^2}\right)}}{2\left(1 + A\frac{l}{\bar{r}}\right)}, \tag{49.32}$$

or polydispersity factor, is equal to the ratio of the coagulation constants of a polydisperse aerosol having particles of mean radius $\bar{r}$ and an isodisperse aerosol with particles of the same radius.

Rearrangement of the basic equation of coagulation for polydisperse aerosols (49.28) led Todes [425] to conclude that the coagulation constant of coarse mists, where Cunningham's correction to the diffusion coefficient is negligible, tends to become constant when $t \gg 1/Kn_0$ or, according to (49.11), $n \ll n_0$; such aerosols have coagulated to a considerable degree. The limiting value is roughly 10 per cent higher than the coagulation constant for isodisperse mists, irrespective of the initial droplet size distribution. Pshenai-Severin [426] solved equation (49.28) by successive approximation for aerosols with an initial droplet mass distribution

$$f'(m) = (f_0'/\sigma) \exp(-m/\sigma)$$

where σ is a constant. He found the droplet concentration in a coagulating mist to be 12 per cent less than the figure obtained by the simplified calculation with K constant.

The theoretical predictions will now be compared with some experimental results. For the reasons given above, and also in § 50, it is best to use experiments with mists or smokes, such as stearic acid smoke, with spherical primary particles giving rounded, more or less dense aggregates.

The main experimental fact established by many investigations is the straight-line relationship between time and the reciprocal of the particle concentration. This is in agreement with equation (49.11) and implies that the coagulation constant in aerosols (including smokes) does not change with time; this has been even confirmed when the particle concentration falls to one-tenth or thirtieth of the initial value during the course of an experiment (see Fig. 71, taken from Derjaguin and Vlasenko [277]).

Corresponding to such changes in concentration the mean particle size increases by a factor of two or three, and the coagulation constant of the aerosols with which the experiments were conducted (initial mean radius of 0·1 to 0·3 μ) must have decreased quite noticeably (see Table 28); the effect of increasing the mean particle size must therefore have been compensated by the simultaneous increase in polydispersity of the coagulating aerosol. The first factor is greater when the initial particle radius is small and, in fact, aerosols with a very small weight concentration (1 to 5 mg.m^{-3}) and a mean particle radius of 0·5 to 0·7 $\times$ 10^{-5} cm exhibit an appreciable diminution in the coagulation constant during the process [427].

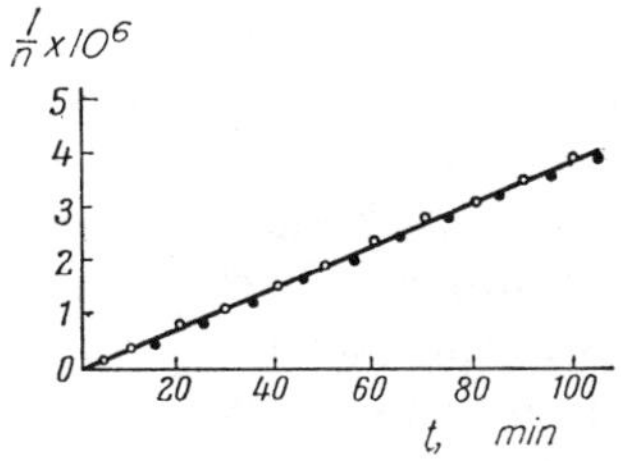

FIG. 71. The kinetics of thermal coagulation.

A careful comparison of experimental and theoretical values of the coagulation constant has been made by Tikhomirov, Tunitskii and Petryanov [424]. Their experiments were conducted with mists of mineral oil, tricresyl phosphate and sulphuric acid, the mean particle radii being about 0·3 μ. About half way through an experiment a sample of mist was taken from the chamber and diluted with air to stop coagulation. The particle size distribution in the sample was then determined by the vertical electric field method. The quantities $\bar{r}$, $\overline{\left(\frac{1}{r}\right)}$ and $\overline{\left(\frac{1}{r^2}\right)}$ were found and the polydispersity factor was calculated as described above. Its value in all the experiments was only a few per cent (a maximum of ten) greater than one. Values of the coagulation constant calculated from this were systematically 20–30 per cent below experimental. For example, in an experiment with mineral oil mist:

$$\bar{r} = 2{\cdot}63 \times 10^{-5}\ \text{cm}; \quad \overline{\left(\frac{1}{r}\right)} = 0{\cdot}378 \times 10^{5}\ \text{cm}^{-1}; \quad \overline{\left(\frac{1}{r^2}\right)} = 0{\cdot}145 \times 10^{10}\ \text{cm}^{-2};$$

$$\vartheta = 1{\cdot}02; \quad K_{\text{theor}} = 3{\cdot}95 \times 10^{-10}\ \text{cm}^3\ \text{sec}^{-1}; \quad K_{\text{exp}} = 4{\cdot}78 \times 10^{-10}\ \text{cm}^3\ \text{sec}^{-1}.$$

Although the number of droplets (15–20) used in these experiments to determine the mean sizes was clearly inadequate, and there is usually a tendency to select the brighter or coarser particles in making a visual estimation of particle sizes, there is little doubt that the experimental values exceed the theoretical ones.

In the experiments of Patterson and Cawood [428] a reproducible value of 5·2 $\times$ 10^{-10} cm^3 sec^{-1} was obtained for the coagulation constants of aerosols of stearic and oleic acids. Theoretically, without consideration of the polydispersity effect, this should have been about $4{\cdot}6 \times 10^{-10}$ at the beginning of the experiment and somewhat less as the particles became coarser. The discrepancy can be explained on the basis of the theoretical results of Todes and Pshenai-Severin. Another explanation is given below on page 308. It is highly desirable that measurements should be made of the coagulation constant in homogeneous mists, obtained by the nucleation method [6], in which the polydispersity factor is practically one.

Nolan and Kennan [437] measured the rate of coagulation of aerosols produced by volatilizing platinum in filtered air; this is the only information available for highly dispersed aerosols having $r < 0{\cdot}1\,\mu$ and, with certain reservations, it can be used to compare the theory of coagulation with experiment. The mean particle radius, which varied from $0{\cdot}5 \times 10^{-6}$ to $2{\cdot}6 \times 10^{-6}$ cm, was found by the diffusion method (see page 206) and the particle concentration was measured with a Pollak counter [292], the number of condensation nuclei being established by a light extinction method after condensing water vapour on them. The mean value of K/rD in these experiments was 33, deviations from the mean being large although no systematic variation with increase in r was observed. According to (49.12) the value should be $8\pi \approx 25$ for monodisperse aerosols, without correction for the concentration discontinuity, but high values of K would be likely on account of the polydispersity of the platinum aerosol. It appears, then, from these experiments that the correction for the concentration discontinuity is unnecessary, although this is not conclusive. In experiments with horizontal ducts (see page 206) the authors found values of 0·4 to 0·7 for the apparent density of the platinum particles, showing that the aerosol consisted of coarse aggregates of a large number of primary particles and hence that the degree of polydispersity was very large. It is also possible, on the basis of the discussion on pages 18 and 42, that an appreciable proportion of the particles had elongated shapes. It is clear from §§ 49 and 50 that a large increase in the coagulation constant must result and compensate for the effect of the discontinuity in concentration. The problem cannot be finally solved until the rate of coagulation of highly dispersed, slightly coagulated aerosols has been studied.

Some thought must be given to the work of Langstroth and Gillespie [361] in which the coagulation constant was determined from the decrease in number concentration of the aerosol due to deposition on the walls and bottom of the chamber.† Since the rates of decrease in number due to deposition and coagulation are proportional to n and n^2 respectively, the coefficients β and K in the equation

$$-\frac{dn}{dt} = \beta n + Kn^2 \tag{49.33}$$

or in its integrated form

$$\ln\left(\frac{1}{n} + \frac{K}{\beta}\right) - \ln\left(\frac{1}{n_0} + \frac{K}{\beta}\right) = \beta t \tag{49.34}$$

were chosen so that the equations expressed the actual decrease in n found experimentally. This gave reasonable values for K, but the method is unreliable, particularly because the linear relation between $\ln n$ and t, which is the basis of the method, does not hold in polydisperse aerosols (page 253). Langstroth and Gillespie therefore

† With particles measuring a few tenths of a micron, in a chamber of several cubic metres, the correction for deposition is very small.

discarded this method, determined β directly (see page 254) and calculated K by difference. The values found were not constant during the experiment but rose and fell; they were also much greater than the theoretical values and others which have been found experimentally. The data of Langstroth and Gillespie are therefore not very reliable.

In conclusion the change in particle size distribution during coagulation will be mentioned. It was studied by Smoluchowski [429] for an initially isodisperse aerosol on the assumption that the coagulation constant $K(r_1, r_2)$ was the same for all the particles in the coagulating aerosol. If n_0 is the initial number of particles in unit volume, the number of aggregates containing ν primary particles, when the total number of particles (both primary and aggregates) is n, is given by

$$n_\nu = \frac{n^2(n_0 - n)^{\nu-1}}{n_0^\nu}. \tag{49.35}$$

The sequence $n_1, n_2, n_3 \ldots$ is thus a geometrical progression with a common ratio $(n_0 - n)/n_0$, n being given by (49.11).

The coagulation of a polydisperse system with an arbitrary initial particle size distribution has been studied by Tunitskii [430]. Let $n(m, t)\,dm$ be the concentration of particles having masses between m and $m + dm$ at time t. The variation of $n(m, t)$ with time is given by the following equation in which t has been omitted from the right hand side in the interest of simplicity.

$$\frac{dn(m, t)}{dt} = \frac{1}{2}\int_0^m K(m_1, m - m_1)\, n(m_1)\, n(m - m_1)\, dm_1$$
$$- n(m)\int_0^\infty K(m, m_1)\, n(m_1)\, dm_1, \tag{49.36}$$

$K(m, m_1)$ is the coagulation constant for particles of mass m and m_1. The first term on the right hand side is the increase in the number of particles of mass m due to those formed from smaller ones; the second term is their decrease owing to coagulation with other particles. Equation (49.36) can only be solved if K is constant, so that the aerosol must not be too polydisperse. The solution is

$$n(m, t) = \frac{1}{2\pi\left(1 + \frac{K}{2} n_0 t\right)} \int_0^\infty e^{-im\xi} \frac{\Phi(\xi)\, d\xi}{1 - K\tau\Phi(\xi)}\, d\xi \tag{49.37}$$

where n_0 is the total initial particle concentration and

$$\tau = \frac{t}{2\left(1 + \frac{K}{2} n_0 t\right)}, \qquad \Phi(\xi) = \int_0^\infty e^{-im\xi}\, n(m, 0)\, dm. \tag{49.38}$$

If the initial distribution of particle masses, $n(m, 0)$, is known, any subsequent distribution can be found by numerical integration although this requires a lot of work. Schumann [583] gave an asymptotic solution to equation (49.36) for constant K when $Kn_0 t \gg 1$:

$$n(m) = \frac{c}{\bar{m}^2} e^{-m/\bar{m}} \tag{49.39}$$

c is the weight concentration of the aerosol and $\bar{m}$ the mean mass of a particle at time t,

$$\bar{m} = \frac{c}{n} = \frac{c(1 + 0{\cdot}5\, Kn_0 t)}{n_0} \approx \frac{Kct}{2}. \tag{49.40}$$

He believed that this formula would hold for any initial particle size distribution but was unable to give a rigorous proof.

Another asymptotic solution of equation (49.36) with a constant value of K is due to Todes [425].

A theory of the coagulation of very fine aerosols is given in Zebel's paper [800]. The author equates the rate of diffusion to kinetic deposition, not at the distance of the apparent mean free path of the particles l_B from the absorbing sphere (see p. 291), but at the actual surface of the latter. He obtains a formula, which for identical particles reduces to

$$K = K_0 \Bigg/ \left(1 + \frac{4D\sqrt{m}}{r\sqrt{6kT}}\right) \tag{49.41}$$

(m is the mass of each particle and K_0 is the coagulation constant without the correction for concentration discontinuity) whereas equation (49.41) can be written as

$$K = K_0 \Bigg/ \left(1 + \frac{3D\sqrt{m}}{r\sqrt{2\pi kT}}\right). \tag{49.42}$$

Zebel's view that his formula is also valid when $3D\sqrt{m}/r\sqrt{2\pi kT} \approx l_B/r \gg 1$ is incorrect; equation (49.22) should be used in these circumstances.

Investigations of the coagulation rate of condensation nuclei in the air of a room ($r = 2$–3×10^{-6} cm) made in Nolan's laboratory by O'Connor [801] yielded the improbably large value of $K/rD = 100$–140.

From theoretical and experimental data on the coagulation of bipolarly charged aerosols and taking into consideration the fact that particles of the size indicated rarely carry more than one elementary charge (see Table 11), it is evident that charges on the particles cannot be responsible for such high values of the coagulation constant. Some serious source of error must have been present in these experiments.

Experiments by Cawood and Whytlaw-Gray [802] on the coagulation constant of aerosols of Fe_2O_3, prepared by the photochemical decomposition of vapour of $Fe(CO)_5$ in air at pressures of 200–760 mm Hg indicated a considerable increase at lower pressures. This is not surprising since in the initial stage of their existence the particles are undoubtedly very small and, in accordance with (49.12a), their coagulation constant increases with the ratio l/r.

Using a differential counter of condensation nuclei which enabled the concentration of nuclei of different sizes, corresponding to different degrees of supersaturation, to be measured, Yaffe and Cadle [803] made a qualitative study of the kinetics of coagulation of very fine aerosols prepared by heating TiO and NaCl. Following theory, the number of particles of the smallest sizes, $r = 0{\cdot}4\text{–}0{\cdot}6 \times 10^{-7}$ cm, rapidly decreased from the start while the coarser particles first increased, then attained a maximum and finally diminished in number.

Melzak [804] proposed a method of solving the basic integro-differential equation of coagulation (49.36) with constant K which is very similar to Tunitskii's method (see p. 299). The solution of this equation with values of K which depend on the mass or volume of the coagulating particles is certainly of much greater importance, but seems to be possible only by numerical methods. The first calculation of this kind, giving the change in size distribution of a coagulating aerosol containing initially a given range of particle sizes, was made by Zebel [805] with the help of an electronic computer.

In Derjaguin's view [806] a theory of coagulation should consider that before two approaching particles can come into contact a gaseous film between them has to be squeezed out. According to Taylor [807], the rate of approach of two spheres of radius r under the action of a force F is equal to

$$V = \frac{Fh}{3\pi\eta r^2}. \tag{49.43}$$

In the deduction of this formula the gap between the spheres h is assumed to be very small compared with r. As can be seen from this formula, the spheres can never come into contact. In fact, experiments with spheres falling through a very viscous liquid [808] show that approaching spheres fall apart instead of sticking together. The reflection of large water drops in air after colliding with one another, due to the presence of an air film, is also well known. Everyday experience shows, however, that hard spheres come into contact when they collide. As regards large particles, the problem will be discussed later in § 56. In the theory of Brownian coagulation, with particles of $r < 1\,\mu$, the solution is much simpler. In this case Taylor's formula is evidently valid only when h is smaller than the mean free path of gas molecules, in which event hydrodynamical formulae are inapplicable to the squeezing out of the gas film.

Dautrebande and his collaborators [482], [809] have claimed an intensive coagulation of silica dust by aerosols of NaCl. According to Avy [810], a perceptible decrease in particle concentration of an aerosol of indigo with $\bar{r} = 0{\cdot}1$ to $0{\cdot}15\,\mu$ and $n_0 = 2 \times 10^5$ was observed when a mist from a sodium chloride solution having $\bar{r} = 2\text{–}3\,\mu$ was sprayed into the aerosol, the concentration of NaCl mist in the chamber being brought to $n = 2 \times 10^4$. When injection of the mist ceased coagulation of the aerosol of indigo proceeded at practically the same rate as in the control experiment without the NaCl aerosol, as was to be expected from the theory of Brownian coagulation.

Le Bouffant [811] showed that NaCl aerosol ($\bar{r} \approx 0{\cdot}1\,\mu$) when mixed with a cloud of coal dust ($\bar{r} = 0{\cdot}25\mu$) of about the same weight concentration caused no appreciable increase in the size of the coal particles.

It has long been known that the hygroscopic nature of NaCl and, consequently, the air humidity, are of particular importance in such experiments. Walkenhorst and Zebel [812, 813] found that a dry aerosol of NaCl ($r \leqslant 0{\cdot}1\,\mu$) had no effect upon the stability of coal dust, but when the humidity of the air was 95–99 per cent a rapid increase in the size and density of the aggregates of coal and NaCl particles, leading to their precipitation, was observed due to the absorption of moisture (see p. 20). Similar observations were made with SiO_2 aerosols. Dautrebande's observation [814] that SiO_2 aerosols ($\bar{r} = 0{\cdot}35\,\mu$) mixed with NaCl aerosols were deposited in the bronchial tubes of test animals, while pure SiO_2 aerosols deposited in the lungs, may be due to inhaled air becoming saturated with water vapour as it passes down the respiratory tract [644].

§ 50. THE COAGULATION OF AEROSOLS HAVING ELONGATED PARTICLES

The coagulation constant of aerosols with spherical particles was calculated from the rate of diffusion of the particles on to an absorbing sphere. A different shape of absorbing surface has to be used for non-spherical particles [431]. As in § 49 diffusion on to the absorbing surface can be considered stationary and the general equation of diffusion in three dimensions (37.18) becomes

$$\Delta n = 0 \tag{50.1}$$

where Δ is the Laplace operator. It is also necessary for n to satisfy the boundary conditions $n = 0$ on the absorbing surface and $n = n_0$ at an infinite distance from it. The function $\psi = n - n_0$ satisfies both the equation $\Delta\psi = 0$ and the conditions $\psi = -n_0$ at the absorbing surface and $\psi = 0$ at an infinite distance from it. Thus ψ has the same form as the potential in the field round a conductor, charged to potential $-n_0$, of the same shape as the absorbing surface. The concentration gradient of an aerosol at an absorbing surface, $(dn/dh)_0$, is therefore equal to the field intensity at the surface of the conductor, which is proportional to the charge density σ on the surface; it thus depends on the curvature of the surface at a given point. The concentration gradient, and the flux of particles diffusing upon unit area, have maxima at the poles of a prolate ellipsoid of revolution and at the equator of an oblate one.

The total number of particles diffusing towards the absorbing surface in unit time is

$$\Phi = \int D\left(\frac{dn}{dh}\right)_0 dS, \tag{50.2}$$

where dS is an element of surface and the integral is taken over the whole surface. Since

$$\left(\frac{dn}{dh}\right)_0 = -\left(\frac{d\psi}{dh}\right)_0 = -4\pi\sigma,$$

then

$$\Phi = -\int D \times 4\pi\sigma\, dS = 4\pi D q = 4\pi D C_E n_0, \tag{50.3}$$

where q is the charge on the surface at a potential $-n_0$ and $C_E = q/n_0$ is its capacitance.

The rate of diffusion of particles on to an absorbing surface is thus proportional to its electrical capacitance. This will be used to find the coagulation constant of ellipsoidal particles having polar and equatorial semi-axes c and a, respectively, with spherical particles of the same volume. The radius of the latter is $r = \sqrt[3]{a^2 c}$. The absorbing surface is the locus of points lying at a distance r from the surface of an ellipsoidal particle and can be replaced, to a first approximation, by an ellipsoid with semi-axes $a + r$ and $c + r$.

The capacitance of a prolate ellipsoid of revolution is [432]

$$C_E = \sqrt{c^2 - a^2}/\ln \frac{c + \sqrt{c^2 - a^2}}{a} \tag{50.4}$$

and of an oblate ellipsoid

$$C_E = \sqrt{a^2 - c^2}/\text{arc cos}\left(\frac{c}{a}\right). \tag{50.5}$$

Using these formulae, the values of the dynamic shape factor given in Table 7 for ellipsoidal particles, and allowing for orientation in all directions, it can be shown, for example, that prolate ellipsoids having an axial ratio of 10:1 have a coagulation constant with spherical particles which is 1·10 times greater than the value for spheres;

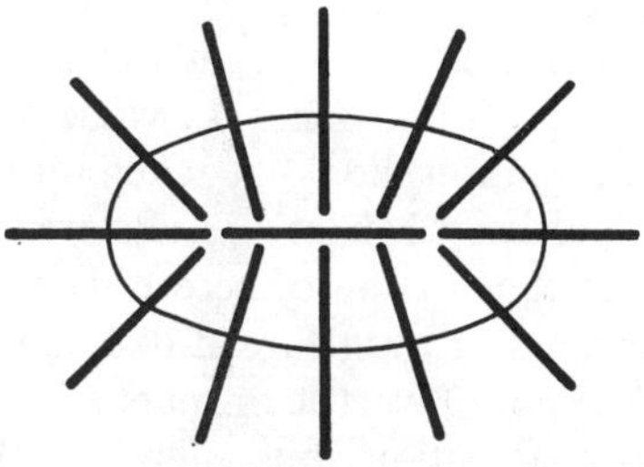

FIG. 72. Coagulation of rod-like particles.

for oblate ellipsoids the figure is 1·04. In this derivation Brownian rotation of the ellipsoidal particles has been neglected; this further increases coagulation and becomes even more important when both the coagulating particles are elongated. Unfortunately the theory of coagulation, taking Brownian rotation into account, presents great mathematical difficulties.

In Müller's theory [431] rod-like particles are assumed to coagulate by colliding mainly at their ends, where the absorbing surface has its maximum curvature, and to move mainly along their axes in the direction of maximum mobility. This leads to coagulation as represented in Fig. 72. The absorbing surface will then be the locus of the centres of the rods which have their ends touching the given rod. It is approximately an ellipsoid with the polar axis $4L$ and the equatorial axis equal to the length of the rod, $2L$. By (50.4) the capacity of this ellipsoid is

$$C_E = \frac{L\sqrt{3}}{\ln(2 + \sqrt{3})} = 1{\cdot}32\,L\dagger. \tag{50.6}$$

The axial mobility of the rods can be calculated if they are assumed to be very elongated ellipsoids, like needles, with an axial ratio β (see (12.11)):

$$B = \frac{\ln 2\beta}{4\pi\eta L}. \tag{50.7}$$

† Müller erroneously put $1{\cdot}49\,L$.

The expression thus obtained for the coagulation constant of rod-like particles is

$$K_s = \frac{2\pi \times 2 \times \ln 2\beta \times kT \times 1{\cdot}32\,L}{4\pi\eta L} = \frac{1{\cdot}32\,kT \ln 2\beta}{\eta}, \tag{50.8}$$

while for spherical particles it is

$$K_0 = \frac{2\pi \times 2\,kT \times 2\,r}{6\,\pi\eta r} = \frac{4\,kT}{3\,\eta}. \tag{50.9}$$

Hence $K_s/K_0 \approx \ln 2\beta$ and for $\beta = 10$, $K_s/K_0 \approx 3$.

However, the scheme of coagulation sketched in Fig. 72 is far from reality. In the first place, the mobilities of ellipsoidal particles along and at right angles to the axis differ only by a factor of two†; secondly, the concentration gradients at the poles and equator of the absorbing surface illustrated in Fig. 72 also differ by only a factor of two. Therefore, it is not permissible to say that rods move only along their axes or that they coagulate only at their ends. In fact the particles may touch at any point although it is true that contact predominates near the ends of the rods, so that the capacity of the absorbing ellipsoid in Fig. 72, and hence the value of K_s/K_0, must be decreased. On the other hand Brownian rotation should be taken into account and this increases the coagulation of rod-like particles. Müller's results are therefore qualitative.

Experimental data on the coagulation of aerosols having elongated particles are scanty. Although very elongated or even thread-like aggregates are found in many smokes they are usually the result of electric or magnetic forces between the particles (see § 52) so that the high coagulation constants in such smokes are not directly related to the topic at present under discussion.

The coagulation of aerosols obtained by subliming dimethyl aminoazobenzene proceeds in an interesting manner [434]. The mist of supercooled droplets formed at first has a normal rate of coagulation but after some time the droplets crystallise into needles and the coagulation constant increases markedly.

Some interesting results were obtained by Artemov [30] who showed that the coagulation constant of ammonium chloride smoke depended largely on the relative humidity of the gas phase. When the humidity was raised from 30 to 40 per cent, K fell from 5 to 3×10^{-10} cm³ sec⁻¹; with further increase in humidity K began to increase, reaching $6{\cdot}5 \times 10^{-10}$ at 50 per cent and 8×10^{-10} at 75 per cent relative humidity.††

Microscopic examination of the settled aggregates showed that as the humidity of the air increased they become more compact and rounded by capillary action of the water deposited on the surface and in the pores of the aggregates. This probably explains the scatter in K, from 4·5 to 9×10^{-10} $cm^3\ sec^{-1}$, in Radushkevich's work [436] for which ammonium chloride smokes were produced under identical conditions except that the humidity of the air in the chamber was not controlled. With stearic acid smokes, which consist of rounded aggregates, values of K are readily reproducible in carefully conducted experiments [428] and lie within narrow limits such as $5{\cdot}0$–$5{\cdot}3 \times 10^{-10}$ $cm^3\ sec^{-1}$.

† Experimental study of the Brownian motion of bacilli having $\beta = 26$ has shown that the Brownian displacement along the axis is only 30 per cent greater than at right angles [433].

†† That ammonium chloride smoke was stabilized by slightly humid air but became less stable at high humidities had previously been noticed by Samokhvalov and Kozhukhova [435].

The decrease in K when the humidity is raised from 30 to 40 per cent can be explained by the rounding of the ammonium chloride aggregates but the increase in K when the humidity is raised further is puzzling. Artemov thought it might be due to a more rapid sedimentation of moistened particles but at 50 per cent humidity ammonium chloride particles cannot have enlarged appreciably and enlargement in any event would cause a slight decrease in the coagulation constant (see Table 28).

Considerable reduction in the coagulation constant in the presence of solvent vapours which also cause rounded aggregates to form has been observed by Artemov in smokes of anthraquinone and nitrosodimethylaniline. In the presence of ammonia, however, which favours the formation of thread-like aggregates from nitrosodimethylaniline, K increased almost 80 per cent. Ammonia accelerates the coagulation of stearic acid smoke, also, because needle-like crystals of ammonium stearate are formed.

Fujitani [815] records that the introduction of ethyl acetate vapour (up to 0·06 millimoles/l) into a mist of a saturated NH_4Cl solution ($\bar{r} = 0{\cdot}25\,\mu$) caused a decrease in the coagulation constant from 8 to 3×10^{-10} cm³ sec⁻¹. A similar effect was obtained with amyl and butyl alcohols. It is unlikely that a decrease in the efficiency of collision could have occurred (see § 56) and, as shown in Table 28, the smaller figure is much closer to the theoretical value of the coagulation constant. Possibly crystallization of NH_4Cl in the droplets and the effects described above occurred in these experiments.

§ 51. THERMAL COAGULATION OF AEROSOLS HAVING CHARGED PARTICLES. THE EFFECT OF MOLECULAR FORCES ON THE RATE OF COAGULATION

The coagulation constant will now be found for equal spherical particles bearing charges q_1 and q_2 (of the same or opposite sign); the electrostatic force between the particles is $\mathbf{F}(\varrho)$, ϱ being the distance between their centres. Diffusion of particles on to the absorbing sphere now proceeds with an ordered radial velocity $\mathbf{V} = B\mathbf{F}$ superimposed on the Brownian motion. Assuming, as before, that the process is stationary and putting $\partial n/\partial t = 0$ in equation (37.19) gives

$$D\Delta n = \operatorname{div}(\mathbf{V}\,n) = B\operatorname{div}(\mathbf{F}n). \tag{51.1}$$

Owing to spherical symmetry this equation in polar coordinates is

$$D\frac{1}{\varrho^2}\frac{d}{d\varrho}\left(\varrho^2\frac{dn}{d\varrho}\right) = B\frac{1}{\varrho^2}\frac{d}{d\varrho}(\varrho^2 Fn)$$

or

$$\frac{d}{d\varrho}\left(D\varrho^2\frac{dn}{d\varrho} - B\varrho^2 Fn\right) = 0, \tag{51.2}$$

whence it follows that

$$4\pi\left(D\varrho^2\frac{dn}{d\varrho} - B\varrho^2 Fn\right) = \text{const} = \Phi. \tag{51.3}$$

The first term on the left hand side of equation (51.3) gives the number of particles diffusing in unit time through a spherical surface concentric with the absorbing sphere, and the second term the number of particles passing through the same surface as a result of the ordered motion. Their sum is the rate of deposition of particles on the absorbing sphere, Φ.

The function $n(\varrho)$ must satisfy the boundary conditions $n = n_0$ at $\varrho = \infty$ and $n = 0$ at $\varrho = 2r$. Integration of equation (51.3) for the first of these conditions yields

$$n(\varrho) = \exp\left(\frac{B}{D}\int_{\infty}^{\varrho} F(\varrho)\,d\varrho\right) \times$$

$$\times \left[n_0 + \frac{\Phi}{4\pi D}\int_{\infty}^{\varrho} \frac{1}{\varrho^2}\exp\left(-\frac{B}{D}\int_{\infty}^{\varrho} F(\varrho)\,d\varrho\right)d\varrho\right]. \tag{51.4}$$

The second condition gives

$$n_0 = \frac{\Phi}{4\pi D}\int_{\infty}^{2r} \frac{1}{\varrho^2}\exp\left(-\frac{B}{D}\int_{\infty}^{\varrho} F(\varrho)\,d\varrho\right)d\varrho = 0. \tag{51.5}$$

Replacing B/D by $1/kT$ and denoting the electrostatic potential, $\int_{\varrho}^{\infty} F(\varrho)\,d\varrho$, by $\psi(\varrho)$ leads to†

$$\Phi = \frac{4\pi D n_0}{\int_{2r}^{\infty} \frac{1}{\varrho^2}\exp\left[\frac{\psi(\varrho)}{kT}\right]d\varrho}. \tag{51.6}$$

For uncharged particles

$$\Phi_0 = 8\pi D r n_0. \tag{51.7}$$

Introducing the new variable $x = 2r/\varrho$ the following expression is obtained for the ratio of the coagulation constants for charged and uncharged particles, $\beta = \Phi/\Phi_0$

$$\beta = 1\Big/\int_0^1 \exp\left[\frac{\psi(2r/x)}{kT}\right]dx. \tag{51.8}$$

If $\psi > 0$ (like charges, repulsion) $\beta < 1$ and a decrease in the coagulation constant results, while for $\psi < 0$ (unlike charges, attraction) the coagulation constant increases.

If the particles are conductors of electricity (see page 55) the force $F(\varrho)$ is given by the series [439]

$$F(\varrho) = \left(1 + 15\frac{r^6}{\varrho^6} + \cdots\right)\frac{q_1 q_2}{\varrho^2}$$

$$-\left(2\frac{r^3}{\varrho^3} + 3\frac{r^5}{\varrho^5} + 4\frac{r^7}{\varrho^7} + \cdots\right)\left(\frac{q_1^2 + q_2^2}{\varrho^2}\right). \tag{51.9}$$

† This equation was derived by Fuchs [438] and later in a more general form by Kramers [302].

The first term of this series $(q_1 q_2/\varrho^2)$ is the force between free charges while the others represent the induction force. If $q_1 = q$ and $q_2 = sq$ ($|s| \geqslant 1$; $s > 0$ in the case of repulsion; $s < 0$ in the case of attraction) the potential ψ is

$$\psi\left(\frac{2r}{x}\right) = \frac{q^2}{2r}\left[s\left(x + \frac{15}{448}x^7 + \cdots\right) - (1 + s^2)\left(\frac{x^4}{16} + \frac{x^6}{64} + \frac{x^8}{256} + \cdots\right)\right]. \qquad (51.10)$$

The relative importance of the first and the remaining terms of this series with regard to β depends mainly on the ratio, $\lambda = q^2/2\,rkT$, of the electric and kinetic energies of the contacting particles. The most important case is that of a stationary charge on an aerosol in a bipolar ionized atmosphere (see page 114), the distribution of charges being given by Boltzmann's formula

$$n(q) = \frac{n_0}{\sqrt{2\,rkT\pi}} \exp\,[-\,q^2/2\,rkT], \qquad (51.11)$$

where n_0 is the total particle concentration and $n(q)\,dq$ is the concentration of particles having charges between q and $q + dq$.

Hence it follows that the mean value $\bar{\lambda} = \overline{q^2}/2\,rkT$ is one half. It can be shown by graphical integration that when $\bar{\lambda} = \frac{1}{2}$ with equal, like charges on the particles ($s = 1$), the error introduced by discarding the induction terms in (51.10) when calculating β is only 1·4 per cent; for $s = 2$ it is 4·3 per cent. With unlike charges the error is even less. The series (51.9) can therefore be terminated after the first (Coulomb) term without introducing any great error, in the case of bipolar charging, and β is given by the simple formula

$$\beta = \frac{\lambda_{12}}{e^{\lambda_{12}} - 1} \qquad (51.12)$$

for repulsion and

$$\beta = \frac{|\lambda_{12}|}{1 - e^{-|\lambda_{12}|}} \qquad (51.13)$$

for attraction, where

$$\lambda_{12} = \frac{q_1 q_2}{2\,rkT}. \qquad (51.14)$$

If the bipolar charging is symmetrical, as it usually is to a good approximation, to every particle with a charge q_1, there is a corresponding one with charge $-\,q_1$ and the change in rate of coagulation on account of the charging can be found in the following way.

The coefficient β for the coagulation of two particles with charges q_2 and q_1 or q_2 and $-\,q_1$ is given by expressions (51.12) and (51.13), respectively. An aerosol charged in a bipolar ionized atmosphere with $\bar{\lambda} = \frac{1}{2}$ makes β equal to 1·271 for unlike charges and 0·770 for like charges, so that particles having unlike charges coagulate appreciably faster than neutral ones while those bearing like charges coagulate more slowly. The arithmetic mean of the two values is 1·02, so that the overall effect of charging is extremely small and could hardly be observed experimentally. Fuchs and Petryanov [166] found it impossible to detect any difference between the coagulation rates of oil mists, whether uncharged or charged by ultraviolet irradiation, using Tyndallometry. The same result was obtained earlier by Whytlaw-Gray and Patterson [440].

Gillespie [360] studied the coagulation of silica dusts carrying comparatively large bipolar charges (see page 255) and found that the coagulation constant increased appreciably with increase in the average charge on the particles. As already stated, these data cannot be accepted as reliable.

Owing to electrostatic dispersion (page 102) the rate of coagulation of unipolar aerosols cannot be determined by the decrease in number concentration of particles but only by the increase in their mean size and is thus rather difficult to measure. It can be proved that the rate of decrease in particle concentration when aerosols are charged unipolarly, i.e. under the simultaneous action of coagulation and electrostatic dispersion, always increases [438]. It is therefore impossible to stabilize aerosols by unipolar charging. Generally speaking, enlargement of the particles is retarded by unipolar charging and can practically stop if the charges are big enough. However, it can be seen from (51.10) that a sufficiently large charge ratio s can cause the potential ψ to become negative, owing to induction forces, so that repulsion changes to attraction, even for like charges. The effect of induction forces is also enhanced when the ratio of the particle sizes increases. Considerable enlargement of heavily charged or very large particles is thus possible, even in unipolar aerosols.

It has been suggested that clouds in the atmosphere may be stabilized by unipolar charges [441] and that precipitation occurs only from clouds containing droplets which are slightly charged, or have charges of both signs. The weak point in this theory, which regards as identical the part played by electric forces in both liquid and gaseous disperse systems, is that at the small number concentration of atmospheric clouds (10^3 to 10^4 cm^{-3}) the rate of thermal coagulation is negligibly small. Furthermore, cloud droplets always have charges of both signs, although one may predominate slightly.

The method set out above for calculating the coagulation constant in the presence of long-range forces between the particles has been used by Tikhomirov, Tunitskii and Petryanov [424] to find out how molecular forces influence the rate of coagulation. If the potential of inter-molecular forces between volume elements dv_1 and dv_2 as a function of their distance apart ϱ, is given by the well-known expression

$$d\psi = -Q\frac{dv_1 dv_2}{\varrho^6}, \tag{51.15}$$

then for plane parallel plates a distance h apart the potential (per 1 cm^2) is given [442] by

$$\psi(h) = -\frac{\pi Q}{12 h^2}, \tag{51.16}$$

and for two identical spherical particles of radius r by

$$\psi(\varrho) = -\frac{\pi^2 Q}{6}\left\{\frac{2r^2}{\varrho^2} + \frac{2r^2}{\varrho^2 - 4r^2} + \ln\left(1 - \frac{4r^2}{\varrho^2}\right)\right\}. \tag{51.17}$$

Arguing as above, the increase in the coagulation constant, β, due to molecular forces is given by (51.8) on substituting in it the expression for ψ,

$$\beta = 1 \Big/ \int_0^1 \exp\left[-\frac{\pi^2 Q f(x)}{6\,kT}\right] dx, \tag{51.18}$$

where

$$f(x) = \left[\frac{x^2}{2} + \frac{x^2}{2(1-x^2)} + \ln(1-x^2)\right]. \tag{51.19}$$

It can be seen from these formulae that the effect of the molecular forces depends not on the particle size but only on the quantity Q/kT. Until very recently there have been no reliable data on the value of the constant Q. From measurements of the adhesion of quartz spheres *in vacuo* Bradley [443] found that Q was between 2 and 5×10^{-13} erg. Hence it can be calculated by graphical integration that $\beta = 1{\cdot}35$ to 1·5. For organic liquid droplets Tikhomirov, Tunitskii and Petryanov assumed $Q = 6 \times 10^{-14}$, from the relationship between Q and surface tension, and found $\beta = 1{\cdot}18$. They concluded that experimental values of the coagulation constant are greater than theoretical values (see § 49) owing to the action of molecular forces.

The effect of molecular forces can be evaluated more reliably from the direct measurements made by Derjaguin and Abrikosova [444] of the molecular attraction between a quartz plate and a lens as a function of their distance apart. These experiments show that the exponent of h in (51.16) must be increased to about three for $h > 5 \times 10^{-6}$ cm, which makes the exponent of ϱ in expression (51.15) equal to seven or thereabouts. Since the derivation of formulae corresponding to (51.16) and (51.17) for an exponent equal to seven is fairly complicated, the value six will be retained and Q calculated for various values of h from (51.16) using the experimental data of Derjaguin and Abrikosova; Q, of course, is no longer constant (Table 31).

TABLE 31. MOLECULAR FORCE BETWEEN QUARTZ PARTICLES

h	(cm)	10^{-5}	2×10^{-5}	3×10^{-5}
$\psi(h)$	(erg cm^{-2})	$2{\cdot}7\times10^{-5}$	$0{\cdot}6\times10^{-5}$	$0{\cdot}2\times10^{-5}$
Q	(erg)	$1{\cdot}1\times10^{-14}$	$0{\cdot}9\times10^{-14}$	$0{\cdot}7\times10^{-14}$

As the value of Q for quartz is almost an order of magnitude less than that taken by Tikhomirov, Tunitskii and Petryanov, it is evident that molecular forces cause the coagulation constant of aerosols to increase no more than one or two per cent.

The effect of charge on the kinematic coagulation of aerosols is dealt with in § 54.

Zebel [800] drew attention to the different effects of unipolar and bipolar charges on aerosols. As seen from (51.12) and (51.13) at large absolute values of λ_{12}, attraction gives $\beta \approx |\lambda_{12}|$ and repulsion $\beta \approx \lambda_{12} e^{-\lambda_{12}}$ so that the effect of a unipolar charge is much greater than that of a bipolar one. With allowance for the concentration discontinuity the parameter β in the theory of coagulation of charged particles is not given by (51.12) but by the expression

$$\beta = \left[\frac{e^{\lambda_{12}} - 1}{\lambda_{12}} + \frac{4\,D}{r}\sqrt{\frac{m e^{\lambda_{12}}}{6\,kT}}\right]^{-1}. \tag{51.20}$$

Contrary to Zebel's view, this is valid only when $l_B/r \ll 1$.

Expanding $e^{\lambda_{12}}$ in (51.12) as a series,

$$\beta = 1\Big/\left(1 + \frac{\lambda_{12}}{2} + \frac{\lambda_{12}^2}{6} + \cdots\right) \tag{51.21}$$

ta formula also given by Gunn [816], but with $\lambda_{12} = q_1 q_2 / r\, kT$.

In the experiments of Benton *et al.* [817] mists with $r = 0{\cdot}5$–$2{\cdot}5\ \mu$ generated by spraying 0·1–1 N solutions of Fe alum and of KCNS were mixed in a chamber. The rate of coagulation Φ_{12} of the droplets with one another was evaluated from the number of coloured droplets which settled to the bottom of the chamber. Φ_{12} was found to increase with the concentration of the solutions. The addition of cation-active detergents to the alum solution and of anion-active ones to the KCNS markedly increased Φ_{12}. A decrease of Φ_{12} occurred when the additions were made the other way round. Since Fe^{3+} and CNS^- ions are adsorbed at the surface of their solutions, as well as the cations of the cation-active and anions of the anion-active detergents, it was supposed that the electrial double layer on the surfaces of the drops had an effect upon their collision efficiency, decreasing it in the case of drops having the same sign on the outside of the layer and increasing it for opposite signs.

In fact, as shown below, free charges on the drops increase collision efficiency, but the mechanism of this phenomenon is such that it cannot operate in the case of an electrical double layer. The electrostatic force between drops carrying these layers is negligible, even when the drops are very close to one another. It was, however, shown by Dukhin and Derjaguin [818] that motion of a drop, and the resulting circulation in it, may cause appreciable displacement of the charges in the double electric layer, which gives rise to polarization of the drop and the formation of an appreciable external field.

It is most likely that the observations of Benton *et al.* are due to a ballo-electric effect. As has been shown by Natanson [819] a ballo-electric effect of the "second kind" is observed when a solution of large ionic strength is sprayed to produce a mist whose droplets are predominantly charged with a sign opposite to that of the ions adsorbed at the surface of the solution. Therefore, the droplets were positive in the case of alum and negative for KCNS. Addition of detergents should still increases this effect. Unfortunately, the charges of the droplets were not measured in these experiments.

§ 52. POLARIZATION COAGULATION OF AEROSOLS

External factors which promote forced coagulation of aerosols include electrical and acoustic fields. Despite their different physical nature these fields have much in common owing to the similar properties of electric and hydrodynamic potentials.

Spherical conducting particles (see page 55). of radius r in an electric field of the strength E have induced in them an electric dipole of moment $P = Er^3$ The force of interaction between two such dipoles having a distance $\varrho \gg r$ between their centres is similar but opposite in sign to that due to fluid motion, which is given by (23.2) and (23.3):

$$F_\varrho = -\frac{3\,E^2 r^6}{\varrho^4}\left(\frac{3}{2}\cos 2\,\theta + \frac{1}{2}\right), \tag{52.1}$$

$$F_\tau = -\frac{3\,E^2\,r^6}{\varrho^4}\sin 2\theta, \tag{52.2}$$

where θ is the angle between the field direction and the line of centres of the particles. The force field between polarized particles, is thus represented by Fig. 23 with the arrows reversed; they attract one another if they are arranged along the field direction and repel when they are at right angles. The potential of this force is

$$\psi(\varrho) = -\frac{E^2 r^6}{\varrho^3}\left(\frac{3}{2}\cos 2\theta + \frac{1}{2}\right) = -\frac{E^2 r^6}{\varrho^3}(2\cos^2\theta - \sin^2\theta), \qquad (52.3)$$

which is notable in that it depends not only on the distance between the particles but also on the direction of their line of centres.

No account has been taken of secondary induction due to the mutual polarization of the particle-dipoles when they are close together. The forces which develop in this way are expressed by complex trigonometric series [445], the use of which to determine the effect of secondary induction on rate of coagulation requires such a tremendous amount of computation that it has been necessary to pass them by. Their effect, however, can be evaluated by another method (see below). An idea of the magnitude of secondary induction forces can be obtained from the interaction potential for particles in contact ($\varrho = 2r$) which is

$$\psi(2r) = -E^2 r^3 (0{\cdot}25\cos^2\theta - 0{\cdot}125\sin^2\theta) \qquad (52.4)$$

according to the approximate formula (52.3) and

$$\psi(2r) = -E^2 r^3 (1{\cdot}404\cos^2\theta - 0{\cdot}099\sin^2\theta) \qquad (52.5)$$

according to the accurate formula.

It is extremely difficult to find a rigorous solution for the diffusion of particles towards an absorbing sphere in a force field given by formula (52.3) which does not possess spherical symmetry. An estimate of the extent to which polarization of the particles affects the rate of coagulation will be made by neglecting tangential movement of the particles and assuming that they move only radially towards the absorbing sphere.

To begin with, the rate of coagulation of a polarized aerosol will be determined in the absence of diffusion or thermal motion of the particles. Two particles so positioned that there is a maximum attraction between them ($\theta = 0$) move towards one another with a velocity

$$V = \frac{6E^2 r^6 B}{\varrho^4} \approx \frac{E^2 r^5}{\pi\eta\varrho^4}. \qquad (52.6)$$

The equation of motion for the particles is therefore

$$-\frac{d\varrho}{dt} = 2V = \frac{2E^2 r^5}{\pi\eta\varrho^4}. \qquad (52.7)$$

Integration shows that particles initially a distance ϱ_0 apart come into contact after

$$t_0 = \frac{\pi\eta\varrho_0^5}{10E^2 r^5} \approx \frac{5\times 10^{-5}\varrho_0^5}{E^2 r^5} \qquad (52.8)$$

seconds.

It is possible to study the course of coagulation experimentally at the number concentrations of about 10^6 to $10^7\ \text{cm}^{-3}$; the mean distance between the particles is then $0{\cdot}5$–$1{\cdot}0 \times 10^{-2}$ cm and ϱ_0 will be taken as $0{\cdot}5 \times 10^{-2}$ cm. Thermal coagulation will be shown to be much accelerated by polarization of the particles when $Er^{3/2} = 4 \times 10^{-6}$; with $r = 10^{-4}$ to 10^{-5} cm this makes $Er^{5/2} = 4 \times 10^{-10}$ to 4×10^{-11}. Inserting these values into (52.8) gives $t_0 = 10^3$ to 10^5 sec. If however the same particles had been moved together until they were $0{\cdot}5 \times 10^{-3}$ cm apart, by some other agency, they would then touch after only 0·01 to 1 sec. It is clear, therefore, that the mechanism of coagulation in an electric field consists in the diffusion of particles towards one another until they are close enough for electric forces to become appreciable. Motion of the particles then assumes an ordered character. This reasoning applies equally to the coagulation of charged particles discussed in the preceding section.

Returning to the radial motion of particles towards an absorbing sphere in the force field described by (52.3), formula (51.8) determines the coefficient β, which is the ratio of the rates of deposition of particles on the absorbing sphere with and without electric forces. β also depends on θ,

$$\beta = 1 \bigg/ \int_0^1 \exp\left[- \frac{E^2 r^3 x^3 \left(\cos^2\theta - \frac{1}{2}\sin^2\theta\right)}{4kT} \right] dx$$

$$= 1 \bigg/ \int_0^1 e^{-\alpha^3 x^3}\, dx = \alpha \bigg/ \int_0^\alpha e^{-y^3}\, dy, \tag{52.9}$$

where

$$\alpha = \left(\frac{E^2 r^3}{4kT}\right)^{1/3} \left(\cos^2\theta - \frac{1}{2}\sin^2\theta\right)^{1/3} = \alpha_1 \left(\cos^2\theta - \frac{1}{2}\sin^2\theta\right)^{1/3}. \tag{52.10}$$

In order to find out how the overall effect of polarization of the particles influences their rate of deposition on the absorbing sphere the mean value of β over the surface of the sphere must be calculated

$$\bar{\beta} = \tfrac{1}{2} \int_0^\pi \beta \sin\theta\, d\theta = \int_0^1 \beta\, d\cos\theta. \tag{52.11}$$

The following values of $\bar{\beta}$ as a function of the dimensionless quantity α_1 (Table 32) have been found by graphical integration.

TABLE 32. EFFECT OF ELECTRIC FIELD ON RATE OF COAGULATION OF MISTS

α_1	1	2	3	5	10	20	Hereafter $\bar{\beta} \approx \frac{1}{3}\alpha_1$
$\bar{\beta}$	1·0	0·95	1·07	1·7	3·4	6·8	

Thus for $\alpha_1 < 3$ the decrease in the rate of deposition in the zone of repulsion ($\cos^2\theta - \frac{1}{2}\sin^2\theta < 0$) is almost exactly counterbalanced by the increase in the zone of attraction. For $\alpha_1 = 5$ deposition is completely halted in the zone of repulsion but in the zone of attraction the rate of deposition increases so much that a noticeable overall increase is produced. This value of α_1 corresponds in air to $Er^{3/2} \approx 4 \times 10^{-6}$, or a field strength of 1400 V cm^{-1} with particles of radius 10^{-4} cm, and 40,000 V cm^{-1} for $r = 10^{-5}$ cm.

The effect of secondary induction forces can be estimated as follows. Their potential, to a first approximation, is proportional to ϱ^{-6} and the additional term kE^2r^q/ϱ^6 is introduced into (52.3), k being assigned a value which makes (52.5) give the correct value $\psi = -1{\cdot}404\,E^2r^3$ when the particles are in contact ($\varrho = 2r$) at $\theta = 0$. It is then possible to find β from the resulting corrected expression for $\psi(\varrho)$. The correction to β is greatest around $\alpha_1 = 1$ and amounts to about 25 per cent, rapidly approaching zero if α_1 deviates from one; for rough calculations it can be entirely neglected.

These figures show that the coagulation of mists is accelerated appreciably only by very strong electric fields. It has been found in practice that a field of 200 V cm^{-1} produces no noticeable enhancement of the coagulation of an oil mist [446]. On the other hand Fuchs and Petryanov [166] reported that a unipolar mist, charged by a corona, coagulated rapidly with the formation of large drops each consisting of several thousand primary droplets. This was largely due to intense polarization of droplets in the vicinity of the discharge electrode where the field strength was very great. Such coagulation also takes place in electrostatic precipitators and greatly increases their efficiency. In thunderclouds, where the field strength reaches about 1000 V cm^{-1}, polarization coagulation of droplets may also be considerable [447].

The coagulation of solid particle aerosols in an electric field is unique. A doublet formed by the coagulation of two primary particles orientates with its long axis parallel to the field (see § 43) and its dipole moment in this position is much greater than that of the primary particles. The doublet coagulates with other particles mainly at the ends and thus grows lengthwise. The dipole moment of such a long aggregate can be calculated approximately by regarding it as a prolate ellipsoid of revolution. By putting $\theta = 0$ in (43.12) and referring to Table 22 it will be seen that the dipole moment of such an ellipsoid is proportional to $v\beta$, where v is the volume of the ellipsoid and β its axial ratio. Both v and β are proportional to the number of primary particles, ν, in the aggregate so that the dipole moment of the aggregate is proportional to ν^2, while that of the potential of interaction between two such aggregates is proportional to ν^4. The coagulation constant therefore increases very rapidly during the course of the process and the aggregates acquire thread- or chain-like shapes (Fig. 73, Fe_2O_3 smoke [448]).

Experiments prove that smokes coagulate in a fairly strong electric field much faster than usual and form linear aggregates [354, 446, 449]. At 5000 V cm^{-1} the rate of coagulation, of ammonium chloride smoke is 10^4 times normal and the effect is noticeable for $E > 200$ V cm^{-1}. The term "directed" has been applied to this type of coagulation which probably takes place when smokes are formed by vaporization in an arc. Nearly all smokes obtained in this way, including magnesium oxide, consist mainly of thread-like aggregates [450, 451] whereas magnesium oxide smoke produced by burning magnesium ribbon contains none [451]. Artemov [30] has observed directed coagulation in an electric field with an ultra-microscope.

Analagous phenomena occur in magnetic fields with smokes of ferro-electric particles, such as the iron smoke obtained from iron pentacarbonyl [452].

Linear aggregates may also form in the absence of an external electric or magnetic field if the particles are permanent electric dipoles or magnets [434, 453]. Although they do not orient themselves in space, apart from a weak action of the earth's magnetic field, on approaching one another the dipoles rotate so that the positive extremity of one particle touches the negative extremity of another, as happens in the formation of linear aggregates. Permanent electric dipoles may result in thermally generated smokes when the crystalline particles of pyroelectric materials are cooled. A typical example is aminoazobenzene, which has pronounced pyroelectric properties [434]. The smoke obtained by vaporizing this substance coagulates 10–100 times faster than a normal aerosol with the formation of rapidly-growing thread-like aggregates. The addition of 10 per cent of paraffin wax suppresses the pyroelectric nature of aminoazobenzene and a smoke produced from this mixture coagulates normally without forming linear aggregates. Pyroelectric properties may disappear if a small amount of decomposition accompanies vaporization; the smokes of some pyroelectric materials such as resorcinol do not exhibit directed coagulation for this reason.

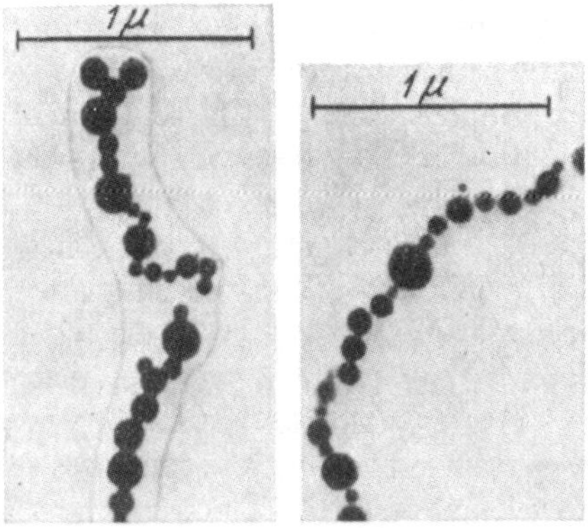

FIG. 73. Aggregates in Fe_2O_3 smoke.

Iron and nickel smokes having particles which are permanent magnets are generated by heating the vapours of the carbonyls in the absence of oxygen [453]. Below the Curie point† (360 °C for nickel and 770 °C for iron) the smokes coagulate to thread-like aggregates but above it rounded aggregates or very short chains are formed. Thread aggregates orient themselves in a magnetic field and turn through 180° when the field is reversed, thus showing that the orientation is not due to induction but to permanent magnetism of the particles. How the magnetism originates is not known. Smokes of iron oxide crystals have similar properties but particles of amorphous iron oxide are not ferromagnetic and do not give linear aggregates without an external field.

A tendency to form linear aggregates is sometimes observed in smokes whose particles cannot possess a permanent electric or magnetic moment (see photograph of aluminium smoke on page 17). It is probably due to contact polarization of the particles or to free charges on them [451]. Dallavalle observed [455] that the coagulation of ammonium chloride smoke with neutral particles yielded more or less rounded aggregates, while strongly charged particles produced linear aggregates, although the average number of primary particles in the aggregates and the rate of coagulation were approximately the same in each case.

† The Curie point is the temperature at which the ferromagnetic properties disappear.

Winkel [820] investigated the coagulation kinetics of aerosols in a chamber where a uniform electric field of $E = 40$–290 V/cm was maintained. With a mist produced by the hydrolysis of $TiCl_4$ vapour the weight concentration changed in accordance with the law $c = c_0 \exp(-KErt)$. The effect of the field was much greater on an aerosol of NH_4Cl, aggregates in the form of long threads being formed. In both cases large aggregates were deposited on the bottom of the chamber rather than on the electrodes, indicating that their free charges were small.

Zebel [821] advanced a theory of the coagulation of aerosols having particles which were electrical or magnetic dipoles; allowance was made for the disorienting action of thermal motion. The significant quantity is $\varrho_p = (P^2/kT)^{1/3}$, where P is the dipole moment, which is equal to the distance between the centres of the dipoles for which the energy of interaction is of the order of kT. The author made calculations for both weak ($\varrho_p \ll 2r$) and strong ($\varrho_p \gg 2{\cdot}8\,r$) interaction. In the first case the coagulation constant was found to increase by a factor $1 + 0{\cdot}095\,(\varrho_p/2r)^6$, and in the second by a factor of $\varrho_p/2r$, approximately.

§ 53. COAGULATION OF AEROSOLS BY ULTRASONIC VIBRATIONS†

The coagulation of aerosols by ultrasonic waves was observed early in the study of ultrasonics [456]. It is easily demonstrated by generating standing ultrasonic waves in a closed glass tube containing the aerosol [131, 135]. If the frequency and intensity of the vibrations are suitable the aerosol coagulates in a few minutes or even seconds, forming annular deposits on the walls of the tube at the antinodes. A few of the finest particles usually remain suspended and coagulate extremely slowly under continued action of the ultrasonic waves.

The theory of ultrasonic coagulation has been the subject of lively discussion [457]. Hiedemann [458] considered the main cause to be the varying degrees to which particles of the polydisperse aerosols commonly encountered were entrained by vibrations of the gas (see page 86). Fine particles executing oscillations of large amplitude comb the aerosol which also contains coarser, almost stationary particles. The absorbing surface is a cylinder whose length is twice the amplitude of oscillation of the fine particles and whose radius is the sum of the radii of the coarse and fine particles. The relative motion of these cylinders and the coarse particles is caused by sedimentation, turbulence etc.

This mechanism is probably important in polydisperse aerosols but it cannot be the only cause of ultrasonic coagulation since coagulation of aerosols occurs at frequencies which are so high that practically all the particles are motionless [459]. Nor can Hiedemann's theory explain why the finest particles, which vibrate most energetically, remain uncoagulated.

Andrade's opinion [460] is that ultrasonic coagulation is due to hydrodynamic attraction between the particles (see page 95). When calculating the rate of coagulation of aerosols in an ultrasonic field he neglected diffusion and committed some other errors, so that the good agreement between the formula he derived and the experimental data

† The usual expression "ultrasonic coagulation" of aerosols is inaccurate since high audible frequencies (1–10 kc/sec) are commonly used for this purpose.

obtained in his laboratory must be regarded as fortuitous. As indicated in the preceding section, it is not permissible to neglect diffusion in electric or sound fields [461] since it is important in bringing particles close enough for the short range forces to operate.

The potential due to the hydrodynamic force between identical particles in a sound field is

$$\varphi(\varrho) = \frac{\pi \gamma_g V_{r_0}^2 r^6}{\varrho^3} (\sin^2 \theta - 2 \cos^2 \theta), \tag{53.1}$$

as can be seen by comparing (23.4), (52.1), and (52.3). In this equation V_{r_0} is the velocity amplitude of the relative motion of particles and medium and θ is the angle between the direction of vibration and the line of centres.

The effect of hydrodynamic forces on the rate of coagulation cannot be determined reliably by the method of the previous section because at short distances between the particles (53.1) is probably inapplicable for reasons given on page 98. The coefficient β, related to the dimensionless quantity

$$\alpha_1 = \left(\frac{\pi \gamma_g V_{r_0}^2 r^3}{8kT} \right)^{1/3} \tag{53.2}$$

is somewhat larger than for coagulation in an electric field.† The effect is noticeable at $\alpha_1 = 3$.

Experimental data on ultrasonic coagulation of aerosols are for the most part no use for comparing with theory, since they have usually been obtained for polydisperse aerosols of unknown particle size and concentration. Typical experimental conditions which give rise to noticeable acceleration of coagulation might be a frequency of 10^4 c/sec and a mean field energy of 100 erg cm^{-3} which corresponds to a maximum amplitude of vibration equal to 0·13 mm. Taking into account that the mean field energy is $\bar{\Omega} = \frac{1}{4} \gamma_\varrho \overline{U_0^2}$ the value of α_1 is

$$\alpha_1 = \left[\frac{\pi (V_{r_0}/U_0)^2 \cdot 4\bar{\Omega}}{8kT} \right]^{1/3} r. \tag{53.3}$$

Using Fig. 20 for the ratio V_{r_0}/U_0 we find that for $r = 0{\cdot}2$, 0·5 and 1 μ, α_1 is 0·6, 4 and 16 respectively. In fact, under such conditions, ultrasound produces appreciable acceleration of coagulation [135] at $r = 0{\cdot}2\,\mu$. However, it must be borne in mind that there is vigorous circulation of air in tubes irradiated by an ultrasonic field (see page 88) which is accompanied by a considerable velocity gradient; the latter is more effective than molecular diffusion in encouraging the approach of the particles.

The theory of hydrodynamic attraction thus permits at least a rough estimate to be made of the conditions necessary for effective ultrasonic coagulation. It also shows that ultrasonics have so little effect upon highly dispersed aerosols [462] on account of the factor $(V_{r_0}/U_0)^{2/3} r$ in (53.3); the rapid increase of the ultrasonic coagulation coefficient during coagulation, as the particles become larger [463], is likewise explained. The frequently-reported ineffectiveness of ultrasonics in aerosols of low number concentration [464] is understandable, since the large mean distance between the particles means that there is little chance of their moving close enough together for hydrodynamic forces to act. The hydrodynamic theory also explains why flat aggre-

† Although the maximum attraction in this case is half the repulsion, the zone of attraction is bigger than that of repulsion (the converse holds in an electric field).

gates, disposed in a plane perpendicular to the direction of vibration, are formed during ultrasonic coagulation [135]; directed coagulation also occurs in this plane ($\theta = \pi/2$).

Coarse, elongated particles, which do not vibrate, are oriented in an ultrasonic field with their long axes perpendicular to the direction of the vibrations [131] in agreement with § 11, but fine particles, which are completely entrained, have their long axes parallel to the vibrations probably because of the orienting action of the velocity gradient in the circulating air flow.

In the calculation above, the energy of the sound vibrations was averaged over the whole volume of aerosol, whereas ultrasonic coagulation occurs mainly near the antinodes. Turbulent mixing proceeds during exposure to more or less intense ultrasonic vibrations so that all the aerosol particles are, on the average, under the same conditions.

Although no special experiments have been set up to compare the rates of ultrasonic coagulation of smokes and mists there does not seem to be any great difference between them. In Hiedemann's laboratory experiments [465] coagulation of an oil mist with $\bar{r} = 0{\cdot}4\,\mu$ was rapid at 10^4 c/s and $\Omega = 25$ erg cm^{-3} while St. Clair [135], working with ammonium chloride smoke having $\bar{r} = 0{\cdot}2\,\mu$ at the same frequency, found that appreciable acceleration of coagulation was reached at $\Omega = 50$ erg cm^{-3}.

There are indications that coarse aggregates (up to $100\,\mu$) develop only when the vibrations are not too intense [135]. Turbulence, in an intense sound field, breaks up the larger aggregates. Aggregates formed during ultrasonic coagulation are generally not very strong and often crumble on deposition. Break-up of aggregates is the reason for the ineffectiveness of ultrasonics for removing fly ash [466]. The addition of water or oil mists to the smoke or dust being coagulated causes the aggregates to stick together and appreciably increases the effectiveness of ultrasonic air cleaning [135, 467].

Regarding the effect of frequency, Hiedemann's theory indicates that the greatest efficiency would be expected at the frequency producing the maximum difference in the amplitudes of vibration of different sized particles. The optimum frequency should therefore increase as the mean particle size decreases.

On the theory of hydrodynamic attraction, at a given degree of entrainment of particles by vibrations of the medium, that is at a given value of V_{r_0}/U_0, efficiency depends only on the value of U_0^2 or on the energy of the sound field. As the frequency is increased coagulation should first increase markedly, but at frequencies where V_{r_0}/U_0 is close to one, when $\omega^2 r \approx 0{\cdot}002$, further increase in frequency at the same field energy can no longer produce a discernible effect and is therefore pointless.

Experimental data on the effect of frequency on coagulation efficiency are extremely scarce. Indications [461] that the minimum effective frequency is 3×10^3 c/s for most aerosols and 7×10^3 c/s for tobacco smoke are of limited value because the mean particle size of these aerosols is not given.

It remains to examine the role played by sound pressure on the particles during ultrasonic coagulation (see § 21), in connection with which the following observation is significant [135].

As long as the vibrations are not intense enough to cause turbulent mixing, concentrated ammonium chloride smoke has a striated appearance right up to the onset of coagulation because the particles are concentrated at the antinodes. Expression (21.2) will be used to see if the pressure produced by standing sound waves on aerosol particles can account for this drift.

Let t_0 be the time required by a particle situated at a wave node to move under the influence of this pressure to the nearest antinode, thus covering a distance $\lambda/4$. For simplicity sin $(4\pi x/\lambda)$ will be put equal to one, suffering the particle to move throughout at its maximum velocity

$$V_{max} = \frac{F_{max}}{6\pi\eta r} = \frac{5\pi\gamma_g V_{r_0}^2 r^2}{18\lambda\eta} \tag{53.4}$$

(U_0 has been replaced by V_{r_0} in formula (21.2) in agreement with the remark on page 89). This gives as a lower limit for t_0

$$t_0 = \frac{\lambda}{4V_{max}} = \frac{0\cdot9\,\lambda^2\eta}{\pi\gamma_g V_{r_0}^2 r^2}. \tag{53.5}$$

Under the above conditions (frequency 10^4 c/s, $\lambda = 3$ cm, $\Omega = 100$ erg cm^{-3}) $t_0 = 3 \times 10^6$, $1\cdot3 \times 10^4$ and 400 sec respectively for $r = 0\cdot2$, $0\cdot5$ and $1\,\mu$. Actually such aerosols coagulate in several tens of seconds, during which time the particles would be able to drift only a negligible part of the distance to the antinodes. The following circumstances must however be taken into account. As pointed out on page 88 recent theoretical work indicates that ultrasonic pressure is actually considerably greater than given by formula (21.2). There are, as well, fairly strong harmonics in an ultrasonic field and up to a certain point the pressure increases with the frequency of vibration. Finally, as coagulation of the aerosol proceeds and the particles get bigger, their drift towards the antinodes under the action of ultrasonic pressure is sharply accelerated.

It is thus quite possible that ultrasonic pressure promotes coagulation of aerosols by propelling the particles towards the antinodes, where hydrodynamic forces between the particles are particularly large, and create high particle concentrations there. It is hardly possible, though, to ascribe an over-riding role to ultrasonic pressure, as St. Clair does [468]. In particular it is difficult to explain from this point of view why the number concentration of the aerosol has such a strong influence on the effect of ultrasonics.

It is also impossible to agree with Kubanskii's belief [469] in the essential role of collisions between particles at the nodes of the ultrasonic field caused by circulation of the aerosol (see page 88). It may be true for very coarse particles, such as the coarse aggregates already formed as a result of coagulation, but cannot be important for primary particles.

The practical application of ultrasonics to depositing industrial aerosols has long been held up by the lack of sufficiently powerful generators of ultrasonic waves. Ultrasonic sirens [462] have been constructed for this purpose. Industrial equipment is available for depositing sulphuric acid mists and other aerosols by ultrasonics at a rate of 1000 m^3 min^{-1}. Up to 90 per cent of the mist can be coagulated in about 4 sec, after which the coarse drops formed are removed in cyclones. A small highly dispersed part of the mist invariably remains uncoagulated and increasing the duration of the treatment does not have much effect on it. As previously stated, ultrasonics are not much use for aerosols of low number concentration.

Very little progress has been made in the past few years in understanding sonic coagulation. In the work of Gudemchuk *et al.* [822], the coagulation of rather monodisperse flowing dioctylphthalate mists with $\bar{r} = 0\cdot28\,\mu$ at the frequency of 13 kc/sec was studied. As shown in Table 13 and Fig. 20, practically complete entrainment of particles by the air vibrations occurred so that coagulation could not be due to Hiedemann's mechanism. When the intensity of the sound was great enough the aerosol coagulated rapidly and visual observations suggested that considerable turbulence was occurring in the gas. A number of parallel plates 90 cm high and spaced 1–2 cm apart were inserted into the coagulation column, which resulted in a relatively small decrease in the amplitude of sonic vibration, but considerably lessened the turbulence. The plates appreciably reduced the rate of coagulation at a given sound intensity, indicating that turbulence of the medium may be an important factor in sonic coagulation.

Since turbulence must interfere with the displacement of particles by sound pressure towards the antinodes of standing waves (see p. 87), and also with the effect of the asymmetry of oscillation of the particles (see p. 89), these effects appear to be relatively unimportant in sonic coagulation.

The hydrodynamic effect of Oseen forces, discussed on p. 102, can only be assessed by special investigations.

§ 54. KINEMATIC (GRAVITATIONAL) COAGULATION OF AEROSOLS

Relative motion between particles of unequal size, owing to the different speeds acquired under the influence of external forces, leads to collisions and coagulation. This process will be called kinematic coagulation† and is particularly important in the case of settling of particles by gravity ("gravitational coagulation" according to Frenkel [470]). Kinematic coagulation also takes place in an electric field, under the action of centrifugal forces, when the velocity of a stream changes, and so on.

The theory of kinematic coagulation is essentially different for aerosols with coarse and fine particles because in the former diffusion can be neglected. Consider a coarse spherical particle falling freely with a velocity V_s through an aerosol consisting of finer particles. If the rate of fall of the latter is negligible the process is equivalent to that discussed in § 34 for the flow of an aerosol moving with a velocity V_s round a stationary sphere. Knowing the appropriate collection efficiency $\ni$, which can in this case be called the capture coefficient, it is easy to calculate the number of fine particles captured in one second by a coarse particle of radius R from the formula

$$\Phi = \pi \ni R^2 V_s. \tag{54.1}$$

If the rate of settling of the fine particles cannot be neglected, then the relative velocity should be used in the calculation, but in this case the velocity distribution in the gas flowing round the coarse particle is different. This results in a change in the capture coefficient; this change increases as the sizes of the coarse and fine particles approach one another. It is very difficult to calculate this correction and it will be passed over. Of much more importance is the correct allowance for the dependence of the capture coefficient on the size of the fine particles: all the methods hitherto given for calculating this are wrong.

† The term "orthokinetic", i.e. "directional", coagulation used for this in colloid-chemical literature has a broader but less specific meaning, including coagulation in electric and sound fields.

The first, crude theory of gravitational coagulation, proposed by Findeisen [471], made no allowance for curvature of the trajectories of the fine particles in the stream flowing round the coarse particle, and the capture efficiency was taken as $\left(\frac{R + r}{R}\right)^2$ where R is the radius of the coarse particle and r the radius of the fine particles. This of course gives values of ∋ which are too high. To check his calculations experimentally Findeisen made a mist of droplets having radii from 5 to 8 μ by adiabatic expansion in a chamber of 60 m³ volume [472]. Analysis of the change in droplet size distribution with time enabled Findeisen to prove the occurrence of coagulation which, in view of the low number concentration of the mist, could only have been gravitational. The rate of coagulation between droplets of similar sizes agreed more or less with calculations, but when the sizes were very different the rate was much less than calculated.

Langmuir [239] made the opposite mistake. Having used a computer to calculate the curved trajectories of the particles (see § 34), he determined the capture coefficient as a function of the parameter l_i/R without taking into account the interception effect of the fine particles, considering them as points. The results he obtained are, of course, much too low and in particular it would appear that particles less than 15 μ in radius do not capture one another at all. Langmuir also used not the relative but the absolute rate of fall of the coarse particles. Finally, as has already been indicated, the method used by Langmuir of interpolating values of ∋ obtained for viscous and potential flow regimes has not been adequately substantiated.

TABLE 33. LIMITING VALUES (*at* $l_i = 0$) OF THE CAPTURE COEFFICIENT IN AEROSOLS WITH SPHERICAL PARTICLES

r/R	1	0·8	0·6	0·4	0·2	0·1	0·05	0·025
For viscous flow	1·25	0·83	0·48	0·22	0·06	0·014	0·0036	0·001
For potential flow	3·50	2·69	1·93	1·25	0·62	0·30	0·15	0·075

Langmuir's calculations of the capture coefficient of water drops for various values of R and r are therefore unreliable and his results are not given here. Shishkin [474] multiplies Langmuir's coefficient by $[(R + r)/r]^2$, which is not a great improvement because the capture coefficient remains zero for droplets less than 15 μ in radius.

In fact, ∋ is not zero for particles of any size. If the reasonable assumption is made that the centre of a particle passing round a sphere moves along a streamline, and the inertia of the particle is neglected, it is possible to calculate from formulae (34.16) and (34.17) minimum values of the coefficient (Table 33) for viscous and potential flow [243]. These values relate to the cross section of the large particle.

The true values of ∋ must always be greater than those given in Table 33. It can be seen from the Table that ∋ falls rapidly as r/R becomes smaller, particularly in viscous flow, which explains Findeisen's observations mentioned above.

In order to calculate the exact value of the capture coefficient, with allowance for inertial displacement, the trajectories of particles of all sizes must be determined. It has already been pointed out on page 141 that dynamical similarity is of limited application in this case. The gravitational coagulation of aerosols with a particle radius not exceeding 15 μ involves only a small inertial displacement. For these, which include natural mists and clouds in which precipitation has not yet begun, Table 33 gives better values than Langmuir's theory for the capture coefficient. On the other hand, kinematic coagulation of comparatively coarse drops with fine ones, at values of Re_f such that the flow of air round the large drops may be considered potential, involves values of ∍ near to these calculated by Langmuir, as confirmed by Gunn's experiments [475].

In these experiments water drops of radius 1·6 mm fell through a tube 3 m high which contained a polydisperse water mist of droplets several tens of microns in radius. The amount of water captured was calculated by the gain in weight of the large drops. By measuring the concentration and droplet size distribution in the mist, and knowing the rate of fall of the large drops at various heights, Gunn calculated the theoretical gain in weight according to formula (34.6) by numerical integration and obtained satisfactory agreement between theory and experiment. The average value of Re_f in these experiments was about 500.

In the work of Telford, Thorndike and Bowen [574] a comparatively isodisperse water mist having droplets of radius about 70 μ, obtained from a spinning-disk atomizer, was fed continuously into a vertical tube with transparent walls. Air saturated with water vapour was fed upwards through the tube with a velocity very slightly greater than the velocity of free fall of the droplets. The slowly rising droplets were photographed with short time-exposures on a horizontally moving film and the relative velocities of neighbouring droplets in roughly the same vertical line were determined by measuring the angles between the straight-line segments on the film. Hence the mean relative velocity $\overline{V}_r = 1{\cdot}1$ cm sec^{-1} was calculated. On the photographs both primary droplets and doublets formed by collision and falling downwards were counted, hence giving the concentration of both. From a knowledge of $\overline{V}_r$, the height of the tube, the radius of the primary droplets and the rate of fall of the double droplets, it was possible to calculate the capture coefficient, which was 12·6. Even for strictly rectilinear fall of the droplets the interception effect makes ∍ equal to 4 only.

Such a high value of ∍ is explained by the authors as follows. The streamlines behind the droplet are inclined at a small angle to the vertical axis passing through the centre of the droplet. Another droplet situated behind the first one and nearly in line with it will be attracted by the flow towards the vertical axis which increases ∍. Since the radial velocity of the droplet is small its effect can come into play only when the radial motion lasts for a long enough time, that is when the velocities of the two droplets are about the same.

Similar observations have been made by Sartor [575] in model experiments with water drops ($r = 1$–2 mm) falling through a viscous oil at Reynolds numbers of about 0·02 to 0·1. The capture coefficient was two or three times greater than that calculated from formula (34.7) and it could be seen directly how the following drop moved sideways into the path of the leading drop. Sartor tried to apply the results of his experiments to gravitational coagulation of mists, using the principle of dynamical similarity, but he failed to keep the Stokes number constant. Another point is that when water droplets move through a viscous fluid a circulation develops in the droplets (see page 241) which may have an appreciable effect on the capture coefficient. Sartor's experiments have therefore a qualitative nature.

Longitudinal interaction between falling droplets was also shown on photographs obtained by Telford *et al.* The foremost droplet maintained a constant velocity right up to the moment of collision, while the one behind began to accelerate when it was still about forty diameters from the point of contact owing to the flow set up behind the other droplet. The Reynolds number in these experiments was 6.

At small Re, when the flow in front of and behind the droplet is symmetrical, it was shown in § 13 that the decrease in resistance of the medium was the same for each droplet; as Re increases to several units this symmetry is lost and the influence of the leading droplet on the following one predominates.

Gravitational coagulation is extremely important in rainfall from clouds. The size of cloud droplets formed by the condensation of vapour is about 10 μ and their rate of fall is small compared with the velocity of vertical currents in the atmosphere. Such droplets cannot reach the surface of the earth; to do so they would have to be at least 100 μ in radius. The mechanism by which cloud droplets grow has long been the subject of speculation. Each of the proposed mechanisms – hydrodynamic attraction between falling droplets, thermal, turbulent and kinematic coagulation, coagulation under the influence of the atmosphere's electric field, isothermal distillation from fine to coarse droplets – has proved insufficient on its own to account for the fall of heavy rain.

In 1935 Bergeron put forward the hypothesis that the formation of rain began in the uppermost regions of clouds where the temperature was below zero and supercooled water droplets were found. Some of these droplets freeze, after which the rapid process of isothermal distillation from droplets to ice crystals can begin. While the enlarged crystals are falling through the cloud they continue to grow by kinematic coagulation and finally emerge as raindrops or snowflakes. Bergeron's hypothesis evidently accounts for rain formation in temperate and polar regions, but in the tropics rain often falls from clouds whose temperature is above zero at every point; droplet growths can then occur only by coagulation.

A theory of the growth of cloud droplets by gravitational coagulation has been developed by Findeisen [471], Langmuir [239], Frenkel [476] and Shishkin [477]. According to Langmuir's calculations, a droplet falling through the whole thickness of a cloud and ending up as a coarse raindrop ($r \approx 1$ mm) would have to start off appreciably greater than the observed mean size of cloud droplets. Langmuir's opinion is that a small number of extra large droplets are formed accidentally which fall through the cloud until they reach a critical size at which they break up, owing to air resistance (see page 44). The comparatively coarse spray so formed is carried upwards by convection currents which reach velocities of several metres a second in cumulus clouds. These droplets then grow in turn, fall and once more break up so that, on Langmuir's model, the process of rain formation in clouds is a chain reaction.

According to Shishkin's calculations [477] rain formation in heavy convective clouds results from condensation of vapour on to the droplets during a single passage through the cloud, thus making Langmuir's hypothesis unneccessary. Shishkin agreed with Langmuir that only droplets greater than 15 μ radius can grow as a result of gravitational coagulation.

If interception and horizontal displacement of the particles are taken into account rain formation by coagulation is encouraged still more. In addition, the effect of charged droplets on gravitational coagulation must be considered (see p. 325).

The suppression of aerosols by water sprays depends on kinematic coagulation. Two cases can be distinguished: (1) when the water droplets fall through the aerosol on account of their weight alone, and (2) when the droplets are projected into the aerosol at a velocity considerably greater than their rate of sedimentation. Looking at the first case, Table 34 shows values for the coefficient of capture of aerosol particles, having various sizes and a density of two, by water droplets of radius R which are falling at their terminal velocity V_s. V_s was found from Table 9 and the values of τ required for calculating Stk were taken from Table 13. The flow of air round the falling droplets was assumed to be potential and was calculated by (34.6). Interception was neglected since r/R is small. For every centimetre of its path a drop captures $\pi R^2 \ni n$ particles, n being the number concentration of the aerosol, so that 1 cm³ of atomized water captures $\pi R^2 \ni n / \frac{4}{3} \pi R^3 = 0{\cdot}75 \ni n/R$ particles and the ratio $\ni/R$ is a measure of the efficiency of the process. The values of $\ni$ on the left hand side of the table are too high because the flow round the falling drops cannot be considered potential; no accurate values of $\ni$ are available for this flow region.

TABLE 34. COEFFICIENT OF CAPTURE OF AEROSOLS BY WATER SPRAY

r, μ		R, cm						
		10^{-3}	2×10^{-3}	5×10^{-3}	10^{-2}	2×10^{-2}	5×10^{-2}	10^{-1}
		$(V_s/2R) \times 10^{-3}$, sec.$^{-1}$						
		0·6	1·2	2·3	3·5	4·0	4·0	3·3
0·5	Stk	$4{\cdot}2 \times 10^{-3}$	$8{\cdot}4 \times 10^{-3}$	16×10^{-3}	24×10^{-3}	28×10^{-3}	28×10^{-3}	23×10^{-3}
	$\ni$	0	0	0	0	0	0	0
	$\ni/R$	0	0	0	0	0	0	0
0·7	Stk	$0{\cdot}8 \times 10^{-2}$	$1{\cdot}6 \times 10^{-2}$	$3{\cdot}1 \times 10^{-2}$	$4{\cdot}7 \times 10^{-2}$	$5{\cdot}4 \times 10^{-2}$	$5{\cdot}4 \times 10^{-2}$	$4{\cdot}4 \times 10^{-2}$
	$\ni$	0	0	0	0·03	0·06	0·06	0·03
	$\ni/R$	0	0	0	3	3	1·2	0·3
1	Stk	$1{\cdot}5 \times 10^{-2}$	$3{\cdot}1 \times 10^{-2}$	6×10^{-2}	$9{\cdot}1 \times 10^{-2}$	$10{\cdot}4 \times 10^{-2}$	$10{\cdot}4 \times 10^{-2}$	$8{\cdot}5 \times 10^{-2}$
	$\ni$	0	0	0·07	0·17	0·21	0·21	0·16
	$\ni/R$	0	0	14	17	10	4	1·6
2	Stk	0·06	0·12	0·23	0·35	0·40	0·40	0·33
	$\ni$	0·07	0·24	0·41	0·54	0·58	0·58	0·53
	$\ni/R$	70	120	80	54	29	12	5·3
5	Stk	0·35	0·7	1·4	2·1	2·4	2·4	2·0
	$\ni$	0·54	0·7	0·85	0·89	0·90	0·90	0·89
	$\ni/R$	540	350	170	89	45	18	9

It can be seen from Table 34 that the efficiency with which aerosols are collected by water sprays depends very much on the size of their particles; for a density of two they are only collected if $r > 0{\cdot}5\,\mu$, regardless of the size of the water drops. This result is in good agreement with experimental data for dusts [478, 479]. The most suitable water drop radius is 20 μ for $r = 2\,\mu$, 50 to 100 μ for $r = 1\,\mu$ and 100 to 200 μ for $r = 0{\cdot}7\,\mu$, thus increasing as the particle size of the aerosol decreases.

When atomized water is projected at a high velocity into an aerosol quite different behaviour is exhibited, because for a given droplet velocity the capture coefficient, and the efficiency with which the atomized water acts, increases steadily as the droplet size is decreased. Capture efficiency can be large even for very fine particles. For example, the coefficient of capture of particles of density 2 by water droplets 50 μ in radius at an injection velocity of 16 m sec^{-1} is 0·4 for $r = 0{\cdot}2\,\mu$ and 0·12 for $r = 0{\cdot}1\,\mu$. Avy [480] produced a considerable reduction in the concentration of indigo smoke having $r = 0{\cdot}1$–$0{\cdot}15\,\mu$ by atomizing a 10 per cent solution of sodium chloride directly in the chamber. However, the velocity of injected droplets falls rapidly as they travel, and a high efficiency of coagulation is possible only in the vicinity of the atomizer.

Kinematic deposition of aerosols by atomized water should not be confused with thermal coagulation of aerosols by water mists, which obeys quite different laws (see § 49), its rate depending mainly on the number concentration of the mist; if it is not high enough no effect is obtained. In the experiments of Avy mentioned above, after atomization of the sodium chloride solution had been stopped the number concentration of the indigo smoke fell at the same rate as in a control experiment with no mist. On the theory of thermal coagulation this would be expected since the concentration of the mist was about 10^4 cm^{-3}. On the other hand Dautrebande [482] claimed a considerable coagulation effect when mists of eosin or sodium chloride solutions having $r = 0{\cdot}2$–$0{\cdot}3\,\mu$ were introduced into quartz dust of about the same particle size. Large aggregates containing both silica and dissolved material were obtained. Dautrebande emphasizes that to obtain a noticeable effect a large amount of dispersed water must be introduced but unfortunately he does not state the number concentration of the mist in the chamber. It can be deduced from the weight concentration and mean particle size of the sodium chloride that the sodium chloride aerosol concentration in these experiments was 10^6–10^7 cm^{-3}, so that thermal coagulation of silica particles with sodium chloride particles must have taken place fairly rapidly.

In dust suppression practice, using water sprays, all three processes take place at the same time – kinematic coagulation near the atomizer, gravitational and thermal elsewhere; the first two processes are responsible for depositing the coarse fractions and the third for the fine fractions.

Recently Venturi scrubbers for dust removal have attracted much attention. They consist of Venturi tubes into which water is sprayed at the throat [483]. Since the water drops cannot immediately acquire the velocity of the gas in the throat, high relative velocities are established between the aerosol particles and the droplets and intense kinematic coagulation proceeds. Recent data [483] indicate that capture of particles by the droplets occurs not only in the throat but also over a distance of more than ten throat diameters down the diffuser. Gradient coagulation, which is discussed in the next section, may also take place in the Venturi. As would be expected the efficiency of Venturi scrubbers improves with increase in the amount of water spray, the velocity of the gas and the size of the aerosol particles; decrease in the size of the water drops is also favourable [485].

For particles of radius $0{\cdot}1$ to $0{\cdot}2\,\mu$ the efficiency is low [484, 485] but it is increased considerably if the gas is first saturated with water vapour. Owing to the reduction in pressure and adiabatic cooling caused by the accelerated gas flow in the throat, vapour condenses on the aerosol particles and enlarges them. By this means it was possible to deposit 99·9 per cent of dioctyl phthalate particles having $r = 0{\cdot}07$–$0{\cdot}20\,\mu$ [486].

Boucher's assertion [250] that for a Venturi scrubber to have maximum efficiency the size distributions of the mist droplets and aerosol particles must be as nearly the same as possible will be considered later.

When diffusion of the particles is important in kinematic coagulation the problem is equivalent to diffusion on to a sphere immersed in a viscous flow. Equation (39.17) can be used with the reservation stated on page 210. The approximate solution proposed by Müller [487] will not withstand criticism. He starts from the incorrect boundary condition that the concentration gradient at the surface of a sphere is perpendicular to the surface and neglects tangential transfer of particles in the flow. As a result Müller finds an expression for the normal concentration gradient at the surface which turns out to be a function of the azimuth, thus contradicting the original assumption.

The range of particle sizes over which both diffusion and relative motion due to settlement have to be considered will now be studied. In aerosols having a small degree of polydispersity the number of particles captured in one second by one of the coarser particles is about $\pi V_s r^2 n$ since $\ni$ is of the order of one (see Table 33). In one second the number of particles coming into contact with a given particle, owing to diffusion, is $2\,Kn = 16\pi\, r D n$. Remembering that as r increases the product $r\,V_s$ increases rapidly while D decreases, it is clear that the transition from thermal coagulation to kinematic coagulation must be quite rapid. It can be verified from the values of D and V_s given in Table 13 that gravitational coagulation becomes insignificant for $r < 0{\cdot}5\,\mu$ and thermal coagulation for $r > 2\,\mu$. Things are rather different in highly polydisperse systems. Since the rate of thermal coagulation of coarse particles with fine ones is high (see § 49) while the rate of gravitational coagulation of these particles is small, the extent of the transition region of mixed coagulation is extended considerably.

Coagulation of mists by coarser, charged drops falling through them under gravity is of great interest in the theory of rainfall from clouds. Levin [488] has investigated the case when the mist droplets are also charged and simple coulomb forces act between them and the large drops. The inertia term in the differential equation of motion of the fine droplets can be neglected when

$$\sigma_1 = \frac{3\,|Qq|}{4\pi\gamma r^3 R V_s^2} \gg 1, \tag{54.2}$$

where Q, q are the charges and R, r the radii of the large and small drops, respectively. V_s is the terminal velocity of the large drops. Neglect of inertia is thus permissible if the droplets are small and carry large charges. When the large drops fall under viscous conditions they will capture oppositely charged fine droplets with an efficiency given by

$$\ni = \frac{2\,|Qq|}{3\pi\eta r(R^2 - r^2)V_s} = \frac{3\,|Qq|}{\pi\gamma r R^2(R^2 - r^2)\,g}. \tag{54.3}$$

Capture of neutral droplets by falling charged drops, on account of induced charges, has been studied by Levin [488] and by Pauthenier and Cochet [489]. If the radius of the falling droplets is less than 10 to 15 μ and the charge greater than 4×10^{-4} e.s.u. it will be shown below that the inertia of the droplets and the hydrodynamic curvature of their trajectories on approaching one another may be neglected; this means that the medium surrounding the falling droplets may be regarded as stationary. Then, according to Pauthenier and Cochet, the expression for the capture coefficient when Stokes's law of resistance holds is

$$\ni = \frac{1}{R^2}\left[\frac{45\,\lambda\, r^2 Q^2}{16\,\gamma g\,(R^2 - r^2)}\right]^{2/5}, \tag{54.4}$$

where $\lambda = (\varepsilon_k - 1)/(\varepsilon_k + 2)$ is the induction factor, which is about one for water droplets. As R increases with the growth of the charged drop formula (54.4) becomes inaccurate and the inertia of the droplets and curvature of the streamlines near them has to be taken into account. The problem can then only be solved numerically. Curves calculated by Pauthenier and Cochet for the dependence of ∋ on R for water droplets with a charge of 4×10^{-4} e.s.u. are shown as continuous lines on Fig. 74 for a series of values of r. The dot-and-dash lines were obtained from formula (54.4) and the dashed lines are for uncharged drops.

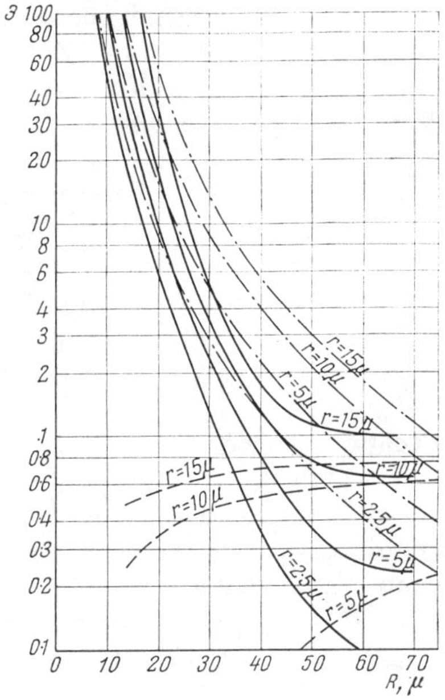

FIG. 74. Capture coefficient of neutral droplets by charged drops.

Similar results were obtained by Levin, who allowed for the curved trajectories but neglected the inertia of the particles, which is permissible when

$$\sigma_2 = \frac{3\,\lambda Q^2}{2\pi\gamma R^4 V_s^2} \gg 1. \tag{54.5}$$

Assuming that the charge on droplets in natural clouds is 10^{-3} R e.s.u. Levin obtained the values of ∋ shown in Table 35.

TABLE 35. CAPTURE OF NEUTRAL DROPLETS BY FALLING CHARGED DROPS

R, μ	$Q \times 10^7$ (e.s.u.)	σ_2	r, μ	∋	r, μ	∋	r, μ	∋
2·5	2·5	1250	1·25	2·32	0·8	1·48	0·5	0·89
3·75	3·75	110	1·87	1·06	1·20	0·69	0·75	0·40
5·00	5·00	20	2·5	0·57	1·6	0·38	1·0	0·22
6·25	6·25	5·1	3·12	0·34	2·0	0·23	2·25	0·13

Unfortunately there are as yet no reliable data on the relationship between the sizes and charges of droplets in clouds, so it is difficult, at present, to assess the effect of charged droplets on rain formation. Concerning the use of charged water droplets for inducing precipitation from the clouds, Pauthenier and Cochet calculated that drops $15\,\mu$ in radius carrying a charge of 4×10^{-4} e.s.u. had the same effect as uncharged drops $60\,\mu$ in radius having sixty-four times the mass.

The effect of charge on gravitational coagulation of water droplets similar in size has been studied experimentally by Telford *et al.* [574]. Artificial bipolar charging of the drops to about 10^{-4} e.s.u., produced a rate of coagulation several times faster than when the drops carried only the natural charge produced by atomization. Coagulation practically stopped with unipolar charging.

A little space will be devoted to the possibility of coagulating aerosols by hydrodynamic interaction between settling particles. This effect has, in the past, been assigned fundamental importance in the formation of rain. Insufficiently accurate determinations of droplet sizes in rain and mist led to distribution curves having several maxima [490]. The radii of the droplets corresponding to these maxima were in the approximate ratios $1^{1/3}:2^{1/3}:4^{1/3}:8^{1/3}$ etc. and the volumes as 1:2:4:8 etc. It was hastily concluded that primary mists were isodisperse, that droplets falling with the same speed attracted one another and coalesced, the same occuring with double droplets and so on. New, more accurate measurements, however, have shown these observations to be erroneous; the droplet size distribution has the form shown in Fig. 2 and isodisperse mists do not occur in nature. It follows from what has been stated in § 23 that it is possible for falling particles to coagulate by hydrodynamic interaction only at high Reynolds numbers, that is with large particles. In this case, however, even when the sizes of the particles are not very different their relative velocity is so great that the effect of hydrodynamic interaction cannot manifest itself. Hydrodynamic forces may be of significance, however, in fluidized beds (see § 58).

The great importance of gravitational coagulation in atmospheric precipitation has resulted in much attention during recent years. At present, it is firmly established that Langmuir's model of gravitational coagulation, in which a large spherical particle falls through an aerosol of smaller particles, which exert no influence on its motion, at a velocity equal to the difference between the actual fall velocities of these particles, represents only a limiting case for a very large difference in the sizes of the particles. In general it is necessary to allow for hydrodynamical interaction between the particles, which depends not only on their relative velocities but also on their velocities relative to the air.

Gravitational coagulation has been studied recently in three different ways: (A) by theoretical computation of the field of flow around a falling drop and hence of its interaction with the other particles. (B) By model experiments in a viscous liquid which reproduce the trajectories and velocities of two or more sedimenting spheres. (C) By examining the motion and the collection efficiency of droplets suspended in an ascending air stream.

By making use of expressions which he deduced for the hydrodynamic forces $F_{\parallel}$ and $F_{\perp}$ (see p. 99) to Stokes' approximation, Hocking [658] employed an electronic computer to calculate the trajectories and collection efficiencies of freely falling water drops. In this case, $\ni > 0$ only when the radius of the larger drop $R > 18\,\mu$ (this conclusion is at variance with experiment, see below). With an increase in R both $\ni$ and the range of the values of the ratio r/R at which $\ni > 0$ are increased (at $r/R = 1$, on Stokes' approximation, as was shown on p. 100 both particles have an identical velocity and do not approach each other). Thus, at $R = 20\,\mu$ this range extends from

0·4 to 0·8 and at $R = 25\,\mu$, from 0·2 to 0·9. The maximum values of $\ni$ (at $r/R \approx 0{\cdot}7$) in these two cases are 0·25 and 0·90 respectively.

Although the sedimentation of drops of this size corresponds to Re_f values of only 0·10–0·30, calculation of the collection efficiency on Stokes' approximation can result in grave errors, since it involves the determination of the trajectories of the drops, beginning with a large distance ϱ between them (up to 50 R) at which $\beta = V_s\,\varrho/\nu > 1$ (see p. 101).

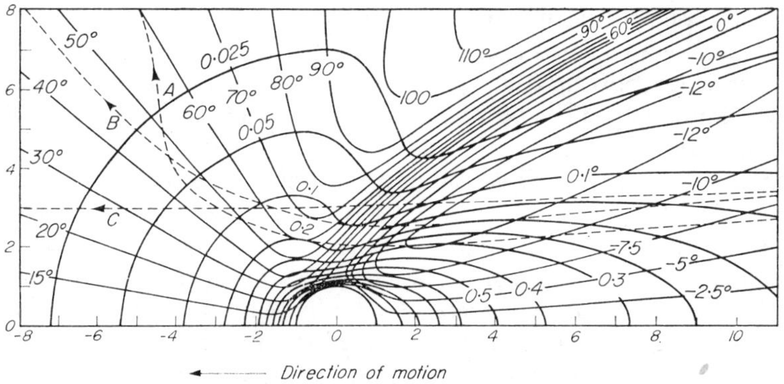

FIG. 74a. Field of flow around a sphere ($\mathrm{Re}_f = 4$).

Pearcey and Hill [823] considered this problem using Oseen's approximation; simplified forms of Oseen's equations for the field of flow around a moving sphere, applicable at large ϱ/R only, were used. The error then increases as the surface of the sphere is approached. The authors determined the field of flow produced by the motion of two spherical drops by adding the fields produced by each drop separately, making use of the zero approximation (see (13.1)), which does not satisfy strictly the boundary conditions on the surfaces of the drops. This error, also, grows as the drops approach each other. The view of the authors that the drops ran through the zone in which the errors are large so quickly that there is little effect on the final result is far from convincing.

The field of flow around a falling water drop having $R = 73\mu$ ($\mathrm{Re}_f = 4$) is shown in Fig. 74 a. The solid lines connect points relating to identical flow velocities, expressed in fractions of the velocity of fall of the drop, and points at which the flowlines have a definite slope θ to the axis of motion. $\theta < 0$, if the lines approach the axis. The graph shows that a wake is formed behind the drop in which the flow velocity decreases slowly with an increase in the distance from the drop and the flow lines are directed towards the axis. In front of the drop, and to the side, the flow velocity diminishes much more rapidly and the flow lines are directed sideways at large angles. Consequently, drops moving behind the given drop of radius R will not only overtake it (see p. 102) but will also be drawn into its wake and approach the axis of flow. The trajectories of drops overtaking a given drop of radius R are shown in Fig. 74a by broken lines. The radii of these drops are 0·1 per cent (A), 1·1 per cent (B) and 11 per cent (C) larger than R. Initially, when at a large distance from the given drop, these drops were in the same place. The smaller the difference in the sizes and fall velocities, the more time is available for both drops to approach each other.

Values of ∋ as a function of the ratio $(R - r)/r$ for various values of R, as calculated by the authors, are shown in Fig. 74 b. The rapid rise of ∋ with a decrease of this ratio has just been explained. The growth of ∋ with increasing Re_f is due to rising asymmetry of the field of flow; and the greater Re_f, the slower proceeds the reduction of the flow velocity in the wake with increasing distance from the drop. The data in Fig. 74 b are calculated on the assumption that initially the drops R and r are an infinite distance apart. In practice, due to horizontal displacement of the drops caused by turbulence and convection currents, the continuous mutual approach of the drops begins with a finite distance between them so that the values of ∋ in Fig. 74 b are too large.

This fact is of particular importance in the case of small drops ($r < 10\,\mu$) as their mutual approach, when the difference in size is small, is extremely slow. Consequently, the curves for such drops shown in Fig. 74 b are of little practical interest.

It is evident from the foregoing that the experimental results of Telford *et al.* agree fairly well with the estimates made by Pearcey and Hill.The values obtained by Telford, ∋ $= 12$ for drops of $r \approx 70\,\mu$ ignoring the interception effect, and ∋ $= 3$ allowing for it, correspond, according to Fig. 74 b, to $(R - r)/r = 0{\cdot}01$–$0{\cdot}1$, i.e. to the conditions of Telford's experiments.

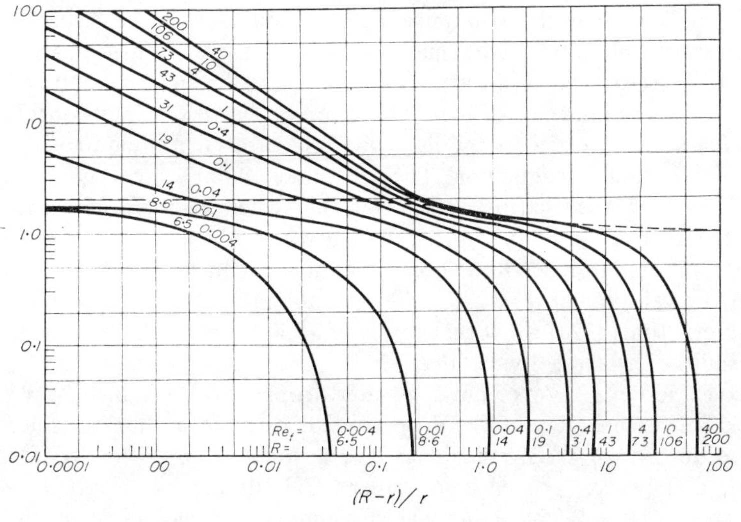

FIG. 74 b. Gravitational coagulation of water drops (R in μ).

Schotland [824] studied the motion of two (in most cases identical) steel spheres of $r = 0{\cdot}7$–7 mm, sedimenting in a viscous liquid having a viscosity of 20 poise. He did not give the direct results of his experiments, but used dynamical similarity to apply them to water drops falling in air with the same value of Stk. He considered it unnecessary to maintain Re constant, assuming Re in his experiments to be very small (for the above mentioned spheres Re varied from 0·05 to 5). Unfortunately, there is an obvious error in the calculations and it is impossible to ascertain from his papers what spheres were used.

Schotland observed the mutual approach of two identical sedimenting spheres, but there was no real contact between them owing to the viscous resistance of the intervening liquid (see p. 301). After approaching one another the spheres separated as predicted by Pearcey and Hill. It was found that $\ni$ was about 1·7 for identical spheres and much less for spheres of different sizes. It seems from this that Re in Schotland's experiments was not particularly small.

Kinzer and Cobb [825] suspended a water drop (R) at a constant height by adjusting the velocity of an ascending air stream to equal V_s of the drop. The air contained mist droplets of mean radius $\bar{r} = 5\text{–}7\,\mu$; the stationary drop was examined with a telescope. The increase in R caused by the capture of mist droplets was determined from the increase in V_s and $\ni$ was estimated in this manner. These calculations were made on the basis of V_s rather than from the relative velocity of the drops, the small drops being assumed to be stationary with respect to the air, hence the values of $\ni$ calculated for near values of r and R are too small.

The ($\ni$, R) curve obtained by the authors is of a rather complicated shape. At $R \approx r \approx 6\,\mu$, $\ni \approx 1$ and with an increase in R (that is a decrease in r/R) it diminishes down to about 0·15 at $R = 30\,\mu$, which is in qualitative agreement with Pearcey and Hill; $\ni$ then rises again to 0·9 at $R = 200\,\mu$. This portion of the curve does not differ greatly from that calculated for potential flow; its shape can be explained by the fact that at constant r the Stk number is proportional to V_s/R and as shown in Table 9 increases with R. With further increase in R, $\ni$ again diminishes to 0·15 at $R = 1{\cdot}5$ mm. As seen from Table 9, in this size range V_s/R diminishes with increasing R.

Kinzer and Cobb noticed that small droplets in the wake of the suspended drops move in small circles located in vertical planes passing through the centre of the drop, evidently under the influence of vortices formed in the wake. They consider the turbulence generated by vortices in the wake of all the drops in a cloud to be the principal factor in gravitational coagulation. Their extremely complicated and not very clear theory will not be discussed beyond noting that when a sphere moves in a viscous medium the formation of the vortices equal to its diameter does not occur until Re $\approx$ 100, and a still higher Re is necessary for detachment [826]. Thus, the theory can hardly be valid for drops of $R \leqslant 300\,\mu$ although it might hold with large drops. Oakes' observations [827] show that drops of $R = 1{\cdot}8$ mm falling through a smoke chamber agitate the aerosol violently.

The collection efficiency of moving water drops was determined by Walton and Woolcock [828] who allowed the drops to fall through a chamber filled with an isodisperse aerosol of methylene blue having spherical particles ($\bar{r} = 1{\cdot}25\text{–}2{\cdot}5\,\mu$). This method, however, proved inconvenient and the authors resorted to another. A water drop of $R = 0{\cdot}25\text{–}1{\cdot}0$ mm was suspended by a glass capillary in a vertical tube through which the aerosol was drawn upward at a rate corresponding to V_s of the drop. Re_f ranged from 70 to 870. The values of $\ni$ obtained were somewhat lower than those calculated by Fonda and Herne for potential flow past a sphere (Fig. 48c).

Walton and Woolcock also calculated the dust suppressing efficiency of a water spray projected into a dust cloud at high velocity, allowing for the gradual reduction in velocity. As would be expected, the capture of fine dust particles can be considerably increased in this way.

The wash-out of radioactive atmospheric aerosols by rain has lately acquired great importance. The foregoing discussion shows that it is possible to use the curve for potential flow, shown in Fig. 48c, for these calculations without serious error.

May [829] released *lycopodium* spores of $\bar{r} = 16\,\mu$ marked with I^{131} into the atmosphere from a height of 3·3 m during a period of rain. A number of Petri dishes were used as collectors of the spores, whether freely sedimenting or washed out by rain, and were placed in a circle of the radius = 10 m around the tower from which the spores were dispersed. Some of the dishes were shielded from the rain and these collected only the spores deposited spontaneously, and not those washed out by rain drops. The number of spores collected by rain drops was estimated by difference. The sizes of the rain drops ($R = 0{\cdot}2$–2 mm) were not determined, but taken from the tables of Best [830] for a given intensity of rain. A fair agreement between the experimental results and theoretical calculations of deposition in potential flow was obtained.

Johnstone *et al.* [831] deduced the following empirical formula for the efficiency of Venturi scrubbers

$$\ni = 1 - \exp(-K\beta\,\mathrm{Stk}^{1/2}) \tag{54.6}$$

where β is the ratio of the volume of the water sprayed to that of the gas cleaned and K is a constant. Brink and Contant [832] confirmed this formula in undustrial experiments with phosphoric acid mists of $r = 0{\cdot}25$–$0{\cdot}75\ \mu$ finding that $\ni$ varied from 0·8 to 0·99. Theoretically, the ($\ni$, Stk) curve shown in Fig. 48c for potential flow past a sphere would be expected; however, this curve can be well approximated over the given range of $\ni$ by (54.6), if a suitable value is chosen for K.

At the first glance, the conclusions of Pearcey and Hill and the experiments of Telford *et al.* support Boucher's view, mentioned on p. 324. These conclusions, however, certainly do not hold for particle sizes of the order of $1\ \mu$.

As regards the effect of electric charges on gravitational coagulation, the theoretical paper of Pauthenier *et al.* [833] should be mentioned; equation (54.4) was generalized to include a vertical electric field of strength E acting on the particles in addition to gravity. A term $3QE/4\pi R$ has to be added to the term $\gamma g(R^2 - r^2)$ where $E > 0$, if the field is directed downwards. The estimates of Gunn [834], who apparently knew nothing about Levin's work (see p. 326), are very primitive and are not considered here.

From time to time accounts appear of the beneficial influence of the electric charging of water sprays upon the efficiency of aerosol deposition in a chamber [835, 836], although there are as yet no accurate quantitative data. The charging of the aerosol itself by a corona discharge only slightly increases its deposition in a Venturi scrubber. It seems that no attempt has been made to charge the water drops in the scrubber.

The attractive force between a spherical particle of radius r and a drop upon which vapour is condensing, according to (16.29), is $6\pi\eta rRD_g(c_\infty - c_s)M'/c'\varrho^2M$ and this varies with the distance like a Coulomb force; Levin's formula (54.3), as was noted by Dukhin and Derjaguin [643], can thus be used in evaluating the rate of gravitational coagulation of an aerosol whose drops are growing in a supersaturated atmosphere but $6\pi\eta rRD_g(c_\infty - c_s)M'/c'M$ must be substituted for $|Qq|$.

With evaporating drops the rate of gravitational coagulation must evidently be reduced. The precipitation of aerosols by water sprays is therefore more effective the higher the humidity of the gas.

§ 55. COAGULATION OF AEROSOLS BY STIRRING AND IN TURBULENT FLOW

Experimental study of the rate of coagulation during stirring is made difficult by the fact that stirring, as pointed out on page 255, accelerates the deposition of particles on the chamber walls. The rate of deposition increases with particle size so that the assessment of the course of coagulation from the increase in mean particle size is complicated.

Langstroth and Gillespie [361] tried to solve this problem by separating the constants of coagulation and deposition. Experiments were conducted with ammonium chloride smoke ($r = 0{\cdot}5\,\mu$, $n = 2 \times 10^5$ cm^{-3}) in a 1 m^3 chamber which was fanned with broad blades reaching almost to the walls of the chamber and oscillating about a horizontal axis passing across the centre. The mean velocity $\bar{U}$ of the air in the chamber was determined with a hot-wire anemometer. The number and weight concentrations of the smoke were measured, the latter being found to vary according to the logarithmic law

$$\ln (c/c_0) = -\beta' t, \tag{55.1}$$

in which the coefficient β' remained almost constant throughout the whole of the experiment, which lasted several hours. The results are given in Table 36.

TABLE 36. COAGULATION OF AMMONIUM CHLORIDE SMOKE DURING STIRRING

$\bar{U}$, cm sec^{-1}	0	18	48	67	83
$K \cdot 10^{10}$, cm^3 sec^{-1}	3·7	4·0	5·0	7·5	8·7
$\beta \cdot 10^5$, sec^{-1}	5·5	9·7	11·0	14·5	16·3
$\beta' \cdot 10^5$, sec^{-1}	5·5	12·0	13·8	18·3	21·7

That the fall in weight concentration due to deposition is faster than the fall in number concentration ($\beta' > \beta$) is explained by the polydispersity of the aerosol (see page 257).

As already pointed out, this method of determining the coagulation constant is unreliable for polydisperse aerosols so that these experiments do not prove conclusively that the coagulation of aerosols having particle radii of about $0{\cdot}5\,\mu$ is accelerated by stirring. In more recent work by the same authors [357] β was determined directly (see page 254) and the coagulation constant was found from the difference. It would appear from the results that K first decreases as the stirring rate is increased and then begins to increase but these strange results probably arise from errors in the measurements.

Richardson [491] studied the effect of stirring on rate of coagulation by measuring the optical density of an aerosol. He made no allowance for deposition of the particles and obtained absurdly high values for the coagulation constants in both stirred and unstirred aerosols. Since the aerosol contained a coarse fraction, with particle radii up to $5\,\mu$, which deposited very rapidly on the walls and bottom of the chambers, the reason is clear. He also generated an aerosol with particle radii of 3 to 4 μ, by atomizing a solution of methylene blue, which was passed through a wind tunnel having a screen to produce turbulence. The mean particle size doubled in 3 sec. The flow velocity and aerosol concentration were not indicated.

Teverovskii [407] measured the number concentration in smoke from a ground-level source and also calculated it theoretically (see § 47). Assuming that the difference was due to coagulation he obtained coagulation constants ranging from 50×10^{-10} $cm^3 sec^{-1}$ at a wind speed of 1·15 m sec^{-1} to $17{,}700 \times 10^{-10}$ $cm^3 sec^{-1}$ at a wind speed of 5·9 m sec^{-1}. Since very large errors are unavoidable when measuring the concentration of such smoke clouds these results, which indicate that atmospheric turbulence can increase the coagulation constant several thousand times, must be regarded with great caution.

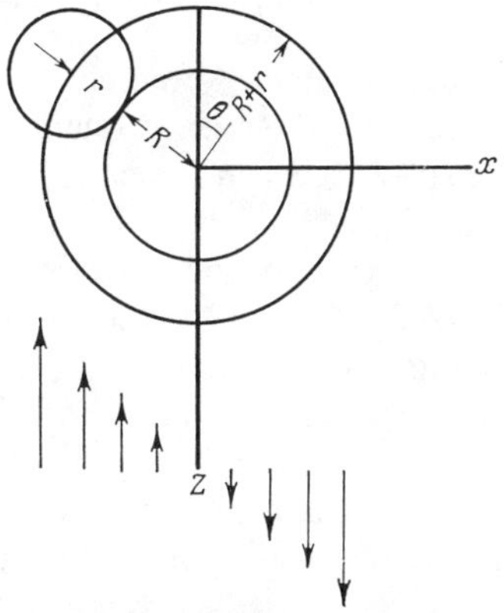

FIG. 75. Theory of gradient coagulation.

Experimental study of coagulation in aerosols during stirring and turbulent flow has thus been very limited. A theoretical approach can begin with Smoluchowski's formula [492] for the gradient coagulation which occurs when a transverse velocity gradient exists in the flow. Smoluchowski neglected diffusion and curvature of the streamlines near the particles thus making the problem easy to solve.

Taking the flow in the direction of the z-axis let the velocity gradient be Γ along the x-axis (Fig. 75). One particle of radius R is taken as the stationary absorbing sphere of radius $R + r$ where r is the radius of the other particles. The flow velocity relative to the chosen particle is $U = \Gamma x = \Gamma(R + r) \sin \theta$. The number of particles passing between the coordinates x and $x + dx$ in 1 sec which touch the absorbing sphere is

$$d\Phi = nU\,2(R + r) \cos \theta \, dx = 2\, n\Gamma(R + r)^3 \cos^2 \theta \sin \theta \, d\theta \tag{55.2}$$

and hence the total number of particles reaching the absorbing sphere, from above and below, is

$$\Phi = 2 \int_{\theta=0}^{\theta=\frac{\pi}{2}} d\Phi = \tfrac{4}{3} n\Gamma(R + r)^3 \tag{55.3}$$

or, for a monodisperse aerosol,

$$\Phi = \tfrac{32}{3} n\Gamma r^3. \tag{55.4}$$

If curvature of the streamlines near the particles is allowed for and the relative velocity, and hence the inertial displacement, of particles passing close to one another is very small, the numerical coefficients in formulae (55.3) and (55.4) should be reduced considerably, according to Table 33.

Model experiments [493] on the aggregation of glass spheres having $r = 60–70\,\mu$, due to a velocity gradient in a viscous syrup at $\Gamma = 0{\cdot}4–0{\cdot}8\ \text{sec}^{-1}$, gave very good agreement with formula (55.4), however. It is possible that rotation of the spheres in a velocity gradient (see page 36) compensates for the effect of the curved trajectories to some extent.

The ratio of the rates of gradient and thermal coagulation is

$$\frac{\Phi}{\Phi_0} = \frac{32\,n\Gamma r^3}{3\times 16\pi n D r} = \frac{2\,\Gamma r^2}{3\pi D}. \tag{55.5}$$

It is clear that coagulation in a velocity gradient cannot cause the effect described by Langstroth and Gillespie. In their experiments the gradient must have been greatest in the 10 cm-wide space between the ends of the blades and the walls of the chamber. Since the greatest linear velocity of the blades was about $2\ \text{m sec}^{-1}$ the maximum value of the gradient must have been $20\ \text{sec}^{-1}$, the average being much less. For $r = 0{\cdot}5\,\mu$ and $\Gamma = 20\ \text{sec}^{-1}$ formula (55.5) gives $\Phi/\Phi_0 = 0{\cdot}4 \times 10^{-2}$; the effect of gradient coagulation is therefore negligible.

Coarsely dispersed aerosols are different. For $r = 10\,\mu$ and $\Gamma = 20\ \text{sec}^{-1}$ the ratio of gradient to thermal coagulation is approximately 16, according to equation (55.5), and coagulation is accelerated noticeably even with small velocity gradients. Gradient coagulation can also be very active in the boundary layer of turbulent flow, as can be verified in the following way. The velocity gradient at the wall, Γ_0, can be found from the formula $\tau_0 = \eta\Gamma_0$ where τ_0 is the tangential drag of the gas on 1 cm^2 of the wall which is related in turn to the friction velocity $U^* = \sqrt{\tau_0/\gamma_g}$. Hence $\Gamma_0 = U^{*2}/\nu$. Nikuradse's experiments [389] indicate that the ratio $\bar{U}/U^*$ lies between 14 and 28 for flow through pipes at Reynolds' numbers from 300 to 100,000 so that $\Gamma_0/\bar{U}^2$ lies between 0·008 and 0·03. If, for example, $\bar{U} = 10\ \text{m sec}^{-1}$ then $\Gamma_0 \approx 10^4$ sec^{-1}; the coagulation of an aerosol having particles of radius $0{\cdot}5\,\mu$ can therefore be accelerated appreciably in the wall layer.

In his theoretical work on the coagulation of aerosols in turbulent flow Levich [494] discusses coagulation effected by fluctuations having a scale of the same order as the particle size, which is appreciably less than the internal scale of turbulence λ_0. There are no reliable experimental data on these fluctuations; fluctuation velocities calculated from Heisenberg's theory [367] are considerably less than those of Obukhov and Yaglom. They [364] derived the following expression for the root mean square relative fluctuation velocity of two particles of a turbulent medium along their line of centres:

$$u_\lambda = \sqrt{\frac{1}{15}\frac{\varepsilon}{\nu}}\,\lambda. \tag{55.6}$$

This was used by Levich as the basis of his calculations. In this formula λ is the distance between the particles, ν the kinematic viscosity of the medium, and ε the rate of dissipation of turbulent energy per gram of the medium, related to λ_0 by

$$\lambda_0 \approx \left(\frac{\nu^3}{\varepsilon}\right)^{1/4}. \tag{55.7}$$

An expression for the coefficient of diffusion due to fluctuations on a scale λ follows,

$$D_\lambda \approx \lambda u_\lambda \approx \frac{1}{4}\sqrt{\frac{\varepsilon}{\nu}}\,\lambda^2 = \beta\lambda^2. \tag{55.8}$$

The rate of diffusion of two particles towards one another in a turbulent medium therefore decreases as they approach. Steady state flow of particles diffusing towards an absorbing sphere of radius $2r$ is expressed by [see formula (38.32) on page 203]

$$\Phi = 4\pi\varrho^2 D_\varrho \frac{dn}{d\varrho} = 4\pi\beta\varrho^4 \frac{dn}{d\varrho} = \text{const.} \tag{55.9}$$

The boundary conditions $n = 0$ at $\varrho = 2r$ and $n = n_0$ at $\varrho = \infty$ give

$$\Phi = 32\pi r^3 \beta n_0 \approx 25\sqrt{\frac{\varepsilon}{\nu}}\,r^3 n_0. \tag{55.10}$$

A similar formula, with a different numerical coefficient, can be obtained in another way by assuming, following Tunitskii [495], that velocity gradient coagulation occurs in turbulent flow due to the relative fluctuation velocity at right angles to the line of centres of the particles. Obukhov and Yaglom use formula (55.6) for this velocity with the coefficient 1/15 replaced by 2/15. The gradient of the fluctuation velocity is thus

$$\frac{du_\lambda}{d\lambda} \approx \sqrt{\frac{2\varepsilon}{15\nu}}, \tag{55.11}$$

while formula (55.4) leads to

$$\Phi \approx \frac{32}{3}\sqrt{\frac{2\varepsilon}{15\nu}}\,r^3 n_0 \approx 4\sqrt{\frac{\varepsilon}{\nu}}\,r^3 n_0 \tag{55.12}$$

for the flow of particles towards a given particle due to fluctuations.

The ratio of the constants of turbulent and thermal coagulation is

$$\varkappa = \frac{\alpha\sqrt{\dfrac{\varepsilon}{\nu}}\,r^3 n_0}{16\pi D r n_0} = \frac{\alpha}{16\pi}\sqrt{\frac{\varepsilon}{\nu}}\,\frac{r^2}{D}, \tag{55.13}$$

where α is either 4 or 25 depending on whether formula (55.10) or (55.12) is used. For particles 1 μ in radius $r^2/D \approx 0{\cdot}1$. Hence it follows that for the coagulation of such particles to be accelerated noticeably by turbulence $\sqrt{\varepsilon/\nu}$ must be about 100 and λ_0 according to formula (55.7) about 0·1 cm.

It can be seen from page 258 that for this a very high flow velocity is necessary. For particles of radius 0·1 μ the effect of turbulent coagulation is negligible while for $r = 10\,\mu$ it is very large. The value of ε becomes large near the walls but here the gradient of the average velocity is also large, which is evidently the main reason for the rapid coagulation of the aerosol in Richardson's experiments mentioned above.

It should be noted that the possibility of coagulating aerosols by small-scale turbulent fluctuations was first suggested by Teverovskii [407] but his calculation of the effect was incorrect and he obtained exaggerated values for the turbulent coagulation constant of particles having a radius of 0·3 μ in the atmosphere.

In another paper [496] Levich discussed the coagulation of aerosols which resulted from the different extents to which particles of different sizes are entrained by turbulence. The expressions obtained by Levich for the root mean square velocity of the particles relative to the medium is

$$v_r = a\tau, \tag{55.14}$$

where τ is the relaxation time of a particle and a the root mean square acceleration in the turbulent flow. This formula can easily be derived by treating the fluctuations as harmonic oscillations of angular frequency ω. According to (20.14)

$$v_r = \frac{u\omega\tau}{\sqrt{1+\omega^2\tau^2}}, \tag{55.15}$$

where u is the root mean square fluctuation velocity.

As shown in § 45, in this case $v_r/u \ll 1$ in

$$v_r = u\omega\tau. \tag{55.16}$$

For harmonic oscillations $a = \omega u$ so that $v_r = a\tau$. The relative velocity of two particles whose distance apart is small in comparison with the scale of fluctuations is $v_{r1} - v_{r2} = a(\tau_1 - \tau_2)$. Since relative motion of particles and medium is possible only with fluctuations on a scale $< \lambda_0$ whose period t_p according to Obukhov and Yaglom does not depend on the scale, the maximum acceleration is given by

$$a = \frac{u_{\lambda_0}}{t_p} = \frac{u_{\lambda_0}^2}{\lambda_0} \tag{55.17}$$

or, according to formulae (55.6) and (55.7),

$$a \approx \frac{\varepsilon^{3/4}}{\nu^{1/4}}. \tag{55.18}$$

Gillespie [837] estimated the coagulation rate of charged aerosols in a velocity gradient. For the rate of coagulation of spherical particles having the radius r_1 and the charge q_1, with those of radius r_2 and charge q_2 he obtained†

$$\Phi_{12} = \tfrac{4}{3}\left[\Gamma(r_1 + r_2)^3 - 3\beta\right] n_2 \tag{55.19}$$

where

$$\beta = \frac{q_1 q_2 (r_1 + r_2)}{6\pi\eta r_1 r_2}.$$

In the absence of charges (55.19) reduces to the Smoluchowski equation (55.4).

Recent papers on the theory of coagulation in turbulent flow include the work of Saffman and Turner [838] which does not differ in principle from that of Levich although their computations are somewhat more accurate, allowance being made, for instance, for the variations of relative particle velocity, assuming a Gaussian distribution.

† Gillespie has -6β in this formula.

For the rate of turbulent coagulation by the first mechanism of Levich due to the difference in fluctuation velocity of the medium at two points a distance $2r$ apart, the formula of Levich and Tunitskii (55.10) with the coefficient 10·4 was obtained. For coagulation by the second mechanism, due to the difference in the fluctuation velocity of particles caused by unequal inertia, they found

$$\Phi = 5{\cdot}7\, n_0(r_1 + r_2)^2(\tau_1 - \tau_2)\, \varepsilon^{3/4}\, \nu^{-1/4} \qquad (55.20)$$

where τ_1 and τ_2 are the relaxation times of coagulating particles. For the mean square value of relative acceleration in turbulent flow an expression due to Batchelor was used,

$$\overline{\left(\frac{dU'}{dt}\right)^2} = 1{\cdot}3\, \varepsilon^{3/2}\, \nu^{-1/2}. \qquad (55.21)$$

Equation (55.20) shows that the second mechanism of coagulation predominates as the difference in size of coagulating particles increases. The authors estimate that in clouds with $\varepsilon = 5$ cm²/sec³, which is about right for stratus clouds with a weak wind, the second mechanism predominates over the first when $r_1 - r_2 > 2{\cdot}3\,\mu$; when $\varepsilon = 1000$ cm²/sec³ (cumulus with strong convection) already at $r_1 - r_2 > 0{\cdot}6\,\mu$.
A similar approach can be found in a paper by East and Marshall [576] who did not make use of an expression like (55.21) for the mean square value of the acceleration of the medium and therefore failed to obtain an expression for the rate of turbulent coagulation.

There is a difference of opinion as to whether the collection efficiency of the particles should be allowed for in coagulation by the second mechanism. East and Marshall use values of $\ni$ estimated by Langmuir, but Saffman and Levich [839] assume that $\ni \approx 1$. Saffman proceeds in his work from the experiments of Manley and Mason mentioned on p. 334, who studied coagulation in a gradient flow, but these experiments refer only to turbulent coagulation by the first mechanism as treated by Tunitskii (see p. 336). Levich believed that fine particles were carried past large ones and at the same time diffused towards them by the first mechanism. The expression for $\ni$ deduced from this model is, in Levich's view, of the order of unity. If, however, $R = 10^{-3}$ cm and $\varepsilon = 1000$ cm²/sec³ are substituted in this expression $\ni$ comes out to be only 0·05. The magnitude of $\ni$ seems to have a great effect upon the coagulation rate by the second mechanism. On the other hand, the value of $\ni$ in this case probably differs greatly from the figure for gravitational coagulation.

Frisch [840] assumes the mean square relative displacement of two particles under the action of turbulence to be equal to the sum of mean square values of their absolute displacements; from this it would follow that the coefficient of relative turbulent diffusion of two particles was equal to the sum of the diffusion coefficients of each. The theory of Smoluchowski could then be applied to turbulent coagulation with only slight modifications. These assumptions are obviously at variance with the theory of turbulent diffusion and, on this account, Frisch's conclusions are not given here.

Levich [839] proved that formulae of the type (55.10) are valid only when particles approach one another by turbulent diffusion up to actual contact, so that $D_t \gg D$, at any distance between the centres of the particles up to $\varrho = 2r$. According to (55.8) this necessitates

$$2r \gg 2\, D^{1/2}(\nu/\varepsilon)^{1/4}. \qquad (55.22)$$

With $\varepsilon = 1000\,\text{cm}^2/\text{sec}^3$, for example, the condition will be observed for $r > 1\,\mu$. When the inequality (55.22) is reversed, i.e. when $r < 0{\cdot}1\,\mu$, the coagulation rate is determined by the formula of Smoluchowski and the turbulence of the medium plays no part.

§ 56. THE EFFICIENCY OF COLLISIONS BETWEEN AEROSOL PARTICLES

The efficiency of collisions between particles, or the probability of their coalescence or sticking together on collision, is very important in the theory of coagulation. The terms "collision" and "contact" are not synonymous because the cushioning effect of air between particles or large drops may prevent contact.

There are two reasons why it is difficult to study contact efficiency by direct ultramicroscopic observation. The depth of field in such an investigation for a variety of reasons cannot be made less than about $10\,\mu$ so that particles less than this distance apart along the optical axis appear to be in the same plane and the meeting of their images is construed incorrectly by the observer as contact. Stereo-ultramicroscopic observation would overcome this but it has never been attempted. Secondly, the approach of two particles to a distance less than the resolving power of the microscope is also interpreted as a contact. Such particles may separate without touching so that some of the contacts recorded will only be apparent. These considerations have not always been taken into account with the result that all observed contacts of aerosol particles have been claimed to be ineffective [497, 498].

The effectiveness of contact between aerosol particles and a wall can be dealt with by placing a glass or metal mirror at an angle of 10–15° to the optical axis of the microscope [499]. With dark-field illumination the mirror image of each particle is seen together with the particle itself in the field of vision and contact of a particle with the mirror is perceived as contact of the particle with its image. For liquid particles the criterion of attachment is their disappearance because they spread over the mirror; for solid particles it is the cessation of Brownian motion.

Here, also, some of the observed contacts were only apparent, being registered when a particle approached within a distance $\delta/2$ of the wall where δ is the resolving power of the microscope. Let $w(x)$ be the probability that a particle distant x from the wall will move away without touching it to a distance x' at which the movement becomes noticeable to an observer. It follows from the physical meaning of $w(x)$ that

$$w(x) = \tfrac{1}{2}w(x + \Delta x) + \tfrac{1}{2}w(x - \Delta x), \tag{56.1}$$

where Δx is a small displacement. Taylor expansion of the right hand side of this equation gives $\ddot{w}(x) = 0$ so that $w(x)$ is linear. Since $w(0) = 0$ and $w(x') = 1$, clearly $w(x) = x/x'$. In the present case x is $\delta/4$, this being the average distance up to $\delta/2$; under the conditions of the experiment x' was approximately equal to δ. Hence the probability that a contact is only apparent is about 0·25.

Experiments conducted with tobacco smoke (consisting of semi-liquid particles), magnesium oxide and ammonium chloride smokes, having particles of mean radius several tenths of a micron, showed that between 10 and 25 per cent of the contacts appeared to be ineffective. The number was practically the same when the mirror was covered with a thin film of glycerol. Had the ineffective contacts been real their number would most probably have diminished appreciably when the solid surface was replaced by liquid. That ammonium chloride smoke gave a higher percentage of apparently ineffective contacts (20–25 per cent) than tobacco and magnesium oxide smokes (10–15 per cent) is probably explained by particles of the latter having higher electric charges than the former so that mirror forces reduced the likelihood of particles moving away again after having approached the wall. These experiments show that the efficiency of contact between aerosol particles between 0·1 and 1 μ with a solid surface is close, and in all probability equal, to one.

In principle the efficiency of contact between aerosol particles can be found by comparing experimental and theoretical values of the coagulation constant. However, multiple collisions (see page 294) considerably reduce the dependence of the coagulation constant on the efficiency, α, of a single collision so that the ratio $K_{\text{exp}}/K_{\text{theor}}$ must be found fairly accurately. Exact calculation of K_{theor} is at present difficult, because of the effect of polydispersity on the rate of coagulation (see page 295), and the only conclusion that can be derived from measurements of the coagulation constant is that α is near to one. Indirect evidence of high contact efficiency is that coagulation constants of smokes and mists having similar particle size distributions, such as oleic acid mist and stearic acid smoke, are practically the same [428]. Ineffective contact between liquid droplets is unlikely unless solid surface films, such as occur on oxidised or contaminated mercury, are present. Mists, of liquids other than mercury, coagulate by the droplets coalescing rather than sticking together.

The problem under discussion can be approached theoretically by finding the molecular force potential when two spherical particles come into contact. It is given by (51.17), putting $\varrho = 2r + \delta$, where δ is of the order of the molecular diameter. Since $r \gg \delta$ this reduces to

$$\psi(2r + \delta) = -\frac{\pi^2 Q r}{12\,\delta}. \tag{56.2}$$

According to the data on page 309, Q is about 10^{-13} to 10^{-14} erg; it follows that the ratio ψ/kT for particles of radius 10^{-5} cm is about 10^2 to 10^3, and the molecular force potential is appreciably greater than the kinetic energy of the particles. Even perfectly elastic particles might be unable to break away from one another after collision.

A similar conclusion is reached from the expression derived by Bradley [443] and Derjaguin [500] for the force needed to separate spherical particles of radii r_1 and r_2 from one another

$$F = 4\pi\sigma \frac{r_1 r_2}{r_1 + r_2} \tag{56.3}$$

or, for identical particles,†

$$F = 2\pi\sigma r, \tag{56.4}$$

where σ is the free surface energy of the particles.

The energy of Brownian rotation of a doublet formed by two particles about an axis perpendicular to the line of centres is

$$\Omega = \tfrac{1}{2} I\omega^2 = kT, \tag{56.5}$$

† The same expression is obtained for the force of cohesion of wet particles (e.g. in hygroscopic smokes) in which case σ is the surface tension of the liquid (see page 362).

where $I = \frac{64}{15}\pi\gamma r^5$ is the moment of inertia of the doublet with respect to this axis and ω is the mean angular velocity. The centrifugal force tending to break up the doublet is therefore

$$F' = \frac{4}{3}\pi\gamma r^4\omega^2 = \frac{5\,kT}{8\,r}. \tag{56.6}$$

Rupture occurs when $F' > F$ or $r < \left(\frac{5\,kT}{16\pi\sigma}\right)^{\frac{1}{2}}$. Taking σ to be several tens of erg cm^{-2}, aggregates can rupture under the influence of molecular impacts only if $r < 10^{-8}$ cm [502]. For linear aggregates of greater length the centrifugal force is still less.

The theoretical considerations set out indicate that all the contacts between aerosol particles must be effective and this, of course, also applies to contacts between particles and walls. It must be emphasized that this relates exclusively to contacts resulting from Brownian motion. Fairly coarse particles moving under the influence of external forces can rebound after a collision while aggregates can be broken up by air currents, and so on (see below).

Mention must also be made of the large number of researches by Russian authors (Rumyantseva [503], Urazovskii and Kuz'menko [504], Samokhvalov and Kozhukhova [435], Saichuk and Narskikh [505], Andreev and Kibirkshtis [506], Smirnov and Solntseva [507]) on the effect of foreign vapours on the coagulation of aerosols. Both stabilizing and sensitizing effects were noted. Apart from experiments in which a condensation aerosol formed in the presence of foreign vapour which might affect formation and growth of nuclei, the effects mentioned can be explained mainly by the influence of the vapours on the shape and structure of aggregates (see § 50). It has been shown by Radushkevich and Chugunova [508], by Petryanov, Tunitskii and Tikhomirov [509], and by Artemov [510] that in the absence of these complicating factors, foreign vapours have no effect on the rate of coagulation.

This is in complete agreement with the ideas discussed above. An adsorbed mono- or polymolecular layer on their surfaces cannot lower the molecular force potential between particles in contact enough for it to become comparable with the energy of Brownian motion. However, not everything is clear in this question. For instance, it is not yet understood why the rate of coagulation of ammonium chloride smoke first decreases then increases (see page 304) as the humidity of the air is increased. Nor is it clear why linear aggregates are formed in the presence of some vapours etc.

The rebound of large water drops on collision in air is an interesting phenomenon which has been known for a long time. It is caused by a film of air between the colliding drops preventing them from touching. A similar film is formed when fairly large drops fall on to the flat surface of a liquid and may result in their floating for a long time on the surface before coalescing with it. Large raindrops drag a layer of air with them as they fall into water, which later escapes as bubbles.

The first serious studies of the rebound of water drops were made by Rayleigh [242]. The most interesting of his experiments were with two narrow jets of water which were broken up by sound vibrations and met at various angles. When the angle was 180°, so that the drops were moving in opposite directions, they were flattened on impact and immediately coalesced.† At smaller angles of incidence a bridge was

† Rayleigh observed the colliding drops through a rotating perforated disk and the sketches which he made are very similar to the ciné-films made seventy years later.

formed between the drops which gradually became thinner as the drops moved apart. Finally, it either broke into fine secondary drops or contracted again, when coalescence occurred.

Such experiments are difficult to reproduce. The presence of specks of dust or grease on the surface raises the percentage of effective collisions. An electric field greatly encourages coalescence. In Rayleigh's opinion drops coalesce if some projection on the surface can pierce the air cushion between them during the short duration of the collision; a charge on the surface of the drops promotes the rapid growth of projections formed by chance. Extraneous particles projecting above the surface also assist the piercing of the air cushion and the formation of a bridge.

According to Rayleigh's observations, an atmosphere containing a considerable proportion of a gas which is readily soluble in water (sulphur dioxide, carbon dioxide, water vapour) enhances the collision efficiency, evidently because the gas layer is made thinner by dissolution of gas in the drops. Hydrogen produces a similar effect, owing to its low viscosity, being easily extruded from between the drops.

Newall [257] determined the thickness of layer of the air between colliding water drops by means of interference fringes. The thickness is of the order of 1 μ and decreases in the direction of flow. Increasing either the velocity of the jets or the angle between them leads to a decrease in the thickness.

Kaiser's experiments [287] with soap bubbles must be mentioned since they are directly related to the problem under discussion. He pressed bubbles 3–6 cm in radius with a force of 20–60 mg upon a flat soap film. After 3–25 sec the air layer between them disappeared and coalescence occurred. It was shown by Newton's rings that the thickness of the layer, which was less in the middle ($\sim 1\,\mu$) than at periphery, decreased steadily, at first fairly quickly and then more slowly, to 0·08 μ when coalescence occurred. The squeezing out of the layer is described satisfactorily by Stefan's equation for movement of a viscous fluid between approaching parallel disks

$$\frac{1}{h^2} - \frac{1}{h_0^2} = \frac{4Ft}{3\pi\eta R^4} \tag{56.7}$$

where h_0 is the initial distance between the disks, h is the distance at time t, F is the compressing force and R the radius of the disks.

When an electric potential difference exists between the soap bubble and the film the air is squeezed out of the gap much more rapidly and coalescence occurs sooner when the thickness is still 0·3–0·4 μ. This is due to electrostatic attraction between the surfaces which form a parallel-plate condenser. In Kaiser's experiments this attractive force was several times larger than the mechanical compression applied and, of course, it increased as the air layer became thinner.

Mahajan [291] has studied liquid drops floating on the surface of the same liquid, a commonly observed phenomenon. Observation of Newton's rings again established that the thickness of the air layer decreased steadily and this determined the life time, τ, of the drop. Generally speaking τ is also proportional to the viscosity of the liquid but with glycerin, mercury and saturated solutions of sugar in water τ was so small that it could not be measured. For water τ is 1–2 sec, for aniline 2–3 sec, for a solution of soap in water 4–5 sec and for olive oil several minutes. The effects of dust, surface contaminants and electric fields are the same as in Rayleigh's experiments. Drops will not float unless dropped from a certain height which, for solutions of soap in water, is $> 1{\cdot}8$ cm, corresponding to a velocity at impact exceeding 60 cm sec^{-1}. When the air pressure is reduced τ falls rapidly.

Derjaguin and Prokhorov [515, 516, 582] showed that liquids of considerable vapour pressure, such as hexane, octane and ether, yielded a stable air layer between stationary drops pressed together lasting for an indefinite time. This makes it much easier to study the layer, which was found to be lens-shaped, the thickness being about 1 μ in the centre and several tenths of a micron at the thinnest part. The mean curvature of the drop surface $\frac{1}{2}\left(\frac{1}{R_1}+\frac{1}{R_2}\right)$ decreases towards the middle of the layer where it becomes negative. Laplace's formula shows, therefore, that the excess gas pressure in the layer increases from the periphery towards the centre where it reaches a value of the order of 1000 dyne cm^{-2}. The gas, in consequence, must flow continuously out of the layer.

For a non-volatile liquid this flow results in the air layer thinning rapidly and coalescence of the drops takes place. When the vapour pressure of the liquid is high enough, vapour diffuses from the layer into the surrounding space in addition to the hydrodynamic flow. A gradient of the partial pressure of the vapour is thus set up in the layer, together with a gradient of the partial air pressure which is about the same in absolute value but in the opposite direction. Air therefore diffuses into the layer. Since the rate of diffusion is proportional to the thickness of the layer, while the efflux velocity varies as the cube of the thickness, a steady state is reached as the layer becomes thinner, when diffusion of air into the layer is compensated by its flow outwards and, in fact, only the vapour of the liquid leaves the film. The higher the vapour pressure of the liquid, or, in the case of water drops, the less humid the air, the greater is the steady thickness of the film and its stability.† Derjaguin and Prokhorov confirmed their quantitative theory of this phenomenon by means of model experiments.

Aganin [511] allowed water drops of radius 0·5–1·2 mm to fall upon a glass plate set at a certain angle θ to the horizontal and covered with a uniform film of water. With weakly-charged drops, formed by earthing the capillary jet from which they issued, coalescence occurred only when they fell from more than a certain critical height, otherwise they rebounded. The experiments were readily reproducible. With drops of radius 0·75 mm falling on to a horizontal plate the critical height was 1·96 cm, corresponding to a velocity at impact of 62 cm sec^{-1}. The critical velocity increased with the angle between the velocity of the drop and the normal to the surface at the point of contact. The product $V \cos \theta$, i.e. the critical value of the normal component of the velocity and, constant at 70 cm sec^{-1} for drops of radius 0·5 mm, and 46 cm sec^{-1} for a radius of 1·2 mm. According to Aganin's observations no change in the charge on the drops resulted from ineffective collisions, showing that there was no contact between the water drop and the film of water.

When the drops carried large charges (of the order of 10^{-2} e.s.u.) a bridge was formed between the drop and the film of water, as in Rayleigh's work, which broke as the drop rebounded, leaving part of the drop on the film. The size of this residue increased with the charge on the drop until complete coalescence occurred when the charge reached a critical value. Experiments by Aganin, with weakly-charged drops

† A similar phenomenon occurs with water drops on a hot plate, which has been termed the spheroidal state.

falling on to a stationary drop on the end of a vertical capillary, gave similar results. With $r = 0{\cdot}83$ mm all the collisions were effective for $V \geqslant 21$ cm sec^{-1} and all were ineffective for $V \leqslant 18$ cm sec^{-1}.

In the experiments of Gorbatchev and Mustel [512] water drops of radius 0·65 mm collided with a velocity of 100 cm sec^{-1}, their trajectories forming an angle of 2°. The planes of motion of the drops were parallel and a controlled distance Δ apart. The angle of impact, θ, between the relative velocity of the drops and the line of centres, decreased as Δ was increased. The mean impact efficiency fell from 100 per cent for $\Delta = 0$ to zero for $\Delta = r$. There was no exchange of water from one drop to the other at ineffective collisions.

Gorbatchev and Nikiforowa [513] experimented with drops of radius 0·5 to 0·75 mm. One drop was suspended on the end of a glass capillary fixed to a long pendulum and the other drop was situated on a waxed table. By moving the table it was possible to obtain a collision at any desired angle. In this case an upper critical velocity, V_{kr}, was discovered above which the collisions were ineffective. This differed from the lower critical velocity in that it became smaller with increase in the angle of impact from 80 cm sec^{-1} at $\theta = 10°$ to 50 cm sec^{-1} at 50° and 20 cm sec^{-1} at 70°. The drop radius was 0·75 mm. Some transfer of liquid from one drop to the other was observed during ineffective collisions.

Tverskaya carried out experiments [514] with water drops of radius 0·5 to 1·5 mm falling from a capillary tube on to a stationary drop located on a fine metal ring connected electrically to the capillary. The dependence of the upper critical velocity, V_{kr}, on angle of impact was measured. The values V'_{kr} below which all the collisions were effective and V''_{kr} above which all the collisions were ineffective were determined, both types of collision being observed in the range V'_{kr} to V''_{kr}. This was believed to be mainly due to oscillations of the falling drops which thus had various shapes at the moment of impact. For drops of radius 1·5 mm, in air having a relative humidity of 50 per cent, the values of V'_{kr} and V''_{kr} were 40 and 61 cm sec^{-1} for $\sin\theta = 0{\cdot}6$, 60 and 76 cm sec^{-1} for $\sin\theta = 0{\cdot}4$, and 76 and 90 cm sec^{-1} for $\sin\theta = 0{\cdot}2$. Ciné-films of colliding drops taken by Tverskaya completely confirmed Rayleigh's observations. At high velocities and large angles of impact the width of the bridge formed between the drops, and its lifetime, is extremely small.

Tverskaya and Yudin [473] made the important observation that the collision efficiency rises appreciably as the drop size decreases. When r decreased from 1·1 to 0·7 mm V_{kr} increased several times. This is easily explained because the smaller the drops the less they are flattened on collision; furthermore, with the same degree of deformation of the drops, the rate at which the air film is squeezed out according to equation (56.7) increases rapidly as the drop size decreases. The effect of humidity on the coalescence of water drops, mentioned above, was confirmed by the author's establishing that increase in the relative humidity, ranging in the experiments from 36 to 98 per cent, notably increased the collision efficiency, particularly when the angles of impact are not very large. For $r = 1{\cdot}15$ mm and $\sin\theta = 0{\cdot}73$, V''_{kr} was 25 and 50 cm sec^{-1} at humidities of 36 and 93 per cent respectively; for $\sin\theta = 0{\cdot}65$ and 93 per cent humidity V''_{kr} was so large that it could not be determined at all. Similar results were obtained earlier by Prokhorov and Yashin [517, 582].

Prokhorov and Leonov [518, 582] studied the effect of moisture deficiency on the rate of gravitational coagulation of a water mist. The mist was passed upwards through a vertical tube at a velocity of 20 or 40 cm sec^{-1}, droplets of radius greater than

30 or 40 μ, whose rate of fall exceeded the flow velocity, being separated out. The mist was then led through a heat exchanger mounted in the tube which varied the temperature a few degrees either way and so changed the degree of saturation of the air. It was then observed by an ultramicroscope. A constriction in the tube was situated 5 cm above the observation window and when sufficiently coarse drops coagulated in this region the larger resulting drops fell downwards and were recorded by an observer. In order that drops which had become enlarged by condensation of vapour should not fall out, a small local expansion was made in the tube to segregate the coarsest drops with the necessary margin of size. At small supersaturations the number of droplets falling in unit time remained unchanged but, below saturation the number decreased almost in direct proportion to the humidity deficiency, thus indicating a decrease in the efficiency of collisions between the drops. A graph given in the paper shows that already at 83 per cent relative humidity the collision efficiency of droplets having radii less than 40 μ is two to four times less than at 100 per cent humidity. This interesting result, however, needs checking.

Fedoseev, Manakin and Domentianova [519] studied the deposition on the bottom of a smoke chamber of a condensation water mist and a mist obtained by atomizing an electrolyte solution. They found that the mass of deposit obtained when both mists were present was markedly greater than the sum of the deposits due to the individual mists on their own. The natural supposition that condensation of vapour on the droplets of the solution had occurred is rejected by the authors, who put forward the rather improbable hypothesis that the difference in vapour pressure between the two kinds of droplets assists their coagulation by eliminating the film of air. The authors present the following observation as a substantiation of their point of view. If a weak solution of potassium thiocyanate and a strong solution of ferric nitrate are atomized in a chamber all the drops in the deposit are coloured, but if the concentrations of the solutions are equal there are very few coloured drops. This curious phenomenon should be studied more thoroughly because another explanation might be advanced.

It can be seen that experimental material on the efficiency of collisions between drops is exceedingly varied. As yet there is no theory based on the laws of hydrodynamics and capillary physics, and the phenomenon certainly warrants the attention of theoretical physicists.

Now, the influence of the wettability of aerosol particles on their coagulation by water sprays and the effect of adding wetting agents to the water will be dealt with. The opinion is widespread that the efficiency with which poorly wetted dusts, such as coal dust, are coagulated by water mists is low; in its favour is the fact that fused spherical particles of fly ash are caught by water sprays better than unburnt coal particles [520]. However, the pricipal reason for the low efficiency with which coal particles are caught is most probably their flaky shape, which causes them to orient themselves parallel to the surface of the drops round which they are flowing: the motion of the particles towards the drops is thereby retarded and an air film may be formed between the coarse particles and drops (see below). In Drinker's well-known book [215] even the very low absorbability of phosphorus pentoxide smoke by water is ascribed to poor wettability of the particles!

As pointed out above, all the contacts are evidently effective, irrespective of the surface nature of the particles, during thermal coagulation. However, in the kinematic coagulation of more or less coarse solid particles with large drops, an air film of the type formed at collisions between large drops may exist, particularly if a plate-like particle approaches a drop with its flat face. In this case the surface tension of the liquid may exert a considerable influence on the rate of penetration of the film and on the final result of the collision. Wettability of the particles by the liquid is possibly of some importance here. Since poorly wetted particles remain on the surface of the drop, without being drawn in, the whole of the surface of the drop may become covered with a layer of dust, should the dust concentration be high enough [202]. Tearing off of the particles by the air flow may then occur (see § 57).

The advantage to be gained from the addition of wetting agents to the water sprays used for dust suppression is of great practical interest. The effect of wetting agents in wet drilling is considerable, since they facilitate flow of water over the walls of the hole in the form of a continuous film, so that solid particles hitting the wall cannot be blown away as they can from a dry wall. The effect of wetting agents on the precipitation of already formed dusts is less than in drilling. In some cases, such as the deposition of oil mist by water sprays, the effect is zero [521] but in others, such as the deposition of quartz dust, a favourable effect is apparent [580]. Wetting agents can have a double action. Since they lower the surface tension of water considerably finer atomization is achieved which, under the specific conditions analysed in § 54, can raise the efficiency of deposition. In addition, as already pointed out, they can directly increase the collision efficiency between dust particles and water drops.

The important phenomena of deposition and coagulation of aerosols result from two essentially different processes, the approach of the particles to a surface or to one another, and the process of adhesion or coalescence. The mutual approach of particles is described by the differential equations discussed above, and can, in principle, be estimated with any desired degree of accuracy from the field of flow and the external forces. Only mathematical difficulties are encountered. This is not the case for the processes occurring on collision, the mechanism of which is complicated and has been inadequately studied so that the collision efficiency, or the adhesion or coalescence probability, cannot be calculated, even to a first approximation, but can be determined only from experiments.

The mechanism of collision of solid particles with other solid particles or surfaces is essentially different from that of liquids and requires special consideration. Collision between a liquid and a solid is an intermediate case.

During collision between solids the viscous resistance of the thin gaseous interlayer can be neglected. If a spherical particle of radius r moves with velocity V_0 towards a flat surface it can be assumed that the viscous resistance will become appreciable when the distance, h, between the particle and the plane is reduced to a certain value h_0. From that moment the particle is acted upon by a force which is twice as large as the one between two approaching spheres (see p. 301)

$$F = -\frac{6\pi\eta r^2 V}{h}. \tag{56.8}$$

By solving the differential equation of motion of the particle under the action of this force, we find that the particle will stop at a point corresponding to the value

$$h_s = h_0 e^{-V_0/\beta} \tag{56.9}$$

where

$$\beta = 9\eta/2r\gamma.$$

If it is assumed that V_0 is the velocity of a freely falling particle in air and that $h_0 \approx 0{\cdot}05\, r$, h_s, which can easily be calculated, is always less than the mean free path of the gas molecules. At such a value of h_s there can be no question of a viscous resistance of the gaseous interlayer. Moreover, owing to the roughness of all real surfaces, contact will be established when h_s is of the order of the height of the micro-projections. Thus (possibly with the exception of the gravitational coagulation of large lamellar particles oriented parallel to one another when falling) a direct contact seems to be always established between solid particles, or between particles and a surface, in air.

The theory of collisions indicates that the coefficient of restitution, or ratio of relative velocities V_r'/V_r after and before collision, may have any value from one, for completely elastic collision, to zero, depending on the material, size, shape and relative velocity of the colliding bodies. In elastic collision the maximal stress in the region of the contact, H_{max}, should be less than a certain critical value which is many times larger than the elastic limit of the material under static load owing to the rapidity with which the load is applied. For central elastic collision between two equal spheres, or between a sphere and a plane, the maximum value of the repelling force on Herz's theory [841] is

$$F_{max} = Kr^2 V_r^{6/5} \tag{56.10}$$

where K is a constant depending on the elastic properties of the material, and the maximal stress is

$$H_{max} = K' V_r^{2/5} \tag{56.11}$$

which does not depend on r. In fact, at sufficiently small values of V_r collisions are always elastic, as was confirmed by the experiments of Raman [842] and Andrews [843] using aluminium and copper spheres. Very plastic materials, however, like lead, have a critical value of V_r so small that elastic collision cannot be observed.

In the case of oblique collision the foregoing argument must be applied to the normal component of the relative velocity, V_{rn}, directed along the line of centres. Data on the change in the tangential component V_{rt}, caused by the force of friction between the spheres, are very scarce. It will be shown below that the coefficient of restitution V_{rt}'/V_{rt} may be either more or less than V_{rn}'/V_{rn}. It should be noted that oblique collision causes a torque due to friction. If a sphere collides obliquely with a wall while rotating in the same direction in which it would rotate if rolling along the wall, the angle of incidence, θ, between the trajectory of the sphere and the normal to the wall is smaller than the angle of reflection, θ', but when the rotation occurs in the opposite direction $\theta > \theta'$.

Since H_{max} is independent of r these relationships should presumably hold also for coarse aerosol particles with $r > 0{\cdot}1$ mm. At smaller values of r the adhesion force, F_{ad}, which to a first approximation is proportional to r (see p. 340), becomes effective. Since the maximal force of repulsion in elastic collision, F_{max}, is porportion to r^2 (see (56.10)) it is evident that the ratio F_{ad}/F_{max} increases with a decrease in r. Very fine particles therefore stick to a surface, even when elastic collisions would be expected with the particles of such a perfectly elastic substance as quartz. The greater the plasticity of particles, the larger are the values of V_r and r at which they stick. The condition for sticking can be written as

$$\frac{mV_{rn}'^2}{2} < \Omega_{ad} \tag{56.12}$$

where V'_{rn} is the normal velocity of rebound of a particle in the absence of the adhesive force and Ω_{ad} is the work necessary to tear off the particle [844].

The rebounding of solid particles affects the operation of louvre dust separators (see p. 131) and the pneumatic transportation of granular materials. The mean transport velocity of the particles $\overline{V}$ in the latter case is considerably less than the velocity $\overline{U}$ of the air stream carrying them; when particles are carried up a vertical tube it is less than the effective velocity $\overline{U} - V_s$. This was first believed to be due to friction between the grains and the walls of the tube. In horizontal tubes, the major portion of the material tranported moves along the bottom of the tube, and in vertical tubes this can occur due to the air stream having a screwing motion; in these cases friction in the usual sense of the word is present. There is, however, another reason why the particles lag behind the air flow, especially in narrow tubes, namely, the loss of tangential velocity on collision with the wall.

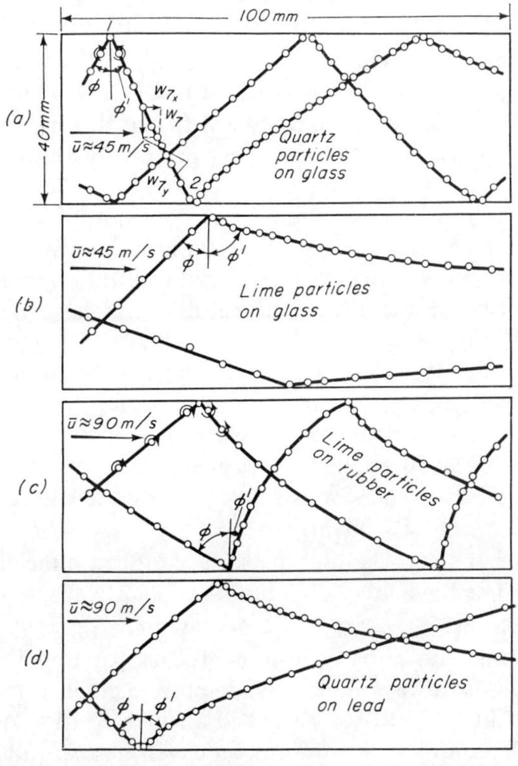

FIG. 75a. Trajectories of quartz and lime particles in pipes made of various materials.

By taking high speed motion pictures, Adam [845] determined the trajectories and velocities of quartz and lime particles, having $r = 0{\cdot}1$–$0{\cdot}5$ mm, in a transparent horizontal tube measuring 4 × 4 cm at flow rates of 40–90 m/sec (Fig. 75 a). Collisions of quartz particles with the glass walls were usually elastic. A slight decrease in the reflection angle θ' is accounted for by the rotation of particles after their collision with the opposite wall (see p. 357). A few non-elastic collisions associated with the damage to the particle or the wall, at contact, were observed in which V_t and V_n were considerably decreased. All collisions of quartz particles with lead walls were non-elastic. The collision of the more plastic lime particles with glass walls resulted in V_t, and especially V_n, being decreased, so that $\theta' > \theta$. The collisions of lime particles with glass walls are therefore much less frequent than those of quartz particles. When lime particles collide with rubber walls, however, V_t is almost completely lost, while V_n is retained, so that $\theta' \ll \theta$; the particles then acquire especially high rotation velocities (up to 700 rev/sec).

Owing to frequent collisions of quartz particles (r = 0·25–0·4 mm) with the walls of the tubes when the particles lose part of their tangential velocity, the mean value of the latter, $\overline{V_t}$, is about equal to 0·5 $\overline{U}$. For lime particles, as would be expected, $\overline{V_t}/\overline{U}$ is much greater (0·85). It is easy to understand how $\overline{V_t}/\overline{U}$ decreases with a reduction in the diameter of the tube.

The resistance of the tube to the flow of dust-air mixture can be resolved into two components corresponding to clean air and dust, the second component λ being approximately proportional to the weight concentration of the dust c and increasing as $\overline{V_t}/\overline{U}$ decreases since, more energy is then expended in accelerating the particles. For quartz particles, λ is almost twice as large for r = 0·25–0·4 mm as for r = 0·03 to 0·05 mm (at the same c); the smaller ones follow the air flow more closely and therefore collide with the walls less often. For CaO particles with r = 0·25–0·4 mm λ is still less.

It is believed by some who have studied the deposition of solid particles on hard surfaces that the particles can stick to a wall and be blown off later by the flow. No doubt this occurs with variable flow rates, but it is rather unlikely at a constant rate. As explained on p. 355 and 357 the critical dynamic flow rate is much less than the static, a much greater flow rate being necessary to dislodge a particle resting on a surface than to prevent one from adhering. Apparently, a particle colliding with a surface will stick to it if it happens to arrive directly at a deep enough potential pit, or if it rolls or slips along the surface until it reaches such a pit. In order to blow the particle off again an amount of energy greatly in excess of the kinetic energy of the particle before collision is needed, so that a higher flow rate, or some other external factor, must be available.

Electric charges on the particles and the surface influence the efficiency of deposition and are responsible for better adhesion of unipolarly charged talc dust to plant leaves and other objects [668]; charges also operate in xerography (see p. 122).

The chief method of increasing deposition efficiency is to cover the surface with a film of grease. This is always done with impingement instruments. In a paper, which was discussed in § 33, Röber described an investigation of the efficiency of various greases by depositing different dusts on cover-glasses, with a konimeter, and then determining the flow rate at which the deposit was blown off by a jet of clean air from the konimeter. In view of the dynamic effect, pointed out above, no conclusions about deposition efficiency can be gained from these experiments, although it is possible that an idea of the relative efficiencies of different greases might be obtained.

It is impossible to blow particles with r below 0·5–1·0 μ from deposits on clean plates, even at U = 200 m/sec, but the adhesion of particles with r = 1–5 μ can be secured only by the use of an adhesive film, the properties of the substance used and the thickness of the layer being of great importance. Grease layers which are either not viscous enough, or else too thick, are blown off by the air jet together with the dust particles resting on them. If the grease layer is too tough particles cannot penetrate and are easily blown from its surface. The best results are obtained with a thickness of 0·1–0·5 μ; thinner layers fail to retain particles with $r = 2{\cdot}5$–$5\,\mu$. When $r > 5\,\mu$, with *lycopodium* spores, for instance, above a certain velocity of impinging jet it is impossible to retain particles with any kind of grease.

Dergachev [846] studied the deposition of coal dust with $r = 50$–$80\,\mu$ at $U = 25$ m/sec. A plate was smeared with Vaseline and placed normal to the flow and it was found that the deposition efficiency grew with the thickness of the layer up to $\sim 70\,\mu$, or about the value of r. There was no further change, an efficiency of 100 per cent being attained.

In the experiments of Rosinski *et al.* [847] particles of ZnS of $r = 10\,\mu$ were deposited on sticky tapes with $U = 4$ m/sec. All particles were found to stick to a tape for which the work of tearing away from a steel plate was $\Omega = 22$ g/cm, but only 11 per cent stuck to another tape for which $\Omega = 340$ g/cm, the latter surface being too hard. It is possible that under instantaneous loading, at the moment of collision, adhesives which are not soft enough behave as solids.

As already pointed out, the process of collision of liquid drops is essentially different from that of solid particles. In the first place, the absence of surface roughness and the slight deformability of drops enables an air interlayer to be formed between them. The greater the relative velocity upon collision, the less is the thickness of the interlayer and the greater is the probability of the formation of a liquid bridge between two drops due to the presence of ripples on adjoining sections of their surface. The existence of a lower critical velocity and the rule of Aganin (see p. 342) are explained in this way.

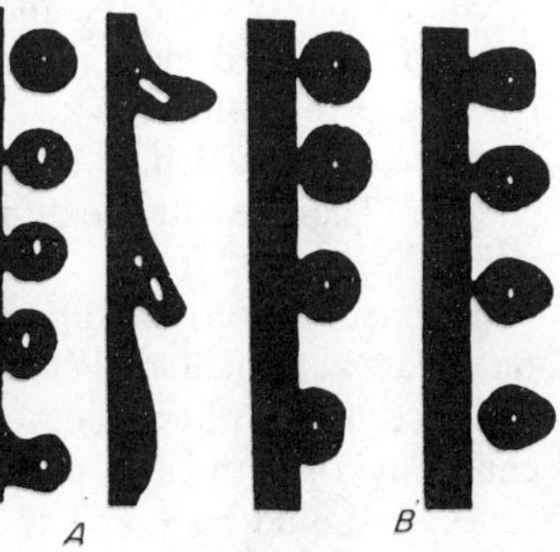

FIG. 75b. Effective (A) and ineffective (B) collisions of dioctyphthalate drops with a vertical glass plate.

Furthermore, when the drops collide, not only is the narrow region in the vicinity of the contact area deformed, but the whole drop changes shape (for illustration see Fig. 75b). The increase in the surface area of the drop and in the free surface energy then act like elastic deformation and elastic energy. The transfer of the latter back into kinetic energy is accomplished by the flattened drops resuming their spherical shape and repelling one another. If a layer of air remains intact between the drops during collision their separation is unimpeded, but should a liquid bridge be formed the outcome of collision depends on whether kinetic energy of the diverging drops exceeds the work of rupturing the bridge. If deformation is large, that is when the relative velocity suffices, the collision will be ineffective. Oblique collisions are less effective than the central because the tangential component of velocity contributing to the rupture of the bridge is retained to a great degree.

It is sometimes stated [848] that for collisions between the drops to be effective it is necessary for the kinetic energy of their relative motion to exceed a certain fraction of their surface energy, but this is not borne out by experiment. It may apply to the collision of drops with unwettable solid particles (see below).

The mechanism of collision between liquid drops and solids is intermediate between the processes considered above. It will be shown that, as a rule, no air interlayer is formed in this case and liquid coalescence is also absent. However, the mechanism of rebound involving the transition of capillary energy to kinetic energy, is the same as with liquid drops. In effective oblique collisions, when the capillary energy is insufficient to detach a drop from the solid, it is usual for the drop to slide along the surface.

In the experiments of Gallili and LaMer [849] a thin vertical jet of homogeneous glycerol mist, with $\bar{r} = 1{\cdot}5$–$5\,\mu$ impinged at a velocity $V = 11$–16 m/sec upon the upper side of a hydrophobized glass plate set at 40° to the horizontal. The width of the deposit of droplets increased with a rise in V and r, whereas the reverse should have taken place had the collisions been effective. It was concluded that the droplets bounced off after their first collision and deposited elsewhere. That there was no slipping of the drops along the surface was proved by having an initial deposit of very fine glycerol droplets on the plate which were not wiped away by the larger impinging drops. Still larger drops with $r = 15$–$20\,\mu$ failed to bounce off after impaction, but slid along as in the experiments of Hartley and Brunskill, described below.

Similar experiments were conducted by Gillespie and Rideal [850] with a mist of dibutylphthalate ($\bar{r} = 2{\cdot}3\,\mu$) on an untreated glass plate. At $V < 2$ m/sec no deposit was formed; an increase in V made the collection efficiency increase very quickly, reaching 100 per cent at 3 m/sec, after which it began to fall equally rapidly to 0·1 at 4 m/sec, when it rose again. These results are difficult to account for and need verification.

In another series of experiments drops of dibutylphthalate ($r \approx 1$ mm) were allowed to fall vertically upon a plate at a small angle to the vertical; a motion picture gave a good illustration of the mechanism of collisions. (See Fig. 75 b, which shows effective and non-effective collisions. In both cases the drops slipped upon the plate over the distance of 5–10 mm). It should be noted that the liquid forms a sharp contact angle with the plate so there can be no continuous air interlayer.

By means of a centrifuge the normal force F needed to detach drops of dibutylphthalate from a surface was also measured. For drops having $r = 6{\cdot}8\,\mu$, which formed lenses on the surface with $r' = 15\,\mu$, F was found to be about 5×10^{-4} dynes, and for drops with $r = 1{\cdot}4\,\mu$ ($r' = 3\,\mu$) about 9×10^{-6} dynes. These values of F are some six orders of magnitude less than the forces calculated by the authors, which are necessary for the tearing off of the drops from the plate without their deformation. In reality, the drops deform, under the action of a normal force, and develop a gradually narrowing neck.

Gillespie [851] prepared mists, by spraying molten stearic acid, having $\bar{r} = 1$–$2\,\mu$, which were deposited on a steel wire with $R = 0{\cdot}085$ mm at flow rates of 6–26 cm/sec; the surface distribution of the deposit was studied as a function of the angle θ from the forward stagnation point. Since the Stk number in these experiments was between 0·01 and 0·1 deposition was due to interception and electric forces. The distribution of droplets changed appreciably when the wire was coated with silicone, hence it is evident that they slid along the surface of the wire owing to the drag of the air stream.

In experiments by Hartley and Brunskill [852] water drops of equal size ($r = 25$ to $200\,\mu$) fell vertically at a velocity V_s on to surfaces oriented at an angle θ to the horizontal. All collisions were effective on smooth surfaces, even when they were made hydrophobic with silicone oil, paraffin etc. Large drops bounced off from rough surfaces, including plates coated with carbon black or MgO and the leaves of plants covered with wax platelets. On pea leaves, all collisions were effective at $r < 50\,\mu$ and ineffective at $r > 125\,\mu$, independent of θ. In the intermediate range $\ni$ decreased with an increase in θ. When the surface tension was decreased, by adding large quantities of methanol or acetic acid to the water, $\ni$ markedly increased, while the velocity of rebounding drops decreased. For plates coated with carbon black $\ni$ was smaller than for leaves, and fragments of carbon black could be seen on the bouncing drops, indicating real contact upon collision. This is confirmed by the absence of any effect on $\ni$ when a reduction in the air pressure down to 10 mm Hg was made.

The work of detaching a drop resting on a surface is proportional to σr^2 and decreases rapidly with an increase in the contact angle, while the kinetic energy of a freely falling drop is proportional to r raised to a higher power; the results above can therefore be well accounted for. Hartley and Brunskill explain the effect of surface roughness by the fact that on poorly wettable rough surfaces the advancing contact angle is very large; the usually measured static value of this angle is probably much below the instantaneous value during impact of the drop.

These considerations apply, also, to the collision of droplets with much larger solid particles. Collision with small solid particles or deposition of the latter upon wet surfaces requires further examination. Although of particular interest to industry, investigation up to now has been confined chiefly to adding wetting agents to water for the purpose of increasing collision efficiency, the mechanism of the process having been studied but little.

Taubman and Nikitina [853] allowed drops of water and solutions of wetting agents of the same size ($r = 1$ mm) to fall through polydisperse dusts of quartz, talc ($r \leqslant 2{\cdot}5\,\mu$) and coal ($r \leqslant 1\,\mu$), and determined the quantity of dust collected by the drops. With some solutions $\ni$ was double the value for water for all three kinds of dust. It is evident that some time is required for hydrophylic quartz to become wetted so that wetting is but poor upon instantaneous contact. No correlation between the static surface tension of the solutions and $\ni$ was detected, but $\ni$ increased with a decrease in the dynamic surface tension, measured on a newly formed liquid surface. It is of interest to note that some surface active substances make the surface of dust particles hydrophobic and decrease $\ni$.

The authors suggest that the molecules of surface active substances migrate over the surface of the particle at the moment of impact, rendering it hydrophobic or hydrophylic. This is possible, however, only in the rather improbable case of the velocity of migration being greater than the relative velocity of the drop and of the particle at the moment of impact.

Ermilov [854] studied the effect of wetting agents on PbO smoke ($r \approx 0{\cdot}3\,\mu$) depositing in a straight through cyclone with wetted walls. At large air speeds, U, at the entrance to the cyclone (15 m/sec) ∍ is large (0·95) and there is little increase in ∍ (up to 0·97–0·98) upon adding wetting agents to water. When U is decreased, ∍, of course, also decreases, but the effect of adding wetting agents is greatly enhanced: at $U = 1$ m/sec, ∍ = 0 for water and can be raised to 0·9 by choosing suitable wetting agents. Again no relationship was found between the effect of wetting agents and the static surface tension of the solution.

Pozin *et al.* [855] studied the wettability of dusts and their retention when bubbled through water and showed that the retention of particles of hydrophobized pyrites and of hydrophylic baryta with $r \geqslant 2{\cdot}5\,\mu$ was about the same: at $r \leqslant 2\,\mu$ retention of pyrites particles was much less. The addition of wetting agents had little effect on ∍, in the case of baryta, but increased the retention of pyrites.

In the experiments of McCully *et al.* [856] a thin jet of aerosol of glass spheres of $r = 2{\cdot}5$–$25\,\mu$ collided with a suspended water drop. Photographs clearly show that some of the spheres bounced off the drop. When the spheres were hydrophobic the proportion of ineffective collisions increased considerably, and the spheres which did not bounce off remained on the surface of the drop, whereas untreated spheres penetrated inside.

It is evident from the foregoing that the effect of wettability of particles becomes apparent only when the kinetic energy of collision is small. A theory of collision between drops and unwettable particles was advanced by Pemberton [857] who considered an absolutely unwettable spherical particle on which a contact angle $\theta = 180°$ was formed by water. As already pointed out, the instantaneous value of θ may be close to 180°. It can be easily shown that the work of immersion of such particle in the water is equal to $\frac{8}{3}\pi r^2 \sigma$ and the author concluded that collision with a large water drop would be effective when the kinetic energy of the component of relative velocity normal to the surface of the drop, V_{rn}, was greater than the work of immersion. By calculating V_{rn} for potential flow round a sphere, by means of an electronic computer, Pemberton plotted the (∍, Stk) curves for the collection of unwettable particles. It is clear from this theory that collision efficiency depends on the dynamic surface tension of the liquid because, at the moment of collision, the surface of the liquid undergoes very rapid extension.

The energy of complete or partial immersion of a particle in a liquid and the (∍, Stk) curve can be calculated in a similar way for any value of the advancing contact angle θ: it is only necessary to take into consideration how θ itself increases with V_{rn}.

Although the use of wetting agents enables the dust concentration in mines during wet drilling to be lowered, the problem can be solved radically only by completely eliminating all access of air into the borehole [858].

CHAPTER VIII

DISPERSAL OF POWDERS AS AEROSOLS

§ 57. THE DETACHMENT AND TRANSPORT OF PARTICLES BY WIND

Previous chapters have covered deposition and coagulation of aerosol particles as they adhere to surfaces or to one another. The reverse processes of detachment of particles and the breaking up of aggregates will now be considered. It was mentioned in the foreword that these processes play an important part in nature and everyday life and some phenomena of direct practical value, such as the fluidization of powders, have been studied fairly thoroughly; the systematic theoretical and experimental study of the field, as a whole, is, however, still in an elementary stage.

The dispersion of powders by air currents is especially important and this problem will be discussed in some detail; difficulties are the absence of reliable data on molecular forces, which oppose the separation of particles, and uncertainty about the force exerted by the air current on the particles. The data in this chapter are therefore, for the most part, qualitative.

In the second half of this section the detachment of particles from surfaces will be dealt with. This is almost always preceded by the particles either slipping or rolling along the surface, principally because, near the wall, the horizontal component of the air current, and consequently of the hydrodynamic force acting on a small particle, is considerably greater than the corresponding normal component. Molecular attraction between a moving particle and a surface is appreciably less than for a stationary particle; furthermore, when particles move along the surface new factors appear which promote breakaway (see below). A start will therefore be made by discussing how particles lying on a surface are carried off tangentially by air currents; this has been studied much more fully than the actual process of detachment. The transport of particles is of interest on its own account since it explains the movement of sand by winds and pneumatic transport of powders and grains.

The simplest example is the transport of particles along a smooth horizontal wall which was studied by Syrkin [522] in a wind tunnel 10 cm wide. His results with corundum powder are given in Table 37, in which r stands for the particle radius and $\bar{U}_{crit}$ the mean air speed in the tunnel at which particles on the iron or glass floor were carried away by the wind. Actually, some particles are carried away at lower speeds and others at greater so that $\bar{U}_{crit}$ is an average.

In the theoretical part of his work Syrkin assumed that transport of particles took place by rolling. In order to overturn a cubical particle of edge $2r$ a force

$$F > 8r^3\gamma g \tag{57.1}$$

must be applied to it at a height r. If the mean velocity of the air flowing round the particle is U_m the hydrodynamic force acting on it is

$$F_M = \psi\, 4\, r^2 \gamma_g\, U_r^2/2, \tag{57.2}$$

where ψ is the drag coefficient. Thus it follows from the inequality (57.1) that the particle will roll if

$$U_m > \left(\frac{4\, r\gamma g}{\gamma_g \psi}\right)^{1/2}; \tag{57.3}$$

This was not compared with experiment. Since the coefficient of friction, $\varkappa$, between solid bodies is less than one, the force $F = 8\, r^3\, \gamma_g \varkappa$ necessary to make the particle slide is less than for rolling so that a cubical particle will slide rather than roll. That $\bar{U}_{\text{crit}}$ is less on a glass wall than on an iron one indicates the large part played by friction in Syrkin's experiments. Corundum particles are usually more or less round so that they require a force several times less than that given by formula (57.1) to make them roll. For calculations we shall take $\varkappa = 0{\cdot}25$, thus discarding the factor 4 in equation (57.3).

TABLE 37. THE PICK-UP OF CORUNDUM PARTICLES BY AN AIR STREAM†

	r, cm							
	10^{-4}	10^{-3}	$3{\cdot}5\times10^{-3}$	5×10^{-3}	8×10^{-3}	10^{-2}	2×10^{-2}	5×10^{-2}
$\bar{U}_{\text{crit}}$ (exp.) on iron wall (cm sec^{-1})	—	—	1140	1060	1080	1090	1270	1630
$\bar{U}_{\text{crit}}$ (exp.) on glass wall (cm sec^{-1})	—	—	860	680	650	650	750	870
$U_{r,\text{crit}}$ (cm sec^{-1})	(0·016)	(1·65)	18·5	30	67	90	204	500
U^*_{crit} (cm sec^{-1})	(5·3)	(16·7)	30	32	37	39	41	41
$\bar{U}_{\text{crit}}$ (theor.) (cm sec^{-1})	(76)	(280)	550	590	710	740	740	780

Assuming turbulent flow (§ 46) along the surface with a laminar layer which has a velocity profile given by (46.11),

$$U = \frac{U^{*2} z}{\nu}. \tag{57.4}$$

Since the value of $U^* z/\nu$ in Syrkin's experiments did not exceed 12, equation (46.12) indicates that the particles in these experiments were immersed in the laminar layer,

† That various factors which cannot be gauged have a strong influence on the results of such experiments is clear from a comparison of this table with the data of Bloomfield and Dalavalle [372].

Their values of the critical speed for quartz dust are:

0·5 cm sec^{-1} for $r = 5\,\mu$, 3 cm sec^{-1} for $r = 12{\cdot}5\,\mu$,
12 cm sec^{-1} for $r = 25\,\mu$, 50 cm sec^{-1} for $r = 50\,\mu$,
125 cm sec^{-1} for $r = 250\,\mu$ and 500 cm sec^{-1} for $r = 500\,\mu$.

so that the velocity of the air flow round them is given by formula (57.4). By introducing the Reynolds number for flow round the particle, $\mathrm{Re} = 2\,r\,U_m/\nu$, the inequality (57.3), with the factor 4 discarded, can be put in the form

$$\psi\,\mathrm{Re}^2 > \frac{4\,r^3 \gamma g}{\nu^2 \gamma_g} \tag{57.5}$$

and the critical values of Re and U_r for corundum ($\gamma = 4$) found by mean of Table 5 (page 32); the critical friction velocities, U^*_{crit}, were determined by putting $z = r$ and $U = U_m$ in (57.4). The maximum value of the dimensionless quantity $U^*_{\text{crit}}\,r/\nu$ (for $r = 5 \times 10^{-2}$ cm) is 12. Finally, the theoretical values of $\overline{U}_{\text{crit}}$ given in the table were calculated from formula (46.29), $U^* \approx 0{\cdot}2\ \overline{U}/\mathrm{Re}_f^{1/8}$. In this derivation the assumption was made that the drag is the same for free particles and for particles lying on the wall. In view of this, and of the low accuracy of measurements of this sort, the agreement between theory and experiment is not unsatisfactory.

According to calculation the critical velocity should diminish steadily as the particle size decreases but in fact, as can be seen from Table 37, there is a noticeable tendency for $\overline{U}_{\text{crit}}$ to increase when the particle radius falls below 50 μ. Although no systematic measurements have been made on fine particles, disconnected observations show that the critical velocity increases as the particles become smaller (experiments with coal particles 25–50 μ in radius [581]). This is also indicated by experiments with impingement instruments (see page 155) and Rumpf's oberservations [523] with a disk placed parallel to a stream of dust having $r = 0{\cdot}5$–$6\,\mu$ are of interest. A thick loose deposit containing particles of all sizes was obtained at low flow velocities while a thin dense deposit containing particles from 0·5 to 1 μ in radius was obtained at high velocities, the coarser particles having been blown away by the wind; at still higher flow velocities there was no deposit at all.

In the above calculations no allowance was made for molecular forces between the particles and the surface. If these are calculated from (56.3), with $r_2 = \infty$, the force of adhesion, $F = 4\pi\sigma r$, is proportional to the particle radius. This formula, however, is applicable only to ideally smooth particles and surfaces and cannot be used for practical calculations. By taking the reasonable value of several tens of dynes per centimetre for σ it suggests that spherical particles several millimetres in radius would be held by molecular forces to the underside of a horizontal surface, which is clearly impossible. The adhesive force is determined not by the macroscopic radius of a particle but by the considerably smaller radii of curvature of the asperities which actually form the contact between a particle and a surface. In any case the force falls much more slowly with decrease in particle size than does the force of gravity; it therefore falls slower than r^3. In fact, when a disk sprinkled with powder is turned over the coarse particles fall away while the fine ones remain. Since the hydrodynamic force is proportional to $U_m r$, and hence to r^2, for sufficiently fine particles, the adhesive force decreases more slowly than r^2.

Syrkin also determined a "dynamic" critical flow velocity at which corundum particles thrown into the air stream failed to stick to the bottom of the tunnel and were carried away by the wind. This velocity, about $2\frac{1}{2}$–3 times less than the static critical velocity, was 3·5–5 m sec^{-1}. According to experiments by Averbukh and Shabalin [524] the critical dynamic velocity for aluminium hydroxide particles 25–30 μ in radius is 2–3 m sec^{-1}.

The motion of particles over the surface of a layer of similar particles is important and arises in the transport of sand and soil by winds and the pneumatic conveyance of friable materials. It was first established by Sokolov [525] that transport is effected in three ways: (1) particles roll along the surface; (2) particles break away from the surface and immediately fall back, thus moving in jumps; (3) particles remain airborne. The first two of these processes were investigated thoroughly by Bagnold [526, 527] in a 30 × 30 cm wind tunnel with glass walls and mean air velocities up to 10 m sec^{-1}. A thick layer of sand was spread on the bottom of the tunnel. Bagnold's experiments were repeated by Chepil [528, 529] with various soils and he obtained completely analogous results. The remarks below concern the experiments of both authors.

In one series of experiments the critical dynamic velocity was determined, a fine stream of sand being poured continuously into the tunnel upstream of the working section. Below the critical velocity of the air sand grains falling on to the layer of sand caused granules on its surface to move as described below but, at a certain distance from the place where the sand was poured in, increasing with the air speed, the movement ceased. Above the critical velocity the movement spread without slackening, throughout the tunnel.

Many of the sand grains move in jumps as illustrated in Fig. 76. It can be seen from the figure that the grains jump almost vertically upwards, attain a considerable horizontal velocity downwind and fall at a very sharp angle to the horizontal. The heights jumped, at the air speeds indicated, attain several tens of centimetres and the falling sand grains produce small hollows in the surface of the sand. They either ricochet from the surface to execute another jump or bury themselves in the sand, transferring their momentum to other particles which in turn roll or jump along. The transport of sand is thus a chain reaction.

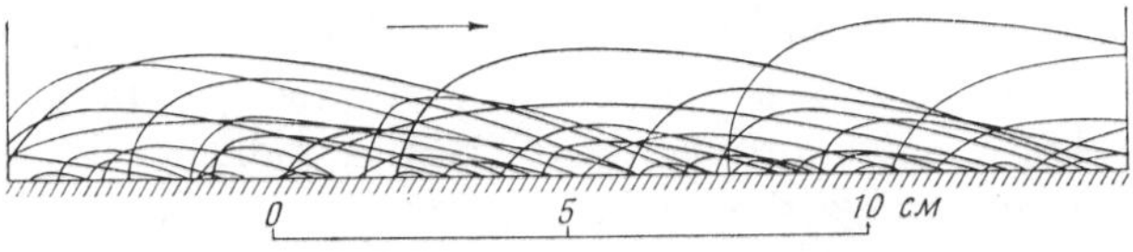

FIG. 76. Trajectories of jumping sand grains.

If no stream of sand is introduced from outside individual grains, protruding from the layer on the floor, begin to roll at a certain air speed but soon stop if they fall into a hollow. An increase in air speed again causes a few particles to roll and stop, and so on. When the air speed is sufficiently great some rolling grains collide with other, coarser grains, protruding above the surface, and receive an upward impulse which causes them to jump. Everything then happens as described above and at an air speed exceeding the static critical velocity the momentum received from the air stream by the jumping sand grains is sufficient for the process of jumping and knocking out of fresh grains to become continuous and to be propagated still further.

Many jumping particles rotate at high speeds (200–1000 rev/sec [528]) which indicates that just before jumping they were rolling along the surface. For sand having $r = 0{\cdot}09$–$0{\cdot}15$ mm four or five times more is conveyed by jumping than by rolling. The factor varies from two to twenty-five for soils depending on their state of dispersion [528]. Coarse sand grains having radii greater than 0·2 or 0·3 mm cannot jump when the wind speed is below 10 m sec^{-1} but only roll along while grains greater than 1 mm in radius remain stationary. The fine fractions of a polydisperse sand are therefore blown away by wind. The very finest fraction may be dispersed as an aerosol (see below).

Bagnold noticed that in experiments without admission of sand from outside, motion of the sand grains always began at the downstream end of the tunnel and was propagated upstream. This is evidently associated with the gradual development of turbulence in the air as it progresses down the tunnel. No screen was used in Bagnold's experiments to promote turbulence.

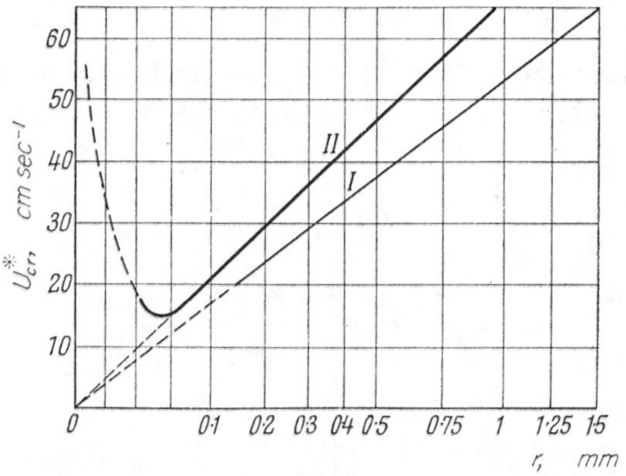

FIG. 77. Critical flow velocities for sand

Since the static critical velocity is greater than the dynamic, the removal of soil by the wind spreads a great distance once it is started. The process begins on small knolls due to vortices and chance gusts. The effectiveness of even slight obstacles, such as grass stems and stones, against wind erosion is explained by the fact that particles of soil cannot pass them to continue the process [528].

The results of critical velocity measurements published by Bagnold [527] for sands of different dispersity are illustrated in Fig. 77. The continuous lines, I for the dynamic and II for the static velocities, are drawn through experimental points while the broken lines are extrapolations. The critical friction velocity is plotted on the ordinate axis against $\sqrt{r}$ on the abscissa. It can be seen that for $r > 50\,\mu$ Z^*_{crit} is proportional to $\sqrt{r}$, namely,

$$\text{Dynamic } U^*_{crit} = 164\sqrt{r} \text{ cm sec}^{-1}, \tag{57.6}$$

$$\text{Static } U^*_{crit} = 208\sqrt{r} \text{ cm sec}^{-1}. \tag{57.5}$$

Below $50\,\mu$ U^*_{crit} increases for finer particles owing to molecular forces between them, in agreement with what has been stated above.

In calculating critical velocities it must be remembered that the wind velocity profile in this case corresponds to flow about a rough surface and is given by

$$U/U^* = 5{\cdot}75 \log\left(\frac{30\,z}{z_0}\right), \tag{57.8}$$

where z_0 is approximately equal to the height of the irregularities, in this case the diameter of a sand grain. In Bagnold's opinion the rolling of sand grains is a result of the friction between the air and the surface of the sand. Assuming that the frictional force acting on a particle lying on a layer of sand (Fig. 78) is the same as for a particle in the upper layer of sand grains, namely $\pi r^2 \tau_0$, the condition for rolling can be found by equating the moment of the gravitational force F_2 about the point of contact A, equal to $(2/3)\,\pi r^4 \gamma g$, to the moment of the hydrodynamic force F_1 about the same point, $(4/3)\,\pi r^3 \tau_0$. The condition for rolling is $\tau_0 > 0{\cdot}5\, r\gamma g$ whence

$$U^*_{\text{crit}} = \left(\frac{r\gamma g}{2\gamma_g}\right)^{1/2}. \tag{57.9}$$

A relationship between critical velocity and sand grain size, appropriate to Bagnold's experiments, has been derived but the absolute value of the static U^*_{crit} is only

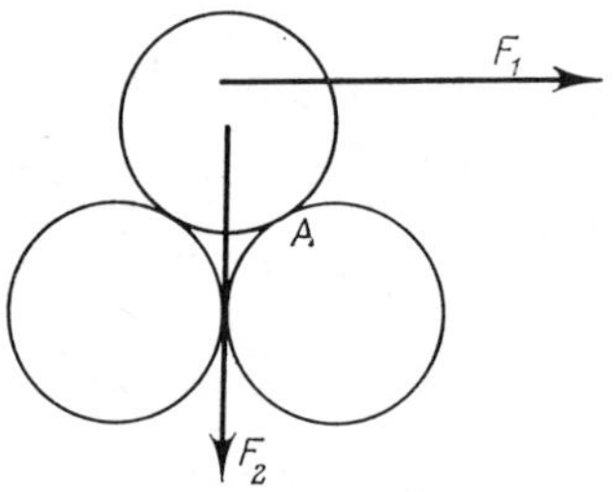

FIG. 78. Diagram showing rolling sand grains.

one over 4·5 of that calculated from formula (57.9). The force on a particle lying on a layer of sand is then greater than the force acting on the same particle if it was at the level of the surface of the layer. In order to calculate U^*_{crit} from the drag on the particle, it will be assumed that the resistance acts at the centre of the particle and that the flow velocity at the level of the centre is given by (57.8) with z put equal to $2\,z_0$ so that $U_m \approx 10\,U^*$. This gives

$$U^*_{\text{crit}} \approx \left(\frac{0{\cdot}013\, r\gamma g}{\psi\gamma_g}\right)^{1/2}. \tag{57.10}$$

The presence of ψ in this expression destroys the proportionality between U^*_{crit} and $\sqrt{r}$. The sand studied most carefully by Bagnold had $\bar{r} = 0{\cdot}12$ mm which makes $\psi = 3{\cdot}9$. It follows from (57.10) that $U^*_{\text{crit}} = 8$ cm sec^{-1} while the experimental static value was 23 cm sec^{-1}. Therefore, as in the case of sliding over a smooth wall, the drag of the particle has been overestimated.

Bagnold's observations in the desert [530], with sand particles of the same size, gave $U^*_{\text{crit}} = 23$ cm sec^{-1}, the same as the static value in the wind tunnel. This corresponds to a wind speed of 5·2 m sec^{-1} at a height of 10 cm and 6·5 m sec^{-1} at a height of 1 m. For the same sand grains Sokolov [531] found a critical wind speed of 4·5–6·7 m sec^{-1} at a height of 10 cm and also found that U_{crit} was proportional to $\sqrt{r}$. According to Genzel (see [532]) motion of sand begins at the following wind speeds (Table 38).

TABLE 38. CRITICAL WIND VELOCITY FOR SANDS OF DIFFERENT PARTICLE SIZE

r (mm)	0·087–0·12	0·12–0·25	0·25–0·5	0·5–1·0
U_{crit} (cm sec^{-1}) (dry sand)	380	480	600	900
U_{crit} (cm sec^{-1}) (sand with 2 per cent moisture)	600	750	950	1200

Here also there is approximate proportionality between U_{crit} and $\sqrt{r}$ for dry sand, so this can be accepted as fairly well established.

According to Chepil [528] soils erode most readily if r = 0·05–0·07 mm. In this case U_{crit} = 3·6–4 m sec^{-1} at a height of 15 cm. Soils with $r < 25\,\mu$ and sands with $r >$ 0·5 mm hardly erode at all, the reason in the former case being the attractive forces between the particles.

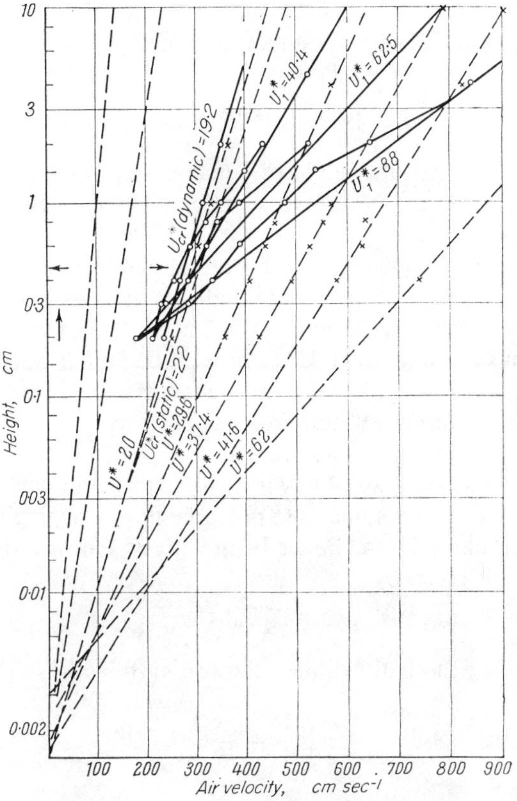

FIG. 79. Effect of jumping sand grains on wind velocity profile.

Jumping sand grains which receive momentum from the air flow decelerate the latter in the layer adjoining the surface. Bagnold made wind-tunnel measurements of the velocity distribution above a layer of moist sand which remained motionless, even above the critical velocity, and above the same layer dried, when the flow began to move the sand again.

The results of these experiments are presented in Fig. 79 in which the flow velocity is plotted along the abscissa and the vertical distance from the surface of the sand is plotted logarithmically along the ordinate axis. The broken lines refer to moist and the continuous ones to dry sand having $r = 0{\cdot}09$–$0{\cdot}15$ mm. The two lines on the extreme left refer to both wet and dry sand because the velocity was below critical. The corresponding friction velocity is shown on each line. It can be seen from the graph that the velocity distribution above moist sand has the logarithmic profile (57.8) with $z_0/30 = 0{\cdot}0015$ cm. The lines come to a focus at the point $z = 0{\cdot}0015$ cm and $U = 0$. At a height of 2 mm above dry sand the flow velocity does not depend on the mean wind speed, $\bar{U}$, in the tunnel, while below 2 mm, as Bagnold points out, the flow velocity even decreases as $\bar{U}$ is increased. In general the flow velocity in the layer adjacent to the surface is considerably less when the sand grains are jumping. This is explained by the large increase in the number of jumping grains as U increases and the corresponding rise in momentum lost by the flow. In this case the focus of the profile lines has coordinates $z = 0.2$ cm and $U = 200$–250 cm sec^{-1}.

According to Bagnold [527] the amount of sand of the particle size mentioned above transported in unit time by the air flow is

$$q = 1{\cdot}5 \times 10^{-9}\,(U - U'_{\text{crit}})^3, \qquad (57.11)$$

where U is the flow velocity at a height of 1 m and U'_{crit} is the critical velocity at the height of the focus i.e. 250 cm sec^{-1}. Measurements made by Bagnold in the Libyan desert gave very similar results.

According to Danovskii [533] the transport of snow begins at wind speeds of 4–5 m sec^{-1} (2 m sec^{-1} on ice) and increases rapidly as the wind strengthens. If the rate of transport of snow is taken to be unity at 5–7 m sec^{-1} then it reaches 10 at 8–10 m sec^{-1}, 30 at 11–14 m sec^{-1} and 70 at 15–18 m sec^{-1}. It is thus approximately proportional to $(U - U_{\text{crit}})^2$. The rapid growth in transport of snow and sand as the wind strengthens explains the terrible nature of these occurrences in violent storms.

The mechanism by which particles are detached from a surface by wind is not yet completely clear. Joukowski's well-known theory can be applied to coarse particles around which the air flow approximates to potential flow. Chaplygin showed [534] that for a particle shaped like a half-cylinder lying with its flat face on a wall the lifting force per unit length is given by

$$F_M = \tfrac{8}{3}\,\gamma_g r U_m^2, \qquad (57.12)$$

where r is the radius of the half-cylinder. According to Jeffreys [535] the analogous formula

$$F_M = \pi\left(\frac{1}{3} + \frac{\pi^2}{9}\right)\gamma_g r U_m^2 \qquad (57.13)$$

is obtained for cylindrical particles. In these equations, as above, U_m denotes the flow velocity on a level with the centre of the particle.

Losiyevskii [536] studied experimentally the lifting force acting on a rectangular plate of height $0{\cdot}4$ cm and area $S = 12$ cm^2 lying on the bottom of a water trough or at a small distance h from the bottom. In this case $F_M = k\gamma_g S U_m^2/2$, k being from $0{\cdot}2$ to $0{\cdot}3$, and as h increases the lift first increases a little but beyond $h = 0{\cdot}5$ mm it falls abruptly. The horizontal force on the plate increases steadily with h, at first very quickly and then more slowly. Einstein and El-Samni [537] measured the pressure at various points on the surface of hemispheres having $r = 3{\cdot}4$ cm lying with their flat faces on the bottom of a channel and established that there was a lifting force proportional to $\gamma_g U_m^2$. The force fluctuates considerably in amplitude and the authors consider that detachment occurs when the pulsating force reaches some critical value.

The origin of the lifting force can be explained as follows. The field of flow round an obstacle placed on a surface or on a wall has the appearance illustrated in Fig. 67, the flow being accelerated at the upper part and decelerated at the lower. At a distance from the obstacle the static pressure of the air in a vertical section near the wall layer is constant, hence, according to Bernoulli's theorem, the air pressure at the top of the obstacle is less than at the bottom and an upward force results. It must be emphasized that this does not apply to fine particles immersed in the laminar layer for which viscous forces exceed inertial and Bernoulli's theorem is not relevant.

A particle rolling along a surface at a velocity less than that of the wind experiences a Magnus force directed outwards the magnitude of which can be obtained by Joukowski's theorem. Some authors (Turitsyn [538], Young [539] *et al.*) consider this to be the main reason for the detachment of particles. The lifting force per unit length of a cylindrical particle having radius r is

$$F_M = 2\pi r \gamma_g U_r V_t, \tag{57.14}$$

where U_r is the velocity of the air relative to the particle and V_t is the tangential velocity of the surface of the rolling particle relative to its axis which is the same as the forward velocity of the particle. Although both U_r and V_t are less than U_m, the numerical coefficient in (57.14) is greater than that of (57.13) with the result that the two formulae give values to F_M of the same order of magnitude. Equation (57.14) does not apply to particles inside the laminar layer.

If the flow velocity required to detach a particle from a horizontal wall is calculated from (57.13) it turns out to be considerably less than the velocity necessary to make particles slide or roll along the surface, in contrast to Bagnold's experiments which show that the particles begin to roll first and then become airborne. This is another indication that formulae (57.12–57.14) are not valid for fine particles.

In very many cases a wind evidently causes particles to leave horizontal surfaces in the following manner. The particles first begin to slide or roll along until their speed is high enough for small bumps on the surface or on the particles themselves to cause them to jump. If a particle jumps out of the laminar layer it is either caught up by vertical turbulent fluctuations or, if it is rolling, by the Magnus force, and is carried clear. Fine particles held mainly by molecular attraction probably break away in a similar manner. If a plate is covered with powder and held vertically the coarser particles roll off while the finer ones stay on the plate. If the plate is then turned with its powdered side downwards the adhering particles still remain on the plate. This shows that the force necessary to remove a particle in a direction perpendicular to the surface is not less than the force required to move a particle along the surface†. Since the hydrodynamic forces along a surface are always greater than those at right angles to it, it is natural to expect that a particle will begin to move by sliding or rolling. This inevitably results in a reduction in the number of points of contact between the particle and the surface and, consequently, in a drop in the force of adhesion. The probability of a particle leaving the surface is thus considerably enhanced.

† It was shown in work by Derjaguin, Ratner and Futran [540] that the force of static friction between freshly drawn quartz threads, in the absence of an external load, is equal to the force required to separate two threads which are in contact.

In addition to jumps, particles may be detached by local vortices which originate at inequalities of the surface and by vortices with vertical axes, or whirlwinds. Low pressure is created inside these by centrifugal force, and air which flows in may entrain particles.

The ability of fine particles to remain attached to surfaces is called adherence and is very important for insecticide powders. Since the surfaces concerned, such as leaves, are inclined at various angles to the horizontal the particles' own weight no longer hinders breakaway but assists it. Adherence is therefore governed exclusively by molecular forces and improves with decrease in particle size. The shape and the nature of the particle and of the surface are also important since the magnitude of σ in formulae of the type (56.3) depends on them. Clearly the plasticity of the particle and of the surface play a large part as well. Molecular forces may cause flattening of the surfaces at points of contact and Derjaguin showed [500] that if there is a plastic deformation it leads to a considerable increase in the force required to detach the particle. Particles of soft substances therefore adhere better than hard ones.

As an example, a metal plate dusted with pyrophyllite powder retains twenty times more after tapping than when bauxite of similar particle size is used, since this substance is much harder [541]. Flat or needle-like particles adhere better than rounded particles of the same volume. Humidity is of great importance. The water meniscus formed at the point of contact of a spherical particle and a wall attracts the particle to the wall with a force of the order $2\pi r\sigma$ where σ is the surface tension of water†. Since the radius of a particle is generally considerably greater than the radii of curvature of the submicroscopic protrusions at which contact actually occurs (see page 355) the adherence rises rapidly with humidity. Experimental results on the adherence of various powders are extremely scarce so that only qualitative ideas can be given.

The most dubious factor in the theory of the detachment and transfer of particles by air streams is the adhesive force between solids. At present this cannot be evaluated theoretically, mainly because the geometry of the surfaces in the contact area is not known. For the same reason it is impossible to extrapolate to $h = 0$ the data on molecular forces between non-contacting bodies obtained by Derjaguin and Abrikossova (see p. 309). Experimental measurements with small (of the order of 1 mm) quartz and glass spheres with surfaces freshly solidified from the melt are, unfortunately, very contradictory. Under clean experimental conditions, *in vacuo* or in completely dry air, so that no adsorbed water layer existed on the surfaces, Stone [402], How *et al.*, [859] and McFarlane and Tabor [501] found that the adhesion force was very small (of the order of some hundredths of a dyne) or zero, while according to the others (Bradley [443]) it amounted to several tens of dynes. There are also some (Harper [860], Tomlinson [861]), who claim that sometimes adhesion is equal to zero and at other times it is quite large. From the results of special experiments almost all the authors came to the conclusion that the electric charges on the spheres had no marked effect upon adhesion.

† Consider the film of liquid between a sphere and plane in contact (Fig. 80). When the radius of the film ϱ is small relative to the radius r of the sphere $h = r(1 - \cos\theta) \approx 0{\cdot}5\, r\,\theta^2$. If both surfaces are wetted by the liquid the radius of curvature of the meniscus is $\sim h/2$, the negative pressure within the film is $2\,\sigma/h$ according to Laplace's formula and the cohesive force is

$$F = (2\,\sigma/h)\,\pi\varrho^2 \approx 4\pi\, r\sigma. \qquad (51.15)$$

For two equal spheres in contact h is twice as large and $F \approx 2\pi r\sigma$ (57.16). MacFarlane and Tabor [501] verified this formula by direct measurements of the cohesive force between wetted spheres.

It is possible that these discrepancies are caused by the presence of dust particles on the surface of the spheres, which interfere with their contact and are difficult to remove. This is indicated by the adhesion between two spheres which have been left lying about for some time usually being zero. The measurements of Bradley, according to which the adhesive force between quartz spheres ($r = 0{\cdot}2$–$0{\cdot}5$ mm) is expressed by (56.3) with $\sigma = 35$ dyne/cm, may therefore be the most reliable. However, the only experimental data in existence on the adhesion of very small particles (see p. 373) makes F_{ad} two orders of magnitude smaller than was found by Bradley.

Adhesion is always greater in moist air. According to McFarlane and Tabor [501] when the humidity of air is increased from 80 per cent to 90 per cent the adhesive force between a glass plate and glass spheres rises from 0 to $4\pi\sigma r$, where σ is the surface tension of water (see p. 338). Such a value of adhesion on plates with rough surfaces, and probably bearing dust particles, is attained only when the thickness of the water film is greater than the height of surface projections or the diameter of the dust particles.

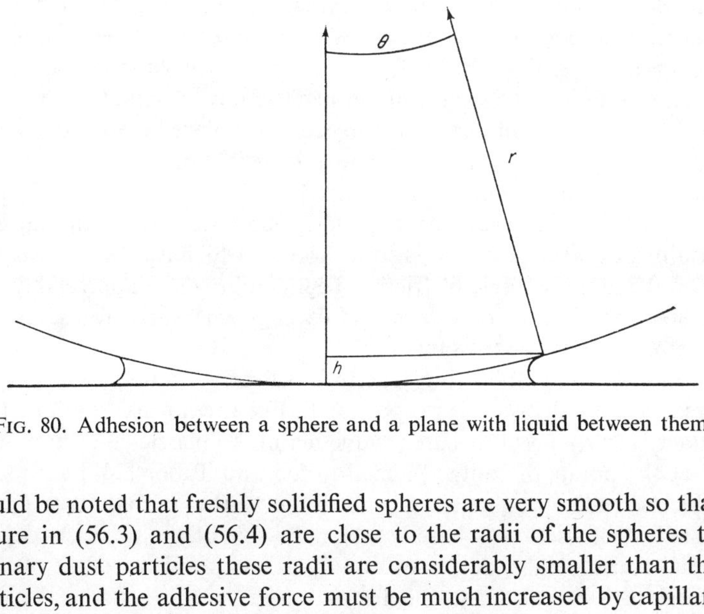

FIG. 80. Adhesion between a sphere and a plane with liquid between them.

It should be noted that freshly solidified spheres are very smooth so that the radii of curvature in (56.3) and (56.4) are close to the radii of the spheres themselves. With ordinary dust particles these radii are considerably smaller than the half-size of the particles, and the adhesive force must be much increased by capillary condensation of moisture in the vicinity of the points of contact.

Larsen [862] used a centrifuge to measure the adhesion between glass spheres, having $r = 67\,\mu$ and $34\,\mu$, and a glass plate in the presence of a liquid interlayer and confirmed (57·15). He then deduced an expression for the adhesion between a sphere and a cylinder in the presence of an interlayer

$$F = 2\pi\sigma r\left\{\frac{\varrho/r}{[(\varrho/r^2) + (R/r)^2]^{1/2}} + \frac{1}{[(\varrho/r)^2 + 1]^{1/2}}\right\}\Big/\left(1 + \frac{r}{R}\right) \qquad (57.17)$$

where R is the radius of the cylinder and ϱ the radius of the interlayer. This expression is also in good agreement with experiment. In experiments with oscillating filaments of oiled glass with $R = 128$ and $420\,\mu$, and spheres of $r = 55\,\mu$ Larsen found that the detachment of the spheres began when the maximal inertial force $ma/(2\pi\nu)^2$ (a is the amplitude and ν the frequency of oscillations, m the mass of the sphere) exceeded the adhesive force calculated by means of (57.17) and was complete after about 1 min.

Experiments with spheres blown off clean glass filaments confirmed a considerable increase (by one order of magnitude) in adhesion with atmospheric humidity. The displacement of the spheres resting on oiled filaments due to air drag was also studied. Unfortunately, there are a number of gaps in Larsen's paper which prevent a proper assessment of the results of his interesting experiments; no numerical data is given about the agreement between theory and experiment nor is it clear how the hydrodynamic forces acting upon the spheres were calculated.

How the detachment of particles by an air stream begins has not yet been settled. Particles may be torn off under the direct action of hydrodynamic lift normal to the wall, or they may begin rolling along the wall, under the action of tangential drag and bounce away after colliding with other particles or excrescences on the surface. Calculation of the lift and drag forces experienced by particles of different shapes, resting on a surface and immersed in, or rising above the boundary layer, is very difficult. The lift on fairly large objects, cubes, spheres etc., which projected well beyond a laminar layer in a water trough was measured at large Re by Goncharov [863] who demonstrated that the lift force, in accordance with the Kutta-Joukowsky theorem was the larger the more enveloping was the flow around the body in the vertical plane. The force diminished if the objects were placed closer together and was much greater upon an object which rose above its neighbours. Unfortunately, no numerical data is included in Goncharov's book.

Calculation of the force necessary to dislodge fine particles on the surface is even more uncertain. For large objects it is static friction which is easy to measure, but already for $r < 0{\cdot}5$ mm the angle of friction begins to increase appreciably with a decrease in r [864]; the scatter of values of this angle measured for several particles of the same size increases at the same time.

The reason may be the following. The force of static friction, which is in fact the tangential resolute of adhesion, is proportional, like the normal resolute, to the true area of contact. The extent of this area is due mainly to plastic deformation of microprojections at the points of contact, and Bowden and Tabor [241] have shown that for large objects it is approximately proportional to the normal force F_n, on which fact the basic law of friction is based. When the size of the object is decreased F_n, which is equal to its weight, decreases rapidly as well, and the effect of the normal force of adhesion, which also causes deformation in the contact zone, becomes more apparent. The real contact area, and consequently the force of friction, is proportional to the sum of these normal forces, which was first pointed out by Derjaguin and Lazarev [109]. As the adhesive force of the particles is nearly proportional to r, the coefficient and the angle of friction increase with a decrease in r.

Differences of roughness in the contact zones of large objects having the same F_n are more or less averaged, but for fine particles the normal resolute of adhesion and, consequently, the contact area depend to a great extent on the microgeometry of the surfaces and therefore vary greatly from particle to particle, from one orientation of a particle to another and from one point on a surface to another. This explains why all experiments on the blowing off of the dust yield curves of flowrate against the fraction of particles blown off in which complete removal calls for a flow several times greater than the one at which the process begins, even when the dust particles are all the same size [844]; similar scatter occurs in measurements of the angle of friction of particles. Experimental study of the blowing of dust from surfaces is therefore troublesome.

Cremer and Conrad [867] avoided the difficulty by measuring the angle of friction, θ, of a uniform layer of dust particles which were fairly uniform in size. Layers of differing thickness were deposited by sieving. In successful experiments the entire layer slid off in one piece and the average value of θ was measured.

The condition of slipping is expressed by the equation

$$Mg \sin \theta = \mu \, Mg \cos \theta + H \tag{57.18}$$

where M is the mass of the dust per cm^2 of the layer, μ the ordinary coefficient of friction and H the tangential adhesive force per cm^2. By plotting $Mg \sin \theta$ against $Mg \cos \theta$ for particles of dolomite or magnesite having $r > 150\,\mu$, Cremer and Conrad obtained a straight line passing through the origin, implying that H was zero. Straight lines have also been obtained for $r < 150\,\mu$, but they intersect with the ordinate axis at a point corresponding to the value of H. The experiments showed that the value Hr for a given dust ($r = 30$–$150\,\mu$) and a given surface was constant, being equal to 1·4 dyne/cm for a magnesite dust on glass, 1·3 on iron and 1·25 on magnesite. After the dusts had been heated H decreased.

If the dust particles in the first layer are close-packed in a cubic lattice there are $1/4r^2$ particles per cm^2 so that a tangential adhesive force $F_{ad} = 4r^2 H \approx 5r$ dynes acts upon each particle; this force, like the normal resolute, is approximately proportional to r†. On the other hand, Patat and Schmid [868], using the same method found for carborundum dust on a steel plate that Hr^2 = const. while $Hr^{0 \cdot 7}$ = const. for alundum on the same plate. In the first case, therefore, F_{ad} did not depend on r, which cannot be correct, and in the second case it was proportional to $r^{1 \cdot 3}$.

Batel [869] found that heating quartz dust and a glass plate in vacuo at 300 °C reduced H to zero; Patat and Schmid, however, using carborundum dust on steel, oberserved an appreciable residual value of H. There is no doubt that a considerable portion of the force F_{ad} is due to the capillary action of adsorbed water, as in the case of normal adhesion.

All who have used the method of Cremer and Conrad point out that the ordinary coefficient of friction, μ, rises appreciably when r is diminished and that it is not possible to resolve the coefficient of friction into macroscopic and adhesion components. The sole conclusion that can be drawn from the extensive experimental material collected by Patat and Schmid is that the phenomenon is of a great complexity.

At the present time, therefore, neither theoretical data nor empirical formulae exist to describe the detachment of particles by an air stream. It might appear that the copious experimental material on the pick-up of particles by flowing water could be applied to air, but, as was emphasized by Bagnold [870], substantial differences exist between the two processes. Some experiments of a qualitative character remain to be described.

The observations of Rumpf [523, 871] suggest that it is more difficult to blow off a deposit of a limestone dust on a vertical metal plate placed parallel to the flow when

† According to the experiments of Beischer (see p. 373), the adhesion between Fe_2O_3 particles of $r = 0 \cdot 25\,\mu$ is equal to 5×10^{-5} dynes, i.e. $F_{ad} \approx 2r$ dynes; in this case, in accordance with theory, the adhesion between two particles must be half that between the particle and a flat surface. The postulate that the forces of adhesion and friction are similar in magnitude for fine particles is thus confirmed.

a higher air speed was used to form it. The author thinks this is due to the greater velocities of particles colliding with the plate. It is possible that plastic deformation of the particles and an increase in the contact area take place. Another possibility is that after collision the particles roll along the surface until they reach a potential pit, where there is a large adhesive force. The greater the velocity of the particle, the deeper the pit may be, and the more firmly does the particle adhere to the surface.

The relationship between size of particle and critical rate of flow has been studied by many. Klimanov [872] found that a coal dust fraction of $r = 37$–$52\,\mu$ gave $U_{crit} = 6$ m/sec, $r = 21$–$37\,\mu$ gave 7 m/sec and $r = 0$–$21\,\mu$, 10 m/sec. In experiments by Jordan [844] 50 per cent of a deposit of quartz dust was blown away in two seconds by a thin air jet normal to the plate with $U = 45$ m/sec and $r = 3{\cdot}5\,\mu$; a velocity of 95 m/sec was required for $r = 2\,\mu$ and 145 m/sec for $r = 1\,\mu$. This confirms that for particles with radii smaller than 50–100 μ U_{crit} increases as r decreases. Bloomfield and Dalavalle (see p. 354) claimed an opposite effect for quartz dust with $r \gg 5\,\mu$, which is hardly correct.

Dawes [873] deposited layers of polydisperse powdered limestone, chalk, plaster of Paris, shale and fly-ash (with $\bar{r} = 16$–$46\,\mu$), measuring 5×17 cm^2 area and 6 mm thick by sieving upon No. 0 emery paper; this was placed on the bottom of a wind tunnel 7·5 cm high. When $\bar{U}$ exceeded 15–20 m/sec blowing off of particles, or erosion, began by a process similar to that studied by Bagnold (see p. 358), the rate of erosion being proportional to $\bar{U}^3$ and depending to a great extent upon the properties of the powder. The property best characterizing the powders proved to be their "cohesion", H, defined as the force per cm^2 cross-section necessary to disrupt a column of the powder. The rate of erosion of powders at $H < 200$ dyne/cm^2 is 4–6 times greater than at $H = 300$ and 10–15 times greater than at $H = 700$. The value of H increases as the mean particle size falls and with increasing spread of sizes, especially if there is a very fine fraction. H is also directly associated with the angle of repose. Study of dispersibility of powders, by a method similar to that of Andreasen (see p. 372), showed that it rose with increasing H in almost all cases. The connection between H and the particle radius is in line with the adhesion between the particles being proportional to their radius (see above).

At the flow rates exceeding the critical value, U'_{crit}, complete removal, or denudation, of the powder, beginning at the leading edge, occurs in less than a second. For different powders U'_{crit} varied from 28 to 40 m/sec (its maximal value in these experiments). On a polished plate U'_{crit} was 30 per cent lower than on emery paper. Vibration of the support greatly reduced U'_{crit} but had no effect upon the rate of erosion which depends very much on the angle of repose of the leading edge, smaller angles favouring an increase in U'_{crit}.

These observations, coupled with measurements of air pressure in the layer indicate that denudation results from an aerodynamic drag at the leading edge. Dawes believes that denudation occurs when the drag is greater than the force of friction F_t between the layer and the support, which was determined in special experiments. In agreement with the experiments of Cremer and Conrad, he proved that $H_t = H_0 + \mu p$ where p is the normal stress and μ and H_0 are constants. However, from the tables given by the author, the correlation between U'_{crit} and H_0 for the same powder is rather poor, and the mechanism of the process may be more complicated.

In Dawes' opinion erosion proceeds in a similar way by the tangential drag upon the surface of the layer exceeding the force of friction of the powder so that some of the layer is displaced and blown away. This hypothesis is but poorly confirmed by the experimental data, the erosion of fine powders actually appearing to proceed by the mechanism described by Bagnold and Chepil.

The phenomena discussed here are of great interest in connection with pneumatic transportation of granular materials. A number of works of applied character are concerned with this important technical process. Its physical principles are, however, incompletely elucidated.

In pneumatic transportation two-phase flow is usually observed in horizontal pipes, a continuous layer of material moving slowly along the bottom of the pipe. Its velocity is determined by the tangential drag acting upon the surface of the layer and by the force of friction against the bottom of the pipe. With coarse materials, a major portion moves by saltation, like natural sands. In the upper part of the pipe the particles are suspended.† Fine, soft particles, in not too narrow pipes, collide with the walls infrequently and their mean longitudinal velocity $\overline{V}$ differs only slightly from the air velocity $\overline{U}$. Large, hard particles, in narrow pipes, travel in a zigzag way from one side to the other (see Fig. 75a) so that $\overline{V}_t$ may be considerably less than $\overline{U}$. The less is $\overline{U}$ the heavier the particles and the higher the concentration of the material being transported, the greater is the proportion moving along the bottom of the pipe. In vertical pipes of circular section the stream usually rotates to some extent and material is thrown out to the walls by centrifugal force; similar phenomena are observed in this case as well.

It is clear that according to the conditions the ratio $\overline{V}_t/\overline{U}$ can vary within a wide range, as has been found in a number of experiments. Unfortunately, only in Adam's paper, mentioned above (see p. 348), is allowance made for such important factors as the hardness of the particles and of the walls. The observed decrease in the ratio $\overline{V}_t/\overline{U}$ with an increase in the particle size and a decrease in the diameter of the pipe are quite well explained by the considerations given above.

§ 58. THE FLUIDIZATION AND DISPERSAL OF POWDERS

In this section the process will be discussed by which materials are converted from an aggregated state to aerosols. In the preceding section a particular case was examined. When air flows at comparatively low velocities over a layer of powder, particles become detached from the surface and are carried upwards by turbulence. This is certainly the principal mechanism by which solid particles are conveyed into the atmosphere in nature. While the particles are being carried away the layer is compacted by the air flow and assumes a streamlined shape [522] until there is no more erosion.

For a renewal of the process to commence a considerably greater flow velocity is required when similar events are repeated. However, at very high flow velocities, not only are individual particles carried away but whole aggregates are torn out and immediately disintegrated by the current of air; as a result hollows are formed in the surface [526], which becomes more and more uneven, thus facilitating the further

† According to the observations of Mehta *et al.* [874] glass spheres of $r = 48\,\mu$ move at $\overline{U} = 5 - 25$ m/sec, mostly by saltation, while the spheres of $r = 18\,\mu$ travel as an aerosol.

tearing out of aggregates. This constitutes a transition from surface to bulk atomization. The latter occurs in nozzles used for injecting coal dust in furnaces, in the pneumatic separation of powders (see page 42), in the dispersal of insecticide dusts and when chemical reactions proceed in the "boiling layer". It is thus of great technical importance.

Understanding of the mechanism by which powders are dispersed has recently progressed considerably owing to the experiments on fluidization of powders, or their transference into the "boiling" state. This is of great interest in chemical technology and a description will be given of the phenomenon in gases and liquids.

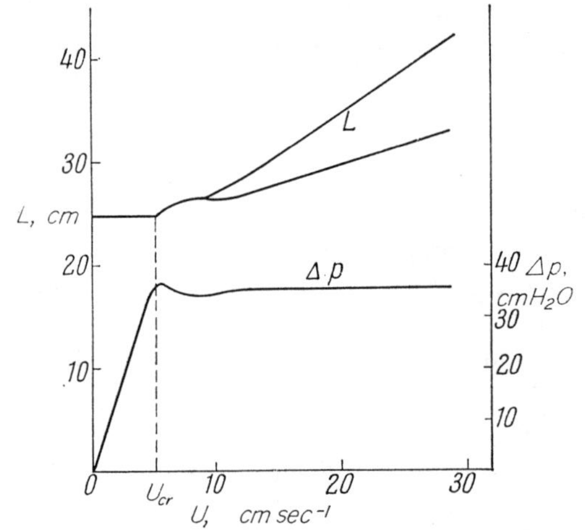

FIG. 81. Relationship between flow velocity, pressure and the height of the layer of powder during fluidization.

Water or a gas flows upwards, with gradually increasing speed, through a layer of powder in a cylindrical vessel which has a porous bottom (see Fig. 81 representing the fluidization by air of glass powder having spherical particles of radius 0·135 mm [542]). At low flow velocities the particles remain stationary; the thickness, L, of the layer and the fraction of volume occupied by particles, φ, are constant and the resistance, Δp of the layer is given by the well-known Kozeny formula:

$$\Delta p = L \operatorname{grad} p = \frac{\varkappa L U \eta \varphi^2}{r^2(1-\varphi)^3}, \tag{58.1}$$

where grad p is the pressure gradient of the medium in the layer, r is the mean particle radius, U is the flow velocity above the layer and $\varkappa$ is a factor depending on the shape of the particles. When the pressure gradient of the medium equals the gradient of hydrostatic pressure of the powder, $g\varphi(\gamma - \gamma_g)$, that is when

$$\Delta p = Lg\varphi(\gamma - \gamma_g) = \frac{\varkappa L U \eta \varphi^2}{r^2(1-\varphi)^3}, \tag{58.2}$$

the resultant of the forces acting on a particle becomes zero. At the flow velocity U_{crit} corresponding to this condition the layer begins to expand and the resistance becomes constant since the product $L\varphi$ is constant.

At this stage of expansion the nature of the fluid and the particle size begin to influence the behaviour. For fluidized powders having $r > 0{\cdot}15$–$0{\cdot}20$ mm in a gas, or considerably smaller in a liquid, the resistance continues to obey formula (58.1) and the expansion of the layer, or the decrease in φ, as the flow velocity increases can be determined from equation (58.2). This shows that under these conditions a powder expands uniformly; contact between neighbouring particles is preserved but the powder assumes a more flocculent structure. In an isodisperse powder, having spherical particles, φ decreases from 0·66 to 0·54, approximately [543]. For close packed spheres $\varphi = 0{\cdot}74$ and for simple cubic packing $\varphi = 0{\cdot}52$, so that the expanded layer must approximate to simple cubic packing.

In finer powders, where forces of cohesion between particles begin to play a noticeable part, deviations from (58.1) are observed. The ratio $U/\Delta p$ increases more rapidly than the formula predicts, being proportional not to $(1 - \varphi)^3/\varphi^2$ but to $[(1 - \varphi)^3/\varphi^2]^s$ where $s > 1$ and increases with decreasing particle size [544]. This shows that the powder is no longer expanding uniformly but is breaking into separate aggregates with slit-like passages between them. Much of the gas goes through these passages so that the flow velocity is greater than calculated by (58.1).

The effect of cohesive forces on the properties of powders is also evident under static conditions without air movement. The so-called bulk specific gravity of a powder, $\varphi\gamma$, does not depend appreciably on particle size in coarse powders since it is determined by the ratio of the weight of the particles to the frictional force between them, which is proportional to the weight. However, as the particle size of a powder is decreased, molecular forces begin to come into play and these increase the friction between particles and favour the formation of a more porous structure; the bulk gravity therefore decreases [545] (see Fig. 82). The angle of repose of very coarse powders is similarly independent of particle size but increases when the particles are smaller [546].

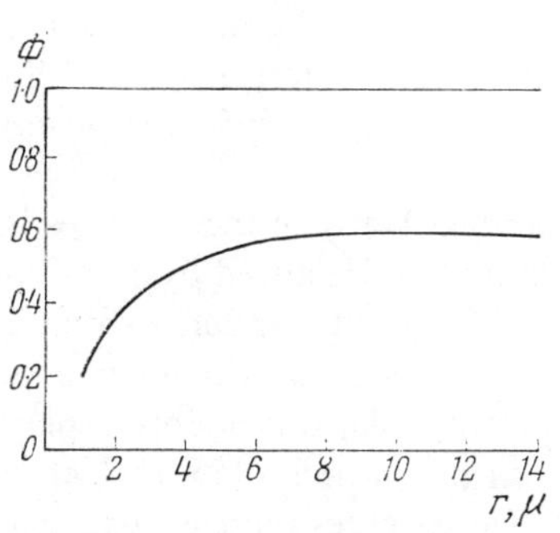

FIG. 82. Particle size and bulk gravity of powders.

The molecular forces between particles in water are so small or the loosening action of water is so large, that uniform expansion occurs even with highly dispersed powders. In gases the work expended in overcoming molecular forces during uniform expansion of a layer of fine particles is appreciably greater than the work necessary to break the powder into separate aggregates, which is why the latter actually happens. As the layer expands its fluidity (measured, for example, by the resistance offered to a stirrer rotated in it) increases by a factor of several tens [547] while the angle of repose decreases considerably, the powder assuming liquid-like properties. In water the angle of repose finally reaches zero showing that contact between particles has ceased. In a gas the angle retains some finite value.

It should be noted that equation (58.2) holds for gases only if the height of the layer is not much greater than the diameter of the tube containing the powder. If this is not so, friction of the powder against the wall of the tube begins to have an effect and expansion of the layer begins when the ratio grad $p/g\varphi\gamma$ is somewhat greater than one [548].

After a certain amount of expansion, which is between 5 and 20 per cent of the original volume depending on the powder, the particles begin to move and the resistance of the layer decreases a little (Fig. 81), probably as a result of reduced friction between the particles. Beyond this point the behaviour in water is quite different from that in gas. In water there is complete fluidization; the layer continues to expand uniformly, particles are suspended freely in the flow and the relationship between φ and the flow velocity is the same as between φ and the rate of settling of a suspension in a restricted space [542] (see § 14).

Whether complete fluidization can occur in a gas is unknown because no experiments have been carried out with powders such as *lycopodium* which are easily fluidized. The conditions for complete fluidization in a gas are extremely unfavourable owing to large molecular forces and the low viscosity of gases. What is observed is therefore as follows. At a certain flow velocity the gas begins to bubble through the uniformly expanded layer of a coarse powder as through a liquid. The height of the layer therefore continues to increase but fluctuates violently (see Fig. 81). Gas bubbles cause energetic mixing of the powder, promoting a circulation rising at the centre of the tube and descending at the walls [544]. In this state the powder is very much like a boiling liquid. As the flow velocity increases particles whose rate of settling is less than the flow velocity above the layer begin to leave the "liquid" phase, being drawn out of the boiling layer into the "gaseous" (aerosol) phase. The particle concentration in the aerosol phase keeps increasing until finally the boundary between the two phases disappears and the powder is completely carried away by the gas [549].

It is shown below that contact between the particles is preserved in the expanded layer but, since the force of gravity is balanced by aerodynamic drag, friction between the particles is very small and fluidized powders exhibit high fluidity. That complete fluidization is nevertheless not observed with coarse powders, in which cohesive forces cannot play an appreciable role, is probably explained by the action of hydrodynamic forces (§ 23) between the particles [550] which prevent the layer of powder from expanding uniformly. These forces increase with particle size.

So-called aggregative fluidization is observed with fine powders. At low flow velocities channels are formed in the layer and the main mass of gas flows through these. When the flow velocity is increased the channels expand and energetic mixing and continuous formation and disintegration of aggregates develop in the layer, accompanied by carriage of individual particles into the aerosol phase. The finer the particles, the stronger are the aggregates and the worse is the fluidization. Since hydrodynamic forces increase and the effect of molecular forces diminishes with increase in particle size, it would be expected that optimum conditions for fluidization occur at some intermediate particle size. In fact there are indications that the most uniform fluidization of powders is observed at $r = 25\,\mu$ [551] or $20\,\mu$ [552], and that the tendency towards bubble formation in the layer increases as *r* increases in coarse powders, and as *r* decreases in fine powders [553].

Since the friction of powders against the wall of the tube falls as the particle size increases, the effect of molecular forces diminishing, the phenomenon of bumping is observed with coarse particles in narrow tubes. The powder is completely lifted by the gas which then breaks through the layer and allows it to fall again.

Fluidized catalysts have the following advantages: free access of the reacting chemicals to the surface of the catalyst particles, rapid removal of heat permitting overheating to be avoided, and the possibility of continuously transferring the catalyst from the reactor to the regenerator and vice versa. The coefficient of heat transfer from a gas flowing in a tube to the walls rises considerably in the presence of fluidized powder owing to active mixing of the powder by gas bubbles [554].

TABLE 39. VOLUME FRACTIONS OCCUPIED BY PARTICLES DURING FLUIDIZATION

	$\bar{r}$, μ	φ_0	φ_{min}
Glass spheres	285	0·67	0·55
	115	0·66	0·54
Rounded sand	185	–	0·58
	80	–	0·56
	42	–	0·53
	25	–	0·46
Angular sand	160	–	0·50
	100	–	0·47
	42	–	0·43
	27	–	0·42
Iron oxide	187	0·50	0·48
	100	0·48	0·44
	35	0·42	0·39

If the flow is decreased gradually the powder remains expanded when the flow is stopped and must be shaken to return it to the original unexpanded state [544]. It follows that contact between particles must be preserved in the expanded layer and the assertion, often made, that the particles are surrounded on all sides by a gaseous film which prevents direct contact is without foundation.

This makes it easy to determine values of φ_{min} for the expanded layer and some are given in Table 39 [555]. The table also includes values of φ_0 for a quiescent unexpanded layer. If φ_{min} is measured by gradually increasing the flow the results fluctuate. In this case φ_{min} depends on the initial packing of the particles [556], since they are not able to rearrange themselves. Some authors have measured φ_{min} when gas was being passed through, obtaining values considerably lower than those in the table because of gas bubbles in the layer. The value of φ_{min} corresponding to the maximum possible porosity, or the maximum possible volume of a given powder while preserving contact between particles, is one of the most important physical characteristics of powders.

It can be seen from Table 39 that porosity increases with decreasing particle size not only in the undisturbed layer but also in the expanded one. This also indicates the continued action of molecular forces, and hence of contact between the particles, after the layer has expanded.

Some powders can be put into the expanded state by careful pouring without passing gas through them [557]. Expanded powders may possess high fluidity and resemble a liquid in some respects. The air between the particles evidently retards their fall and perpetuates their loose structure. Since the force of gravity is unbalanced their fluidity may be due to the smallness of the molecular forces and thus of the friction between the particles. Of all powders, *lyopodium* probably has the greatest fluidity, this being promoted by the uniformity, lightness and spherical form of the particles, and by the presence on their surfaces of ridges which favour looseness of the structure of the powder.

Experiments on fluidization have elucitated only the effect of the fluid (water or gas) and, to some degree, of the fineness of the powder on the ease with which it can be dispersed as an aerosol. The effects of particle shape and of the nature of the material are uncertain. It is well known that some powders raise dust when they are poured while others having the same particle size do not.

Dispersibility of powders is important technically, particularly for insecticidal dusts. Much of the preceding section on adherence is again relevant. Dispersibility depends primarily on the cohesive forces between particles and improves as the particle size increases while depending very much upon the moisture content of the powder. Hydrophobic powders, such as talc, are easier to disperse than hydrophilic ones like quartz and limestone. According to Albinsson [481] the rate at which quartz having $r = 25$–$50\,\mu$ can be sieved depends on the humidity of the air in which it has been kept and falls sharply at relative humidities above 70 per cent. Some reduction in the rate of sieving, however, is observed at humidities below 10 per cent but it can be eliminated by adding traces of radioactive material to the powder which removes the electric charges from the particles. The charges which are produced by sieving a very dry powder encourage the formation of stable aggregates and the adhesion of particles to the sieve.

Powders of soft plastic materials are more difficult to disperse than hard ones; isodisperse powders atomize better than polydisperse because, in the latter, more of the intergranular space is filled so there are more points of contact between particles.

Experimental data on the dispersibility of various powders are extremely scarce and only those obtained by Andreasen [558] are given here. Two cubic centimetres of powder was poured through a narrow slit into a vertical tube of height 250 cm and diameter 4·5 cm. The particles were separated to some extent as they fell through the air and the percentage of powder which had not settled on the bottom of the tube in 6 sec was determined. Since individual particles could not have reached the bottom in this time the author assumed that his figures represented the percentage of dispersed powder, which he called the dispersibility. Actually the unsettled fraction also contained small aggregates. Some results of Andreasen's experiments are given in Table 40.

Dispersibility is of great importance in combating coal-mine explosions with inert powders. Dust, usually limestone, raised by a detonation wave checks the spread of an explosion. Powder used for this purpose must be easily dispersed and it has been found that the addition to pulverized limestone of 0·5 per cent soot, which reduces the cohesive forces between the limestone particles, considerably improves its dispersibility by a detonation wave [559].

As already mentioned, theoretical calculation of the cohesive forces between particles is unreliable and their experimental determination by measuring the force necessary to break thread-like aggregates is of interest (Beischer [560]). Such aggregates, obtained in an electric field from non-magnetic iron oxide smoke having particle radii of $0{\cdot}25\,\mu$, broke under a force of 4 to 6×10^{-5} dyne. If it is assumed,

TABLE 40. DISPERSIBILITY OF SOME POWDERS

	Particle radius limits (μ)	Dispersibility (%)
Lycopodium	12	100
Wood charcoal dust	0–25	85
Wood charcoal dust	0–7	23
Aluminium powder	0–15	66
Talc	0–20	57
Carbon black	0–15(?)	47
Potato starch	0–35	27
Graphite dust	0–25	17
Pulverized slate	0–25	13
Cement	0–45	5·5
Prepared chalk	0–6	1·5
Polydisperse silica dust (coarse)	–	21
Polydisperse silica dust (fine)	–	8
Isodisperse silica dust	11·5	68
	8	83
	5·6	45
Porcelain dust with fine fractions removed	7	50
	2·7	52
	1·1	21
	0·45	12
Porcelain dust without fine fractions removed	–	5

following (56.4), that the cohesive force is proportional to the particle radius, then

$$F = (1{\cdot}6 \text{ to } 2{\cdot}4)r, \quad \text{dyne}. \tag{58.3}$$

Formula (56.4), with a plausible value for σ, gives values of the breaking force which are many times greater than the experimental ones. Aggregates of zinc oxide require about the same breaking force as iron oxide but aggregates obtained in an electric field from smokes of ammonium chloride, arsenic trioxide or anthracene have considerably greater strength.

Beischer explains this by saying that in the more volatile substances cementation of the aggregates occurs by diffusion and recrystallization. Finally, in aggregates formed in the pyroelectric smoke of aminoazobenzene or in magnetic iron oxide smoke, in which there are electrostatic or magnetic forces between the particles as well as molecular ones, the breaking force is 100–1000 times larger than for an aggregate of zinc oxide or of non-magnetic iron oxide.

In conclusion the rupture of aggregates by air currents will be discussed. Two basic mechanisms are possible. The first resembles the rupture of liquid drops (see

Fig. 15). Flat aggregates orient themselves perpendicular to the flow direction and at the centre, where the flow velocity is least, the pressure reaches a maximum according to Bernoulli's theorem and the aggregate deforms like the drop in Fig. 15; a shearing force develops which increases with the size of the aggregate and with the relative velocity of the aggregate and the air. The larger the aggregate the more easily does it rupture when falling through air. The velocity of fall is generally small and more complete disaggregation can be achieved by the momentarily large velocities applied when aggregates are introduced into a rapidly flowing airstream. The action of dust injectors and the atomization of insecticide dusts spread from aircraft probably depend on this kind of process.

The second mechanism of rupture occurs in the presence of fairly large velocity gradients, such as those occurring in impingement dust sampling instruments (see page 152) and needs no further explanation. Its efficiency also increases with the length of the aggregate. This is clearly the way in which aggregates are ruptured during energetic mixing and in turbulent flow.

Data in the literature indicate that when an aerosol of solid particles is subjected to sufficiently intense agitation a stationary state can be reached when the rates of coagulation and disaggregation are equal [502]. Apart from fluidization, the mechanisms of powder dispersal remain almost untouched in scientific literature.

A recurrent problem in the laboratory investigation of aerosols is the dispersion of powders into the aerosol state, free from aggregated particles. Perfect dispersion is usually required either for experimental aerosols having certain properties or for the laboratory reproduction of industrial aerosols with solid particles. In the first case a powder with the desired mean size and size distribution can be prepared by grinding and wet classification; in the second, material can be taken from an industrial dust precipitator. Perfect dispersion of the former is necessary, but it may be desirable to aim at the same degree of aggregation as the original with the industrial aerosol; this can be achieved by allowing a perfectly dispersed aerosol to coagulate.

Coarse powders ($r > 10\,\mu$) can be dispersed efficiently by an air jet, but finer particles present difficulties. The removal of particles having $r < 10\,\mu$, for instance, from an elutriator is very slow, even if they are introduced into the elutriator from outside [875]. It is evident that aggregates are formed instantaneously.

Poor dispersibility of fine powders can be accounted for in the following way. An aggregate located in a velocity gradient is subjected to a difference of flow velocity at two points located a distance $2\,r$ from each other which is equal to $2\,r\Gamma$, where Γ is the velocity gradient. A force results, tending to break apart two particles in contact, which is proportional to r^2

$$F_M \approx 6\pi\eta r\, 2r\Gamma = 12\pi\eta r^2\Gamma \tag{58.4}$$

whereas the adhesive force, as seen above, is approximately proportional to r.

Apart from the particle size, the shape of the particles, their mechanical properties, plasticity and the packing density, or their number of contacts of a particle with its neighbours, are all important. It is especially difficult to disperse powders covering a great range of particle size since their packing density is much greater than in isodisperse powders. A decrease in packing density, or in the bulk density of the powder, on exposure, for example, to sonic vibration, substantially facilitates dispersion.

As mentioned on p. 156, the experiments of Davies *et al.* [218] demonstrated breaking up of coal dust aggregates, with r of the order of $1\,\mu$, when they were blown through a slit of width $2h = 0{\cdot}08$ mm at a velocity $\bar{U} = 170$ m/sec. The mean velocity gradient in the slit being equal to $3/2\ \bar{U}/h$, the separating force, F_M, for particles of $r = 1\,\mu$, (58.4) was $3{\cdot}6 \times 10^{-4}$ dynes. Experimental data obtained by Beischer [560] for the force of adhesion between the particles of this order of size follows the formula (58.3)

$$F_{ad} \approx 2\,r \tag{58.5}$$

whence at $r = 1\,\mu$, $F_{ad} \approx 2 \times 10^{-4}$ dynes, which is of the same order as F_M.

Eadie and Payne [876] designed an apparatus for the complete dispersion of powders which was based on this principle. The powder was coarsely dispersed by an air jet and then forced under a pressure of up to thirty atmospheres through an annular slit 10–250 μ wide. It was shown by sedimentation analysis of the resulting aerosol that when the pressure, or flow velocity, was increased, the size of the particles issuing from the slit decreased only down to a certain limit and then remained constant. This was taken as evidence of perfect disaggregation. However, there appears to have been no microscopic examination of the particles and there is not enough evidence in their short paper to bear out the correctness of their conclusion. It is possible, for example, that the sonic flow velocity was attained in the slit, which does not rise with an increase in pressure. It must be acknowledged, however, that disaggregation in a gradient flow is a most promising approach to the perfect dispersion of powders.

Another method of dispersion is the use of oscillations. Oscillations of sonic frequency affect powders in two ways, increasing their density at small amplitudes and loosening them up at large [877]; the loosening is accompanied by a considerable decrease in cohesion, an increase in the ease of flow and a decrease in the angle of repose etc. Chin-Young-Wen and Simmons [878] dispersed a powder of glass spheres having $r = 15\,\mu$ by sonic oscillation. A more efficient method of dispersion, however, is the simultaneous action of sound and of air flow through the powder.

Morse [879] worked with highly cohesive, poorly dispersible powders of plaster of Paris, marble ($\bar{r} \approx 1\,\mu$) etc. When air was blown through beds of these powders they were not loosened and the air passed through in the form of large, separate bubbles. Application of a sonic field with a frequency of 20–670 c/s and sufficient intensity (> 110 db) under the bed expanded it appreciably (as much as 1·5 times, with plaster of Paris), and fairly perfect fluidization was achieved. At larger velocities of the air flow a disaggregated aerosol would presumably be formed.

It may be possible to increase the dispersibility of powders by the addition of a very small amount of another highly disperse powder. An example is the addition of carbon black to limestone dust, mentioned on p. 372. Addition of MgO is also very effective. The experiments of Craik and Miller [880] show that the addition of 1 per cent MgO of $\bar{r} \approx 0{\cdot}03\,\mu$, to powders of starch, sugar and NaCl with $\bar{r} \approx 5\,\mu$ considerably reduces the angle of repose and, consequently, the cohesion. The particles of MgO and carbon black appear to play the same role here as the dust particles in the experiments on the adhesion between the spheres, which were mentioned on p. 363, decreasing the contact area between the particles of the original powder.

It is easy to understand that a small quantity of moisture, which considerably increases the adhesion between the particles, has an opposite effect. According to the experiments of Parker and Stevens [881] fluidization deteriorates upon addition of some hundredths of 1 per cent of water to a powder of glass spheres with $r = 150\,\mu$.

Two more problems, discussed in the voluminous literature on fluidization, should be mentioned here.

An interesting property of fluidized powders which has found many uses in industry [882] is their ability to flow like a fluid down the pipes which are slightly inclined to the horizontal. The flowing system is quite uniform to begin with, but

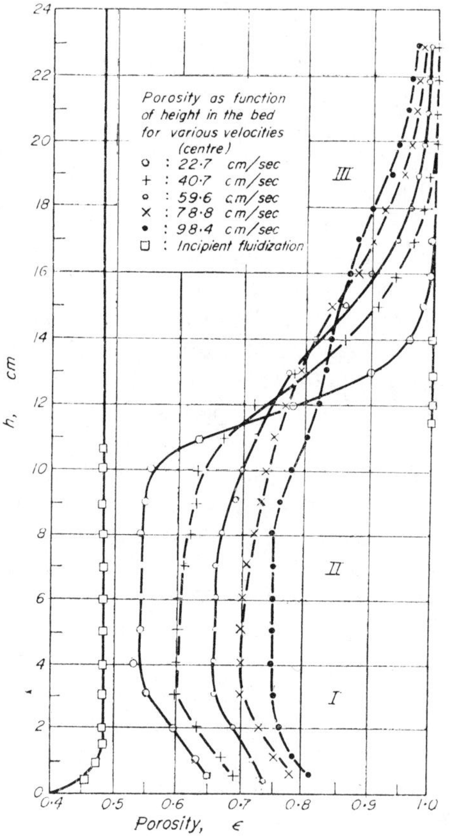

FIG. 82a. Fluidization of glass beads at different flow rates.

after some distance it disintegrates into the phases usually encountered in pneumatic transportation.

It is very difficult to determine visually what occurs in a column of fluidized powder. The introduction of an electrical condenser technique was a great step forward in the investigation of fluidization. A small condenser is introduced at various points and the porosity of the powder at a given point ascertained from its capacitance. An investigation by this method of fluidization of glass spheres with $r = 100\,\mu$, at different air velocities, [883] is shown in Fig. 82a. As seen from the figure, there is a small zone at the bottom of the column with an increased porosity corresponding to the exit of air bubbles from the supporting grid; above this is a zone of a fluidized powder, the porosity of which is enhanced with an increase in U, and, finally, the aerosol zone.

The decrease in aerosol concentration with an increase in h is explained by the coarser particles being thrown up to some height and then falling back, a phenomenon which was thoroughly studied by Zenz and Weil [884] whose theoretical computations are in fair agreement with experiment.

Baumgarten and Pigford [885] used a γ-ray absorption method to prove that the real volume fraction, φ, of the disperse phase between the rising gas-bubbles, in a fluidized bed of fractionated glass beads ($\bar{r}$ = 35–100 μ), was constant and the same as in the expanded bed, φ_{min}.

REFERENCES

1. GIBBS, W., *J. Soc. Chem. Ind. (London)* **51**, 1042 (1932).
2. VIGDORCHIK, YE., *The Retention of Aerosols During Respiration (Zaderzhka aerozolei pri dykhanii)*, LIOT, Leningrad, Chap. 1 (1948).
3. PRZIBRAM, K., *Handb. Phys.*, Berlin, **22** (I), Kap. 4 (1933).
4. *Handb. Experimentalphys.*, Leipzig, **13** (I), Kap. 5 (1929).
5. VONNEGUT, B., *Chem. Rev.*, **44**, 277 (1949).
6. SINCLAIR, D., LAMER, V., *Chem. Rev.*, **44**, 245 (1949).
7. FIGUROVSKII, N., *Coll. Papers on Problems of Kinetics and Catalysis (Sb. Problemy kinetiki i kataliza)*, ONTI, Moscow–Leningrad, **7**, p. 9 (1949).
8. NEIBURGER, M., WURTELE, M., *Chem. Rev.*, **44**, 321 (1949).
9. ROLLER, N., *J. Frankl. Inst.*, **223**, 609 (1937).
10. ROSIN, P., RAMMLER, E., *Koll. Z.*, **67**, 16 (1934).
11. NUKIYAMA, S., TANASAWA, Y., *Trans. Soc. Mech. Engrs., Japan*, **5**, No 18, 63 (1939).
12. ZELENY, J., MCKEEHAN, L., *Phys. Z.*, **11**, 78 (1910).
13. LEVIN, L., *Dokl. Akad. nauk SSSR*, **94**, 1045 (1954).
14. HATCH, T., CHOATE, S., *J. Frankl. Inst.*, **207**, 369 (1929).
15. VOEGTLIN, C., HODGE, H., *Pharmacology and Toxicology of Uranium Compounds*, p. 468, New York (1949).
16. FRIEDMAN, S., *Chem. Eng. Progr.*, **48**, 118 (1952).
17. RANZ, W., WONG, J., *Arch. Ind. Hyg. Occ. Med.*, **5**, 464 (1952).
18. KOTTLER, F., *J. Frankl. Inst.*, **250**, 339, 419 (1950).
19. KOLMOGOROFF, A., *Dokl. Akad. nauk SSSR*, **31**, 99 (1941).
20. EPSTEIN, B., *Ind. Eng. Chem.*, **40**, 2289 (1948).
21. HATCH, T., CHOATE, S., *J. Frankl. Inst.*, **215**, 27 (1933).
22. NOLAN, P.J., O'TOOLE, C.P., *Geofis. Pura e Appl.*, **42**, 117 (1959).
23. ICHIDA, Y., *Nature, Lond.*, **140**, 70 (1937).
24. PLACZEK, G., *Z. Phys.*, **55**, 81 (1929).
25. SANZENBACHER, R., *Z. Phys.*, **39**, 251 (1926).
26. MATTAUCH, J., *Z. Phys.*, **32**, 439 (1925).
27. TRAUNER, O., *Z. Phys.*, **46**, 237 (1928).
28. WASSER, E., *Z. Phys.*, **27**, 226 (1924).
29. BOWLING, R. *et al.*, *Ind. Eng. Chem., News Ed.*, **19**, 965 (1941).
30. ARTEMOV, I., *Coll. Papers on New Ideas in the Study of Aerosols (Sb. Novyye idei v oblasti izucheniya aerozolei)*. Acad. Sci. U.S.S.R., p. 68 (1949).
31. WHYTLAW-GRAY, R., PATTERSON, H.S., *Smoke*, Chap. 10, London (1932).
32. GERLACH, W., *Phys. Z.*, **20**, 298 (1919).
33. REGENER, E., *Ber. Preuss. Akad. Wiss.*, **192**, 632 (1920).
34. WADELL, H., *J. Frankl. Inst.*, **217**, 459 (1934).
35. RAN'KO YU, *Koll. Zh.*, **10**, 42 (1948).
36. HANSON, E., DANIEL, J., *J. Appl. Phys.*, **18**, 439 (1947).
37. EPSTEIN, P., *Phys. Rev.*, **23**, 710 (1924).
38. MILLIKAN, R., *Phys. Rev.*, **21**, 217 (1923).
39. MILLIKAN, R., *Phys. Rev.*, **22**, 1 (1923).
40. WEYSSENHOFF, J., *Ann. Physik*, **62**, 1 (1920).
41. SCHILLER, L., *Handb. Experimentalphys.*, Leipzig, **4** (II), 342 (1932).
42. OSEEN, C., *Neuere Methoden und Ergebnisse in der Hydrodynamik*. Leipzig, § 16 (1927).
43. RYBCZINSKI, W., *Anzeig. Akad. Krakau*, 40 (1911).
44. LAMB, H., *Hydrodynamics*, 6th ed., § 337, Cambridge (1932).
45. LAMB, H., *Hydrodynamics*, 6th ed., § 338, Cambridge (1932).

46. Cunningham, E., *Proc. Roy. Soc.*, **83** A, 357 (1910).
47. Stacy, L., *Phys. Rev.*, **21**, 239 (1923).
48. Wasser, E., *Phys. Z.*, **34**, 257 (1933).
49. Knudsen, M., Weber, S., *Ann. Physik*, **36**, 982 (1911).
50. Reiss, M., *Z. Phys.*, **39**, 623 (1926).
51. Laby, T., *Nature, Lond.*, **150**, 648 (1942).
52. Schiller, L., *Handb. Experimentalphys.*, Leipzig, **4** (II), 339 (1932).
53. Millikan, R., *The Electron*, Chap. 4 and 5, Chicago (1917).
54. Kellström, G., *Nature, Lond.*, **136**, 682 (1935).
55. Goldstein, S., *Proc. Roy. Soc.* **123** A, 225 (1929).
56. Schmiedel, J., *Phys. Z.*, **29**, 593 (1928).
57. Möller, H., *Phys. Z.*, **39**, 57 (1938).
58. *Handb. Experimentalphys.*, Leipzig, **4** (II), Kap. 4, 13 (1932).
59. Davies, C. N., *Proc. Phys. Soc.*, **57**, 259 (1945).
60. Klyachko, L., *Otopl. i ventil.*, No. 4 (1934).
61. Pettijohn, E., Christiansen, E., *Chem. Eng. Progr.*, **44**, 157 (1948).
62. McNown, J., Malaika, J., *Trans. Amer. Geophys. Union*, **31**, 74 (1950).
63. Gans, R., *München. Ber.*, 191 (1911).
64. Kunkel, W., *J. Appl. Phys.*, **19**, 1056 (1948).
65. Lamb, H., *Hydrodynamics*, 6th ed., § 124, Cambridge (1932).
66. Kilb, A., *Forsch. Ing. Wesen*, **58**, 6, 89 (1934).
67. Lyashchenko, P., *Gravitational Methods of Enrichment* (*Gravitatsionnyye metody obogashcheniya*). Gostoptekhizdat, Moscow, p. 40 (1940).
68. Peterlin, A., Stuart, H., *Z. Phys.*, **112**, 1 (1939).
69. Jeffery, G., *Proc. Roy. Soc.*, **102** A, 161 (1923).
70. Trevelyan, B., Mason, S., *J. Coll. Sci.*, **6**, 354 (1951).
71. Oseen, C., *Neuere Methoden und Ergebnisse in der Hydrodynamik*. Leipzig, § 18 (1927).
72. Schiller, L., *Handb. Experimentalphys.*, Leipzig, **4** (II), 367 (1932).
73. Lamb, H., *Hydrodynamics*, 6th ed., § 343, Cambridge (1932).
74. Finn, R., *J. Appl. Phys.*, **24**, 771 (1953).
75. Heiss, J., Coull, J., *Chem. Eng. Progr.*, **48**, 133 (1952).
76. *Coll. Papers on Problems of Removing Dust from Air* (*Sb. Voprosy ochistki vozdukha ot pyli*). Vsesoyuzn. Nauchno-issled. izdat. teplosnabzheniya, otopleniya i ventilyatsii, Moscow, p. 52 (1940).
77. Gonell, H., *Z. Ver. Deutsch. Ing.*, **72**, 27 (1928).
78. Herwig, H., *Feuerungstechnik*, **28**, 122 (1940).
79. Gunn, R., *J. Meteor*, **6**, 243 (1949).
80. Best, A., *Quart. J. Roy. Meteor. Soc.*, **76**, 302 (1950).
81. Blanchard, D., *Trans. Amer. Geophys. Union*, **31**, 836 (1950).
82. Smoluchowski, M., *Proc. Internat. Mathem. Congr. Cambr.* (1912).
83. Burgers, J., *Proc. Amsterd. Acad. Sci.*, **44**, 1177 (1941); **45**, 126 (1942).
84. Prosad, K., Sen, D., *Phil. Mag.*, **25**, 993 (1938).
85. Kermak, W. *et al.*, *Proc. Roy. Soc. Edinb.*, **49**, 170 (1929).
86. Hawksley, P., *Some Aspects of Fluid Flow*, p. 114 (1950).
87. Lewis, E., Bowerman, E., *Chem. Eng. Progr.*, **48**, 603 (1952).
88. Richardson, J., Zaki, W., *Chem. Eng. Sci.*, **3**, 65 (1954).
89. Regener, E., *Z. Phys. Chem.*, **139**, 420 (1928).
90. Millikan, R., *Phil. Mag.*, **19**, 209 (1910).
91. Ehrenhaft, F., *Wien. Ber.*, **119**, IIa (1910).
92. Gerlach, W., *Handb. Phys.*, Berlin, **22** (I), Kap. 1 (1933).
93. Einstein, A., *Z. Phys.*, **27**, 1 (1924).
94. Meyer, E., Gerlach, W., *Elster-Geitel Festschrift*, 196 (1915).
95. Parankiewitz, J., *Phys. Z.*, **19**, 280 (1918).
96. Millikan, R., *The Electron*, Chap. 3, Chicago (1917).
97. Bär, R., *Ann. Physik*, **67**, 157 (1922).
98. Rotzeig, B., Fuchs, N., *Zh. fiz. khim.*, **9**, 35 (1937).
99. Bär, R., *Ann. Physik*, **59**, 393 (1919).
100. Hopper, V., Laby, T., *Proc. Roy. Soc.*, **178** A, 243 (1941).
101. Rubinowitz, S., *Ann. Physik*, **62**, 695 (1920).
102. Hettner, G., *Z. Phys.*, **27**, 12 (1924).

103. HETTNER, G., *Z. Phys.*, **37**, 179 (1926).
104. MATTAUCH, J., *Ann. Physik*, **85**, 967 (1928).
105. EPSTEIN, P., *Z. Phys.*, **54**, 537 (1929).
106. ROSENBLATT, P., LAMER, V., *Phys. Rev.*, **70**, 385 (1946).
107. SAXTON, R., RANZ, W., *J. Appl. Phys.*, **23**, 917 (1952).
108. NIELSEN, R., *Heat. Pip. Air. Lond.*, **12**, 389 (1940).
109. DERJAGUIN, B., LASAREV, V., *Koll. Zh.* **1**, 295 (1935).
110. WATSON, H., *Trans. Inst. Min. Metall.*, **46**, 155 (1937).
111. BACON, L., *J. Frankl. Inst.*, **221**, 251 (1936).
112. WATSON, H., *Trans. Faraday Soc.*, **32**, 1073 (1936).
113. DEGUILLON, F., *C. R. Acad. Sci., Paris*, **231**, 274 (1950).
114. RAY, S., *Ann. Physik*, **66**, 71 (1921).
115. BOUSSINESQ, J., *Theorie Analitique de Chaleur*, t.II, p.224, Paris (1903).
116. LANDAU, L., LIFSHITZ, E., *Fluid Mechanics,* § 24, Pergamon Press, London (1959).
117. BOGGIO, T., *Rendi Conti*, **16**, 613, 730 (1907).
118. VLADIMIRSKII, V., TERLETSKII, YA., *Zh. Eksp. Teor. Fiz.*, **15**, 258 (1945).
119. BERKOWITCH, S., *Helv. Phys. Acta*, **7**, 170 (1934).
120. SCHMIDT, F., *Ann. Physik*, **61**, 633 (1920).
121. LUNNON, R., *Proc. Roy. Soc.*, **110** A, 302 (1926).
122. KHUDYAKOV, G., CHUKHANOV, Z., *Dokl. Akad. nauk SSSR*, **78**, 681 (1951).
123. KHUDYAKOV, G., *Izv. Akad. nauk SSSR, ser. tekh.*, 1022 (1953).
124. LAWS, J., *Trans. Amer. Geophys. Union*, **22**, 709 (1941).
125. HITSCHFELD, W., *Trans. Amer. Geophys. Union*, **32**, 697 (1951).
126. LAPPLE, C., SHEPHERD, C., *Ind. Eng. Chem.*, **32**, 605 (1940).
127. LAMB, H., *Hydrodynamics*, 6th ed., § 356, Cambridge (1932).
128. YU-CHEN-YANG, *Ann. Physik*, **76**, 333 (1925).
129. ISRAEL, H., *Meteor. Z.*, 36 (1942).
130. KÖNIG, W., *Ann. Physik*, **42**, 353 (1891).
131. BRANDT, O., FREUND, H., HIEDEMANN, E., *Z. Phys.*, **104**, 511 (1937).
132. ZERNOV, W., *Ann. Physik*, **26**, 79 (1908).
133. WAGENSCHEIN, M., *Ann. Physik*, **65**, 461 (1921).
134. CASSEL, H., SCHULZ, H., *Air Pollution*, p.634, New York (1952).
135. ST. CLAIR, H., *Ind. Eng. Chem.*, **41**, 2434 (1949).
136. KING, L., *Proc. Roy. Soc.*, **147** A, 212 (1934).
137. RUDNICK, J., *J. Acoust. Soc. Amer.*, **23**, 633 (1951).
138. WESTERVELT, P., *J. Acoust. Soc. Amer.*, **23**, 312 (1951).
139. ANDRADE, E. N. da C., *Proc. Roy. Soc.*, **134** A, 445 (1931).
140. LAMB, H., *Hydrodynamics* 6th ed., § 362–363, Cambridge (1932).
141. SEWELL, C., *Phil. Trans.*, **210** A, 239 (1910).
142. KYNCH, G., *J. Fluid Mechan.* **5**, 193 (1959).
143. EPSTEIN, P., CARHART, R., *J. Acoust. Soc. Amer.*, **25**, 553 (1953).
144. ALTBERG, W., GOLTZMANN, M., *Phys. Z.*, **26**, 149 (1925).
145. LAIDLER, T., RICHARDSON, E., *J. Acoust. Soc. Amer.*, **9**, 217 (1938).
146. KNUDSEN, V., WILSON, J., ANDERSON, N., *J. Acoust. Soc. Amer.*, **20**, 849 (1948).
147. TYNDALL, J., *Sound*, 6th ed., Lecture VII, London (1895).
148. INGÅRD, H., *J. Acoust. Soc. Amer.*, **25**, 405 (1953).
149. SIEG, H., *Elektr. Nachricht. Techn.*, **17**, 193 (1940).
150. CHANDRASEKHAR, S., *Rev. Mod. Phys.*, **15**, 2 (1943).
151. OSWATICH, K., *Phys. Z.*, **42**, 365 (1941).
152. POYNTING, J., THOMSON, J., *Sound.*, p.169, London (1940).
153. OSEEN, C., *Neuere Methoden und Ergebnisse in der Hydrodynamik*. Leipzig, § 14, 21 (1927).
154. SCHILLER, L., *Handb. Experimentalphys.*, Leipzig, **4** (II), 344 (1932).
155. KIRCHHOFF, G., *Mechanik*. Leipzig, § 251 (1877).
156. LAMB, H., *Hydrodynamics* 6th ed., 137, Cambridge (1932).
157. GORBATCHEV, S., SEVERNYI, A., *Koll. Z.*, **73**, 146 (1935).
158. BJERKNES, V., BJERKNES, J., SOLBERG, H., BERGERON, T., *Physikalische Hydrodynamik*, Berlin, (1933).
159. LANDAU, L., LIFSHITZ, E., *Fluid Mechanics,* § 20, Pergamon Press, London (1959).

160. König, W., *Ann. Physik*, **42**, 549 (1891).
161. Cook, S., *Phil. Mag.*, **3**, 471 (1902).
162. Bazilevich, V., *Coagulation Phenomena in Water Clouds and Mists (Koagulyatsionnyye yavleniya v vodnykh oblakakh i tumanakh)*. Dissertation. Geofian (1941–1943).
163. Thomas, G., *Ann. Physik*, **42**, 1079 (1913).
164. Townsend, J., *Phil. Mag.*, **45**, 471 (1898).
165. Wolodkewitch, N., *Z. Phys.*, **84**, 593 (1933).
166. Fuchs, N., Petryanov, I., *Acta physicochim. URSS*, **3**, 827 (1935).
167. Whytlaw-Gray, R., Patterson, H. S., *Smoke*, p. 164, London (1932).
168. Dessauer, F., *10 Jahre Forschung auf dem physikalisch-medizinischen Grenzgebiet*, Leipzig (1931).
169. Chizhevskii, A., *Curing Lung Diseases with Ionized Air (Lecheniye legochnykh boleznei ionizirovannym vozdukhom)*. Vsesoyuz. ob-vo vrachei-gomeopatov. Moscow (1930).
170. Abbott, R., *Phys. Rev.*, **12**, 381 (1918).
171. Fuchs, N., Petryanov, I., *Koll. Z.*, **65**, 173 (1933).
172. Wells, P., Gerke, R., *J. Amer. Chem. Soc.*, **41**, 312 (1919).
173. White, L., Hill, D., *J. Coll. Sci.*, **3**, 251 (1948).
174. Rosenblum, N., *Tech. Phys. U.S.S.R.*, **4**, 1 (1937).
175. *Modern Developments in Fluid Dynamics*, Ed. by S. Goldstein, § 139, Oxford (1943).
176. Chistov, A., *Otopl. i ventil.*, No. 11, 14 (1935); No. 11/12, 27 (1939).
177. Lissovskii, P., *Acta physicochim. URSS*, **13**, 157 (1940).
178. Fuchs, N., *Izv. Akad. nauk SSSR, ser. geogr. geofiz.*, **11**, 341 (1947).
179. Zimmerschied, W., *Koll. Z.*, **72**, 135 (1935).
180. Chapman, S., *Phys. Rev.*, **52**, 184 (1937).
181. Israel, H., *Gerlands Beitr.*, **31**, 173 (1931).
182. *Handb. Experimentalphys.*, Leipzig, **13** (I), Kap. 5 (1929).
183. Petryanov, I., Tunitskii, N., *Zh. fiz. khim.*, **13**, 1131 (1939).
184. Nolan, P. J., Kenny, P., *J. Atmos. Terr. Phys.*, **3**, 181 (1953).
185. Cochet, R., *Rev. Génér. Electricité.*, **62**, 113 (1953).
186. Munden, D., *J. Appl. Chem.*, **2**, 65 (1952).
187. Wells, W., *Amer. J. Publ. Health*, **23**, 58 (1933).
188. Landau, L., Lifshitz, E., *Fluid Mechanics*, § 17, Pergamon Press, London (1959).
189. Shafir, A., *Gig. i sanit.*, No. 7/8, 1 (1944).
190. Sawyer, K., Walton, W., *J. Sci. Instr.*, **27**, 272 (1950).
191. Smukhnin, P., Kouzov, P., *Centrifugal Cyclone Dust Separators (Tsentrobezhnyye pyleotdeliteli – tsiklony)*. Stroiizdat, Moscow (1935).
192. Kouzov, P., *Removing Dust from Air in Cyclones (Ochistka vozdukha ot pyli v tsiklonakh)*. Liot, Leningrad (1938).
193. Larcombe, H., *Mining Mag.*, **77**, 137, 208, 273, 346 (1947).
194. Rosin, P., Rammler, E., Intelmann, W., *Z. Ver. Deutsch. Ing.*, **76**, 433 (1932).
195. Muhlrad, W., *Genie civile*, **124**, 152 (1947).
196. Lissmann, M., *Chem. Met. Eng.*, **37**, 630 (1930).
197. Shepherd, C., Lapple, C., *Ind. Eng. Chem.*, **31**, 972 (1939).
198. Meldau, R., *Z. Ver. Deutsch. Ing.*, **76**, 1189 (1932).
199. Belyayev, N., *Khim. prom.*, No. 5 (1949).
200. Briggs, W., *Trans. Amer. Inst. Chem. Eng.*, **42**, 511 (1946).
201. Farbar, L., *Trans. Amer. Soc. Mech. Eng.*, **75**, 953 (1953).
202. Zalogin, N., Shukher, S., *Cleaning Flue Gases (Ochistka dymovykh gazov)*. Gosenergoizdat, Moscow–Leningrad (1948).
203. McCabe, L., *Ind. Eng. Chem.*, **45**, No. 12, 105 A (1953).
204. Sell, W., *VDI-Forschungsheft*, No. 347 (1931).
205. Albrecht, F., *Phys. Z.*, **32**, 48 (1931).
206. Mazin, I., *Trud. Tsentr. Aerolog. Observator.*, No. 7, 39 (1952).
207. Syrkin, S., *Theory of Modelling the Trajectories of Solid Particles in Curvilinear Flow (Teoriya modelirovaniya trayektorii tverdykh chastits v krivolineinom potoke)*. Kotloturbinnyi institut (1934).
208. Rosin, P., Kayser, H., Rammler, E., *Z. tech. Phys.*, **16**, 205 (1935).
209. Barth, R., Barth, W., *Rauch und Staub*, **22**, 93 (1932).
210. Volkov, P., *Development of a Method for Studying the Motion of Solid Particles in Models (Razrabotka metodiki issledovaniya dvizheniya tverdykh chastits na modelyakh)*. Kotloturbinnyi institut (1936).

211. WALTON, W., *Symposium on Particle Size Analysis*, 136, London (1947).
212. DAVIES, C. N., *Symposium on Particle Size Analysis*, p. 25, London (1947).
213. MAY, K., DRUETT, H., *Brit. J. Ind. Med.*, **10**, 142 (1953).
214. HIRST, J., *Ann. Appl. Biol.*, **39**, 257 (1952).
215. DRINKER, P., HATCH, T., *Industrial Dust* (1936).
216. BOURDILLON, R., *J. Hygiene*, **41**, 197 (1941).
217. MAY, K., *J. Sci. Instr.*, **22**, 187 (1947).
218. DAVIES, C. N., AYLWARD, M., LEACEY, D., *Arch. Ind. Hyg. Occ. Med.*, **4**, 354 (1951).
219. DAVIES, C. N., AYLWARD, M., *Proc. Phys. Soc.*, **64** B, 889 (1951).
220. LEVIN, L., *Dokl. Akad. nauk SSSR*, **91**, 1329 (1953).
221. WILCOX, J., *Arch. Ind. Hyg. Occ. Med.*, **7**, 376 (1953).
222. RANZ, W., WONG, J., *Ind. Eng. Chem.*, **44**, 1371 (1952).
223. PIK, V., SHURCHILOV, L., *Gig. truda i bezop.*, No. 4, 22, (1933).
224. JORDAN, D., *Brit. J. Appl. Phys. Suppl.*, No. 3, 194 (1954).
225. BIRSE, E., ROBERTS, J., *Trans. Faraday Soc.*, **44**, 273 (1948).
226. FICKLEN, J., GOOLDEN, L., *Science*, **85**, 587 (1937).
227. FAIRS, G., *J. Soc. Chem. Ind. (London)* **60**, 141 (1941).
228. LAMB, H., *Hydrodynamics*, 6th ed., § 68, Cambridge (1932).
229. LAMB, H., *Hydrodynamics*, 6th ed., § 343, Cambridge (1932).
230. LAMB, H., *Hydrodynamics*, 6th ed., § 96, Cambridge (1932).
231. LAMB, H., *Hydrodynamics*, 6th ed., § 337, Cambridge (1932).
232. LAMB, H., *Hydrodynamics*, 6th ed., § 171, Cambridge (1932).
233. LAMB, H., *Hydrodynamics*, 6th ed., § 342, Cambridge (1932).
234. LANGMUIR, I., BLODGETT, K., Army Air Force, Tech. Rep., No. 5418 (1946).
235. LAPPLE, C., *Chem. Engr.*, **58**, 144 (1951).
236. LANDAHL, H., HERMANN, K., *J. Coll. Sci.*, **4**, 103 (1949).
237. THOM, A., *Proc. Roy. Soc.*, **141** A, 651 (1933).
238. GEIST, J., YORK, J., GRANGER, G., *Ind. Eng. Chem.*, **43**, 1371 (1951).
239. LANGMUIR, I., *J. Meteor.*, **5**, 175 (1948).
240. TAYLOR, G. I., Aeron. Res. Comm. Rep. No. 4350, London (1940).
241. BOWDEN, F., TABOR, D., *Friction and Lubrication of Solids*, Oxford (1956).
242. RAYLEIGH, J., *Proc. Roy. Soc.*, **34**, 130 (1882); *Phil. Mag.*, **48**, 321 (1899).
243. FUCHS, N., *Dokl. Akad. nauk SSSR*, **81**, 1043 (1951).
244. BRUN, E. *et al.*, *C.R. Acad. Sci.*, *Paris* **224**, 1518 (1947).
245. DESSENS, H., *Quart. J. Roy. Meteor. Soc.*, **75**, 23 (1949).
246. *Modern Developments in Fluid Dynamics*, Ed. by S. GOLDSTEIN, § 48, Oxford (1943).
247. SCHLICHTING, H., *Grenzschicht-Theorie*. Karlsruhe, chap. 9, § 4; chap. 10, § 2 (1951).
248. YEOMANS, A. *et al.*, *J. Econ. Entom.*, **42**, 591 (1949).
249. KHIMACH, M., SHISHKIN, N., *Trudy* GGO, No. 31 (93); 29 (1951).
250. BOUCHER, P., *Chal. et Industrie*, **33**, 363 (1952).
251. FRANK, P., MISES, R., *Differential- und Integralgleichungen der Mechanik und Physik*, Vol. II, Chap. 13, New York (1943).
252. CAWOOD, W., *Trans. Faraday Soc.*, **32**, 1068 (1936).
253. LEONTOVICH, M., *Statistical Physics (Statisticheskaya fizika)*. Gostekhizdat, Moscow-Leningrad, Chap. 6 (1944).
254. EINSTEIN, A., *Ann. Physik*, **17**, 549 (1905).
255. FRANK, P., *Ann. Physik*, **52**, 323 (1917).
256. SMOLUCHOWSKI, M., *Ann. Physik*, **21**, 756 (1906).
257. NEWALL, H., *Phil. Mag.*, **20**, 30 (1885).
258. ORNSTEIN, L., *Proc. Amsterd. Acad. Sci.*, **21**, 96 (1918).
259. BÄCKER, B., *Z. Phys.*, **38**, 609 (1926).
260. ZEILINGER, F., *Ann. Physik*, **75**, 403 (1924).
261. FÜRTH, R., *Ann. Physik*, **59**, 409 (1919).
262. LANDMANN, S., *Z. Phys.*, **27**, 237 (1924).
263. SCHRÖDINGER, E., *Phys. Z.*, **16**, 289 (1915).
264. FLETCHER, H., *Phys. Rev.*, **4**, 440 (1914).
265. SMOLUCHOWSKI, M., *Phys. Z.*, **16**, 320 (1915).
266. SCHMID, E., *Phys. Z.*, **22**, 438 (1921).
267. SCHMID, E., *Wien. Ber.*, **129**, IIa, 813 (1920).

268. WINKEL, A., WITZMANN, H., *Z. Elektrochem.*, **46**, 181 (1940).
269. SMOLUCHOWSKI, M., *Ann. Physik*, **48**, 1104 (1915).
270. KOLMOGOROFF, A., *Grundbegriffe der Wahrscheinlichkeitsrechnung*, Berlin (1933).
271. KOLMOGOROFF, A., LEONTOWITSCH, M., *Phys. Z. Sowjetunion*, **4**, 1 (1933).
272. FRANK, P., MISES, R., *Differential- und Integralgleichungen der Mechanik und Physik*, Vol. II, Chap. 13, § 2, New York (1943).
273. SMOLUCHOWSKI, M., *Phys. Z.*, **17**, 588 (1916).
274. FRANK, P., MISES, R., *Differential- und Integralgleichungen der Mechanik und Physik*, Vol. II, Chap. 13, § 3, 1, New York (1943).
275. FRANK, P., MISES, R., *Differential- und Integralgleichungen der Mechanik und Physik*, Vol. II, Chap. 14, § 2, 5 New York (1943).
276. BUCHWALD, E., *Ann. Physik*, **66**, 1 (1921).
277. DERJAGUIN, B., VLASENKO, G., *Dokl. Akad. nauk SSSR*, **63**, 155 (1948).
278. RADUSHKEVICH, L., *Zh. fiz. khim.*, **13**, 1322 (1939).
279. DAVIES, C. N., *Proc. Roy. Soc.*, **200** A, 100 (1949).
280. SMOLUCHOWSKI, M., *Z. Phys. Chem.*, **92**, 140 (1917).
281. JAKOB, M., *Heat Transfer*, **1**, p. 267, New York (1949).
282. GRAETZ, L., *Ann. Physik*, **18**, 79 (1883).
283. NUSSELT, W., *Z. Ver. Deutsch. Ing.*, **54**, 1154 (1910).
284. TOWNSEND, J., *Phil. Trans.*, **193**, 129 (1900).
285. GORMLEY, P., KENNEDY, M., *Proc. Roy. Irish Acad.*, **52** A, 163 (1949).
286. LEVEQUE, M., *Ann. Mines*, **13**, 276 (1928).
287. KAISER, E., *Ann. Physik*, **53**, 667 (1894).
288. RADUSHKEVICH, L., *Acta physicochim. URSS*, **6**, 161 (1937).
289. NOLAN, J. J., GUERRINI, N., *Trans. Faraday Soc.*, **32**, 1175 (1936).
290. FRÖSSLING, N., *Gerlands Beitr.*, **52**, 170 (1938).
291. MAHAJAN, L., *Koll. Z.*, **65**, 20 (1933); **69**, 16 (1934); *Phil. Mag.*, **7**, 247 (1929); **10**, 383 (1930); *Z. Phys.*, **79**, 389 (1932); **81**, 605 (1933); **84**, 676 (1934); **90**, 663 (1934).
292. POLLAK, L., MURPHY, T., *Arch. Meteor. Geophys. Bioklim.* **5**, 100 (1952).
293. LEVICH, V., *Physico-Chemical Hydrodynamics (Fiziko-khimicheskaya gidrodinamika)*, Acad. Sci., U.S.S.R., Moscow (1952), § 14.
294. ROBERTS, O., *Proc. Roy. Soc.*, **104** A, 640 (1923).
295. WITZMANN, H., *Z. Elektrochem.*, **46**, 313 (1940); *VDI Verfahrenstechnik*, No. 4, 107 (1940).
296. BORN, H., ZIMMER, K., *Naturwiss.*, **28**, 447 (1940).
297. KATZ, S., MACRAE, D., *J. Phys. Coll. Chem.*, **52**, 695 (1948).
298. KUCHERUK, V., SEROV, V., *Industrial Dust and Dust-removing Ventilation (Promyshlennaya pyl' i obespylivayushchaya ventilyatsiya)*. Stroiizdat, Moscow (1941).
299. WEISCHHAUS, L., *Chem. Eng.*, **54**, No. 8, 113 (1947).
300. SILVERMAN, L., *Chem. Eng. Progr.*, **47**, 462 (1951).
301. PRING, R., *Air Pollution*, p. 280, New York (1952).
302. KRAMERS, H., *Physica*, **7**, 290 (1940).
303. ROWLEY, F., JORDAN, R., *Heat. Pip. Air Cond.*, 469 (1938).
304. KAUFMANN, A., *Z. Ver. Deutsch. Ing.*, **80**, 593 (1936).
305. *Handbook on Aerosols*, Washington (1950).
306. LAMER, V., *Air Pollution*, p. 607, New York (1952).
307. SMITH, W., STAFFORD, E., *Air Pollution*, p. 264, New York (1952).
308. CLAYTON, A. *et al.*, *Ind. Eng. Chem.*, **46**, 176 (1954).
309. PEROT, J., *Paper Trade J.*, **124**, No. 25, 54 (1947).
310. RAMSKILL, E., ANDERSON, W., *J. Coll. Sci.*, **6**, 416 (1951).
311. EISENBART, H., *Die Gasmaske*, **8**, 72 (1936).
312. STAFFORD, E., SMITH, W., *Ind. Eng. Chem.*, **43**, 1346 (1951).
313. ROSSANO, A., SILVERMAN, L., *Heat. Ventilating*, **51**, No. 5, 102 (1954).
314. WEBER, R., *Angew. Chem.*, **20** B, 335 (1948).
315. DAUTREBANDE, L. *et al.*, *C.R. Acad. Sci., Paris*, **205**, 156, 240 (1937).
316. DAUTREBANDE, L. *et al.*, *C.R. Acad. Sci., Paris*, **205**, 329 (1937).
317. TACHIBANA, T., *Chem. Abstr.*, **41**, 3343f (1947).
318. GOYER, G., GRUEN, R., LAMER, V., *J. Phys. Chem.*, **58**, 137 (1954).
319. FAIRS, G., *J. Soc. Chem. Ind.* (*London*) **60**, 141 (1941).
320. DAUTREBANDE *et al.*, *J. Ind. Hyg. Tox.*, **30**, 103 (1948).

321. WILSON, J., LAMER, V., *J. Ind. Hyg. Tox.*, **30**, 265 (1948).
322. LANDAHL, H. *et al.*, *Arch. Ind. Hyg. Occ. Med.*, **3**, 359 (1951).
323. LANDAHL, H., BLACK, S., *J. Ind. Hyg. Tox.*, **29**, 269 (1947).
324. DAVIES, C. N., *Proc. Roy. Soc.*, **133** B, 282 (1946).
325. GUTMANN, M., *Die Pollenallergie*, München (1929).
326. DAUTREBANDE, L., *Physiol. Rev.*, **32**, 214 (1952).
327. BUCKLAND, F. *et al.*, *Nature, Lond.*, **166**, 354 (1950).
328. SHOSHKES, M. *et al.*, *Arch. Ind. Hyg. Occ. Med.*, **1**, 20 (1950).
329. VAN WIJK, A., PATTERSON, H., *J. Ind. Hyg. Tox.*, **22**, 31 (1940).
330. BROWN, J., HATCH, T., *Amer. J. Publ. Health*, **40** A, 450 (1950).
331. DAVIES, C. N., *Brit. J. Ind. Med.*, **9**, 120 (1952).
332. LIFSHITS, I., LYKHINA, YE, ERENBURG, G., *Gig. i sanit.* No. 10, 17 (1948).
333. FINDEISEN, W., *Arch. Gesamt. Physiol.*, **236**, 367 (1935).
334. LANDAHL, H., *Bull. Mathem. Biophys.*, **12**, 43 (1950).
335. WILSON, J., *J. Coll. Sci.*, **2**, 27 (1947).
336. DAUTREBANDE, L., *Aérosols Médicamenteux*, Paris (1946).
337. TUNITSKII, N., TIKHOMIROV, M., PETRYANOV, I., *Zh. tekh. fiz.*, **10**, 1723 (1940).
338. LEVICH, V., *Physico-chemical Hydrodynamics (Fiziko-khimicheskaya gidrodinamika)*. Acad. Sci. U.S.S.R, Moscow, Chap. 6 (1952).
339. KREVELEN, D., *Chem. Eng. Progr.*, **46**, 29 (1950).
340. REMY, H., SEEMANN, W., *Koll. Z.*, **72**, 3 (1935).
341. REMY, H., RUHLAND, K., *Z. anorg. Chem.*, **139**, 51 (1924).
342. REMY, H., *Trans. Faraday Soc.*, **32**, 1185 (1936).
343. REMY, H., BEHRE, C., *Koll. Z.*, **71**, 129 (1935).
344. REMY, H., FINNERN, H., *Z. anorg. Chem.*, **159**, 241 (1926).
345. LUCHNSKII, G., *Zh. fiz. khim.*, **13**, 302 (1939).
346. ALEKSEEVA, B., ANDRONOV, M., *Lab. Prakt. U.S.S.R.*, **16**, No. 1, 18 (1941).
347. BRONSIKY, D., DIWOKY, F., *Chem. Met. Eng.*, **47**, 541 (1940).
348. BICKMORE, J., LEVY, M., HALL, J., *Photogr. Sci. Eng.*, **4**, 37 (1960).
349. GANS, R., *Ann. Physik*, **86**, 628 (1928).
350. PERRIN, F., *C. R. Acad. Sci., Paris*, **181**, 514 (1925).
351. COHN, E., *Das elektromagnetische Feld.* Leipzig, S. 110 (1900).
352. BLOCH, E., *C. R. Acad. Sci., Paris*, **146**' 970 (1908).
353. ZEEMANN, P., HOOGENBOOM, C., *Phys. Z.*, **13**, 913 (1912).
354. WALMSLEY, H., *Phil. Mag.*, **7**, 1097 (1927).
355. O'KONSKI, C., THACHER, C., *J. Phys. Chem.*, **57**, 955 (1953).
356. PRANDTL, L., *Führer durch Strömungslehre*, Braunschweig, Kap. 5, § 18 (1949).
357. GILLESPIE, T., LANGSTROTH, G., *Canad. J. Chem.*, **29**, 133, 201 (1951).
358. VIGDORCHIK, E., *Trudy Leningr. inst. okhrany truda*, **6**, No. 7 (1933).
359. SHIFRIN, K., GORDON, I., FAINSHTEIN, M., *Izv. Akad. nauk SSSR, ser. geogr. geofiz.*, **3**, 238 (1949).
360. GILLESPIE, T., *Proc. Roy. Soc.*, **216** A, 569 (1951).
361. LANGSTROTH, G., GILLESPIE, T., *Canad. J. Research*, **25** B, 455 (1947).
362. KOLMOGOROFF, A., *Dokl. Akad. nauk SSSR*, **30**, 299 (1941).
363. SIMMONS, L., SALTER, C., *Proc. Roy. Soc.*, **165** A, 73 (1938).
364. OBUKHOV, A., YAGLOM, A., *Prikl. mat. mekh.*, **15**, 1 (1951).
365. OBUKHOV., A., *Dokl. Akad. nauk SSSR*, **67**, 643 (1949).
366. PRIESTLEY, C., SHEPPARD, P., *Quart. J. Roy. Meteor. Soc.*, **78**, 488 (1952).
367. HEISENBERG, W., *Z. Phys.*, **124**, 628 (1948).
368. MACCREADY, P., *J. Meteor.*, **10**, 434 (1953).
369. SCHMIDT, W., *Der Massenaustausch in freier Luft und verwandte Erscheinungen* (1925).
370. *Izv. Akad. nauk SSSR, ser. tekh.*, 1742, 1831, 1848 (1952).
371. GREGORY, P., *Ann. Appl. Biol.*, **38**, 357 (1951).
372. BLOOMFIELD, J., DALAVALLE, J., *Publ. Health Bull.* No. 217 (1935).
373. VANONI, V., *Proc. Amer. Soc. Civil Eng.*, **70**, 793 (1944).
374. KALINSKE, A., PIEN, C., *Trans. Amer. Geophys. Union*, **24**, 530 (1943).
375. KALINUSHKIN, M., *Trudy Z A G I*, No. 266 (1936).
376. SHERWOOD, T., WOERTZ, B., *Ind. Eng. Chem.*, **31**, 1034 (1939).
377. LINDGREN, E., *Arkiv Fysik*, **7**, 293 (1954).

378. FILATOV, B., *Koll. Zh.*, **16**, 65 (1954).
379. TER LINDEN, A., *Chemie-Ing-Techn.*, **25**, 328 (1952).
380. FUCHS, N., *Zh. tekh. fiz.*, **21**, 704 (1951).
381. BARENBLATT, G., *Prikl. mat. mekh.*, **17**, 261 (1953); **19**, 61 (1955).
382. FAGE, F., TOWNEND, H., *Proc. Roy. Soc.*, **135** A, 656 (1932).
383. MINSKII, E., *Dokl. Akad. nauk SSSR*, **49**, 329 (1945).
384. SCHMIDT, W., *Ind. Eng. Chem.*, **41**, 2428 (1949).
385. PRANDTL, L., *Führer durch Strömungslehre*, Braunschweig, Kap. **3**, § 5 (1949).
386. PRANDTL, L., *Führer durch Strömungslehre*, Braunschweig, Kap. **3**, § 4 (1949).
387. *Modern Developments in Fluid Dynamics*, Ed. by S. GOLDSTEIN, § 280, Oxford (1943).
388. LEVICH, V., *Physico-chemical Hydrodynamics (Fiziko-khimicheskaya godrodinamika). Acad. Sci. U.S.S.R.*, Moscow Chap. 3 (1952).
389. *Modern Developments in Fluid Dynamics*, Ed. by S. GOLDSTEIN, § 154, 155, Oxford (1943).
390. ALEXANDER, L., COLDREN, C., *Ind. Eng. Chem.*, **43**, 1325 (1951).
391. TOWNEND, H., *Proc. Roy. Soc.*, **145** A, 180 (1934).
392. SCHUBAUER, G., *J. Appl. Phys.*, **25**, 188 (1954).
393. ZHUKOVSKII, N., *Snow Drifts and Silting of Rivers (O snezhnykh zanosakh i zailenii rek).* Kiev (1911).
394. ASSET, G., PURY, D., *Arch. Ind. Hyg. Occ. Med.*, **9**, 273 (1954).
395. SHELEIKHOVSKII, G., *Smoke in Towns (Zadymleniye gorodov)*. Izd. Ministerstva kommun. khoz-va, Moscow-Leningrad (1949).
396. ANDREEV, P., *The Spreading in Air of Gas Emitted by Industrial Undertakings (Rasseyaniye v vozudukhe gazov, vybrasyvayemykh promyshlennymi predpriyatiyami)*. Gosstroiizdat, Moscow (1952).
397. SUTTON, O., *Micrometeorology*. New York (1953).
398. SUTTON, O., *Atmospheric Turbulence*. London (1949).
399. DOBSON, G., Nat. Advis. Comm. Aeronaut., Rep, No. 67 (1949).
400. RICHARDSON, L., *Proc. Roy. Soc.*, **104** A, 640 (1923).
401. YUDIN, M., *Dokl. Akad. nauk SSSR*, **51**, 99 (1946).
402. STONE, N., *Phil. Mag.* **9**, 610 (1930)
403. SUTTON, O., *Proc. Roy. Soc.*, **135** A, 143 (1932).
404. CALDER, K., *Quart. J. Mech. Appl. Math.*, **2**, 153 (1949).
405. SUTTON, O., *Quart. J. Roy. Meteor. Soc.*, **73**, 257, 426 (1947).
406. LAIKHTMAN, D., CHUDNOVSKII, A., *The Physics of the Atmosphere near the Earth's Surface (Fizika prizemnogo sloya atmosfery)*. Gostekhizdat, Leningrad-Moscow, Chap. 5 (1949).
407. TEVEROVSKII, E., *Coll. Papers on New Ideas in the Study of Aerosols (Sb. "Novyye idei v oblasti izucheniya aerozolei")*. Acad. Sci. U.S.S.R., Moscow, p. 108 (1949).
408. CALDER, K., *Air Pollution*, p. 787 New York (1952).
409. DEACON E., *Quart. J. Roy. Meteor. Soc.*, **75**, 89 (1949).
410. PASQUILL, E., *Proc. Roy. Soc.*, **202** A, 150 (1950).
411. FROST, R., *Proc. Roy. Soc.*, **186** A, 20 (1946).
412. BUDYKO, M., LAIKHTMAN, D., TIMOFEYEV, M. *Meteor. i Gidrol.*, No. 3, 27 (1953).
413. BOSANQUET, C., PEARSON, J., *Trans. Faraday Soc.*, **32**, 1249 (1936).
414. GOSLINE, C., *Chem. Eng. Progr.*, **48**, 165 (1952).
415. THOMAS, D., HILL, G., *Ind. Eng. Chem.*, **41**, 2403 (1949).
416. CHURCH, P., *Ind. Eng. Chem.*, **41**, 2753 (1949).
417. JOHNSTONE, H. F., WINSCHE, W., SMITH, L., *Chem. Rev.*, **44**, 353 (1949).
418. DAVIES, D. R., *Proc. Cambr. Phil. Soc.*, **46**, 500 (1950).
419. SMOLUCHOWSKI, M., *Z. Phys. Chem.*, **92**, 129 (1917).
420. HARPER, W., *Trans. Faraday Soc.*, **30**, 636 (1934).
421. FUCHS, N., *Z. Phys. Chem.*, **171** A, 199 (1934).
422. SMOLUCHOWSKI, M., *Z. Phys. Chem.*, **92**, 151 (1917).
423. MÜLLER, H., *Kolloidchem. Beih.*, **26**, 257 (1928).
424. TIKHOMIROV, M., TUNITSKII, N., PETRYANOV, I., *Acta physicochim. URSS*, **17**, 185 (1942).
425. TODES, O., *Symposium on Problems of Kinetics and Catalysis (Sb. Problemy kinetiki i kataliza)*. ONTI, Moscow–Leningrad, **7** 137 (1949).
426. PSHENAI-SEVERIN, S., *Dokl. Akad. nauk SSSR.*, **94**, 865 (1954).
427. WHYTLAW-GRAY, R., PATTERSON, H. S., *Smoke*, p. 48, London (1932).
428. PATTERSON, H. S., CAWOOD, W., *Proc. Roy. Soc.*, **136** A, 538 (1932).

429. Smoluchowski, M., *Z. Phys. Chem.*, **92**, 144 (1917).
430. Tunitskii, N., *Zh. Eksp. Teor. Fiz.*, **8**, 418 (1938).
431. Müller, H., *Kolloidchem. Beih.*, **27**, 223 part 3, (1928).
432. Kottler, F., *Handb. Phys.*, Berlin, **12**, Kap. 4, § 86–88 (1927).
433. Przibram, K., *Handb. Experimentalphys.*, Leipzig, **8** (II) 703 (1929).
434. Beischer, D., Winkel, A., *Z. Phys. Chem.*, **176** A, 1 (1936).
435. Samokhvalov, M., Kozhukhova, O., *Zh. fiz. khim.*, **8**, 420, (1936).
436. Radushkevich, L., *Zh. fiz. khim.*, **9**, 6 (1937).
437. Nolan, P. J., Kennan, E., *Proc. Roy. Irish Acad.*, **52** A, 171 (1949).
438. Fuchs, N., *Z. Phys.*, **89**, 736 (1934).
439. Kottler, F., *Handb. Phys.*, Berlin, **12**, § 82 (1927).
440. Whytlaw-Gray, R., Patterson, H. S., *Smoke*, p. 166, London (1932).
441. Wiegand, A., Frankenberger, E., *Phys. Z.*, **31**, 204 (1930).
442. Hamaker, H., *Physica*, **4**, 1058 (1937).
443. Bradley, R., *Phil. Mag.*, **13**, 853 (1932).
444. Derjaguin, B., Abrikosova, I., *Disc. Faraday Soc.* **18**, 36 (1954).
445. Krasny-Ergen, W., *Ann. Physik*, **27**, 459 (1936).
446. Winkel, A., *VDI Verfahrenstechnik*, 25 (1941).
447. Gunn, R., *J. Geophys. Res.*, **55**, 171 (1950).
448. Schweckendiek, O., *Z. Naturforsch.*, **5** a, 397 (1950).
449. Winkel, A., *Z. Angew. Chem.*, **54**, 152 (1941).
450. Beischer, D., *Z. Elektrochem.*, **44**, 375 (1938).
451. Whytlaw-Gray, R., Patterson, H. S., *Smoke*, Chap. 8, London (1932).
452. Winkel, A., Haul, R., *Z. Elektrochem.*, **44**, 823 (1938).
453. Beischer, D., Winkel, A., *Naturwiss.*, **25**, 420 (1937).
454. Faxen, H., *Arkiv Mat., Astr. Fys.* **17**, No. 27 (1923).
455. Dallavalle, J. *et al.*, *Brit. J. Appl. Phys. Suppl.*, No. 3, 198 (1954).
456. Wood, W., Loomis, A., *Phil. Mag.*, **4**, 433 (1927).
457. *Trans. Faraday Soc.* **32**, 1122 (1936).
458. Brandt, O., Freund, H., Hiedemann, E., *Koll. Z.*, **77**, 103, 168 (1936).
459. Parker, R., *Trans. Faraday Soc.*, **32**, 1115 (1936).
460. Andrade, E. N. daC., *Trans. Faraday Soc.*, **32**, 1111 (1936).
461. Clair, H. St., Bur. Mines, Rep. Investig., No. 3400 (1938).
462. Danser, H., Neumann, F., *Ind. Eng. Chem.*, **41**, 4439 (1949).
463. Brandt, O., Hiedemann, E., *Trans. Faraday Soc.*, **32**, 1101 (1936).
464. Stokes, C., *Chem. Eng. Progr.*, **46**, 423 (1950).
465. Brandt, O., *Koll. Z.*, **76**, 272 (1936).
466. O'Mara, R., *Combustion*, **21**, No. 4, 40 (1950).
467. Soderberg, C., *Iron Steel Eng.*, **29**, 87 (1952).
468. Clair, H. St., *Air Pollution*, p. 382, New York (1952).
469. Kubanskii, P., *Zh. tekh. fiz.*, **24**, 1044 (1954).
470. Frenkel, J. *The theory of Electrical Phenomena in the Atmosphere (Teoriya yavlenii atmosfernogo elektrichestva)*. Chap. 3. Gostekhizdat, Leningrad–Moscow, (1949).
471. Findeisen, W., *Meteor Z.*, **56**, 356 (1939).
472. Findeisen, W., *Gerland Beitr.*, **35**, 295 (1932).
473. Tverskaya, N., Yudin, I., *Trud. Leningr. gidromet. inst.*, No. 5–6 (1956).
474. Shishkin, N., Inform, sborn, No. 1: Gidrometsl., Gidrometizdat, Moscow, 47 (1951).
475. Gunn, R., Hitschfeld, W., *J. Meteor.*, **8**, 7 (1951).
476. Frenkel, J., Shishkin, N., *Izv. Akad. nauk SSSR, ser. geogr. geofiz.*, **10**, 301 (1946).
477. Shishkin, N., *Clouds, Rainfall and Thundercloud Electricity (Oblaka, osadki i grozovoye elektrichestvo)*. Gostekhizdat, Moscow, Chap. 6 (1954).
478. Dautrebande, L. *et al.*, *J. Ind. Hyg. Tox.*, **30**, 108 (1948).
479. Dautrebande, L., Capps, R., *Arch. Internat. Pharmacodyn. Therap.*, **82**, 512 (1950).
480. Avy, A., *Arch. Malad. Profess.*, **14**, 342 (1953).
481. Albinsson, A., *Arkiv Kemi*, **6**, 293 (1953).
482. Dautrebande, L. *et al.*, *Arch. Internat. Pharmacodyn. Therap.*, **80**, 413 (1949).
483. Johnstone, H. F., Roberts, M., *Ind. Eng. Chem.*, **41**, 2417 (1949).
484. Johnstone, H. F., *et al.*, *Ind. Eng. Chem.*, **46**, 1601 (1954).
485. Ekman, F., Johnstone, H., *Ind. Eng. Chem.*, **43**, 1358 (1951).

486. SCHAUER, P., *Ind. Eng. Chem.*, **43**, 1532, (1951).
487. MÜLLER, H. *Kolloidchem. Beih.*, **27**, 223, part 5 (1928).
488. LEVIN, L., *Dokl. Akad. nauk. SSSR*, **94**, 467 (1954).
489. PAUTHENIER, M., COCHET, R., *Rev. Génér. Electricité*, **62**, 255 (1953).
490. KÖHLER, H., *Meteor. Z.*, **42**, 137 (1925).
491. RICHARDSON, E., *Acustica*, **2**, 141 (1952).
492. SMOLUCHOWSKI, M., *Z. Phys. Chem.*, **92**, 156 (1917).
493. MANLEY, R., MASON, S., *J. Coll. Sci.*, **7**, 354 (1952).
494. LEVICH, V., *Dokl. Akad. nauk SSSR*, **99**, 809 (1954).
495. TUNITSKII, N., *Zh. fiz. khim.*, **20**, 1136 (1946).
496. LEVICH, V., *Dokl. Akad. nauk SSSR*, **99**, 1041 (1954).
497. DADY, G., *C. R. Acad. Sci., Paris*, **225**, 1349 (1947).
498. SWINBANK, W., *Nature, Lond.*, **159**, 849 (1947).
499. FUCHS, N., *Acta physicochim. URSS*, **3**, 819 (1935).
500. DERYAGIN, B., *Koll. Z.*, **69**, 155 (1934).
501. MACFARLANE, J., TABOR, D., *Proc. 7th Int. Congr. Appl. Mech.* p. 335, (1948).
502. BRADLEY, R., *Trans. Faraday Soc.*, **32**, 1088 (1936).
503. RUMYANTSEVA, E., *Zh. fiz. khim.*, **2**, 283 (1931).
504. URAZOVSKII, S., KUZ'MENKO, S., *Zh. fiz. khim.*, **6**, 897 (1935).
505. SAICHUK, V., NARSKIKH, O., *Koll. Zh.*, **2**, 841 (1936).
506. ANDREEV, N., KIBIRKSHTIS, S., *Zh. obshch. khim.*, **6**, 1698 (1936).
507. SMIRNOV, L., SOLNTSEVA, V., *Koll. Zh.*, **4**, 401 (1938).
508. RADUSHKEVICH, L., CHUGUNOVA, O., *Zh. fiz. khim.*, **12**, 34 (1938).
509. PETRYANOV, I., TUNITSKII, N., TIKHOMIROV, M., *Zh. fiz. khim.*, **15**, 811 (1941).
510. ARTEMOV, I., *Dokl. Akad. nauk SSSR*, **50**, 289 (1945).
511. AGANIN, M., *Zh. geofiz.*, **5**, 408 (1935); *Izv. Akad. nauk SSSR, ser. geogr. geofiz.*, **3**, 305 (1940).
512. GORBATCHEV, S., MUSTEL, E., *Koll. Z.*, **73**, 21 (1935).
513. GORBATCHEV, S., NIKIFOROVA, V., *Koll. Z.*, **73**, 14 (1935).
514. TVERSKAYA, N., *Trudy Glavnoi geofiz. oberservatorii*, No. 47, 112 (1954).
515. DERYAGIN, B., PROKHOROV, P., *Dokl. Akad. nauk SSSR*, **54**, 511 (1946).
516. PROKHOROV, P., *Zh. fiz. khim.*, **21**, 1045 (1947).
517. PROKHOROV, P., YASHIN, V., *Koll. Zh.*, **10**, 122 (1948).
518. PROKHOROV, P., LEONOV, L., *Koll. Zh.*, **14**, 66 (1952).
519. FEDOSEEV, V., MANAKIN, B., DOMENTIANOVA, Z., *Koli. Zh.*, **14**, 274, 470 (1952).
520. GONELL, H., *Koll. Z.*, **71**, 31 (1935).
521. BURGOYNE, J., RASBACH, D., *Fuel*, **28**, 281 (1949).
522. SYRKIN, S., *Study of the Conditions for Pneumatic Conveyance of Dust through Pipes (Issledovaniye uslovii pnevmaticheskogo transporta pyli po trubam)*. Kotloturbinnyi Inst. (1935).
523. RUMPF, H., *Chemie-Ing-Techn.*, **25**, 317 (1953).
524. AVERBUKH, YA., SHABALIN, K., *Khim. prom.*, 290 (1947).
525. SOKOLOV, N., *Dunes, their Formation, Development and Internal Structure (Dyuny, ikh obrazovaniye, razvitiye i vnutrenneye stroyeniye)*. St. Petersburg (1894).
526. BAGNOLD, A., *Proc. Roy. Soc.*, **157** A, 594 (1936).
527. BAGNOLD, A., *The Physics of Blown Sand and Desert Dunes*, London (1941).
528. CHEPIL, W., *Soil Sci.*, **60**, 305, 395 (1945).
529. CHEPIL, W., *Soil Sci.*, **71**, 141 (1951).
530. BAGNOLD, A., *Proc. Roy. Soc.*, **167** A, 282 (1938).
531. SOKOLOV, N., Quoted by H. JEFFREYS. *Proc. Cambr. Phil. Soc.*, **25**, 272 (1929).
532. YAKUBOV, T., *Wind Erosion of Soil (Vetrovaya eroziya pochvy)*. Sel'khozgiz (1946).
533. DANOVSKII, L., *Snow Fences on the Railoads of Western Siberia (Snegozashchitnyye zabory na dorogakh Zapadnoi Sibiri)*. Novosibirsk (1950).
534. CHAPLYGIN, S., *Collected works (Sobrannyye sochineniy)* Gostekhizdat, Moscow–Leningrad, Chap. 2, p. 537 (1948).
535. JEFFREYS, H., *Proc. Cambr. Phil. Soc.*, **25**, 272 (1929).
536. LOSIYEVSKII, A., *Trud. Nauchn.-issled. inst. vodnogo transporta*, No. 121 (1934).
537. EINSTEIN, A., EL-SAMNI. *Rev. Mod. Phys.*, **21**, 520 (1949).
538. TURITSYN, B., *Otopl. i ventil.*, No. 1 (1935).
539. YOUNG, R., *Heat. Vent. Eng.*, **26**. 469 (1953).
540. DERJAGUIN B., RATNER, S., FUTRAN, M., *Dokl. Akad. nauk SSSR*, **92**, 1137 (1953).

541. MILLER, J. *et al.*, *Science*, **107**, 144 (1948).
542. WILHELM, R., KWAUK, M., *Chem. Eng. Progr.*, **44**, 195 (1948).
543. ERGUN, S., ORNING, A., *Ind. Eng. Chem.*, **41**, 1179 (1949).
544. LEVA, M. *et al.*, *Chem. Eng. Progr.*, **44**, 511, 619 (1948).
545. DALLAVALLE, J., *Micromeritics*, p. 118, New York (1943).
546. HAHN, F., *Dispersoidanalyse*, Dresden, § 42 (1927).
547. MATHESON, G., HERBST, W., HOLT, H., *Ind. Eng. Chem.*, **41**, 1099 (1949).
548. LEWIS, W., GILLILAND, E., BAUER, W., *Ind. Eng. Chem.*, **41**, 1104 (1949).
549. PARENT, J. *et al.*, *Chem. Eng. Progr.*, **43**, 429 (1947).
550. MORSE, R., BALLOU, C., *Chem. Eng. Progr.*, **47**, 199 (1952).
551. WICKE, E., HEDDEN, K., *Chemie-Ing-Techn.*, **24**, 82 (1952).
552. MATHESON, G. *et al.*, *Ind. Eng. Chem.*, **41**, 1099 (1949).
553. BRÖTZ, W., *Chemie-Ing-Techn.*, **24**, 60 (1952).
554. LEVA, M., GRUMMER, M., *Chem. Eng. Progr.*, **48**, 307 (1952).
555. LEVA, M., *Ind. Eng. Chem.*, **41**, 1206 (1949).
556. VAN HEERDEN, C., NOBEL, A., VAN KREVELEN D., *Chem. Eng. Sci.*, **1**, 37 (1951).
557. BRANDT, O., *Koll. Z.*, **85**, 24 (1938).
558. ANDREASEN, A., *Koll. Z.*, **86**, 70 (1939).
559. TIDESWELL, F., WHEELER, R., *Trans. Inst. Min. Eng.*, **97**, 176 (1938).
560. BEISCHER, D., *Koll. Z.*, **89**, 214 (1939); *Z. Angew. Chem.*, **52**, 514 (1939).
561. AOI, T., *J. Phys. Soc. Japan*, **10**, 119 (1955).
562. PREINING, O., *Staub.* Heft 39, 45 (1955).
563. ROHATSCHEK, H., *Acta Phys. Austriaca*, **9**, 151 (1955).
564. STASZEWSKI, W., *Acta Phys. Polon.*, **13**, 208 (1953).
565. DAVIES, C.N., *Proc. Inst. Mech. Eng.*, **1** B, 185 (1952).
566. WONG, J., RANZ, W., JOHNSTONE, H., *J. Appl. Phys.*, **26**, 244 (1955).
567. CHEN, C.Y., *Chem. Rev.*, **55**, 595 (1955).
568. KOVASZNAY, L., *Proc. Cambr. Phil. Soc.*, **44**, 58 (1948).
569. BLASEWITZ, A., JUDSON, B., *Chem. Eng. Progr.*, **51**, 6 J (1955).
570. SILVERMAN, L., FIRST, M., *Ind. Eng. Chem.*, **44**, 2777 (1952).
571. WHITE, C., *Proc. Roy. Soc.*, **186** A, 472 (1946).
572. GREEN, H., THOMAS, D., *Proc. Inst. Mech. Eng.*, **1** B, 203 (1952).
573. CROZIER, W., SEELY, B., *Trans. Amer. Geophys. Union*, **36**, 42 (1955).
574. TELFORD, J., THORNDIKE, N., BOWEN, E., *Quart. J. Roy. Meteor. Soc.*, **81**, 241 (1955).
575. SARTOR, D., *J. Meteor.*, **11**, 91 (1954).
576. EAST, T., MARSHALL, J., *Quart. J. Roy. Meteor. Soc.*, **80**, 26 (1954).
577. STIMSON, M., JEFFERY, G., *Proc. Roy. Soc.*, **111** A, 110 (1926).
578. SILVERMAN, L., CONNERS, E., ANDERSON, D., *Ind. Eng. Chem.* **47**, 952 (1955).
579. MAGONO, C., *J. Meteor.*, **11**, 77 (1954).
580. CUMMINGS, W. *et al.*, *J. Appl. Chem.*, **2**, 413 (1952).
581. BRYAN, A., ,SMELLIE, I., *J. Royal. Techn. College, Glasgow*, **4**, 406 (1937).
582. PROKHOROV P., *Disc. Faraday Soc.*, **18**, 41 (1954).
583. SCHUMANN, T., *Quart. J. Roy. Met. Soc.*, **66**, 195 (1940).
584. HERDAN, G., *Small Particle Statistics*, Elsevier (1953).
585. JUNGE, C., *J. Meteor.*, **12**, 13 (1955).
586. *Brit. J. Appl. Phys. Suppl.*, No 3 (1954).
587. HEYWOOD, H., *Chem. and Ind.*, **56**, 149 (1937).
588. DALLAVALLE. J., ORR, C., *Brit. J. Appl. Phys. Suppl.*, No 3, 198 (1954).
589. TSIEN, Z., *J. Aeronaut. Soc.*, **13**, 653 (1946).
590. KRZYWOBLOWSKI, M., *Acta Phys. Austriaca*, **9**, 216 (1955).
591. GOKHALE, N., GATHA, K., *Indian J. Phys.*, **32**, 521 (1958).
592. SCHMITT, K., *Z. Naturforsch.*, **14**a, 870 (1959).
593. TREVELYAN, B., MASON, S., *J. Coll. Sci.*, 6., 354 (1951).
594. SHARKEY, T., *Brit. J. Appl. Phys.*, **7**, 52 (1956).
595. MANLEY, R., MASON, S., *J. Coll. Sci.*, **7**, 354 (1952).
596. GARSTANG, T., *Proc. Roy. Soc.*, **142**A, 491 (1933).
597. TOLLERT, H., *Chemie-Ing-Techn.*, **26**, 141 (1954).
598. CRUIZE, A., *Engineering*, **183**, 366 (1957).
599. *Brit. J. Appl. Phys. Suppl.*, No 3, p. 93, 94. (1954).

600. TIMBRELL, V., *Brit. J. Appl. Phys. Suppl.*, No 3, p. 86.
601. HAMILTON, R. J., *Brit. J. Appl. Phys. Suppl.*, No 3, p. 90, 93 95.
602. TIMBRELL, V., *Brit. J. Appl. Phys. Suppl.*, No 3, p. 86.
603. LUDWIG, J., *Chem. Z.*, **79**, 774 (1955).
604. EVESON, G., HALL, E., WARD, S., *Brit. J. Appl. Phys.*, **10**, 43 (1959).
605. FAXÉN, H., *Z. Angew. Mathem. Mechan.*, **7**, 79 (1927).
606. GUREL, S., WARD, S., WHITMORE, R., *Brit. J. Appl. Phys.*, **6**, 83 (1955).
607. CHOWDHURY, K., FRITZ, W., *Chem. Eng. Sci.*, **11**, 92, (1959).
608. VAN DER LEEDEN, P. *et al.*, *Appl. Sci. Res.*, **5** A, 338 (1955).
609. HINZE, J., *Appl. Sci. Res.*, **1** A, 18 (1949).
610. HAPPEL, J., *Amer. Inst. Chem. Eng. J.*, **4**, 197 (1958).
611. KUWOBARA, S., *J. Phys. Soc. Japan*, **14**, 527 (1959).
612. BRINKMAN, H., *Appl. Sci. Res.*, **1** A, 27 (1949).
613. HAWKSLEY, P., *Brit. J. Appl. Phys. Suppl.*, No 3, 1 (1954).
614. VERSCHOOR, H., *Appl. Sci. Res.*, **2** A, 155 (1951).
615. NODA, H., *Bull. Chem. Soc.* Japan, **30**, 495 (1957).
616. CHENG, P., SCHACHTMAN, H., *J. Polym. Sci.*, **16**, 19 (1955).
617. HANRATTY, T., BANDUKWALA, A., *Amer. Inst. Chem. Eng. J.*, **3**, 293 (1957).
618. WHITMORE, R., *Brit. J. Appl. Phys.*, **6**, 239 (1955).
619. WILSON, B., *Austral. J. Appl. Res.*, **3**, 252 (1952).
620. STRAUBEL, H., *Z. Aerosolforsch. Ther.*, **4**, 385 (1955).
621. WUERKER, R., SHELTON, H., LANGMUIR, R., *J. Appl. Phys.*, **30**, 342 (1959).
622. WALDMANN, L., *Z. Naturforsch.*, **14**a, 589 (1959).
623. BAKANOV, S., DERYAGIN, B., *Koll. Zh.*, **21**, 377 (1959).
624. WALKENHORST, W., *Beitr. Silikoseforsch.*, Heft 18 (1952).
625. FUCHS, N., YANKOVSKII, S., *Dokl. Akad. nauk. SSSR*, **119**, 1177 (1958).
626. THÜRMER, H., *Staub*, **20**, 6 (1960).
627. WATSON, H., *Brit. J. Appl. Phys.*, **9**, 78 (1958).
628. SCHADT, C., CADLE, R., *J. Coll. Sci.*, **12**, 356 (1957).
629. BRUN, E., LEROUX, J., MARTY. P., *J. Rech. Centre Nat. Rech. Scient.*, No. 31, 256 (1955).
630. FUCHS, N., YANKOVSKII, S., *Koll. Zh.*, **21**, 133 (1959).
631. ZERNIK, W., *Brit. J. Appl. Phys.*, **8**, 117 (1957).
632. SCHMIDT, E., BECKMANN, W., *Techn. Mechan. und Thermodynamik*, **1**, 341 (1930).
633. KRAUS, W., *Messung des Temperatur- und Geschwindigkeitsfeldes bei freier Konvektion*, Karlsruhe (1955).
634. WATSON, H., *Trans. Faraday Soc.*, **32** 1073 (1936).
635. MAITROT, M., *Ann. Chim.* (Paris), **8**, 657 (1953).
636. TAUZIN, P., *C. R. Acad. Sci., Paris*, **235**, 1119, 1632 (1952).
637. ROHATSCHEK, H., *Acta Phys. Austriaca*, **9**, 151 (1955).
638. AITKEN, J., *Trans. Roy. Soc. Edinb.*, **32** (1883).
639. FACY, L., *Arch. Meteor. Geophys. Bioklim.*, **8** A, 269 (1955); *C. R. Acad. Sci., Paris.* **246**, 102, 3161 (1958). *Geof. Pura e Appl.*, **40**, No 2, (1958).
640. FUCHS, N., *Evaporation and Droplet Growth in Gaseous Media*, § 2, Pergamon Press (1959).
641. HAPPEL, J., PFEFFER, R., *Amer. Inst. Chem. Eng. J.*, **6**, 129 (1960).
642. SEMRAU, K., MARYNOWSKI, C., LUNDE, K., LAPPLE, C., *Ind. Eng. Chem.*, **50**, 1615 (1958).
643. DERJAGUIN, B., DUKHIN, S., *Dokl. Akad. nauk. SSSR.* **111**, 613 (1956).
644. VERZAR, F., HÜGIN, F., MASSION, W., *Pflüger's Arch. Ges. Physiol.*, **261**, 219 (1955).
645. DERJAGUIN, B., BAKANOV. S., *Dokl. Akad. nauk. SSSR*, **117**, 959 (1957).
646. PROKHOROV. P., LEONOV, L., *Disc. Faraday Soc.*, **30**, 124 (1960).
647. DERJAGUIN, B., DUKHIN, S., *Dokl. Akad. nauk SSSR*, **106**, 851 (1956).
648. DUKHIN, S., DERJAGUIN, B., *Dokl. Akad. nauk SSSR*, **112**, 407 (1957).
649. FEDOSEEV, V., POLYANSKII, A., *Trud. Odessk. univ., Fiz. mat. fakul't.*, **5**, 95 (1954).
650. DERJAGUIN, B., DUKHIN, S., *Izv. Akad. nauk SSSR, ser. geofiz.*, 779 (1957).
651. TCHEN CHAN-MOU, *Mean Value and Correlation Problems Connected with the Motion of Small Particles Suspended in a Turbulent Fluid*, The Hague (1947).
652. PEARCEY, T., HILL, G., *Austral. J. Phys.*, **9**, 19 (1956).
653. SERAFINI, J., Nat. Adv. Comm. Aeron. Rep. No. 1159 (1954).
654. INGEBO, R., Nat. Adv. Comm. Aeron. Techn. Note No. 3762 (1956).
655. GUCKER, F., DOYLE, G., *J. Phys. Chem.*, **60**, 989 (1956).

656. Dukhin, S. *Koll. Zh.*, **22**, 128 (1960).
657. Zink, J., Delsasso, L., *J. Acoust. Soc. Amer.*, **30**, 765 (1958).
658. Hocking, L., *Quart. J. Roy. Meteor. Soc.*, **85**, 44 (1959).
659. Faxén, H., *Arkiv Mat. Astr. Fys.*, **19** A, No. 22 (1925).
660. Smoluchowski, M., *Bull. Acad. Sci. Cracov.*, **1** A, 28 (1911).
661. Pshenai-Severin, S., *Izv. Akad. nauk SSSR, ser. geofiz.*, 1045 (1957).
662. Pshenai-Severin, S., *Izv. Akad. nauk SSSR, ser. geofiz.*, 1254 (1958).
663. Pshenai-Severin, S., *Dokl. Akad. nauk SSSR*, **125**, 775 (1959).
664. Dörr, W., *Acustica*, **5**, 163 (1955).
665. Foster, W., *Brit. J. Appl. Phys.*, **10**, 206 (1959).
666. Drozin, W., LaMer, V., *J. Coll. Sci.*, **14**, 74 (1959).
667. Dunskii, V., Kitayev. A., *Koll. Zh.*, **22**, 159 (1960).
668. Göhlich, H., *VDI-Forschungsheft*, No. 467 (1958).
669. Robinson, A., *Comm. Pure Appl. Mathem.*, **9**, 69 (1956).
670. Levin, L., *Izv. Akad. nauk SSSR, ser. geofiz.*, 914 (1957).
671. Gladkova, J., Natanson, G., *Zh. fiz. khim.*, **32**, 1160 (1958).
672. Prengle, R., Rotfus, R., *Ind. Eng. Chem.*, **47**, 379 (1955).
673. Gillespie, T., Langstroth, G., *Canad. J. Chem.*, **30**, 1056 (1952).
674. Hinkle, B., Clyde Orr, Dallavalle, J., *J. Coll. Sci.*, **9**, 70 (1954).
675. Yoshikawa, H., Swartz, G., MacWaters, J., Fite, W., *Rev. Sci, Instr.*, **27**, 359 (1956).
676. Nolan, P. J., O'Connor, T., *Proc. Roy. Irish Acad.*, **57** A, 161 (1955).
677. Goetz, A., *Geof. Pura e Appl.*, **36**, 49 (1957).
678. Feifel, E., *Forsch. Ing. Wesen*, A **9**, 68 (1938).
679. Barth, W., *Brennstoff-Wärme-Kraft*, **8, I** (1956).
680. Solbach, W., *Tonind. Z.*, **80**, 38, 298 (1956); **82**, 474 (1958).
681. Maslov, V., Marshak, Yu., *Teploenergetika*, No. 6, 63 (1958).
682. Walter, E., *Staub*, **18**, 3 (1958).
683. Rumpf, H., *Staub*, Heft **47**, 634 (1956).
684. Daniels, T., *Engineer, Lond.*, **203**, 358 (1957).
685. Stairmand, C., *J. Inst. Fuel*, **29**, 58 (1956).
686. Yaffe, C., Hosey, A., Chambers, J., *Arch. Ind. Hyg. Occ. Med.*, **5**, 62 (1952).
687. Smith, J., Goglia, M., *Trans. Amer. Soc. Mech. Eng.* **78**, 389 (1956).
688. Walton, W., *Brit. J. Appl. Phys. Suppl.*, No 3, 29, (1954)
689. Watson, H., *Amer. Ind. Hyg. Ass. Quart.*, **15**, 21 (1954).
690. Badzioch, S., *Brit. J. Appl. Phys.*, **10**, 26 (1959).
691. Walter, E., *Staub*, Heft 53, 880 (1957).
692. Dennis, R., Samples, W., Anderson, D., Silverman, L., *Ind. Eng. Chem.* **49**, 294 (1957).
693. Röber, R., *Staub*, Heft 48, 41, Heft 49, 273, Heft 50, 418 (1957).
694. Lippmann, M., *Amer. Ind. Hyg. Ass. J.*, **20**, 406 (1959).
695. Brun, R., Lewis, W., Perkins, P., Serafini, J., Nat. Adv. Comm. Aeron., Rep. No. 1215 (1955).
696. Brun, R., Nergler, H., Nat. Adv. Comm. Aeron., Techn. Note No. 2904 (1953).
697. Tribus, M., Young, G., Boelter, L., *Trans. Amer. Soc. Mech. Eng.* **70**, 977 (1948).
698. Dorsh, R., Brun, R., Gregg, J., Nat. Adv. Comm. Aeron., Techn. Note No. 3099 (1954).
699. Brun, R., Gallagher, H., Vogt, D., Nat. Adv. Comm. Aeron., Techn. Note No. 3047 (1953).
700. Bergrun, N., Nat. Adv. Comm. Aeron., Rep. No. 1107 (1952).
701. Mazin, I. *The Physical Principles of Aircraft Icing (Fizicheskiye osnovy obledeneniya samoletov)*. Gidrometeorizdat (1957).
702. Hacker, P., Brun, R., Boyd, B., Nat. Adv. Comm. Aeron., Techn. Note No. 2999 (1953).
703. Brun, R., Serafini, J., Gallagher, H., Nat. Adv. Comm. Aeron., Techn. Note No. 2903 (1953).
704. Herne, H., *Aerodynamic Capture of Particles*, p. 26, Pergamon Press (1960).
705. Tribus, M., Guibert, A., *J. Aeron. Sci.*, **19**, 391 (1952).
706. Levin, L., *Physics of Coarse Aerosols*, § 8 Acad. sci. U.S.S.R. Moscow (1961).
707. Natanson, G., *Dokl. Akad. nauk SSSR*, **116**, 109 (1957).
708. Levin, L., *Izv. Akad. nauk SSSR, ser. geofiz.*, 422 (1959).
709. Davies, C. N., Peetz, V., *Proc. Roy. Soc.*, **234** A, 269 (1956).
710. Davies, C. N., *Proc. Phys. Soc.*, **63** B, 288 (1950).
711. Lewis, J., Ruggieri, R., Nat. Adv. Comm. Aeron., Techn. Note No. 4092 (1957).

712. IGNAT'EV, V., *Teploenergetika*, No. 3 (1958).
713. AMELIN, A., BELYAKOV, M., *Koll. Zh.*, **18**, 385 (1956); Soobshch. nauchn. rabot NIUIF, No. 9, p. 58 (1958).
714. AMELIN, A., BELYAKOV, M., *Zav. lab.*, **21**, 1463 (1955).
715. JARMAN, R., *Chem. Eng. Sci.*, **10**, 268 (1959).
716. SIMMONS, L., COWDEREY, C., Aero. Res. Counc. Rep. Mem. No. 2276 (1945).
717. KRAEMER, H., JOHNSTONE, H., *Ind. Eng. Chem.*, **47**, 2426 (1955).
718. NATANSON, G., *Dokl. Akad. nauk SSSR*, **112**, 696 (1957).
719. GILLESPIE, T., *J. Coll. Sci.*, **10**, 299 (1955).
720. DUKHIN, S., DERJAGUIN, B., *Koll. Zh.*, **20**, 326 (1958).
721. LEVIN, L., *Izv. Akad. nauk SSSR, ser. geofiz.*, 1073 (1959).
722. FUCHS, N., *Evaporation and Droplets Growth in Gaseous Media*, § 5, Pergamon Press (1959).
723. TODOROV, I., SHELUDKO, A., *Koll. Zh.*, **19**, 496 (1957).
724. POLLAK, L., O'CONNOR, T., *Geof. Pura e Appl.*, **31**, 66 (1955).
725. POLLAK, L., O'CONNOR, T., METNIEKS, A., *Geof. Pura e Appl.*, **34**, 177 (1956); **36**, 70 (1957).
726. FÜRTH, R., *Geof. Pura e Appl.*, **31**, 80 (1955).
727. RICHARDSON, J., WOODING, E., *Chem. Eng. Sci.*, **7**, 51 (1957).
728. GORMLEY, P., *Proc. Roy. Irish Acad.*, **45** A, 59 (1938).
729. KENNEDY, M., Quoted by NOLAN, P. J., KENNY, P., *J. Atmos. Terr. Phys.*, **3**, 181 (1953).
730. DEMARCUS, W., THOMAS, J., U. S. Atom. Commiss., ORNL-1413 (1952).
731. THOMAS, J., *J. Coll. Sci.*, **10**, 246 (1955).
732. THOMAS, J., U. S. Atom. Commiss., ORNL-1648 (1954).
733. THOMAS, J., *J. Coll. Sci.*, **11**, 107 (1956).
734. POLLAK, L., METNIEKS, A., *Geofiz. Pura e Appl.*, **37**, 183 (1957).
735. GOYER, G., PIDGEON, F., *J. Coll. Sci.*, **11**, 697 (1956).
736. CHAMBERLAIN, A., MEGAW, W., WIFFEN, R., *Geofiz. Pura e Appl.*, **36**, 223 (1957).
737. FRIEDLANDER, S., *Amer. Inst. Chem. Eng. J.*, **3**, 11 (1957).
738. AKSEL'RUD, G., *Zh. fiz. khim.*, **27**, 1445 (1953).
739. GARNER, F., SUCKLING, R., *Amer. Inst. Chem. Eng. J.*, **4**, 114 (1958).
740. GARNER, F., KEEY, R., *Chem. Eng. Sci.*, **9**, 119 (1958).
741. NATANSON, G., *Dokl. Akad. nauk SSSR*, **112**, 100 (1957).
742. DOBRY, R., FINN, R., *Ind. Eng. Chem.*, **48**, 1540 (1956).
743. CHEN, C.Y., *Chem. Rev.*, **55**, 595 (1955).
744. RADUSHKEVICH, L., *Zh. fiz. khim.*, **32**, 282 (1958).
745. TAMADA, K., FUJIKAWA, H., *Quart. J. Mech. Appl. Math.*, **10**, 425 (1957).
746. KUWOBARA, S., *J. Phys. Soc. Japan*, **14**, 527 (1959).
747. LANDT, E., *Staub*, Heft 48, 9 (1957).
748. STAIRMAND, C., *Heat. Vent. Eng.*, **26**, 343 (1953).
749. FRIEDLANDER, S., *Ind. Eng. Chem.*, **50**, 1161 (1958).
750. GALLILI, I., *J. Coll. Sci.*, **12**, 161 (1957).
751. THOMAS, J., YODER, R., *Arch. Ind. Health*, **13**, 545 (1956).
752. WONG, J., RANZ, W., JOHNSTONE, H., *J. Appl. Phys.*, **27**, 161 (1956).
753. HUMPHREY, A., GADEN, E., *Ind. Eng. Chem.*, **47**, 924 (1955).
754. DAVIES, C. N., *J. Inst. Heat. Vent. Eng.*, **20**, 55 (1952).
755. LEERS, R., *Staub*, Heft 50, 402 (1957).
756. FAIRS, G., *Trans. Inst. Chem. Eng.*, **36**, 476 (1958).
757. LAMER, V., DROZIN, V., *Proc. 2nd Intern. Congr. Surf. Activ.*, **3**, 600, London (1957).
758. HASENCLEVER, D., *Staub*, **19**, 37 (1959).
759. ROSSANO, A., SILVERMAN, L., *Heat. Ventilattng*, **51**, No. 5, 102 (1954).
760. WINKEL, A., *Staub*, Heft 41, 469 (1955).
761. THOMAS, D., *J. Inst. Heat. Vent. Eng.*, **20**, 35 (1952).
762. GILLESPIE, T., *J. Coll. Sci.*, **10**, 299 (1955).
763. BILLINGS, C., DENNIS, R., SILVERMAN, L., U. S. Atom. Commiss., NYO-1592 (1954).
764. FITZGERALD, J., DETWEILER, C., *Arch. Ind. Health*, **15**, 3 (1957).
765. WALKENHORST, W., *Staub*, **19**, 69 (1959).
766. SNYDER, C., PRING, R., *Ind. Eng. Chem.*, **47**, 960 (1955).
767. ELISEEV, N., *Collected Data on Dust Removal in Non-ferrous Metallurgy (Sbornik materialov po pyleulavlivaniyu v tsvetnoi metallurgii)* p. 251, Moscow (1951).
768. THOMAS, J., YODER, R., *Arch. Ind. Health*, **13**, 550 (1956).

769. MORROW, P., MEHRHOF, E., CASARETT, L., MORKEN, D., *Arch. Ind. Health*, **18**, 292 (1958).
770. ALTSHULER, B. *et al.*, *Arch. Ind. Health*, **15**, 293 (1957).
771. DAUTREBANDE, L., BECKMANN, H., WALKENHORST, W., *Arch. Ind. Health*, **16**, 179 (1957).
772. POZIN, M., MUKHLENOV, I., TARAT, E., *The Foam Method of Cleaning Gases of Dust, Smoke and Mist (Pennyi sposob ochistki gazov ot pyli, dyma i tumana)*. Goskhimizdat (1953).
773. O'KONSKI, C., GUNTHER, R., *J. Coll. Sci.*, **10**, 563 (1955).
774. BATCHELOR, G., TOWNSEND, A., *Surveys in Mechanics*, p.352, Cambridge (1956).
775. SOO, S., *Chem. Eng. Sci.*, **5**, 57 (1956).
776. VI-CHENG LIU, *J. Meteor.*, **13**, 399 (1956).
777. SOO, S., TIEN, C., KADAMBI, V., *Rev. Sci. Instr.*, **30**, 821 (1959).
778. KUBYNIN, N., *Izv. Vsesoyuzn. teplotekhnich. inst.*, No. 1 (1951).
779. HOFFMANN, W., *VDI-Ber.*, **7**, 15 (1955).
780. TCHERNOV, A., *Izv. Akad. nauk Kazakhsk. SSR, ser. energetich.*, No. 8 (1955).
781. DAWES, J., SLACK, A., Safety in Mines Res. Est., Res. Rep., No. 105 (1954).
782. LIN, C., MOULTON, R., PUTNAM, G., *Ind. Eng. Chem.*, **45**, 636 (1953).
783. DEISSLER, R., Nat. Adv. Comm. Aeron., Rep. No. 1210 (1955).
784. FRIEDLANDER, S., JOHNSTONE, H., *Ind. Eng. Chem.*, **49**, 1151 (1957).
785. SCHLICHTING, H., *Grenzschichttheorie*, Kap. XVII, § 3, Karlsruhe (1951).
786. LAUFER, J., Nat. Adv. Comm. Aeron., Rep. No. 1174 (1954).
787. PERELES, E., Safety in Mines. Res. Est., Res. Rep. No. 144 (1958).
788. OWEN, P., *Aerodynamic Capture of Particles*, p. 8, Pergamon Press (1960).
789. KOEPE, F., *Staub*, Heft 28, 39 (1952).
790. MADISETTI, A., *Staub*, Heft 47, 652 (1956).
791. FRENKIEL, F., KATZ, I., *J. Meteor.*, **13**, 388 (1956).
792. GIFFORD, F., *J. Meteor.*, **14**, 410 (1957).
793. TRAPPENBERG, R., *Staub*, Heft 39, 5; Heft 40, 189 (1955);
DIEM, M., TRAPPENBERG, R., *Forsch. Ber. Nordrhein-Westf.*, 380 (1957).
794. CZANADY, G., *Austral. J. Phys.*, **8**, 546 (1955); **10**, 558 (1957).
795. ROUNDS, W., *Trans. Amer. Geophys Union*, **36**, 395 (1955).
796. GODSON, W., *Arch. Meteor. Geophys. Bioklim.*, **10** A, 305 (1958).
797. FORTAK, H., *Z. Meteor.*, **11**, 19 (1957).
798. DENISSOV, A., *Izv. Akad. nauk SSSR, ser. geofiz.*, 834 (1957).
799. GANDIN, L., SOLOVEITCHIK, R., *Trudy Glavn. geofiz. observat.*, No. 77 (1958).
800. ZEBEL, G., *Koll. Z.*, **157**, 37 (1958).
801. O'CONNOR, T., *Geof. Pura e Appl.*, **31**, 107 (1955).
802. CAWOOD, W., WHYTLAW-GRAY, R., *Trans. Faraday Soc.*, **32**, 1059 (1936).
803. YAFFE, I., CADLE, R., *J. Phys. Chem.*, **62**, 510 (1958).
804. MELZAK, Z., *Quart. Appl. Math.*, **11**, 231 (1953).
805. ZEBEL, G., *Koll. Z.*, **156**, 102 (1958).
806. DERJAGUIN, B., *Dokl. Akad. nauk SSSR*, **109**, 967 (1956).
807. Quoted from HARDY, W., BIRCUMSHAW, I., *Proc. Roy. Soc.*, **108** A, 12 (1925).
808. SCHOTLAND, R., *J. Meteor.*, **14**, 381 (1957).
809. DAUTREBANDE, L., BECKMANN, H., WALKENHORST, W., *Beitr. Silikoseforsch.*, Heft 31 (1954).
810. AVY, A., *Staub*, Heft 32, 147 (1953).
811. LEBOUFFANT, L., *Rev. Ind. Miner.*, **36**, 62 (1955).
812. WALKENHORST, W., ZEBEL, G., *Beitr. Silikoseforsch.*, Heft 44, 53 (1956).
813. WALKENHORST, W., *Z. Aerosolforsch. Ther.*, **4**, 538 (1955).
814. Quot. from WALKENHORST, W., *Z. Aerosolforsch. Ther.*, 7, 157 (1958).
815. FUJITANI, J., *Bull. Chem. Soc. Japan*, **30**, 683 (1957).
816. GUNN, R., *J. Coll. Sci.*, **10**, 107 (1955).
817. BENTON, D., ELTON, G., PEACE, E., PICKNETT, R., *Intern. J. Air Pollut.*, **1**, 44 (1958).
818. DUKHIN, S., DERJAGUIN, B., *Dokl. Akad. nauk SSSR*, **121**, 503 (1958).
819. NATANSON, G., *Koll. Zh.*, **20**, 705 (1958).
820. WINKEL, A., *Staub*, Heft 41, 469 (1955).
821. ZEBEL, G., *Staub*, **19**, 381 (1959).
822. GUDEMCHUK, V., PODOSHEVNIKOV, B., TARTAKOVSKII, B., *Akustich. Zh.*, **5**, 246 (1959).
823. PEARCEY, T., HILL, G., *Quart. J. Roy. Meteor. Soc.*, **83**, 77 (1957).
824. SCHOTLAND, R., *J. Meteor.*, **14**, 381 (1957); *Artificial Stimulation of Rain*, p.170, Pergamon Press (1957).

825. Kinzer, G., Cobb, W., *J. Meteor.*, **13**, 295 (1956); **15**, 138 (1958). *Artificial Stimulation of Rain*, p.167, Pergamon Press (1957).
826. Torobin, L., Gauvin, W., *Canad. J. Chem. Eng.*, **37**, 167 (1959).
827. Oakes, B., *Aerodynamic Capture of Particles*, p. 179, Pergamon Press (1960).
828. Walton, W., Woolcock, A., *Aerodynamic Capture of Particles*, p. 129, Pergamon Press (1960).
829. May, F., *Quart. J. Roy. Meteor. Soc.*, **84**, 451 (1958).
830. Best, A., *Quart. J. Roy. Meteor. Soc.*, **76**, 16 (1950).
831. Johnstone, H.F., Field, R., Tassles, H., *Ind. Eng. Chem.*, **46**, 1601 (1954).
832. Brink, J., Contant, C., *Ind. Eng. Chem.*, **50**, 1157 (1958).
833. Pauthenier, M., Cochet, R., Dupuy, J., *C.R. Acad. Sci. Paris*, **246**, 2233 (1958).
834. Gunn, R., *J. Meteor.*, **12**, 511 (1955).
835. Tkach, V., *Coll. Papers on Problems of Hygiene and Occupational Illnesses in the Mining, Chemical and Engineering Industries* (Sb. *Voprosy gigiyeny truda i profzabolevanii v gornorudnoi, khimicheskoi i mashinostroitzel'noi promyshlennosti*) p.9, Kiev (1958).
836. Chizhov, I., *Collected Data on Dust Removal in Non-ferrous Metallurgy (Sbornik materialov po pyleulavlivaniyu v tsvetnoi metallurgii)* p.361, Moscow (1957).
837. Gillespie, T., *Proc. Roy. Soc.*, **216** A, 569 (1953).
838. Saffman, P., Turner, J., *J. Fluid Mechan.*, **1**, 16 (1956).
839. Levich, V., *Physico-chemical Hydrodynamics (Fizikokhimicheskaya Gidrodinamika)*, Chap. 5, 2nd ed., Moscow (1959).
840. Frisch, H., *J. Phys. Chem.*, **60**, 463 (1956).
841. Dinnik, A., *Selected Works (Izbrannyye trudy)*, **1**, Chap.5, Akademizdat (1952).
842. Raman, C., *Phys. Rev.*, **12**, 442 (1918).
843. Andrews, J., *Phil. Mag.*, **9**, 593 (1930).
844. Jordan, D., *Brit. J. Appl. Phys. Suppl.*, No 3, 194 (1954).
845. Adam, O., *Chemie-Ing-Techn.*, **29**, 151 (1957).
846. Dergachev, I., *Izv. Vsesoyuzn. teplotekhnich. inst.* No. 6 (1949).
847. Rosinski, J., Nagamoto, C., Ungar, A., *Koll. Z.*, **164**, 26 (1959).
848. Browne, G., Palmer, N., Wormell, T., *Quart. J. Roy. Meteor. Soc.*, **80**, 291 (1954).
849. Gallili, I., LaMer, V., *J. Phys. Chem.*, **62**, 1295 (1958).
850. Gillespie, T., Rideal, E., *J. Coll. Sci.*, **10**, 281 (1955).
851. Gillespie, T., *J. Coll. Sci.*, **10**, 266 (1955).
852. Hartley, C., Brunskill, R., *Surface Phenomena in Chemistry and Biology*, p.224, Pergamon Press (1958).
853. Taubman, A., Nikitina, S., *Dokl. Akad. nauk SSSR*, **110**, 600, 816 (1956); **116**, 113 (1957).
854. Ermilov, P., *Uchen. zapiski Yaroslavsk. tekhnolog. inst.*, **1**, 111 (1956); *Izv. Vyssh. uchebn. zaved., khim. i khim. tekh.*, **2**, 134 (1959).
855. Pozin, M., Mukhlenov, N., Demshin, V., *Zh. prikl. khim.*, **28**, 841 (1955).
Mukhlenov, N., Demshin, V., *Zh. prikl. khim.*, **28**, 922 (1955).
856. McCully, C. *et al.*, *Ind. Eng. Chem.*, **48**, 1512 (1956).
857. Pemberton, C., *Aerodynamic Capture of Particles*, p. 168, Pergamon Press (1960).
858. Landwehr, M., Kortner, R., Walkenhorst, W., Bruckmann, E., *Staub*, Heft 53, 841 (1957).
859. How, P., Benton, D., Puddington, I., *Canad. J. Chem.*, **33**, 1375 (1955).
860. Harper, W., *Proc. Roy. Soc.*, **231** A, 388 (1955).
861. Tomlinson, G., *Phil. Mag.*, **6**, 695 (1928).
862. Larsen, R., *Amer. Ind. Hyg. Ass. Quart.*, **19**, 265 (1958).
863. Goncharov, V., *Investigation of the Dynamics of Channel Flows (Issledovaniye dinamiki ruslovykh potokov)*. Gidrometeorizdat (1954).
864. Fowler, R., Chodziesner, W., *Chem. Eng. Sci.* **10**. 157 (1959).
865. Landau, L., Lifshits, E., *Fluid Mechanics*, p. 611, Pergamon Press (1959).
866. Bourot, J-M., Brun, R., Morillon, B., *C.R. Acad. Sci. Paris*, **250** 2118 (1960).
867. Cremer, E., Conrad, F., Kraus, J., *Angew. Chem.*, **64**, 10 (1952).
Cremer, E., *Staub*, Heft 37, 427 (1954).
868. Patat, F., Schmid, W., *Chemie-Ing-Techn.*, **32**, 8 (1960).
869. Batel, W., *Chemie-Ing-Techn.*, **31**, 343 (1959).
870. Bagnold, A., *The Physics of Blown Sands and Desert Dunes*, 2nd ed., p.72, 163, Methuen (1954).
871. Rumpf, H., *VDI-Ber.*, **6**, 17 (1955).
872. Klimanov, A. *Sborn. Nauchn. trud. Mosk. gorn. inst.*, No. 16 (1955).

873. DAWES, J., Safety in Mines Res. Est., Res. Rep. No. 36 and 49 (1952).
DAWES, J., WYNN, A., Safety in Mines Res. Est., Res. Rep. No. 46 (1952).
874. MEHTA, N., SMITH, J., COMMINGS, E., *Ind. Eng. Chem.*, **49**, 986 (1957).
875. REIF, A., WHYTE, C., GIBLIN, M., *Arch. Ind. Health*, **14**, 442 (1956).
876. EADIE, F., PAYNE, R., *Iron Age*, No. 10, 99 (1954).
877. KROLL, W., *Chemie-Ing-Techn.*, **27**, 33 (1955).
878. CHIN-YOUNG WEN, SIMMONS, H., *Amer. Inst. Chem. Eng. J.*, **5**, 264 (1959).
879. MORSE, R., *Ind. Eng. Chem.*, **47**, 1170 (1955).
880. CRAIK, D., MILLER, B., *J. Pharm. Pharmacol.*, **10**, Suppl., p.136 (1958).
881. PARKER, H., STEVENS, W., *Amer. Inst. Chem. Eng. J.*, **5**, 314 (1959).
882. D'ARCY-SMITH, F., *Ind. Chemist.*, **33**, 181 (1957).
883. BAKKER, P., HEERTJES, P., *Brit. Chem. Eng.*, **3**, 241 (1958).
884. ZENZ, F., WEIL, N., *Amer. Inst. Chem. Eng. J.*, **4**, 472 (1958).
885. BAUMGARTEN, P., PIGFORD, R., *Amer. Inst. Chem. Eng. J.*, **6**, 115 (1960).
886. TAUZIN, P., *Cahiers Phys.*, No. 19, 1; No. 20, 16 (1944).

NAME INDEX

(Only the name of the first author of two or more collaborators is indexed. The numbers in italics relate to reference numbers. The numbers in ordinary type are page numbers.)

SUBJECT INDEX

THE BOOK OF BEASTS: Being a Translation from a Latin Bestiary of the Twelfth Century, T. H. White. Wonderful catalog real and fanciful beasts: manticore, griffin, phoenix, amphivius, jaculus, many more. White's witty erudite commentary on scientific, historical aspects. Fascinating glimpse of medieval mind. Illustrated. 296pp. 5⅜ × 8¼. (Available in U.S. only) 24609-4 Pa. $5.95

FRANK LLOYD WRIGHT: ARCHITECTURE AND NATURE With 160 Illustrations, Donald Hoffmann. Profusely illustrated study of influence of nature—especially prairie—on Wright's designs for Fallingwater, Robie House, Guggenheim Museum, other masterpieces. 96pp. 9¼ × 10¾. 25098-9 Pa. $7.95

FRANK LLOYD WRIGHT'S FALLINGWATER, Donald Hoffmann. Wright's famous waterfall house: planning and construction of organic idea. History of site, owners, Wright's personal involvement. Photographs of various stages of building. Preface by Edgar Kaufmann, Jr. 100 illustrations. 112pp. 9¼ × 10. 23671-4 Pa. $7.95

YEARS WITH FRANK LLOYD WRIGHT: Apprentice to Genius, Edgar Tafel. Insightful memoir by a former apprentice presents a revealing portrait of Wright the man, the inspired teacher, the greatest American architect. 372 black-and-white illustrations. Preface. Index. vi + 228pp. 8¼ × 11. 24801-1 Pa. $9.95

THE STORY OF KING ARTHUR AND HIS KNIGHTS, Howard Pyle. Enchanting version of King Arthur fable has delighted generations with imaginative narratives of exciting adventures and unforgettable illustrations by the author. 41 illustrations. xviii + 313pp. 6⅛ × 9¼. 21445-1 Pa. $5.95

THE GODS OF THE EGYPTIANS, E. A. Wallis Budge. Thorough coverage of numerous gods of ancient Egypt by foremost Egyptologist. Information on evolution of cults, rites and gods; the cult of Osiris; the Book of the Dead and its rites; the sacred animals and birds; Heaven and Hell; and more. 956pp. 6⅛ × 9¼. 22055-9, 22056-7 Pa., Two-vol. set $21.90

A THEOLOGICO-POLITICAL TREATISE, Benedict Spinoza. Also contains unfinished *Political Treatise*. Great classic on religious liberty, theory of government on common consent. R. Elwes translation. Total of 421pp. 5⅜ × 8½. 20249-6 Pa. $6.95

INCIDENTS OF TRAVEL IN CENTRAL AMERICA, CHIAPAS, AND YUCATAN, John L. Stephens. Almost single-handed discovery of Maya culture; exploration of ruined cities, monuments, temples; customs of Indians. 115 drawings. 892pp. 5⅜ × 8½. 22404-X, 22405-8 Pa., Two-vol. set $15.90

LOS CAPRICHOS, Francisco Goya. 80 plates of wild, grotesque monsters and caricatures. Prado manuscript included. 183pp. 6⅝ × 9⅝. 22384-1 Pa. $4.95

AUTOBIOGRAPHY: The Story of My Experiments with Truth, Mohandas K. Gandhi. Not hagiography, but Gandhi in his own words. Boyhood, legal studies, purification, the growth of the Satyagraha (nonviolent protest) movement. Critical, inspiring work of the man who freed India. 480pp. 5⅜ × 8½. (Available in U.S. only) 24593-4 Pa. $6.95

HOW TO WRITE, Gertrude Stein. Gertrude Stein claimed anyone could understand her unconventional writing—here are clues to help. Fascinating improvisations, language experiments, explanations illuminate Stein's craft and the art of writing. Total of 414pp. 4⅜ × 6⅜. 23144-5 Pa. $5.95

ADVENTURES AT SEA IN THE GREAT AGE OF SAIL: Five Firsthand Narratives, edited by Elliot Snow. Rare true accounts of exploration, whaling, shipwreck, fierce natives, trade, shipboard life, more. 33 illustrations. Introduction. 353pp. 5⅜ × 8½. 25177-2 Pa. $7.95

THE HERBAL OR GENERAL HISTORY OF PLANTS, John Gerard. Classic descriptions of about 2,850 plants—with over 2,700 illustrations—includes Latin and English names, physical descriptions, varieties, time and place of growth, more. 2,706 illustrations. xlv + 1,678pp. 8½ × 12¼. 23147-X Cloth. $75.00

DOROTHY AND THE WIZARD IN OZ, L. Frank Baum. Dorothy and the Wizard visit the center of the Earth, where people are vegetables, glass houses grow and Oz characters reappear. Classic sequel to *Wizard of Oz*. 256pp. 5⅜ × 8. 24714-7 Pa. $4.95

SONGS OF EXPERIENCE: Facsimile Reproduction with 26 Plates in Full Color, William Blake. This facsimile of Blake's original "Illuminated Book" reproduces 26 full-color plates from a rare 1826 edition. Includes "The Tyger," "London," "Holy Thursday," and other immortal poems. 26 color plates. Printed text of poems. 48pp. 5¼ × 7. 24636-1 Pa. $3.50

SONGS OF INNOCENCE, William Blake. The first and most popular of Blake's famous "Illuminated Books," in a facsimile edition reproducing all 31 brightly colored plates. Additional printed text of each poem. 64pp. 5¼ × 7. 22764-2 Pa. $3.50

PRECIOUS STONES, Max Bauer. Classic, thorough study of diamonds, rubies, emeralds, garnets, etc.: physical character, occurrence, properties, use, similar topics. 20 plates, 8 in color. 94 figures. 659pp. 6⅛ × 9¼. 21910-0, 21911-9 Pa., Two-vol. set $15.90

ENCYCLOPEDIA OF VICTORIAN NEEDLEWORK, S. F. A. Caulfeild and Blanche Saward. Full, precise descriptions of stitches, techniques for dozens of needlecrafts—most exhaustive reference of its kind. Over 800 figures. Total of 679pp. 8⅛ × 11. Two volumes. Vol. 1 22800-2 Pa. $11.95
Vol. 2 22801-0 Pa. $11.95

THE MARVELOUS LAND OF OZ, L. Frank Baum. Second Oz book, the Scarecrow and Tin Woodman are back with hero named Tip, Oz magic. 136 illustrations. 287pp. 5⅜ × 8½. 20692-0 Pa. $5.95

WILD FOWL DECOYS, Joel Barber. Basic book on the subject, by foremost authority and collector. Reveals history of decoy making and rigging, place in American culture, different kinds of decoys, how to make them, and how to use them. 140 plates. 156pp. 7⅞ × 10¾. 20011-6 Pa. $8.95

HISTORY OF LACE, Mrs. Bury Palliser. Definitive, profusely illustrated chronicle of lace from earliest times to late 19th century. Laces of Italy, Greece, England, France, Belgium, etc. Landmark of needlework scholarship. 266 illustrations. 672pp. 6⅛ × 9¼. 24742-2 Pa. $14.95

SUNDIALS, Albert Waugh. Far and away the best, most thorough coverage of ideas, mathematics concerned, types, construction, adjusting anywhere. Over 100 illustrations. 230pp. 5⅜ × 8½. 22947-5 Pa. $4.50

PICTURE HISTORY OF THE NORMANDIE: With 190 Illustrations, Frank O. Braynard. Full story of legendary French ocean liner: Art Deco interiors, design innovations, furnishings, celebrities, maiden voyage, tragic fire, much more. Extensive text. 144pp. 8⅜ × 11¾. 25257-4 Pa. $9.95

THE FIRST AMERICAN COOKBOOK: A Facsimile of "American Cookery," 1796, Amelia Simmons. Facsimile of the first American-written cookbook published in the United States contains authentic recipes for colonial favorites—pumpkin pudding, winter squash pudding, spruce beer, Indian slapjacks, and more. Introductory Essay and Glossary of colonial cooking terms. 80pp. 5⅜ × 8½. 24710-4 Pa. $3.50

101 PUZZLES IN THOUGHT AND LOGIC, C. R. Wylie, Jr. Solve murders and robberies, find out which fishermen are liars, how a blind man could possibly identify a color—purely by your own reasoning! 107pp. 5⅜ × 8½. 20367-0 Pa. $2.50

THE BOOK OF WORLD-FAMOUS MUSIC—CLASSICAL, POPULAR AND FOLK, James J. Fuld. Revised and enlarged republication of landmark work in musico-bibliography. Full information about nearly 1,000 songs and compositions including first lines of music and lyrics. New supplement. Index. 800pp. 5⅜ × 8¼. 24857-7 Pa. $14.95

ANTHROPOLOGY AND MODERN LIFE, Franz Boas. Great anthropologist's classic treatise on race and culture. Introduction by Ruth Bunzel. Only inexpensive paperback edition. 255pp. 5⅜ × 8½. 25245-0 Pa. $5.95

THE TALE OF PETER RABBIT, Beatrix Potter. The inimitable Peter's terrifying adventure in Mr. McGregor's garden, with all 27 wonderful, full-color Potter illustrations. 55pp. 4¼ × 5½. (Available in U.S. only) 22827-4 Pa. $1.75

THREE PROPHETIC SCIENCE FICTION NOVELS, H. G. Wells. *When the Sleeper Wakes, A Story of the Days to Come* and *The Time Machine* (full version). 335pp. 5⅜ × 8½. (Available in U.S. only) 20605-X Pa. $5.95

APICIUS COOKERY AND DINING IN IMPERIAL ROME, edited and translated by Joseph Dommers Vehling. Oldest known cookbook in existence offers readers a clear picture of what foods Romans ate, how they prepared them, etc. 49 illustrations. 301pp. 6⅛ × 9¼. 23563-7 Pa. $6.50

SHAKESPEARE LEXICON AND QUOTATION DICTIONARY, Alexander Schmidt. Full definitions, locations, shades of meaning of every word in plays and poems. More than 50,000 exact quotations. 1,485pp. 6½ × 9¼. 22726-X, 22727-8 Pa., Two-vol. set $27.90

THE WORLD'S GREAT SPEECHES, edited by Lewis Copeland and Lawrence W. Lamm. Vast collection of 278 speeches from Greeks to 1970. Powerful and effective models; unique look at history. 842pp. 5⅜ × 8½. 20468-5 Pa. $11.95

A CONCISE HISTORY OF PHOTOGRAPHY: Third Revised Edition, Helmut Gernsheim. Best one-volume history—camera obscura, photochemistry, daguerreotypes, evolution of cameras, film, more. Also artistic aspects—landscape, portraits, fine art, etc. 281 black-and-white photographs. 26 in color. 176pp. 8⅜ × 11¼. 25128-4 Pa. $12.95

THE DORÉ BIBLE ILLUSTRATIONS, Gustave Doré. 241 detailed plates from the Bible: the Creation scenes, Adam and Eve, Flood, Babylon, battle sequences, life of Jesus, etc. Each plate is accompanied by the verses from the King James version of the Bible. 241pp. 9 × 12. 23004-X Pa. $8.95

HUGGER-MUGGER IN THE LOUVRE, Elliot Paul. Second Homer Evans mystery-comedy. Theft at the Louvre involves sleuth in hilarious, madcap caper. "A knockout."—Books. 336pp. 5⅜ × 8½. 25185-3 Pa. $5.95

FLATLAND, E. A. Abbott. Intriguing and enormously popular science-fiction classic explores the complexities of trying to survive as a two-dimensional being in a three-dimensional world. Amusingly illustrated by the author. 16 illustrations. 103pp. 5⅜ × 8½. 20001-9 Pa. $2.25

THE HISTORY OF THE LEWIS AND CLARK EXPEDITION, Meriwether Lewis and William Clark, edited by Elliott Coues. Classic edition of Lewis and Clark's day-by-day journals that later became the basis for U.S. claims to Oregon and the West. Accurate and invaluable geographical, botanical, biological, meteorological and anthropological material. Total of 1,508pp. 5⅜ × 8½.
21268-8, 21269-6, 21270-X Pa. Three-vol. set $25.50

LANGUAGE, TRUTH AND LOGIC, Alfred J. Ayer. Famous, clear introduction to Vienna, Cambridge schools of Logical Positivism. Role of philosophy, elimination of metaphysics, nature of analysis, etc. 160pp. 5⅜ × 8½. (Available in U.S. and Canada only) 20010-8 Pa. $2.95

MATHEMATICS FOR THE NONMATHEMATICIAN, Morris Kline. Detailed, college-level treatment of mathematics in cultural and historical context, with numerous exercises. For liberal arts students. Preface. Recommended Reading Lists. Tables. Index. Numerous black-and-white figures. xvi + 641pp. 5⅜ × 8½.
24823-2 Pa. $11.95

28 SCIENCE FICTION STORIES, H. G. Wells. Novels, *Star Begotten* and *Men Like Gods,* plus 26 short stories: "Empire of the Ants," "A Story of the Stone Age," "The Stolen Bacillus," "In the Abyss," etc. 915pp. 5⅜ × 8½. (Available in U.S. only)
20265-8 Cloth. $10.95

HANDBOOK OF PICTORIAL SYMBOLS, Rudolph Modley. 3,250 signs and symbols, many systems in full; official or heavy commercial use. Arranged by subject. Most in Pictorial Archive series. 143pp. 8⅛ × 11. 23357-X Pa. $5.95

INCIDENTS OF TRAVEL IN YUCATAN, John L. Stephens. Classic (1843) exploration of jungles of Yucatan, looking for evidences of Maya civilization. Travel adventures, Mexican and Indian culture, etc. Total of 669pp. 5⅜ × 8½.
20926-1, 20927-X Pa., Two-vol. set $9.90

THE ART NOUVEAU STYLE BOOK OF ALPHONSE MUCHA: All 72 Plates from "Documents Decoratifs" in Original Color, Alphonse Mucha. Rare copyright-free design portfolio by high priest of Art Nouveau. Jewelry, wallpaper, stained glass, furniture, figure studies, plant and animal motifs, etc. Only complete one-volume edition. 80pp. 9⅜ × 12¼. 24044-4 Pa. $8.95

ANIMALS: 1,419 COPYRIGHT-FREE ILLUSTRATIONS OF MAMMALS, BIRDS, FISH, INSECTS, ETC., edited by Jim Harter. Clear wood engravings present, in extremely lifelike poses, over 1,000 species of animals. One of the most extensive pictorial sourcebooks of its kind. Captions. Index. 284pp. 9 × 12. 23766-4 Pa. $9.95

OBELISTS FLY HIGH, C. Daly King. Masterpiece of American detective fiction, long out of print, involves murder on a 1935 transcontinental flight—"a very thrilling story"—NY Times. Unabridged and unaltered republication of the edition published by William Collins Sons & Co. Ltd., London, 1935. 288pp. 5⅜ × 8½. (Available in U.S. only) 25036-9 Pa. $4.95

VICTORIAN AND EDWARDIAN FASHION: A Photographic Survey, Alison Gernsheim. First fashion history completely illustrated by contemporary photographs. Full text plus 235 photos, 1840–1914, in which many celebrities appear. 240pp. 6½ × 9¼. 24205-6 Pa. $6.00

THE ART OF THE FRENCH ILLUSTRATED BOOK, 1700–1914, Gordon N. Ray. Over 630 superb book illustrations by Fragonard, Delacroix, Daumier, Doré, Grandville, Manet, Mucha, Steinlen, Toulouse-Lautrec and many others. Preface. Introduction. 633 halftones. Indices of artists, authors & titles, binders and provenances. Appendices. Bibliography. 608pp. 8⅜ × 11¼. 25086-5 Pa. $24.95

THE WONDERFUL WIZARD OF OZ, L. Frank Baum. Facsimile in full color of America's finest children's classic. 143 illustrations by W. W. Denslow. 267pp. 5⅜ × 8½. 20691-2 Pa. $5.95

FRONTIERS OF MODERN PHYSICS: New Perspectives on Cosmology, Relativity, Black Holes and Extraterrestrial Intelligence, Tony Rothman, et al. For the intelligent layman. Subjects include: cosmological models of the universe; black holes; the neutrino; the search for extraterrestrial intelligence. Introduction. 46 black-and-white illustrations. 192pp. 5⅜ × 8½. 24587-X Pa. $6.95

THE FRIENDLY STARS, Martha Evans Martin & Donald Howard Menzel. Classic text marshalls the stars together in an engaging, non-technical survey, presenting them as sources of beauty in night sky. 23 illustrations. Foreword. 2 star charts. Index. 147pp. 5⅜ × 8½. 21099-5 Pa. $3.50

FADS AND FALLACIES IN THE NAME OF SCIENCE, Martin Gardner. Fair, witty appraisal of cranks, quacks, and quackeries of science and pseudoscience: hollow earth, Velikovsky, orgone energy, Dianetics, flying saucers, Bridey Murphy, food and medical fads, etc. Revised, expanded In the Name of Science. "A very able and even-tempered presentation."—The New Yorker. 363pp. 5⅜ × 8. 20394-8 Pa. $6.50

ANCIENT EGYPT: ITS CULTURE AND HISTORY, J. E Manchip White. From pre-dynastics through Ptolemies: society, history, political structure, religion, daily life, literature, cultural heritage. 48 plates. 217pp. 5⅜ × 8½. 22548-8 Pa. $4.95